PROGRESS IN BRAIN RESEARCH
VOLUME 49

Recent volumes in PROGRESS IN BRAIN RESEARCH

PROGRESS IN BRAIN RESEARCH
VOLUME 49

THE CHOLINERGIC SYNAPSE

EDITED BY

STANISLAV TUČEK

Institute of Physiology,
Czechoslovak Academy of Sciences,
14220 Prague, Czechoslovakia

CO-EDITORS:

ERIC A. BARNARD

Department of Biochemistry,
Imperial College of
Science and Technology,
London SW7 2AZ,
United Kingdom

BRIAN COLLIER

Department of Pharmacology,
McGill University,
Montreal H3G 1Y6,
Canada

JOHN S. KELLY

MRC Neurochemical
Pharmacology Unit,
Medical School,
Cambridge CB2 2QD,
United Kingdom

ROGER M. MARCHBANKS

Institute of Psychiatry,
London SE5 8AF,
United Kingdom

HUMPHREY P. RANG

Department of Pharmacology,
St. George's Hospital
Medical School, London SW17 0QT,
United Kingdom

ANN SILVER

ARC Institute of Animal
Physiology,
Babraham,
Cambridge CB2 4AT,
United Kingdom

ELSEVIER SCIENTIFIC PUBLISHING COMPANY

AMSTERDAM/NEW YORK/OXFORD

1979

ISBN 0-444-80104-9 (Series)
ISBN 0-444-80105-7 (Vol. 49)

with 214 Illustrations and 49 Tables

Published by:

Elsevier/North-Holland Biomedical Press
335 Jan van Galenstraat, P.O. box 211
Amsterdam, The Netherlands

Sole distributions for the U.S.A. and Canada:

Elsevier North-Holland, Inc.
52 Vanderbilt Avenue
New York, N.Y. 10017

Library of Congress Cataloging in Publication Data

  Main entry under title:

  The cholinergic synapse.

    (Progress in brain research ; vol. 49)
    1. Synapses--Congresses. 2. Cholinergic receptors--
Congresses. 3. Neural transmission--Congresses.
4. Myathenia gravis--Congresses. I. Tuček, S.
II. Barnard, Eric A., 1927-   III. Series.
[DNLM: 1. Synapses--Congresses. 2. Acetylcholine--
Congresses. 3. Receptors, Cholinergic--Congresses.
4. Neuromuscular junction--Congresses. Wl PR667J v.
49 / WL102.8 I61c 1978]
QP376.P vol. 49 [QP364] 612'.82'08s [591.1'88]
  ISBN 0-444-80105-7 (v. 49)   79-577

PRINTED IN THE NETHERLANDS

Preface

The articles and abstracts in this volume are based on the proceedings of an international symposium — The Cholinergic Synapse, which was organized by the Czechoslovak Medical Society and the Institute of Physiology of the Czechoslovak Academy of Sciences in the Castle at Žinkovy, about 130 km South-West of Prague, on May 15–19, 1978. Two important features were characteristic of Žinkovy Symposium and both contributed greatly to its success: (a) its interdisciplinarity, permitting personal encounters and exchange of views between electron microscopists, neurochemists, electrophysiologists, cell biologists, neuropathologists and representatives of other disciplines; (b) its multinational nature offering participants an opportunity to establish many new scientific and personal contacts.

The scientific papers contained in this volume demonstrate the great progress that has been achieved in the understanding of the physiology and pathology of cholinergic synaptic transmission. As Professor R.D. O'Brien put it in a postsymposial letter: The scientific content of the Symposium . . . served to illustrate what astonishing progress has been made in recent years, so that one can talk for three days about one kind of junction, and still not have it all said.

The book cannot, unfortunately, transmit the stimulating atmosphere that prevailed at Žinkovy and all the scientific information that became available during formal and informal discussions. It also cannot show the important contribution of the chairmen and co-chairmen of the sessions — Drs. Ann Silver, J.C. Szerb, F.C. MacIntosh, L.G. Magazanik, R.D. O'Brien, S. Thesleff, V.I. Skok, Edith Heilbronn and V.P. Whittaker, all of whom I should like to thank for their invaluable help.

In spite of its shortcomings, the present volume of Progress in Brain Research will undoubtedly serve as an important source of information to all those interested in various aspects of cholinergic synaptic transmission. Its publication would not have been possible without the tireless effort, ingenuity and sacrifice of time by the co-editors, E.A. Barnard, B. Collier, J.S. Kelly, R.M. Marchbanks, H.P. Rang and Ann Silver and without the devoted secretarial assistance of Mrs. Blanka Buchtová. My thanks are due to all of them, to the authors of all contributions, and also to the members of the Department of Neuromuscular Physiology who participated in the organization of the Symposium at Žinkovy.

Stanislav Tuček

List of Contributors

S. ALEMA, Laboratory of Cellular Biology, C.N.R., Roma, Italy.

M.J. ANDERSON, Neurobiology Department, The Salk Institute, LaJolla, CA, U.S.A.

S. ANGIELSKI, Department of Clinical Biochemistry, Institute of Pathology, Medical Academy, 80-211 Gdańsk, Poland.

J. BAJGAR, Purkyně Medical Research Institute, 50260 Hradec Králové, Czechoslovakia.

M. BAUMANN, Département de Pharmacologie, Ecole de Médecine, CH-1211 Genève 4, Switzerland.

G. BIERKAMPER, Institute of Veterinary Pharmacology and Technology, University, Utrecht, The Netherlands.

V. BIGL, Paul Flechsig Brain Research Institute, Karl-Marx-University, 7039 Leipzig, G.D.R.

P. BOKSA, Department of Pharmacology and Therapeutics, McGill University, Montreal, Quebec H3G 1Y6, Canada.

S. BON, Laboratoire de Neurobiologie, Ecole Normale Supérieure, 75230 Paris, France.

R.J. BRADLEY, The Medical Center, University of Alabama, Birmingham, AL 35294, U.S.A.

P.D. BREGESTOVSKI, Institute of Biological Physics, Academy of Sciences, 142292 Pushchino, U.S.S.R.

M. BRZIN, Institute of Pathophysiology, University, 61105 Ljubljana, Yugoslavia.

G. BURNSTOCK, Department of Anatomy and Embryology, University College, London WC1E 6BT, United Kingdom.

D.J. CARD, Department of Pharmacological and Physiological Sciences, University of Chicago, Chicago, IL, U.S.A.

S. CARSON, Laboratoire de Neurobiologie, Ecole Normale Supérieure, 75230 Paris, France.

B. CECCARELLI, Department of Pharmacology, C.N.R. Centre of Cytopharmacology, University, 20129 Milan, Italy.

J.-P. CHANGEUX, Laboratoire de Neurobiologie Moléculaire, Institut Pasteur, 75723 Paris, France.

M. CHRÉTIEN, Départment de Biologie, C.E.A., Centre d'Etudes Nucléaires de Saclay, 91190 Gif-sur-Yvette, France.

M.V. COHEN, Department of Physiology, McGill University, Montreal, Quebec H3G 1Y6, Canada.

B. COLLIER, Department of Pharmacology and Therapeutics, McGill University, Montreal, Quebec H3G 1Y6, Canada.

S. CONSOLO, Istituto di Ricerche Farmacologiche "Mario Negri", 20157 Milan, Italy.

J.Y. COURAUD, Départment de Biologie, C.E.A., Centre d'Etudes Nucléaires de Saclay, 91190 Gif-sur-Yvette, France.

R.A. DEMEL, Laboratoire de Neurobiologie Moléculaire, Institute Pasteur, 75724 Paris, France.

M.J. DENNIS, Department of Physiology, University of California Medical Center, San Francisco, CA 94143, U.S.A.

P.N. DEVREOTES, Department of Biochemistry, University of Chicago, Chicago, IL, U.S.A.

L. DI GIAMBERARDINO, Département de Biologie, C.E.A., Centre d'Etudes Nucléaires de Saclay, 91190 Gif-sur-Yvette, France.

R. DINGLEDINE, Epilepsy Unit and Department of Physiology, Duke University, Durham, NC 27710, U.S.A.

J. DODD, MRC Neurochemical Pharmacology Research Unit, Department of Pharmacology, Medical School, Cambridge CB2 2QD, United Kingdom.

M.J. DOWDALL, Department of Biochemistry, The Medical School, University, Nottingham, United Kingdom.

F. DREYER, II. Physiologisches Institut, Universität des Saarlandes, D-6650 Homburg/Saar, F.R.G.

B. DROZ, Département de Biologie, C.E.A., Centre d'Etudes Nucléaires de Saclay, 91190 Gif-sur-Yvette, France.

Y. DUNANT, Département de Pharmacologie, Ecole de Médecine, CH-1211 Geneva, Switzerland.

D. DWYER, The Medical Center, University of Alabama, Birmingham, AL 35294, U.S.A.

L. EDER, Département de Pharmacologie, Ecole de Médecine, CH-1211 Genève 4, Switzerland.

A.G. ENGEL, Department of Neurology and Neuromuscular Research Laboratory, Mayo Clinic and Mayo Foundation, Rochester, MN 55901, U.S.A.

D.M. FAMBROUGH, Department of Embryology, Carnegie Institution of Washington, Baltimore, MD 21210, U.S.A.

T. FARKAS, Institute of Biochemistry, Hungarian Academy of Sciences, H-6701 Szeged, Hungary.

V.V. FEDOROV, Sechenov Institute of Evolutionary Physiology and Biochemistry, 194223 Leningrad, U.S.S.R.

C. FROISSART, C.N.R.S. Neurochemical Centre and Institute of Biological Chemistry, Louis Pasteur University, 67085 Strasbourg, France.

B.W. FULPIUS, Department of Biochemistry, University, CH-1211 Geneva 4, Switzerland.

J.M. GARDNER, Department of Embryology, Carnegie Institution of Washington, Baltimore, MD 21210, U.S.A.

V.P. GEORGIEV, Institute of Physiology, Bulgarian Academy of Sciences, 1113 Sofia, Bulgaria.

R.E. GIBSON, Section of Radiochemistry and Radiopharmacology, George Washington Unversity Medical Center, Washington, DC 20037, U.S.A.

V. GISIGER, Laboratoire de Neurobiologie, Ecole Normale Supérieure, 75230 Paris, France.

J. GODLEWSKA-JĘDRZEJCZYK, Institute of Biostructure, Medical Academy, 02-004 Warsaw, Poland.

V. GOLDA, Department of Neurosurgery, Medical Faculty, Charles University, Hradec Králové, Czechoslovakia.

A.M. GOLDBERG, Department of Environmental Health Sciences, The Johns Hopkins University, School of Hygiene and Public Health, Baltimore, MD 21205, U.S.A.

R. GOMENI, Istituto di Ricerche Farmacologiche "Mario Negri", 20157 Milan, Italy.

F. GROHOVAZ, Department of Pharmacology, C.N.R. Centre of Cytopharmacology, University, 20129 Milan, Italy.

I. HANIN, Western Psychiatric Institute and Clinic, Department of Psychiatry, University of Pittsburgh School of Medicine, Pittsburgh, PA 15261, U.S.A.

A.J. HARRIS, Department of Physiology, University of Otago Medical School, Dunedin, New Zealand.

T. HATTORI, Kinsmen Laboratory of Neurological Research, Department of Psychiatry, University of British Columbia, Vancouver, B.C. V6T 1W5, Canada.

E. HEILBRONN, National Defence Research Institute, Department 4, S-172 04 Sundbyberg, Sweden.

J.E. HEUSER, Department of Physiology, University of California School of Medicine, San Francisco, CA 94143, U.S.A.

B. HOLMGREN, Department of Neurophysiology, Centro Nacional de Investigaciones Científicas, La Habana, Cuba.

V. HRDINA, Purkyně Medical Research Institute, 50260 Hradec Králové, Czechoslovakia.

W.P. HURLBUT, Department of Pharmacology, C.N.R. Centre of Cytopharmacology, University, 20129 Milan, Italy.

N. IANCHEVA, Regeneration Research Laboratory, Bulgarian Academy of Sciences, Sofia, Bulgaria.

V.I. IL'IN, Institute of Biological Physics, Academy of Sciences, 142292 Pushchino U.S.S.R.

M. ISRAËL, Department of Neurochemistry, Laboratory of Cellular Neurobiology, C.N.R.S., 91190 Gif-sur-Yvette, France.

I. JIRMANOVÁ, Institute of Physiology, Czechoslovak Academy of Sciences, 14220 Prague, Czechoslovakia.

R. JONES, Department of Biochemistry, University, Birmingham B15 2TT, United Kingdom.

J. JUNTUNEN, Institute of Occupational Health, Section of Neurology and Clinical Neurophysiology, SF-00290 Helsinki, Finland.

O. KADLEC, Institute of Pharmacology, Czechoslovak Academy of Sciences, 12800 Prague, Czechoslovakia.

M.A. KAMENSKAYA, Department of Human and Animal Physiology, Moscow State University, 117234 Moscow, U.S.S.R.

P. KÁSA, Central Research Laboratory, Medical University, H-6720 Szeged, Hungary.

G. KATO, Institut Batelle, Geneva, Switzerland.

J.S. KELLY, MRC Neurochemical Pharmacology Research Unit, Department of Pharmacology, Medical School, Cambridge CB2 2QD, United Kimgdom.

G.E. KEMP, The Medical Center, University of Alabama, Birmingham, AL 35294, U.S.A.

L. KESZTHELYI, Institute of Biochemistry, Biological Research Center, Hungarian Academy of Sciences, H-6701 Szeged, Hungary.

Kh.S. KHAMITOV, Department of Physiology, Medical Institute, 420012 Kazan, U.S.S.R.

T. KIAUTA, Institute of Pathophysiology, University, 61105 Ljubljana, Yugoslavia.

H. KILBINGER, Department of Pharmacology, University, D-6500 Mainz, F.R.G.

H.L. KOENIG, Laboratoire de Neurocytologie, Université P. et M. Curie, 75005 Paris, France.

J. KOENIG, Laboratoire de Neurocytologie, Université P. et M. Curie, 75005 Paris, France.

U. KOPP, Western Psychiatric Institute and Clinic, Department of Psychiatry, University of Pittsburgh School of Medicine, Pittsburgh, PA 15261, U.S.A.

M.E. KRIEBEL, Upstate Medical Center, Syracuse, NY 13210, U.S.A.

H.J. KSIĘŻAK, Medical Research Centre, Polish Academy of Sciences, 00-784 Warsaw, Poland.

M.J. KUHAR, Departments of Pharmacology and Experimental Therapeutics, Psychiatry and the Behavioral Sciences, The Johns Hopkins University School of Medicine, Baltimore, MD 212-5, U.S.A.

D.J. KUPFER, Western Psychiatric Institute and Clinic, Department of Psychiatry, University of Pittsburgh School of Medicine, Pittsburgh, PA 15261, U.S.A.

H. LADINSKY, Istituto di Ricerche Farmacologiche "Mario Negri", 20157 Milan, Italy.

N.D. LAMBADJIEVA, Institute of Physiology, Faculty of Biology, Sofia University, 1113 Sofia, Bulgaria.

L.T. LANDMESSER, Department of Biology, Yale University, New Haven, CT 06520, U.S.A.

W.A. LARGE, Department of Pharmacology, St. George's Hospital Medical School, London SW17, United Kingdom.

L. LEHOTAI, Central Research Laboratory, Medical University, H-6720 Szeged, Hungary.

B. LESBATS, Department of Neurochemistry, Laboratory of Cellular Neurobiology, C.N.R.S., 91190 Gif-sur-Yvette, France.

F. LLADOS, Upstate Medical Center, Syracuse, NY 13210, U.S.A.

S. LOVAT, Department of Pharmacology and Therapeutics, McGill University, Montreal, Quebec H3G 1Y6, Canada.

H.-J. LÜTH, Paul Flechsig Institute of Brain Research, Karl-Marx-University, 701 Leipzig, G.D.R.

J. LYLES, Department of Biochemistry, Imperial College of Science and Technology, London SW7, United Kingdom.

W. ŁYSIAK, Institute of Pathology, Medical Academy, 80-211 Gdańsk, Poland.

J. MACHOVÁ, Institute of Experimental Pharmacology, Slovak Academy of Sciences, 88105 Bratislava, Czechoslovakia.

L.G. MAGAZANIK, Sechenov Institute of Evolutionary Physiology and Biochemistry, 194223 Leningrad, U.S.S.R.

K.L. MAGLEBY, Department of Physiology and Biophysics, University of Miami School of Medicine, Miami, FL 33101, U.S.A.

D. MALTHE-SØRENSSEN, Norwegian Defence Research Establishment, Division for Toxicology, N-2007 Kjeller, Norway.

R. MANARANCHE, Department of Neurochemistry, Laboratory of Cellular Neurobiology, C.N.R.S., 91190 Gif-sur-Yvette, France.

S.P. MANN, A.R.C. Institute of Animal Physiology, Babraham, Cambridge CB2 4AT, United Kingdom.

S. MANOLOV, Regeneration Research Laboratory, Bulgarian Academy of Sciences, 1431 Sofia, Bulgaria.

P. MANTOVANI, Department of Pharmacology, University, 50134 Florence, Italy.

R.M. MARCHBANKS, Department of Biochemistry, Institute of Psychiatry, De Crespigny Park, London SE5 8AF, United Kingdom.

R. MASSARELLI, C.N.R.S. Neurochemical Centre and Institute of Biological Chemistry, Louis Pasteur University, 67085 Strasbourg, France.

J. MASSOULIÉ, Laboratoire de Neurobiologie, Ecole Normale Supérieure, 75230 Paris, France.

K. MAŠEK, Institute of Pharmacology, Czechoslovak Academy of Sciences, 12800 Prague, Czechoslovakia.

D.R. MATTESON, Upstate Medical Center, Syracuse, NY 13210, U.S.A.

E.G. McGEER, Kinsmen Laboratory of Neurological Research, Department of Psychiatry, University of British Columbia, Vancouver, Canada.

P.L. McGEER, Kinsmen Laboratory of Neurological Research, Department of Psychiatry, University of British Columbia, Vancouver, Canada.

U. MEYER, Institute of Anatomy, Humboldt University, 104 Berlin, G.D.R.

R. MILEDI, Department of Biophysics, University College London, London WC1E 6BT, United Kingdom.

P.C. MOLENAAR, Department of Pharmacology, Leiden University Medical Centre, Leiden, The Netherlands.

N. MOREL, Department of Neurochemistry, Laboratory of Cellular Neurobiology, C.N.R.S., 91190 Gif-sur-Yvette, France.

B.J. MORLEY, The Medical Center, University of Alabama, Birmingham, AL 35294, U.S.A.

A.M. MUSTAFIN, Department of Biophysics, Kazakh State University, 480091 Alma-Ata, U.S.S.R.

K.-D. MÜLLER, II. Physiologisches Institut, Universität des Saarlandes, D-6650 Homburg/Saar, F.R.G.

A. NAGY, Institute of Biochemistry, Hungarian Academy of Sciences, H-6701 Szeged, Hungary.

G.A. NASLEDOV, Sechenov Institute of Evolutionary Physiology and Biochemistry, 194223 Leningrad, U.S.S.R.

J.F. NEIL, Western Psychiatric Institute and Clinic, Department of Psychiatry, University of Pittsburgh School of Medicine, Pittsburgh, PA 15261, U.S.A.

C. NEVAR, Western Psychiatric Institute and Clinic, Department of Psychiatry, University of Pittsburgh School of Medicine, Pittsburgh, PA 15261, U.S.A.

J. NEWSON-DAVIS, Department of Neurological Science, Royal Free Hospital, London NW3 2QG, United Kingdom.

A. NORDBERG, Department of Pharmacology, University, S-751 23 Uppsala, Sweden.

R.D. O'BRIEN, Section of Neurology and Behavior, Cornell University, Ithaca, NY 14853, U.S.A.

S.H. OH, The Medical Center, University of Alabama, Birmingham, AL 35294, U.S.A.

W. OVTSCHAROFF, Department of Anatomy, Histology and Embryology, Medical Academy, 1431 Sofia, Bulgaria.

M. PÉCOT-DECHAVASSINE, Laboratoire de Cytologie, Université P. et M. Curie, 75230 Paris, France.

K. PEPER, II. Physiologisches Institut, Universität des Saarlandes, D-6650 Homburg/Saar, F.R.G.

G. PEPEU, Department of Pharmacology, University, 50134 Florence, Italy.

G. PILAR, Physiology Section, Biology Sciences Group, University of Connecticut, Storrs, CT 06268, U.S.A.

R.L. POLAK, Medical Biological Laboratory T.N.O., Rijswijk-Z.H., The Netherlands.

G.I. POLETAEV, Department of General Biology and Biophysics, Medical Institute, 420012 Kazan, U.S.S.R.

J.-L. POPOT, Laboratoire de Neurobiologie Moléculaire, Institut Pasteur, 75724 Paris, France.

T.L. RADZYUKEVICH, Sechenov Institute of Evolutionary Physiology and Biochemistry, 194223 Leningrad, U.S.S.R.

Z. RAKONCZAY, Institute of Biochemistry, Biological Research Center, Hungarian Academy of Sciences, H-6701 Szeged, Hungary.

H.P. RANG, Department of Pharmacology, St. George's Hospital Medical School, London SW17 O9T, United Kingdom.

J. ROBERT, C.N.R.S. Neurochemical Centre and Institute of Biological Chemistry, Louis Pasteur University, 67085 Strasbourg, France.

G. ROBINSON, The Medical Center, University of Alabama, Birmingham, AL 35294, U.S.A.

S.R. SALPETER, Department of Physiology, University of California School of Medicine, San Francisco, CA 94143, U.S.A.

I. von SCHWARZENFELD, Pharmakologisches Institut, Universität, Mainz, F.R.G.

A.A. SELYANKO, Bogomolets Institute of Physiology, 252601 Kiev, U.S.S.R.

L. SERVETIADIS, Département de Pharmacologie, Ecole de Médecine, CH-1211 Geneva, Switzerland.

I. SILMAN, Department of Biochemistry, Imperial College of Science and Technology, London SW7 2AZ, United Kingdom.

V.K. SINGH, Kinsmen Laboratory of Neurological Research, Department of Psychiatry, University of British Columbia, Vancouver, Canada.

V.I. SKOK, Bogomolets Institute of Physiology, 252601 Kiev, U.S.S.R.

V.A. SNETKOV, Sechenov Institute of Evolutionary Physiology and Biochemistry, 194223 Leningrad, U.S.S.R.

A. SOBEL, Laboratoire de Neurobiologie Moléculaire, Institut Pasteur, 75724 Paris, France.

G.T. SOMOGYI, Department of Pharmacology, Semmelweis University of Medicine, H-1085 Budapest, Hungary.

F. SOUYRI, Département de Biologie, C.E.A., Centre d'Etudes Nucléaires de Saclay, 91190 Gif-sur-Yvette, France.

D.G. SPIKER, Western Psychiatric Institute and Clinic, Department of Psychiatry, University of Pittsburgh School of Medicine, Pittsburgh, PA 15261, U.S.A.

R. STERZ, II. Physiologisches Institut, Universität des Saarlandes, D-6650 Homburg/Saar, F.R.G.

G. SUDLOW, Max-Planck-Institut für biophysikalische Chemie, Abt. Neurochemie, D-3400 Göttingen, F.R.G.

K. SUMIKAWA, Section of Neurobiology and Behavior, Cornell University, Ithaca, NY 14853, U.S.A.

A. SUNDWALL, AB KABI, Preclinical Research, S-104 25 Stockholm, Sweden.

J.B. SUSZKIW, Biological Sciences Group, University of Connecticut, Storrs, CT 06268, U.S.A.

I. SYROVÝ, Institute of Physiology, Czechoslovak Academy of Sciences, 14220 Prague, Czechoslovakia.

J.C. SZERB, Department of Physiology and Biophysics, Dalhousie University, Halifax, N.S. B3H 4H7, Canada.

A. SZUTOWICZ, Institute of Pathology, Medical Academy, 80-211 Gdańsk, Poland.

S. ŠTOLC, Institute of Experimental Pharmacology, Slovak Academy of Sciences, 881 05 Bratislava, Czechoslovakia.

S. THESLEFF, Department of Pharmacology, University, S-223 62 Lund, Sweden.

T.L. TÖRÖK, Department of Pharmacology, Semmelweis University of Medicine, H-1085 Budapest, Hungary.

E.G. ULUMBEKOV, Department of Histology, Medical Institute, 420012 Kazan, U.S.S.R.

R. URBÁ-HOLMGREN, Department of Neurophysiology, Centro Nacional de Investigaciones Científicas, La Habana, Cuba.

K. VACA, Physiology Section, Biological Sciences Group, University of Connecticut, Storrs, CT 06268, U.S.A.

L.L.M. VAN DEENEN, Biochemical Laboratory, University, Utrecht, The Netherlands.

L. VENKOV, Regeneration Research Laboratory, Bulgarian Academy of Sciences, Sofia, Bulgaria.

B.N. VEPRINTSEV, Institute of Biological Physics, Academy of Sciences, 142292 Pushchino, U.S.S.R.

M. VIGNY, Laboratoire de Neurobiologie, Ecole Normale Supérieure, 75230 Paris, France.

A. VINCENT, Department of Neurological Science, Royal Free Hospital, London NW3 2QG, United Kingdom.

E.S. VIZI, Department of Pharmacology, Semmelweis University of Medicine, H-1085 Budapest, Hungary.

E.M. VOLKOV, Department of General Biology and Biophysics, Medical Institute, 420012 Kazan, U.S.S.R.

C.A. VULFIUS, Institute of Biological Physics, Academy of Sciences, 142292 Pushchino, U.S.S.R.

F. VYSKOČIL, Institute of Physiology, Czechoslovak Academy of Sciences, 14220 Prague, Czechoslovakia.

P.R. WELDON, Department of Physiology, McGill University, Montreal, Quebec H3G 1Y6, Canada.

H. WENK, Institute of Anatomy, Humboldt University, 104 Berlin, G.D.R.

G. WAHLSTRÖM, Department of Pharmacology, University, S-901 87 Umeå, Sweden.

A. WERNIG, Max-Planck-Institut für Psychiatrie, München, F.R.G.

V.P. WHITTAKER, Max-Planck-Institut für biophysikalische Chemie, Abt. Neurochemie, D-3400 Göttingen, F.R.G.

P. WOOD, Laboratory of Preclinical Pharmacology, St. Elizabeth Hospital, Washington, DC 20032, U.S.A.

T.Y. WONG, C.N.R.S. Neurochemical Centre and Institute of Biological Chemistry, Louis Pasteur University, 67085 Strasbourg, France.

S. WONNACOTT, Department of Biochemistry, Institute of Psychiatry, De Crespigny Park, London SE5 8AF, United Kingdom.

O.P. YURCHENKO, Laboratory of Nerve Cell Biophysics, Institute of Biological Physics, Academy of Sciences, 142292 Pushchino, U.S.S.R.

A.L. ZEFIROV, Department of Physiology, State University, 42008 Kazan, U.S.S.R.

L.N. ZEFIROV, Department of Physiology, State University, 42008 Kazan, U.S.S.R.

J. ZELENÁ, Institute of Physiology, Czechoslovak Academy of Sciences, 14220 Prague, Czechoslovakia.

H. ZIMMERMANN, Max-Planck-Institut für biophysikalische Chemie, Abt. Neurochemie, D-3400 Göttingen, F.R.G.

E. ZORYCHTA, Department of Pathology, McGill University, Montreal, Quebec H3G 1Y6, Canada.

A.D. ZURN, Department of Biochemistry, University, CH-1211 Geneva, Switzerland.

Fig.: A group of participants of the Symposium at Žinkovy.

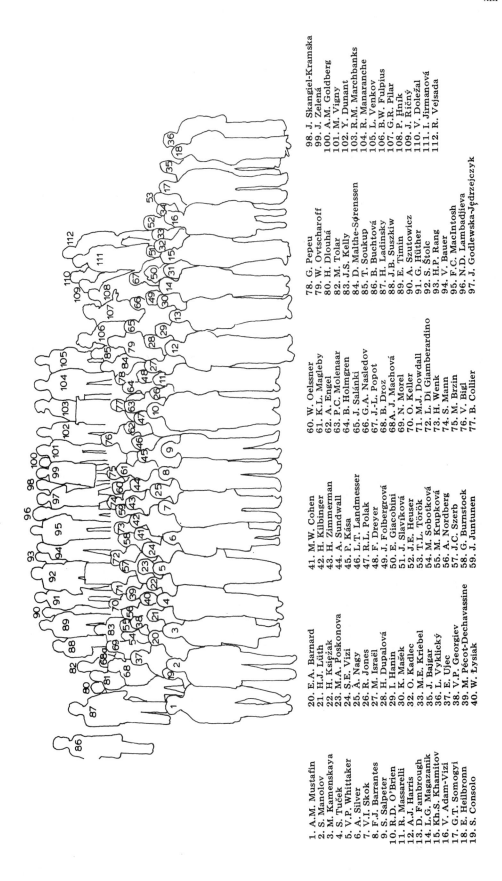

1. A.M. Mustafin
2. S. Manolov
3. M. Kamenskaya
4. S. Tuček
5. V.P. Whittaker
6. A. Silver
7. V.I. Skok
8. F.J. Barrantes
9. S. Salpeter
10. R.D. O'Brien
11. R. Massarelli
12. A.J. Harris
13. D. Fambrough
14. L.G. Magazanik
15. Kh.S. Khamitov
16. V. Adam-Vizi
17. G.T. Somogyi
18. E. Heilbronn
19. S. Consolo

20. E.A. Barnard
21. H.J. Lüth
22. H. Księżak
23. M.A. Poskonova
24. S.E. Vizi
25. Á. Nagy
26. R. Jones
27. M. Israël
28. H. Dupalová
29. I. Hanin
30. K. Mašek
31. J. Slavíková
32. J.E. Heuser
33. M.E. Kriebel
34. M. Sobotková
35. J. Bajgar
36. L. Vyklický
37. E. Ujec
38. V.P. Georgiev
39. M. Pécot-Dechavassine
40. W. Lysiak

41. M.W. Cohen
42. H. Kilbinger
43. H. Zimmerman
44. A. Sundwall
45. P. Kasa
46. L.T. Landmesser
47. R.L. Polak
48. F. Dreyer
49. J. Folbergrová
50. E. Giacobini
51. J. Slavíková
52. J.E. Heuser
53. T.L. Török
54. M. Sobotková
55. M. Krupková
56. A. Nordberg
57. J.C. Szerb
58. G. Burnstock
59. J. Juntunen

60. W. Oelssner
61. K.L. Magleby
62. A. Engel
63. P.C. Molenaar
64. B. Holmgren
65. J. Salánki
66. G.A. Nasledov
67. J.-L. Popot
68. B. Droz
68A.J. Machová
69. N. Morel
70. O. Keller
71. M.J. Dowdall
72. L. Di Giambernardino
73. H. Wenk
74. S. Mann
75. M. Brzin
76. V. Bigl
77. B. Collier

78. G. Pepeu
79. W. Ovtscharoff
80. H. Dlouhá
82. M. Tolar
83. J.S. Kelly
84. D. Malthe-Sørenssen
85. T. Soukup
86. B. Buchtová
87. H. Ladinsky
88. J.B. Suszkiw
89. E. Timin
90. A. Szutowicz
91. G. Hüther
92. S. Stolc
93. H.P. Rang
94. V. Bauer
95. F.C. MacIntosh
96. N.D. Lambadjieva
97. J. Godlewska-Jędrzejczyk

98. J. Skangiel-Kramska
99. J. Zelená
100. A.M. Goldberg
101. M. Vigny
102. Y. Dunant
103. R.M. Marchbanks
104. R. Manaranche
105. L. Venkov
106. B.W. Fulpius
107. G.R. Pilar
108. P. Hník
109. J. Říčný
110. V. Doležal
111. I. Jirmanová
112. R. Vejsada

Abbreviations

Abbreviations are explained in the text when the first time used in the articles. The most frequently used abbreviations are listed below.

acetyl-CoA acetyl-coenzyme A
ACh acetylcholine
AChE acetylcholinesterase
AChR acetylcholine receptor
ATP adenosine triphosphate
ATPase adenosine triphosphatase
α-BgTX α-bungarotoxin
BoTX botulotoxin
BuChE butyrylcholinesterase
CCh carbamylcholine
Ch choline
ChAT choline acetyltransferase
ChE cholinesterase
DA dopamine
dpm disintegrations per minute
DTT dithiothreitol
EAMG experimental autoimmune myasthenia gravis
EM electron microscopy
EPC end-plate current
EPP end-plate potential
GABA γ-aminobutyric acid
HACU high affinity choline uptake
HC-3 hemicholinium-3
IgG γ-immunoglobulin
MEPC miniature end-plate current
MEPP miniature end-plate potential
MG myasthenia gravis
MP membrane potential
NA noradrenaline
NEM *N*-ethylmaleimide
6-OHDA 6-hydroxydopamine

PG	prostaglandin
SCG	superior cervical ganglion
SDHACU	sodium dependent high affinity choline uptake
SDS	sodium dodecyl sulphate
S.E.M.	standard error of the mean
SRA	specific radioactivity
TEA	tetraethylammonium
TLC	thin-layer chromatography
TMA	tetramethylammonium
TTX	tetrodotoxin

Contents

Part I – Ultrastructure of Cholinergic Junctions; Synthesis of Acetylcholine

Part II – Storage and Release of Acetylcholine

Part III – Postsynaptic Action of Acetylcholine: Electrophysiology

Part IV – Postsynaptic Action of Acetylcholine: Biochemistry

Part V – Development of Cholinergic Synapses; Neurotrophic Regulation; Intercelluar Relations

Part VI – Pathology

Part VII – Abstracts

PART I

ULTRASTRUCTURE OF CHOLINERGIC JUNCTIONS
SYNTHESIS OF ACETYLCHOLINE

The Ultrastructure of Autonomic Cholinergic Nerves and Junctions

G. BURNSTOCK

*Department of Anatomy and Embryology, University College London,
London, WC1E 6BT (United Kingdom)*

INTRODUCTION

Cholinergic neurones are widely distributed in the autonomic nervous system. The preganglionic neurones of both parasympathetic and sympathetic systems are cholinergic (see Dale, 1937). The postganglionic parasympathetic neurones are mainly cholinergic and, while the majority of postganglionic sympathetic fibres are adrenergic, a minority are cholinergic — e.g., those supplying the sweat glands, vessels in skeletal muscles of some species and in muscles of the tongue (see Gabella, 1976). Many cholinergic neurones are also present in intramural ganglia in the walls of visceral organs.

ULTRASTRUCTURE OF AUTONOMIC PREGANGLIONIC CHOLINERGIC ENDINGS

Preganglionic fibres originating from neurones in the thoracic and sacral spinal cord and the brain stem penetrate into the ganglia and after extensive branching they synapse on dendrites and cell bodies of ganglion neurones and on somatic and dendritic spines. The bulbous endings of preganglionic fibres are similar to cholinergic endings in other parts of the nervous system, but they are very variable in size (diameter of the sectioned profiles varies between 0.3 and 2.7 μm). Terminal and "en passage" contacts occur.

The nerve endings contain large numbers of small agranular vesicles (40–60 nm in diameter) and a small number (about 1%) of large vesicles (60–150 nm in diameter) with a granular core. Dense projections are present in the presynaptic membrane and synaptic vesicles are clustered around them (Figs. 1a, b).

ULTRASTRUCTURE OF AUTONOMIC POSTGANGLIONIC CHOLINERGIC NERVES

(1) Terminal varicosities

Postganglionic autonomic cholinergic nerves, in common with other autonomic nerves, differ from cholinergic motor nerves supplying skeletal muscle in

4

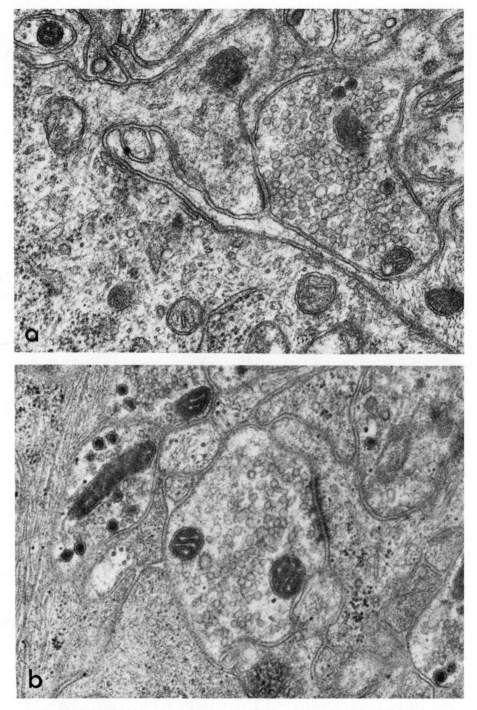

Fig. 1. (a) A cholinergic preganglionic nerve ending forming a synapse on a small process arising from the soma of a ganglion cell in the superior cervical ganglion of the rat. ×50,000. (b) A cholinergic nerve ending forming a synapse on a dendrite originating from an intra-mural neurone in the myenteric plexus of the guinea-pig ileum. ×40,000. (Both a and b by courtesy of Giorgio Gabella, University College London.)

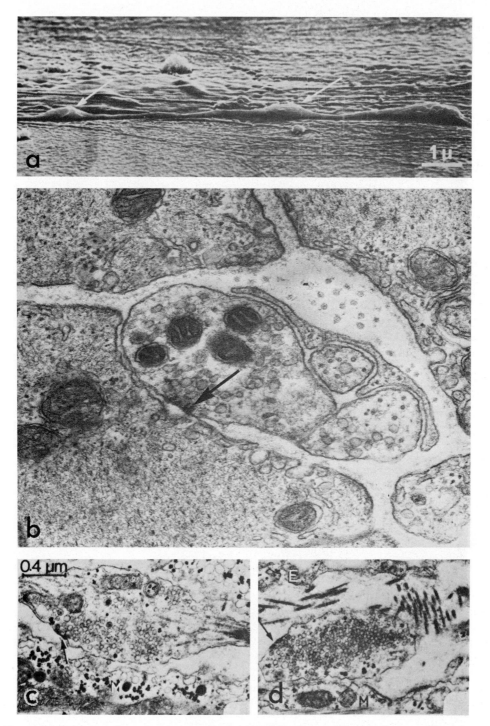

Fig. 2. (a) Scanning electron micrograph of varicosities (arrows) in a single nerve fibre growing in a culture of newborn guinea-pig sympathetic ganglia. Photographed at an angle of
70°C. (Courtesy of J.H. Chamley, C. Hill and G. Burnstock in Burnstock, 1975b.)
(b) A cholinergic varicosity in close contact with a smooth muscle cell of the sphincter
pupillae of the guinea-pig. A dense projection with a cluster of vesicles is present in the prejunctional membrane (arrow). ×48,000. (Courtesy of Giorgio Gabella, University College
London.)
(c, d) Cholinergic varicosities in postganglionic axons in the frog heart. Note that some
vesicles are clustered next to the plasma membrane. The membrane and adjacent cytoplasm
in the areas (arrows) are very dense resembling a synaptic "thickening". In both cases the
varicosities are only a few tens of nanometres from muscle fibres (M). The membrane
"thickening" (arrow) in (c) faces the muscle fibre but in (d) it faces away from the muscle
and toward the endothelium (E). (From McMahan and Kuffler, 1971.)

6

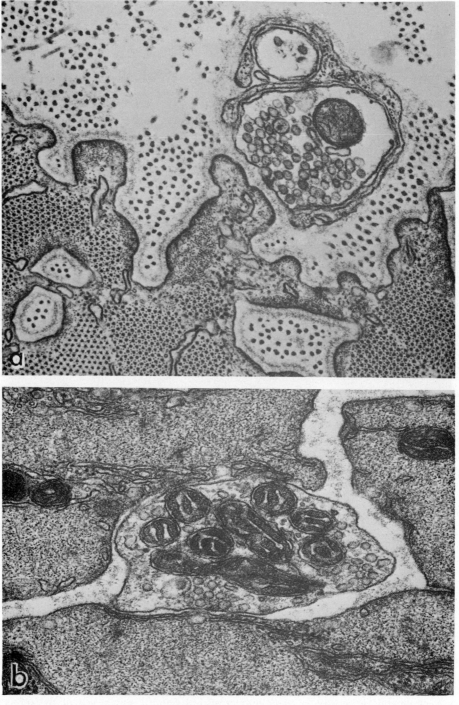

Fig. 3. (a) Myoseptal neuromuscular junction at a muscle-tendon junction in amphibian skeletal muscle: tail muscle of axolotl. Transverse section. Glutaraldehyde fixation. ×55,000. (From Uehara, Campbell and Burnstock, 1976.)
(b) Neuromuscular junction between a cholinergic varicosity and smooth muscle cells of the sphincter pupillae of the guinea-pig. ×50,000. (Courtesy of Giorgio Gabella, University College London.)

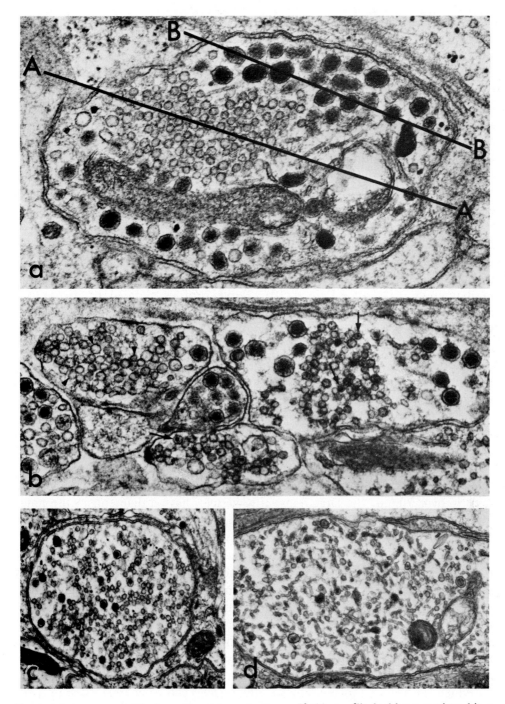

Fig. 4. (a) A nerve profile from the guinea-pig ileum. If this profile had been sectioned in a different plane, then it might have appeared to contain predominantly agranular vesicles (A—A) or predominantly large granular vesicles (B—B). ×54,100.
(b) Two axon profiles containing small granular vesicles (→) and a profile containing small agranular vesicles (▶). Guinea-pig ileum. ×45,600.
(c) Axon profile containing mostly small agranular vesicles and some large granular vesicles. Guinea-pig ileum. ×24,800.
(d) An axon profile containing round and flattened agranular vesicles and some elongated vesicles. Guinea-pig caecum. ×18,900. (From Cook and Burnstock, 1976.)

that they have extensive, varicose terminal fibres in effector organs from which transmitter is released "en passage" (Fig. 2; Burnstock, 1970). Short areas of varicosity membranes (about 0.1 μm) are thickened and associated with clusters of vesicles, suggesting that they may represent sites of transmitter release (Figs. 2b, c, d; McMahan and Kuffler, 1971).

(2) Intraxonal vesicles

Cholinergic autonomic varicosities are similar to motor and preganglionic nerve terminals in the composition of their vesicles, with a predominance of small agranular vesicles, 40—60 nm in diameter, and a small number of large granular vesicles (Figs. 3a, b, 4c, 6a, 9a).

Cholinergic autonomic nerve profiles can usually be clearly distinguished from other autonomic nerve types. Adrenergic profiles are characterised by a predominance of small granular vesicles (40—60 nm in diameter) and a variable number (3—24% depending on location) of large granular vesicles (60—120 nm in diameter) (Figs. 4b, 8d, 9c, d, 10). Nerve profiles claimed to be sensory (Burnstock and Iwayama, 1971) are packed with small mitochondria and contain few, if any, vesicles. A third type of profile representing non-adrenergic, non-cholinergic nerves contains a predominance of "large opaque vesicles" (80—200 nm in diameter). These profiles have been claimed to be either "purinergic" (Burnstock, 1972) or "peptidergic" (Baumgarten et al., 1970; Pearse, Polak and Bloom, 1977).

Analysis of the size and characteristics of the large granular vesicles in the cholinergic nerves shows that they differ in some respects from those seen in adrenergic nerves.

Large dense-core vesicles amount to less than 1% of the total vesicle population in autonomic cholinergic endings. In the cholinergic endings in the guinea-pig iris over 90% of the large dense-core vesicles have a diameter in the range 70—92 nm: in terms of size they constitute, therefore, a more uniform population than the large vesicles of adrenergic endings. The "large granular vesicles" in adrenergic nerves store noradrenaline and probably dopamine-β-hydroxylase (Burnstock and Costa, 1975). However, little is known of the role of the "large granular vesicles" in cholinergic nerves (Burnstock and Iwayama, 1971).

Although there are good grounds for concluding that cholinergic axons contain a predominance of agranular vesicles, there is no proof that all axons with this characteristic are cholinergic. For example, Chan-Palay (1975) has claimed that some of the 5-hydroxytryptamine-containing neurones in the CNS contain predominantly small agranular vesicles. No specific technique so far exists by which the acetylcholine content of the vesicles can be characterised in tissue sections, comparable to the technique using the capacity of adrenergic vesicles to concentrate electron-opaque noradrenaline analogs such as 5- or 6-hydroxy-dopamine (Fig. 9d; Tranzer and Thoenen, 1968).

The innervation of the gastrointestinal tract is extremely complex and up to 9 different types of axonal profiles containing distinctive vesicular populations have been identified (Gabella, 1972; Cook and Burnstock, 1976). Apart from nerve profiles typical of cholinergic nerves with a homogeneous population of round agranular vesicles (Figs. 4b, c), profiles containing many flattened agranular vesicles have been observed (Fig. 4d). Although it has been shown at motor end plates that flattened vesicles may arise from round vesicles as an effect of fixation (Korneliussen, 1972), it was suggested (Cook and Burnstock, 1976) that the location of these vesicles adjacent to profiles containing only round vesicles argues against this explanation in the gut. Further, in the CNS, axons thought to be inhibitory contain predominantly flattened vesicles (Uchizono, 1968). A further complication of nerve identification is that the entire vesicle population of a varicosity is not visible in a thin section. For example, if the nerve profile in the intestine shown in Fig. 4a were sectioned through A—A, it would be considered cholinergic, but if sectioned through B—B it would not. The possibility must also be raised that profiles of this kind containing a mixture of vesicle types (and also those shown in Fig. 4b) may represent nerves storing multiple transmitters (see Burnstock, 1976 and p. 15).

CHOLINERGIC AUTONOMIC NEUROMUSCULAR JUNCTIONS

Electron microscope and histochemical studies of the relationship of nerves to smooth muscle cells combined with electrophysiological studies of transmission have led to the development of a generalised model of the autonomic neuromuscular junction (Fig. 5a; Burnstock and Iwayama, 1971). The essential features of this model are that (1) transmitter is released "en passage" from large numbers of terminal varicosities (Fig. 2); (2) the effector is a muscle bundle rather than a single smooth muscle cell and individual cells within muscle effector bundles are connected by low resistance pathways, represented by "gap junctions" (Fig. 5b). In most organs, some, but not all, muscle cells are "directly innervated"; cells adjoining these are coupled electrotonically and when the muscle cells in an area of an effector bundle become depolarised, an all-or-none action potential is initiated which propagates through the tissue.

(1) Junctional cleft

The minimum width of the cleft between nerve varicosities and effector cells varies considerably in different tissues.

Nerve-muscle separation in the regions of closest apposition in the vas deferens and sphincter pupillae is about 15—20 nm (Figs. 2b, 3b, 6c). From an electron microscopic analysis of semi-serial sections (Merillees, 1968) combined with an electrophysiological study of the neuro-environment of single cells in the vas deferens, it was concluded that transmitter released from varicosities further than about 100 nm away would be unlikely to have a significant effect on muscle cells (Bennett and Merillees, 1966). This conclusion was supported by the experiments of Robinson (1969), who showed that about 15—20% of the axon profiles in the vas deferens had heavy positive staining for acetylcholinesterase (AChE), but that only those muscle membranes within 120 nm of these profiles showed matching AChE staining (Fig. 7). Furthermore, Schwann cell processes intervened between muscle membranes in 80% of all the cases where nerves were separated from muscle by distances greater than 110 nm.

10

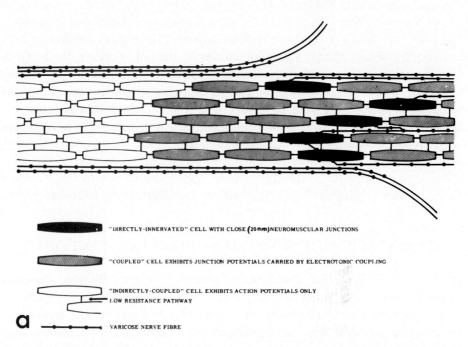

"DIRECTLY-INNERVATED" CELL WITH CLOSE (20 nm) NEUROMUSCULAR JUNCTIONS

"COUPLED" CELL EXHIBITS JUNCTION POTENTIALS CARRIED BY ELECTROTONIC COUPLING

"INDIRECTLY-COUPLED" CELL EXHIBITS ACTION POTENTIALS ONLY
LOW RESISTANCE PATHWAY

a

VARICOSE NERVE FIBRE

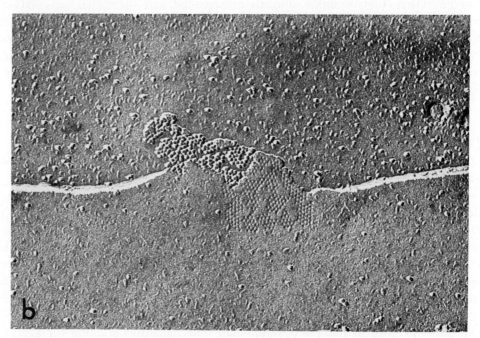

b

Fig. 5. (a) Schematic representation of autonomic innervation of smooth muscle. For explanation see text. (From Burnstock and Iwayama, 1971.)
(b) Freeze-fracture preparation of the circular muscle coat of the guinea-pig ileum. A gap-junction ("nexus") between two smooth muscle cells shown partly on the P-face of one cell (top) partly on the E-face of the other cell (bottom). ×100,000. (Courtesy of Giorgio Gabella, University College London.)

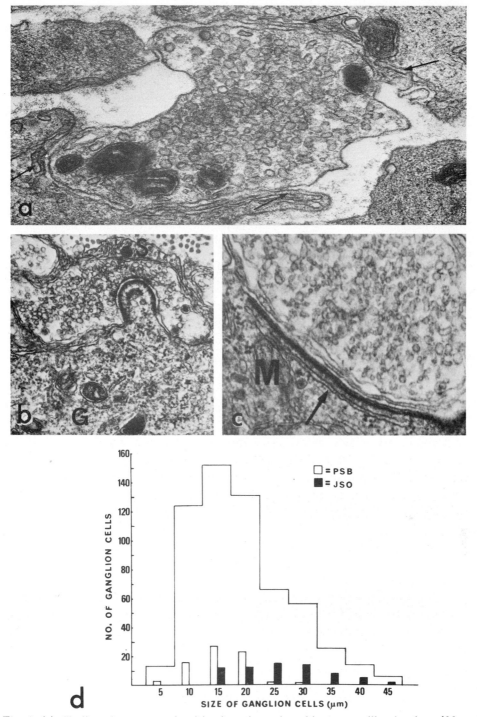

Fig. 6. (a) Cholinergic nerve varicosities in guinea-pig sphincter pupillae in close (20 nm) apposition with smooth muscle. Note the subsynaptic cysternae (arrows). ×50,000. (Courtesy of Giorgio Gabella, University College London.)

(b) A "post-synaptic bar" (PSB) in a somatic spine arising from a small sympathetic ganglion cell (G) from the frog. S, satellite sheath. ×26,000.

(c) A "junctional subsurface organ" (JSO) separated from a presynaptic nerve ending on a frog sympathetic ganglion neurone by a thin process of satellite cell. The JSO is accompanied by an underlying cisterna of endoplasmic reticulum (arrow) and a mitochondrion (M). ×26,000.

(d) A histogram showing the size distribution of 550 frog sympathetic ganglion cells. The horizontal and vertical axes represent the diameter and numbers of measured ganglion cells. Thin open columns show the size distribution of 72 different cells associated with the PSB, and black columns show 68 cells associated with the JSO. The PSBs occur mostly in the small cells (less than 22 μm), while the JSOs occur in larger cells (up to 45 μm). (b, c and d from Watanabe and Burnstock, 1978.)

12

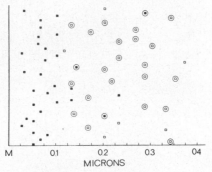

M   0.1   0.2   0.3   0.4

MICRONS

Fig. 7. Distribution of all the AChE-positive axons in the guinea-pig vas deferens which approached within 0.4 μm of a muscle cell membrane. The vertical line at the left of the figure (M) represents muscle membrane, and the horizontal scale indicates the distance in microns between the axon membrane and the muscle cell membrane at the closest point between them. Each square represents one stained axon. The full squares (■) indicate that the membrane of both muscle cell and axon were stained; the empty squares (□) indicate that only the axon was stained. A ring around the square indicates that a Schwann cell was present between axon and smooth muscle cell. The muscle membrane was stained in every approach in which the distance at the closest point was less than 110 nm, but in only a few approaches with a distance greater than this. (From Robinson, 1969.)

In blood vessels, the closest apposition between varicosities in the perivascular plexus at the adventitial-medial border and smooth muscle cells varies about 50 to 80 nm in small muscular arteries (Fig. 8d) and large arterioles to as much as 1–2 μm, in large elastic arteries (see Burnstock, Gannon and Iwayama, 1970; Bevan and Su, 1974; Burnstock, 1975a, 1978a). In the uterine artery of the guinea-pig in late pregnancy where the minimum junctional cleft is about 80 nm, it was suggested on the basis of AChE staining, that the maximum separation for transmitter released from varicosities to be effective was 1000 nm (Bell, 1969).

(2) Postjunctional specialisations

In densely-innervated tissues such as the vas deferens and iris, subsynaptic cysternae are often found at close (20 nm) neuromuscular junctions (Figs. 3b, 6a). No postjunctional specialisations have been consistently found for wider neuromuscular junctions.

GANGLIONIC SYNAPSES

The synaptic cleft in autonomic ganglia is of fairly uniform width, about 25 nm (Fig. 1, 6a, 8b, c).

Most postsynaptic membranes are reinforced by electron-dense material (Fig. 1). Occasionally in sympathetic ganglion cells of mammals there is a sub-synaptic specialization in the form of electron-dense bars or a row of electron-dense dots (Taxi, 1965; Gabella, 1976). These subsynaptic specializations are more prominent in sympathetic ganglia of amphibia (Taxi, 1976) together with another postsynaptic specialization in the form of a wide cisterna of endoplas-

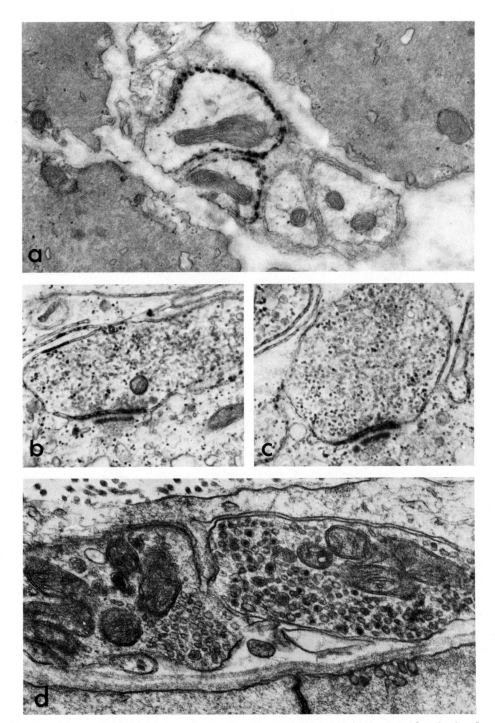

Fig. 8. (a) Two AChE-positive axons in a bundle with two unstained axons. Incubation for acetylcholinesterase, 10 min. Guinea-pig vas deferens. ×30,000. (From Robinson, 1969.)

(b, c) AChE staining in frog sympathetic ganglia. Note that positive staining is confined to region of synapse adjacent to subsynaptic bars. (Courtesy of J. Houchin and P. Robinson, Melbourne University.)

(d) Axon profiles near the anterior cerebral artery of a rat injected with 6-OHDA (250 mg/ kg) 1 h before decapitation. One axon (adrenergic) has many small vesicles with distinct electron-opaque cores and also large vesicles with dense granulation. The other axon (cholinergic) contains no small granular vesicles but many small agranular vesicles and a few large vesicles with moderately granular cores. Osmium fixation. ×30,000. (From Iwayama, Furness and Burnstock, 1970.)

14

mic reticulum, termed a "junctional subsurface organ" or JSO (Watanabe and Burnstock, 1976). Postsynaptic bars or PSB's (Figs. 6b, 8b, c) and junctional subsurface organs (Fig. 6c) do not co-exist in the same ganglion cell, but they are found in different subpopulations of ganglion cells (Fig. 6d); JSO's are found in the larger neurones (up to 45 μm), PSB's in the smaller neurones (less than 22 μm). It has been suggested that these types of postsynaptic specialization might be used to distinguish cholinergic excitatory and inhibitory synapses in frog sympathetic ganglia (Watanabe and Burnstock, 1978); slow excitatory postsynaptic potentials appear to be associated with "junctional subsurface organs" while slow inhibitory postsynaptic potentials appear to be associated with "postsynaptic bars".

LOCALISATION OF ACETYLCHOLINESTERASE AND CHOLINE ACETYLTRANSFERASE

Histochemical localisation of AChE has been widely used for the identification of cholinergic nerves (Burnstock and Robinson, 1967; Fig. 8a), on the basis that such nerves contain high levels of this enzyme relative to other nerve types (see Koelle, 1963). Although AChE activity can also be demonstrated in some adrenergic nerves (Fig. 10; Eränko et al., 1970; Barajas and Wang, 1975) and in nonnervous tissue (Gerebtzoff, 1959; Koelle, 1963), AChE-positivity can be employed as a criterion for cholinergic nerves provided that the reaction intensity is similar to that of known cholinergic elements from the same animal.

Association of postjunctional AChE with an adjacent AChE-positive axon seems to be a useful criterion for a functional cholinergic synapse, since in the uterine artery, postsynaptic AChE was seen consistently in tissues from animals investigated during pregnancy, when the cholinergic vasomotor nerves are functional, but was sparse or absent in tissues from nonpregnant animals when the arterial muscle is insensitive to ACh (Bell, 1969). Since the function of AChE on the smooth muscle membrane is likely to be to degrade neurally liberated ACh, its restriction to a small area of membrane opposite the axon suggests that transmitter action is also limited to a small region. However, it is not yet known whether this region is analogous to the skeletal neuromuscular junction in terms of increased local density of receptors. Histochemical studies of the relationship between cholinergic terminal varicosities and their effectors in the heart (Hirano and Ogawa, 1968) salivary gland (Hand, 1972; Bogart, 1971) and sympathetic neurones (Figs. 8b, c) have also revealed focal areas of AChE activity.

While a histochemical method for localising choline acetyltransferase has been applied with some success to cholinergic motor nerves (Kasa et al., 1970; Burt, 1970) the activity of this enzyme in autonomic cholinergic nerves appears to be too low to allow histochemical localisation.

CHOLINERGIC NEUROMODULATION

Neuromodulation, defined as the regulation of release of transmitter from nerves by the action of neurohumoral agents on prejunctional receptors is a relatively recent concept which is rapidly becoming recognized to be of physiological significance (Story et al., 1975; Hedqvist and Fredholm, 1976; Langer, 1977). Thus, ACh has been shown to inhibit release of noradrenaline from adrenergic nerves via prejunctional muscarinic receptors (Löffelholz and Muscholl, 1969; Story et al., 1975) and noradrenaline released from sympathetic nerve terminals reduces the release of ACh from cholinergic nerves (Paton and Vizi, 1969; Vizi and Knoll, 1971). Ultrastructural demonstration of close apposition of adrenergic and cholinergic nerve varicosities often enclosed within the same Schwann sheath (Fig. 8d) provides morphological support for this concept (Graham et al., 1968; Iwayama et al., 1970; Nelson and Rennals, 1970; Edvinsson et al., 1973; Burnstock and Costa, 1975).

ACETYLCHOLINE AS A CO-TRANSMITTER

(1) Acetylcholine and noradrenaline in developing sympathetic nerves

There is convincing evidence that sympathetic neurones during development both in vitro and in vivo synthesise both tyrosine hydroxylase (a precursor of noradrenaline) and choline acetyltransferase (the precursor of ACh) and that under certain conditions, some of these neurones release both ACh and noradrenaline (see Hill et al., 1976; Furshpan et al., 1976; Patterson and Chun, 1977; Bunge et al., 1978). Just before or after birth (depending on the system and the animal) most sympathetic neurones under the influence of factors produced by the effector organ (including NGF) and transynaptic induction of enzyme activity, are programmed to become either adrenergic or cholinergic (Black et al., 1971, 1976; Thoenen et al., 1972; Le Douarin et al., 1975; Hill and Henry, 1977). In sympathetic nerves destined to become adrenergic, choline acetyltransferase activity is decreased, while tyrosine hydroxylase activity is increased; this is associated with an increase in small granular vesicles. In sympathetic nerves destined to become cholinergic, the reverse is the case and small agranular vesicles predominate (Landis, 1976; Bunge et al., 1978). The efficiency of the amine-uptake pump is reduced in cholinergic sympathetic nerves, but it seems likely that some transport still persists in the adult, since chronic guanethidine treatment produces damage to cholinergic as well as adrenergic neurones in the superior cervical ganglion of the rat, although parasympathetic cholinergic neurones are unaffected (Heath and Burnstock, 1977). Whether some sympathetic nerves in the adult retain the ability to synthesise and release both ACh and noradrenaline is still an open question (see Burn and Rand, 1959; Burnstock, 1976, 1978b). It is particularly interesting that some small vesicles remain agranular even after loading of sympathetic nerves with 5- or 6-hydroxydopamine (Fig. 9), and that some adult nerves containing a predominance of small granular vesicles are AChE-positive (Fig. 10).

(2) Acetylcholine and ATP

Adenosine triphosphate (ATP), a putative transmitter in non-adrenergic, non-cholinergic ("purinergic") nerves in the gastrointestinal tract, bladder, lung

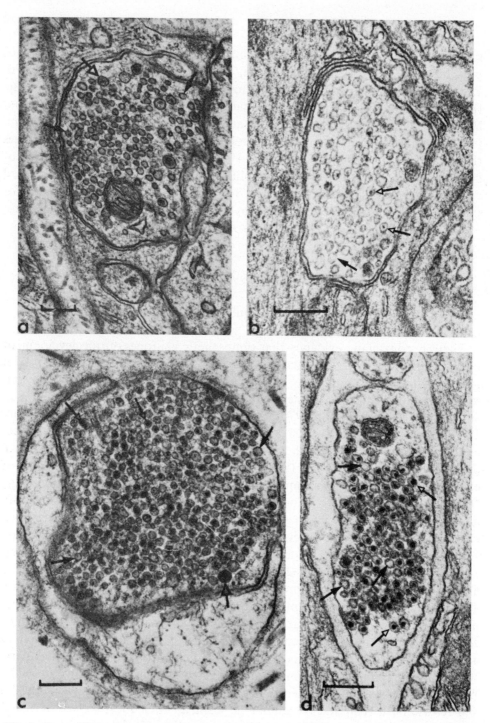

Fig. 9. Proportions of small agranular vesicles, 40–60 nm (↑) and small granular vesicles, 40–60 nm (↑) in sympathetic nerves. Note also the presence of some larger granular vesicles 80–120 nm (⚐). Horizontal calibration: 300 nm.

(a) Cholinergic nerve profile in guinea-pig ureter. OsO_4-glutaraldehyde-OsO_4 fixation.

(b) Nerve profile penetrating "inside" a smooth muscle cell: rat vas deferens. Note the small number of small granular vesicles. Glutaraldehyde fixation. (Courtesy of J. Heath, University of Melbourne.)

(c) Nerve varicosity in sheep ureter. Transverse section. Glutaraldehyde fixation.

(d) Adrenergic varicosity in mammalian smooth muscle: mouse vas deferens. Transverse section. Preincubation in 6-hydroxydopamine (250 mg/kg⁻¹) prior to OsO_4-glutaraldehyde-OsO_4 fixation. (From Furness et al., 1970.) (a and c from Burnstock, 1978b.)

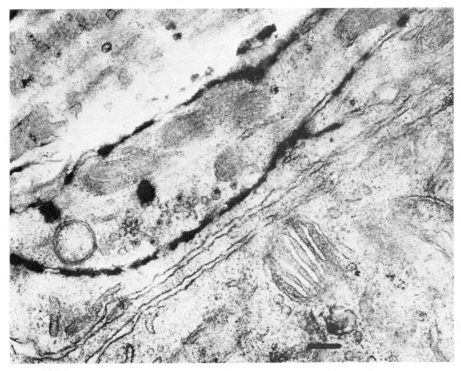

Fig. 10. Distribution of AChE in the pineal gland of rat. Fixation in formaldehyde–glutaraldehyde mixture. The reaction product around the axon containing small granular vesicles is situated between the axon membrane and the pineal cell membrane. The length of the calibration line is 200 nm. (From Eränkö et al., 1970.)

and probably other organs (Burnstock, 1972, 1975c), has been reported to be stored and released together with ACh during stimulation of cholinergic nerves supplying the rat diaphragm (Silinsky and Hubbard, 1973; Silinsky, 1975) and electric organ of torpedine rays (Whittaker et al., 1972; Dowdall et al., 1974; Israël et al., 1975). However, there is no evidence that ATP is released together with ACh from autonomic cholinergic nerves. For example, stimulation of the vagal roots (containing preganglionic fibres supplying intramural purinergic neurones in the toad stomach) resulted in increased nucleoside efflux, but stimulation of the cervical sympathetic branch (containing cholinergic fibres) which joins the vagus nerve did not (Burnstock et al., 1970).

SUMMARY

(1) Terminal varicosities of cholinergic autonomic nerves contain a predominance of small agranular vesicles (40–60 nm) and a small number (about 1%) of large granular vesicles (60–120 nm), which differ in some respects from those found in adrenergic nerves and from "large opaque vesicles" found in non-adrenergic, non-cholinergic ("purinergic" or "peptidergic") nerves.

(2) In the gastrointestinal tract, some nerve profiles contain flattened agranular vesicles which may represent another type of cholinergic nerve; some

profiles contain mixtures of small agranular vesicles and other vesicle types perhaps indicating co-transmitters.

(3) A generalised model of the cholinergic neuromuscular junction is presented, which emphasises "en passage" release of transmitter from extensive terminal, varicose fibres and electrotonic spread of activity via "gap junctions" between neighbouring smooth muscle cells within effector bundles.

(4) The minimum width of the junctional cleft between cholinergic varicosities and effector cells varies between 15—20 nm in densely-innervated tissues, such as vas deferens and iris and 1—2 μm in large elastic arteries.

(5) In densely innervated tissues, sub-synaptic cysternae in smooth muscle are often found at close (20 nm) neuromuscular junctions. In frog sympathetic ganglia, inhibitory and excitatory cholinergic junctions may be recognisable according to postsynaptic specialisations, namely "postsynaptic bars" and "junctional subsurface organs" respectively.

(6) Acetylcholinesterase is a useful marker for cholinergic nerves if used with care, but activity is also present in some adrenergic nerves and non-neuronal tissues. Localisation of postjunctional AChE on effector cells is a useful criterion for a functional cholinergic junction.

(7) Close association of adrenergic and cholinergic nerve varicosities provides morphological support for neuromodulation of ACh release from cholinergic nerves by noradrenaline and of noradrenaline release from adrenergic nerves by ACh.

(8) Some developing sympathetic neurones synthesise and release both noradrenaline and ACh. Around birth most of these neurones are programmed to become either adrenergic or cholinergic, but it is possible that some retain the ability to utilise both transmitters.

ACKNOWLEDGEMENTS

I much appreciate the help of my colleague Giorgio Gabella in the preparation of this manuscript. Some of the work reported was supported by grants from the Medical Research Council and the Wellcome Trust.

REFERENCES

Barajas, L. and Wang, P. (1975) Demonstration of acetylcholinesterase in adrenergic nerves of renal glomerular arterioles. *J. Ultrastruc. Res.*, 53, 244—253.

Baumgarten, H.G., Holstein, A.F. and Owman, C.H. (1970) Auerbach's plexus of mammals and man: electron microscopic identification of three different types of neuronal processes in myenteric ganglia of the large intestine from Rhesus monkeys, guinea-pigs and man. *Zellforsch.*, 106, 376—97.

Bell, C. (1969) Transmission from vasoconstrictor and vasodilator nerves to single smooth muscle cells of the guinea-pig uterine artery. *J. Physiol. (Lond.)*, 205, 695—708.

Bennett, M.R. and Merrillees, N.C.R. (1966) An analysis of the transmission of excitation from autonomic nerves to smooth muscle. *J. Physiol. (Lond.)*, 185, 520—535.

Bevan, J.A. and Su, C. (1974) Variation of intra and perisynaptic adrenergic transmitter concentrations with width of synaptic cleft in vascular tissue. *J. Pharmacol. exp. Ther.*, 190, 30—8.

Black, I.B., Bloom, E.M. and Hamill, R.W. (1976) Central regulation of sympathetic neuron development. *Proc. nat. Acad. Sci. (Wash.)*, 73, 3575—3578.

19

Black, I.B., Hendry, I.A. and Iversen, L.L. (1971) Transynaptic regulation of growth and development of adrenergic neurones in a mouse sympathetic ganglion. *Brain Res.*, 34, 229–240.

Bogart, B.I. (1971) Fine structural localization of cholinesterase activity in the rat submandibular gland. *J. Histochem. Cytochem.*, 18, 730–739.

Bunge, R., Johnson, C. and Ross, D. (1978) Nature and nurture in development of the autonomic neuron. *Science*, 199, 1409–1416.

Burn, J.H. and Rand, M.J. (1959) Sympathetic postganglionic mechanism. *Nature (Lond.)*, 184, 163–165.

Burnstock, G. (1970) Structure of smooth muscle and its innervation. In *Smooth Muscle*. E. Bülbring, A. Brading, A. Jones and T. Tomita (Eds.), Edward Arnold, London, pp. 1–69.

Burnstock, G. (1972) Purinergic nerves. *Pharmacol. Rev.*, 24, 509–581.

Burnstock, G. (1975a) Innervation of vascular smooth muscle: histochemistry and electronmicroscopy. In *Physiological and Pharmacological Control of Blood Pressure. Clin. exp. Pharmacol. Physiol.* Suppl. 2, 7–20.

Burnstock, G. (1975b) Ultrastructure of autonomic nerves and neuroeffector junctions; analysis of drug action. In *Methods in Pharmacology, Vol. III, Smooth Muscle*, E.E. Daniel and D.M. Paton (Eds.), Plenum Press, pp. 113–137.

Burnstock, G. (1975c) Comparative studies of purinergic nerves. *J. exp. Zool.*, 194, 103–133.

Burnstock, G. (1976) Do some nerve cells release more than one transmitter? *Neuroscience*, 1, 239–248.

Burnstock, G. (1978a) Cholinergic and purinergic regulation of blood vessels. In *Handbook of Physiology (Vascular Smooth Muscle)*, D. Bohr, M.D. Somlyo and H.V. Sparks (Eds.), *Amer. Physiol. Soc.*, Williams and Wilkins, Baltimore, Md., in press.

Burnstock, G. (1978b) Do some sympathetic neurones release both noradrenaline and acetylcholine? *Progr. Neurobiol.*, in press.

Burnstock, G. and Costa, M. (1975) *Adrenergic Neurons: their organisation, function and development in the peripheral nervous system*, Chapman and Hall, London, 225 pp.

Burnstock, G., Gannon, B. and Iwayama, T. (1970) Sympathetic innervation of vascular smooth muscle in normal and hypertensive animals. In *Symposium on Hypertensive Mechanisms, Circ. Res.*, 27, (Suppl. II), 5–24.

Burnstock, G. and Iwayama, T. (1971) Fine structural identification of autonomic nerves and their relation to smooth muscle. In *Histochemistry of Nervous Transmission, Progr. Brain Res.*, Vol. 34, O. Eränkö (Ed.), Elsevier, Amsterdam, pp. 389–404.

Burnstock, G. and Robinson, P.M. (1967) Localization of catecholamines and acetylcholinesterase in autonomic nerves. In *American Heart Association Monograph (No. 17), Circ. Res.*, 21, (Suppl. 3), 43–55.

Burt, A.M. (1970) A histochemical procedure for the localization of choline acetyltransferase activity. *J. Histochem. Cytochem.*, 18, 408–415.

Chan-Palay, V. (1975) Fine structure of labelled axons in the cerebellar cortex and nuclei of rodents and primates after intraventricular infusions with tritiated serotonin. *Anat. Embryol.*, 148, 235–265.

Cook, R.D. and Burnstock, G. (1976) The ultrastructure of Auerbach's plexus in the guinea-pig. I. Neuronal elements. *J. Neurocytol.*, 5, 171–194.

Dale, H. (1937) Transmission of nervous effects by acetylcholine. *Harvey Lect.*, (1936, 1937) 32, 229–245.

Dowdall, M.J., Boyne, A.F. and Whittaker, V.P. (1974) Adenosine triphosphate: a constituent of cholinergic synaptic vesicles. *Biochem. J.*, 140, 1–12.

Edvinsson, L., Nielsen, K.C. and Owman, C.H. (1973) Cholinergic innervation of choroid plexus in rabbits and cats. *Brain Res.*, 63, 500–503.

Eränkö, O., Rechardt, L., Eränkö, L. and Cunningham, A. (1970) Light and electron microscopic histochemical observations on cholinesterase – containing sympathetic nerve fibres in the pineal body of the rat. *Histochem. J.*, 2, 479–489.

Furness, J.B., Campbell, G.R., Gillard, S.M., Malmfors, T., Cobbs, J.L.S. and Burnstock, G. (1970) Cellular studies of sympathetic denervation produced by 6-hydroxydopamine in the vas deferens. *J. Pharmacol. exp. Ther.*, 174, 111–122.

Furshpan, E.J., MacLeish, P.R., O'Lague, P.H. and Potter, D.D. (1976) Chemical transmission between rat sympathetic neurons and cardiac myocytes developing in neurocultures – evidence for cholinergic, adrenergic and dual-function neurons. *Proc. nat. Acad. Sci. (Wash.)*, 73, 4225–4229.

20

Gabella, G. (1972) Fine structure of myenteric plexus in the guinea-pig ileum. *J. Anat.*, 111, 69–97.

Gabella, G. (1976) *Structure of The Autonomic Nervous System*, Chapman and Hall, (London), 211 pp.

Gerebtzoff, M.A. (1959) *Cholinesterases: A Histochemical Contribution to The Solution of Some Functional Problems*, Pergamon Press, London, 242 pp.

Graham, J.D.P., Lever, J.D. and Spriggs, T.L.B. (1968) An examination of adrenergic axons around pancreatic arterioles of the cat for the presence of acetylcholinesterase by high resolution autoradiographic and histochemical methods. *Brit. J. Pharmacol. Chemother.*, 33, 15–20.

Hand, A.R. (1972) Adrenergic and cholinergic nerve terminals in the rat parotid gland. Electron microscopic observations on permanganate-fixed glands. *Anat. Rec.*, 173, 131–139.

Heath, J.W. and Burnstock, G. (1977) Selectivity of neuronal degeneration produced by chronic guanethidine treatment. *J. Neurocytol.*, 6, 397–405.

Hedqvist, P. and Fredholm, B.B. (1976) Effects of adenosine on adrenergic neurotransmission — prejunctional inhibition and postjunctional enhancement. *Naunyn-Schmiedlberg's Arch. Physiol.*, 293, 217–223.

Hill, C.E. and Hendry, I.A. (1977) Development of neurons synthesizing noradrenaline and acetylcholine in the superior cervical ganglion of the rat in vivo and in vitro. *Neuroscience*, in press.

Hill, C., Purves, R.D., Watanabe, H. and Burnstock, G. (1976) Specificity of innervation of iris musculature by sympathetic nerve fibres in tissue culture. *Pflüg. Arch. (Ges. Physiol.)*, 361, 127–134.

Hirano, H. and Ogawa, K. (1968) Ultrastructural localisation of cholinesterase activity in nerve endings in the guinea-pig heart. *J. Electron Microsc.*, 16, 313–321.

Israël, M., Lesbats, B., Marsal, J. and Meunier, F.M. (1975) Oscillations of adenosine-triphosphate and acetylcholine levels during stimulation of electric organ of torpedo fish. *C.R. Acad. Sci. Paris*, 280, 905–908.

Iwayama, T., Furness, J.B. and Burnstock, G. (1970) Dual adrenergic and cholinergic innervation of the cerebral arteries of the rat. An ultrastructural study. *Circ. Res.*, 26, 635–646.

Kása, P., Mann, S.P. and Hebb, C. (1970) Localisation of choline acetyltransferase. *Nature (Lond.)*, 226, 812–815.

Koelle, G.B. (Ed.), (1963) Cholinesterase and anticholinesterase agents. *Handbuch der experimentellen Pharmakologie (Ergänzungswerk)*, Vol. 15, Springer-Verlag, Berlin, Heidelberg, p. 298.

Korneliussen, H. (1972) Elongate profiles of synaptic vesicles in motor and plates. Morphological effects of fixation variations. *J. Neurocytol.*, 1, 279–296.

Landis, S.C. (1976) Rat sympathetic neurons and cardiac myocytes developing in microcultures — correlation of fine structure of endings with neurotransmitter function in single neurons. *Proc. nat. Acad. Sci. (Wash.)*, 73, 4220–4224.

Langer, S.Z. (1977) Presynaptic receptors and their role in the regulation of transmitter release. *Brit. J. Pharmacol.*, 60, 481–498.

LeDouarin, N.M., Renaud, D., Teillet, M.A. and LeDouarin, G.H. (1975) Cholinergic differentiation of presumptive adrenergic neuroblasts in interspecific chimeras after heterotropic transplantations. *Proc. nat. Acad. Sci. (Wash.)*, 72, 728–732.

Löffelholz, K. and Muscholl, E. (1969) A muscarinic inhibition of the noradrenaline release evoked by post-ganglionic sympathetic stimulation. *Naunyn-Schmiedeberg's Arch. Pharmacol.*, 265, 1–15.

McMahan, U.J. and Kuffler, S.W. (1971) Visual identification of synaptic boutons on living ganglion cells and of varicosities in postganglionic axons in the heart of the frog. *Proc. roy. Soc. Lond. (Biol.)*, 177, 485–508.

Merrillees, N.C. (1968) The nervous environment of individual smooth muscle cells of the guinea-pig was deferens. *J. Cell. Biol.*, 37, 794–817.

Nelson, E. and Rennels, M. (1970) Innervation of intracranial arteries. *Brain Res.*, 93, 475–490.

Paton, W.D.M. and Vizi, E.S. (1969) The inhibitory action of noradrenaline and adrenaline on acetylcholine output by guinea-pig ileum longitudinal muscle strip. *Brit. J. Pharmacol.*, 35, 10–28.

Patterson, P.H. and Chun, L.L.Y. (1977) Induction of acetylcholine synthesis in primary cultures of dissociated rat sympathetic neurons. I. Effects of conditioned medium. *Develop. Biol.*, 56, 263–280.

Pearse, A.G.E., Polak, J.M. and Bloom, S.R. (1977) The newer gut hormones: Cellular sources, physiology, pathology, and clinical aspects. *Progr. Gastroenterol.,* 72, 746–761.

Robinson, P.M. (1969) A cholinergic component in the innervation of the longitudinal smooth muscle of the guinea-pig vas deferens: the fine structural localization of acetylcholinesterase. *J. Cell. Biol.,* 41, 462–476.

Silinsky, E.M. (1975) On the association between transmitter secretion and the release of adenine nucleotides from mammalian motor nerve terminals. *J. Physiol. (Lond.),* 247, 145–162.

Silinsky, E.M. and Hubbard, J.I. (1973) Release of ATP from rat motor nerve terminals. *Nature (Lond.),* 243, 404–405.

Story, D.F., Allen, G.S., Glover, A.B., Hope, W., McCulloch, M.W., Rand, M.J. and Sarantos, C. (1975) Modulation of adrenergic transmission by acetylcholine. *Clin. exp. Pharmacol. Physiol.,* Suppl. 2, 27–33.

Taxi, J. (1965) Contribution à l'étude des connexions des neurones moteurs du système nerveux autonome. *Ann. Sci. Nat. Zool.,* 7, 413–674.

Taxi, J. (1976) Morphology of the autonomic nervous system. In *Frog Neurobiology,* R. Llinas and W. Precht (Eds.), Springer-Verlag, Berlin-Heidelberg, pp. 93–150.

Thoenen, H., Sano, A., Angeletti, P.U. and Levi-Montalcini, R. (1972) Increased activity of choline acetyltransferase in sympathetic ganglia after prolonged administration of nerve growth factor. *Nature New Biol. (Lond.),* 236, 26–28.

Tranzer, J.P. and Thoenen, H. (1968) An electron microcopic study of selective acute degeneration of sympathetic nerve terminals after administration of 6-hydroxydopamine. *Experientia (Basel),* 24, 155–156.

Uchizono, K. (1968) Morphological background of excitation and inhibition of synapses. *J. Electronmicrosc.,* 17, 55–66.

Uehara, Y., Campbell, G.R. and Burnstock, G. (1976) *An Atlas of The Fine Structure of Muscle and Its Innervation,* Edward Arnold, London, 526 pp.

Vizi, E.S. and Knoll, J. (1971) The effect of sympathetic nerve stimulation and guanethidine on parasympathetic neuroeffector transmission; the inhibition of acetylcholine release. *J. Pharm. Pharmacol.,* 23, 918–925.

Watanabe, H. and Burnstock, G. (1976) Junctional subsurface organs in frog sympathetic ganglion cells. *J. Neurocytol.,* 5, 125–136.

Watanabe, H. and Burnstock, G. (1978) Postsynaptic specialisations at excitatory and inhibitory cholinergic synapses. *J. Neurocytol.,* 7, 119–133.

Whittaker, V.P., Dowdall, M.J. and Boyne, A.F. (1972) The storage and release of acetylcholine by cholinergic nerve terminals: recent results with non-mammalian preparations. *Biochem. Soc. Symp.,* 36, 49–68.

The Importance of Axonal Transport and Endoplasmic Reticulum in the Function of Cholinergic Synapse in Normal and Pathological Conditions

B. DROZ, H.L. KOENIG, L. DI GIAMBERARDINO, J.Y. COURAUD, M. CHRETIEN and F. SOUYRI

Département de Biologie, Commissariat à l'Energie Atomique, C.E.N. de Saclay, BP No. 2, 91190, Gif sur Yvette, and Université Pierre et Marie Curie, Paris (France)

INTRODUCTION

The transmission of nerve impulses from a cholinergic axon to a target cell is achieved by a series of events which take place in the presynaptic neurone and its axon terminals. It is the aim of this chapter to emphasize the crucial role played by axonal transport in supplying cholinergic nerve endings with macromolecules involved in the transmission mechanism.

The study of the renewal of presynaptic macromolecules in cholinergic nerves requires a biological preparation which fulfils the necessary conditions for analyses by high resolution radioautography, cell fractionation or molecular separation. The data on molecular forms of acetylcholinesterase (AChE), reported in this chapter, were derived from chicken sciatic nerve. The chicken ciliary ganglion, a parasympathetic ganglion, was used in radioautographic studies to examine the presynaptic and postsynaptic components of cholinergic synapse. The nerve cell bodies of the preganglionic neurones of the ciliary ganglion are located in the midbrain, in the accessory motor nucleus lining the cerebral aqueduct and corresponding to the Edinger-Westphal nucleus in mammals. The preganglionic nerves contain two types of cholinergic axons. A small proportion of thin axons (about 4 μm in diameter) terminate in synaptic boutons located around small ganglion cells (25–30 μm), but the majority of axons constitute a homogeneous population (length: 10.1 ± 0.8 mm; diameter: 9 ± 1 μm) with each axon terminating in a giant presynaptic calyx (17,000 ± 2300 μm³) which encompasses a large (45–60 μm) postsynaptic ganglion cell (Fig. 2). The ciliary ganglion therefore offers the possibility of studying the turnover of presynaptic constituents either by measuring the local incorporation of labelled precursors into the caliciform nerve endings or by analysing the kinetics of labelled macromolecular products conveyed by axonal transport to the calices after their synthesis by nerve cell bodies in the midbrain.

(I) THE DYNAMIC STATE OF PRESYNAPTIC MACROMOLECULES *

Among the sources able to provide cholinergic nerve endings with new macromolecules, local biosynthesis and axonal transport were investigated, in

---

* Section contributed by B. Droz and H.L. Koenig.

the same material, to estimate the respective parts taken by these processes in compensating for the loss of synaptic constituents.

(A) Local incorporation of labelled amino acids

When [³H]leucine was intravenously administered to chickens, radioautographs of the cliliary ganglion made 5 and 60 min after injection (i.e. during the lag time preceding the arrival of axonally transported proteins) revealed only stray silver grains over the caliciform nerve endings. Meanwhile, an intense reaction was observed over postsynaptic ganglion cells with a more discrete one over Schwann and satellite cells (Koenig and Droz, 1970). A more direct approach was undertaken by in vitro incubation of ciliary ganglia with [³H]leucine (Droz and Koenig, 1971). After a one hour incubation, the ciliary ganglia were washed, fixed with formaldehyde containing cold leucine to prevent an artifactual retention of precursor, and processed for radioautography (Boyenval and Fischer, 1976). The level of radioactivity measured in presynaptic calices was lower than in postsynaptic ganglion cells (Table I); the presynaptic label was distributed mainly to mitochondria and, to the other organelles, to a lesser extent. When a potent inhibitor of protein synthesis, puromycin, was added to the incubation medium, the local incorporation of [³H]leucine was practically abolished in all ganglionic structures including presynaptic calices. In the presence of acetoxycycloheximide, another inhibitor of protein synthesis, the incorporation of [³H]leucine was greatly reduced, but silver grains were still found over mitochondria of pre- and postsynaptic elements.

It could be inferred from these results that mitochondria of cholinergic nerve endings are able to synthesize polypeptides as has been shown in heterogeneous populations of brain synaptosomes (Gambetti et al., 1972). Since presynaptic axons are devoid of cytoplasmic polyribosomes, the question arises whether the local incorporation of labelled amino acids into presynaptic organelles other than mitochondria corresponds to a true synthesis of protein or to the addition of amino acids or peptides to pre-existing polypeptide chains (Uy and Wold, 1977). In either case, the part played by any local presynaptic protein synthesis would account, at most, for less than 2% of the rapidly renewed proteins in calices (Droz, 1973).

TABLE 1

LOCAL INCORPORATION OF PROTEIN AND PHOSPHOLIPID PRECURSORS INTO PRESYNAPTIC CALICES AND POSTSYNAPTIC GANGLION CELL BODIES AFTER IN VITRO INCUBATION OF CILIARY GANGLIA FOR 1 HOUR (EXPRESSED AS NUMBER OF SILVER GRAINS PER 100 μm^2)

Experimental condition	Presynaptic calices	Postsynaptic ganglion cells
* [³H]leucine	49	156
* [³H]leucine + puromycin	4	3
* [³H]leucine + acetoxycycloheximide	8	14
[2-³H]glycerol	5	64
[³H]choline	8	26
[³H]myo-inositol	19	13

* From Droz and Koenig (1971).

(B) Local incorporation of labelled precursors of phospholipids

The possibility that biosynthesis of phospholipids occurs in presynaptic calices was recently investigated by Droz and Koenig (unpublished results, see Table I). After the in vitro incubation of ciliary ganglia with [³H]glycerol (Fig. 1), the level of radioactivity detected in caliciform nerve endings was extremely low and reached hardly 8% of the level of the tracer incorporated into the rough endoplasmic reticulum of postsynaptic ganglion cells. Benjamins and McKahn (1973) have shown that the incorporation of [2—³H]glycerol in nervous tissue reflects the synthesis of new molecules of [³H]glycerophospholipids and not an exchange or recycling of the tracer, hence the lack of significant radioactivity suggests that the presynaptic axon terminals are unable to manufacture phospholipids. When [³H]choline was introduced in the incubation medium, the limited but clear reaction observed over the calices exceeded 30% of the grain concentration counted over the postsynaptic ganglion cells; this significant incorporation of the tracer in the absence of a net synthesis of phospholipids may correspond to a base-exchange reaction with pre-existing phospholipids (Arienti et al., 1976) whereas the incorporation observed in postsynaptic ganglion cells and satellite cells results mainly from a de novo biosynthesis of phospholipids (Table I).

In contrast to results with labelled glycerol or choline, the incubation with [³H]myo-inositol gave rise to a rapid incorporation of the tracer in presynaptic

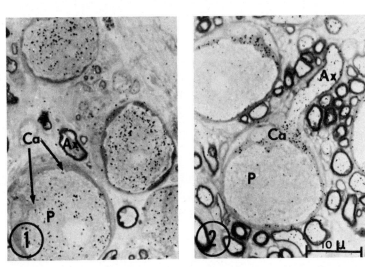

Figs. 1 and 2. Light microscope radioautographs of chicken ciliary ganglion cells after a 15 min incubation with [2-³H]glycerol (Fig. 1) or after an intracerebral injection of [2-³H]glycerol 14 days before (Fig. 2).

In Fig. 1, the virtual lack of silver grains over the preganglionic axons (Ax) and presynaptic calices (Ca) indicates that phospholipids are not significantly synthesized in these structures; in contrast, the radioautographic reaction seen over the perikaryon of postsynaptic ganglion cells (P) points to the nerve cell bodies as active sites of phospholipid biosynthesis.

In Fig. 2, the silver grains seen over the preganglionic axons (Ax) and presynaptic calices (Ca) correspond to the presence of axonally transported phospholipids. The perikaryon of the postsynaptic ganglion cells (P) is almost devoid of label, hence the transsynaptic transfer of [³H]glycerophospholipids may be rated as negligible; compare with [³H]choline (Fig. 9, inset).

calices (Table I). Within one hour, the concentration of radioactivity detected in caliciform nerve endings and preterminal segments of axons was higher than that found in postsynaptic ganglion cells and satellite cells. Gould (1976) has also reported an axonal incorporation of [^{3}H]myo-inositol after the injection of tracer into the sciatic nerve. Thus, in the absence of local biosynthesis of phospholipids, the incorporation of [^{3}H]myo-inositol into presynaptic endings reflects the turnover of the inositol moiety of phosphatidylinositol which probably plays a special role in cholinergic synapses (Larrabee, 1968).

(C) Axonal transport of presynaptic macromolecules to nerve endings

Since the presynaptic calices cannot be supplied with new proteins or phospholipids synthesized in situ, their axonal transport from midbrain cell bodies to axon terminals was investigated. After the injection of a labelled precursor into the cerebral aqueduct, the tracer is taken up by nerve cell bodies of the preganglionic neurones and incorporated into macromolecular products;

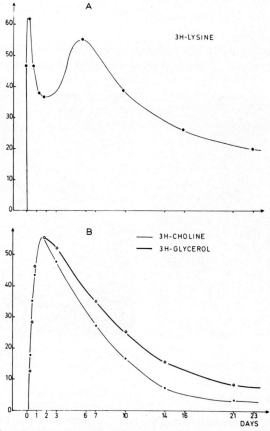

Fig. 3. Kinetics of axonally transported proteins (A) and phospholipids (B) in presynaptic calices. (A) Axonally transported macromolecules arrive in calices with a delay of 1 h after the intracerebral injection of [^{3}H]lysine. The first peak corresponds to the accumulation of rapidly transported proteins, which turn over within a few hours; the second peak results from the accumulation of slowly transported proteins. (B) After the intracerebral injection of [^{3}H]choline or [2-^{3}H]glycerol, the early wave of rapidly transported phospholipids lasts for several days and declines more slowly with [2-^{3}H]glycerol than with [^{3}H]choline.

labelled macromolecules migrate along the preganglionic axons to the ciliary ganglion and accumulate in the presynaptic calices (Fig. 2). Radioautography and cell fractionation have provided information about: (1) the kinetics of the axonal transport of macromolecules to the terminal part of preganglionic axons (Fig. 3); (2) the mean time of sojourn of macromolecules in presynaptic structures and (3) the subcellular distribution of the various macromolecules among the presynaptic organelles (Figs. 4 and 5) (Bennett et al., 1973; Di Giamberardino et al., 1973; Droz et al., 1973; Koenig et al., 1973; Alvarez, 1974).

Membrane constituents

Membrane-bound proteins, glycoproteins and phospholipids were found to arrive in the presynaptic calices of the ciliary ganglion with a delay of 1 to 1.5 h after the intracerebral injection of tracer; the maximal level of radioactivity was reached by 18 or 40 h (Fig. 3, Table II). All these membrane components were transported by fast axonal flow at rates of about 300 mm per day. As Table III shows, it was found that they turn over in presynaptic organelles within times ranging from several hours to weeks and that the same category of macromolecular constituents is renewed at different rates in different synaptic organelles. Since it was also found that (Figs. 4 and 5) proteins,

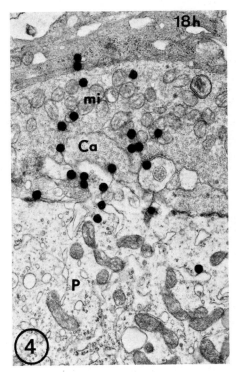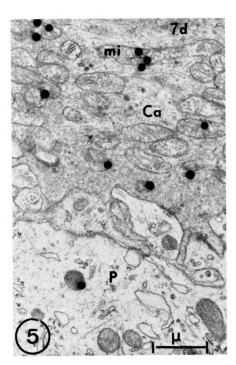

Figs. 4 and 5. Electron microscope radioautographs of presynaptic calices (Ca) at 18 h and 7 days after an intracerebral injection of [³H]choline.

By 18 h (Fig. 4), most silver grains are located over presynaptic plasma membrane and area rich in synaptic vesicles; mitochondria (mi) are still poorly labelled.

After 7 days (Fig. 5), the persisting silver grains are mainly found over mitochondrial profiles. At all times, silver grains are seen over the postsynaptic perikaryon of ganglion cells (P); they are more concentrated over the inner than the outer part of the cytoplasm (see Fig. 9).

TABLE 2

TIME OF MAXIMAL ACCUMULATION OF LABEL IN COMPONENTS OF PRESYNAP-
TIC CALICES OF CILIARY GANGLION AFTER AN INTRACEREBRAL INJECTION OF
LABELLED PRECURSOR

Labelled precursor	Synaptic vesicles and endoplasmic reticulum	Plasma membranes	Mitochondria	Axoplasm
* [^{3}H]lysine	18 h	18 h	144 h	144 h
** [^{3}H]glutamic acid	18 h		144 h	144 h
*** [^{3}H]fucose	18–48 h	18–48 h	very low	very low
[^{3}H]choline	18–40 h	40 h	40 h	very low
[2-^{3}H]glycerol	18 h	40 h	72 h	very low

* From Droz et al. (1973).
** From Alvarez (1974).
*** From Bennett et al. (1973).

glycoproteins and phospholipids turn over within the same organelles at different rates, it is concluded that presynaptic organelles of cholinergic nerve
endings are not renewed as a whole. More probably they result from assembled
constituents which turn over at different rates.

Axoplasmic constituents

The presynaptic axoplasm includes soluble enzymes and a variety of subunits
constituting the axonal cytoskeleton. After the intracerebral injection of
[^{3}H]lysine, the arrival of labelled proteins transported by slow axonal flow was
detected in the caliciform nerve endings after a lag time of about 1.5 day (Droz
et al., 1973). Their rates of transport range from 1 to 10 mm per day and they
reach their maximal concentration at 6 days (Fig. 3; Table II). Once arrived in
synaptic terminals, they stay for about 14 days (Table III).

Mitochondrial components

Multiple processes are responsible for the renewal of mitochondrial constituents in cholinergic endings (Droz, 1975). On one hand, synaptic mitochondria possess an autonomous capacity for synthesizing a limited number of

TABLE 3

MEAN TIMES OF SOJOURN OF VARIOUS PRESYNAPTIC COMPONENTS IN CALICI-
FORM NERVE ENDINGS OF CHICKEN CILIARY GANGLION AFTER AN INTRA-
CEREBRAL INJECTION OF LABELLED PRECURSOR (EXPRESSED IN DAYS)

Labelled precursor	Synaptic vesicles	Plasma membranes	Mito- chondria	Axoplasm
* [^{3}H]lysine	0.7			14
** [^{3}H]fucose	14	6		
[^{3}H]choline	6.5	5	5	
[2-^{3}H]glycerol	7	9	10	

* From Droz et al. (1973).
** From Bennett et al. (1973).

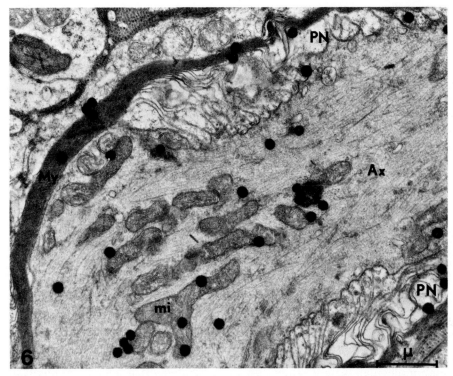

Fig. 6. Electron microscope radioautograph of the last node of Ranvier in a preganglionic nerve fibre 3 days after the intracerebral injection of [^{3}H]choline. In the preganglionic axon (Ax), silver grains are mainly associated with mitochondria (mi) moving down the axon. Silver grains accumulate also over the surrounding myelin sheath (My), especially over the perinodal region (PN).

polypeptides; on the other hand, axonal and synaptic mitochondria receive proteins and phospholipids conveyed by fast axonal transport. In addition, mitochondria moving back and forth (Zelená, 1968) are slowly displaced along the preganglionic axons (Fig. 6) to their terminals (Di Giamberardino et al., 1973; Fig. 5).

(D) The loss of presynaptic macromolecules

The variation in the mean times of sojourn observed in cholinergic calices probably depends on the diversity of mechanisms responsible for the disappearance of synaptic macromolecules (Droz, 1973). First, hydrolytic enzymes present in axon terminals are able to break down protein, peptide, glycoprotein, phospholipid, etc. This catabolic process would account for the presynaptic release of amino acid, sugar or base into the synaptic cleft and their reutilization by postsynaptic ganglion cells or adjacent glial cells (section IV). In axonal and presynaptic phospholipids, a base such as choline may be substituted for another choline molecule without de novo synthesis of phosphatidylcholine, by a base-exchange reaction (Brunetti et al., 1978); the myo-inositol moiety of phosphatidylinositol may also be replaced in situ (Table I). Second, part of the presynaptic constituents are transported in a retrograde direction to the brain and broken down by lysosomal enzymes in the nerve cell bodies (Kristensson and Olsson, 1971). Third, a limited number of protein species, probably associated with synaptic vesicles, could be released during stimulation (Musick and Hubbard, 1972). Fourth, an intracellular transfer or exchange of macro-molecules between presynaptic organelles takes place in caliciform nerve ending (Droz, 1975). In other words, the complexity of the dynamic state of pre-synaptic components in cholinergic endings results from the diversity of processes controlling the influx (local synthesis, addition, exchange, fast or slow axonal flow) and the efflux of macromolecules (catabolic breakdown, retrograde transport, extraneuronal release or transfer from one organelle to another).

(II) MOLECULAR SELECTIVITY AND AXONAL TRANSPORT OF AChE *

When a neuronal enzyme is polymorphic and separable into distinct molec-ular forms, the question arises whether the various molecular forms are con-veyed in bulk along the axons or selectively transported at different rates in different axonal compartments. This problem was recently investigated by Di Giamberardino and Couraud (1978). AChE in chicken, as in other verte-brates (Massoulié and Rieger, 1969), is represented by macromolecules of different molecular weight; they can be separated on a sucrose gradient and characterized by their sedimentation coefficients 5 S, 7 S, 11 S and 20 S. The heaviest form, 20 S, which is particularly abundant in striated muscle and ciliary ganglion, was found to decrease to a very low level after denervation or in acrylamide-induced axonal dystrophy (section V); for this reason, the 20 S AChE is considered as a molecular form mainly associated with cholinergic synapses (Vigny et al., 1976).

In chicken sciatic nerves, 11 S and 20 S forms of AChE, firmly bound to membranes, account for respectively 80 and 3% of the total enzyme activity; the rest is represented by 5 S and 7 S forms which are easily released in solu-tion (Vigny et al., 1976). Axonal transport of the different forms of AChE was studied after transsection of the sciatic nerve: AChE activity was found to accumulate at both proximal and distal sides of the interruption (Lubińska and Niemierko, 1971). In chicken sciatic nerve, the rate of accumulation during the first hour was 108 ± 15 mU/h on the proximal and 26 ± 8 mU/h on the distal side. Di Giamberardino and Couraud (1978) showed by sedimentation analysis performed on a 2 mm segment close to the proximal side that 11 S and 20 S forms piled up rapidly whereas 5 S and 7 S accumulated more slowly (Fig. 7). These results were interpreted as the consequence of the orthograde axonal transport of AChE: 95% of the molecular forms which are in rapid transit to nerve endings correspond to 11 S and 20 S AChE and are probably associated with the membrane of the axonal endoplasmic reticulum. Only 5% of the enzyme movement would depend on the axonal transport of the low molecular weight forms. The analysis of the distal segment indicated that only 11 S and

---

* Section contributed by L. Di Giamberardino and J.Y. Couraud.

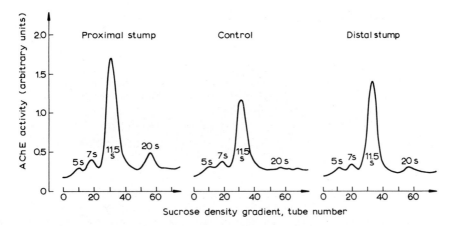

Fig. 7. Sedimentation profiles of AChE from a 2 mm segment of sciatic nerve taken on each side of a transsection performed 6 h before (proximal and distal stump); a 2 mm segment of controlateral intact sciatic nerve is used as control. Note the dramatic increase of the 20 S and 11 S forms in the proximal stump (which reflects orthograde transport) and, to a lesser extent, in the distal stump (which reflects retrograde transport).

20 S forms of AChE move in a retrograde direction back to cell bodies (Fig. 7). Thus the heavy molecular forms of AChE are selectively conveyed by orthograde axonal transport at rates of about 400 mm per day and by retrograde axonal transport at rates of about 150 mm per day; the low molecular forms of AChE would be slowly transported in the orthograde direction at rates of 5—10 mm per day.

In summary, the various molecular forms of AChE are selectively transported to or from cholinergic synapses in different relative amounts and at different rates. The molecular selectivity operated in the axons of cholinergic neurones strongly suggests that the enzyme forms are distributed among distinct axonal compartments.

(III) PRESYNAPTIC COMPARTMENTS AND THE ROLE OF THE ENDOPLASMIC RETICULUM IN THE INTRANEURONAL DISPATCHING OF AXONALLY TRANSPORTED MEMBRANE-BOUND MOLECULES *

The involvement of axonal transport in the supply of specific macromolecules to nerve endings raises the question of how axonally transported macromolecules of definite species are selected and dispatched to strategic sites in which they contribute to the economy of the cholinergic synapse? In the chicken ciliary ganglion as in the optic system (Grafstein, 1969; Cuénod and Schonbach, 1971; Hendrickson, 1972) or the sciatic nerve (Ochs, 1972), electron microscope radioautography and cell fractionation indicate that most of the rapidly transported macromolecules are associated with membrane profiles or particulate fractions (Bennett et al., 1973; Di Giamberardino et al., 1973;

---

* Section contributed by B. Droz and H.L. Koenig.

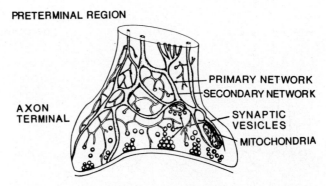

PRETERMINAL REGION

PRIMARY NETWORK

SECONDARY NETWORK

AXON
TERMINAL

SYNAPTIC
VESICLES

MITOCHONDRIA

Fig. 8. Diagram of the smooth endoplasmic reticulum in the presynaptic region of a cholinergic axon. Large tubules of the endoplasmic reticulum anastomose to form the primary
network. The secondary network made of smaller elements originates from wide tubules.
Thin and wide tubules appear to be closely apposed to plasma membrane and mitochondria.
Synaptic vesicles seem to bulge at the ends of thin canaliculi.

Droz et al., 1973; Koenig et al., 1973; Alvarez, 1974). When axonal transport
was impeded by applying light pressure to preganglionic nerves, a dramatic
accumulation of both radioactive molecules and profiles of smooth endoplasmic reticulum was observed in the axon within 3 h (Droz et al., 1973, 1975). In
acrylamide-induced neuropathy, the arrest and the focal accumulation of
labelled proteins coincide with an abnormal extension and disorganization of
the axonal endoplasmic reticulum (Figs. 16 and 17; section V). In other words,
the route followed by molecular aggregates associated with the endoplasmic
reticulum was obstructed and caliciform nerve endings were deficient in
supplies for the renewal of their membrane components. The advent of high-
voltage electron microscope has made it possible to elucidate the spatial configuration of the endoplasmic reticulum in axons and terminals after a specific
impregnation with heavy metal salts (Thiéry and Rambourg, 1976). The
examination of 1–2 μm thick sections of presynaptic calices was facilitated
because most of the presynaptic organelles are arranged in layers. The first
layer, adjacent to the presynaptic membrane, mainly contains synaptic vesicles,
frequently clustered in the active zones (Koenig, 1967). In the second layer,
most of the presynaptic mitochondria are found and the third is occupied by
the axoplasm. In addition, a fourth compartment is represented by the presynaptic endoplasmic reticulum which extends and develops its network
throughout these layers (Fig. 8).

The presynaptic endoplasmic reticulum corresponds to the terminal portion
of channels running along the axon from the nerve cell body. The three-dimen-
sional study of axonal endoplasmic reticulum in presynaptic calices indicates
that large anastomosed tubules give rise to a primary network mainly occupy-
ing the peripheral region of the calyx and frequently enclosing mitochondria
within its meshes (Fig. 8). Since membrane-bound macromolecules are dispatched via the axonal endoplasmic reticulum to nerve endings (Droz et al.,
1975), the transfer of proteins and phospholipids to mitochondria is probably
mediated by contacts with tubules of the endoplasmic reticulum. Thinner
canaliculi originate from large tubules of the primary network and give rise to a
secondary network best developed in the core of the calyx. At the tip of thin

canaliculi, appended vesicles, resembling synaptic vesicles, may be seen. Such a contiguous or continuous relationship probably ensures the exchange of macromolecules between these organelles: proteins, glycoproteins and phospholipids conveyed by fast axonal transport along the channels of the axonal endoplasmic reticulum are indeed rapidly distributed to synaptic vesicles (Fig. 4; Table II; Bennett et al., 1973; Droz et al., 1973; Markov et al., 1976). It is also probable that part of the molecular aggregates making up synaptic vesicles are transported in a retrograde direction to nerve cell bodies. Thus it is inferred that the presynaptic endoplasmic reticulum plays an important role in the assembly and recycling of macromolecular components of cholinergic synaptic vesicles. Occasionally, large tubules and thin canaliculi appear to be adjacent to the axolemmal and presynaptic plasma membranes. Here again the endoplasmic reticulum would constitute a preferential pathway for the molecular traffic responsible for the renewal of plasma membrane components.

Beside the presynaptic endoplasmic reticulum which behaves like cross-roads in dispatching membrane constituents towards specific synaptic sites, the axoplasm of caliciform nerve endings represents the farthest part of a huge compartment continually conveying macromolecules from the perikaryon. Among the constituents of the axoplasmic matrix are tubulin, neurofilament subunits, glycolytic enzymes and an enzyme specific to cholinergic axons, choline acetyltransferase, ChAT. In preliminary experiments, Di Giamberardino and Rossier (unpublished results) combined isotopic tracer and immunochemical procedure and obtained evidence that ChAT associated proteins move with the slow axoplasmic transport in the intact ciliary ganglion system as had been previously shown for ChAT in interrupted nerves by Fonnum et al. (1973) and Tuček (1975).

Cholinergic calices appear to be highly compartmentalized regions of the neuronal cytoplasm in which specialized processes take place simultaneously: synthesis, storage, release or breakdown of acetylcholine. To achieve these distinct mechanisms, the presynaptic nerve endings must be supplied with ChAT, a variety of multimolecular aggregates forming membranes which permit a very rapid sequestration or extrusion of acetylcholine and actin-myosin possibly involved in the movement of the axonal cytoskeleton (Lasek and Hoffman, 1976). All these constituents are manufactured not in situ but in the perikaryon. As emphasized by Couteaux (1974), the coexistence of these processes in a compact area offers a decisive advantage which reduces delay in the supply of acetylcholine. In other words, for the secretion of acetylcholine, cholinergic nerve endings are completely dependent on the supply of soluble and membrane-bound macromolecules which control the secretory process.

(IV) INTERCELLULAR EXCHANGE OF MOLECULES AND TROPHIC INTERACTION *

According to Smith and Kreutzberg (1976), the transcellular interchange of molecules between neurones and glia or pre- and postsynaptic elements is

---

* Section contributed by H.L. Koenig and B. Droz.

34

critical for the functional maintenance of these differentiated cells. The chicken ciliary ganglion offers the opportunity of studying the transfer of molecules from preterminal axons to Schwann cells (Droz et al., 1978) as well as the reciprocal influence exerted by pre- and postsynaptic neurones (Koenig, 1965; Koenig and Droz, 1971; Droz et al., 1973; Chiappinelli et al., 1976; Koenig and Koenig, 1978).

(A) Transfer of molecules from cholinergic axons to myelinating Schwann cells

The supply to Schwann cells of molecules transported along the preganglionic axons by axonal flow was examined by radioautography. After the intra-cerebral injection of labelled amino acids or sugars, the results failed to provide a clearcut evidence for the axon-Schwann cell transfer of axonally transported protein or glycoprotein (Bennett et al., 1973; Droz et al., 1973; Alvarez, 1974). Surprisingly, when preganglionic axons were loaded with transported radio-active phospholipids after the intracerebral injection of [2–^{3}H]glycerol or [^{3}H]choline, the myelin sheath encompassing the axon became rapidly radio-

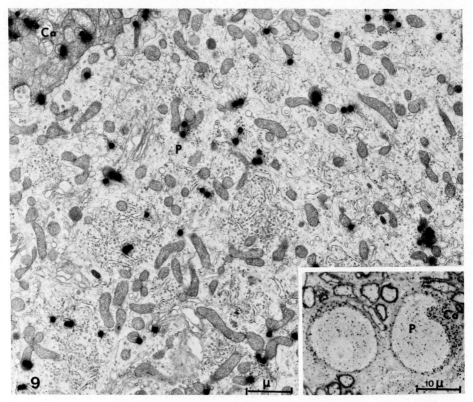

Fig. 9. Transsynaptic passage of label after the intracerebral injection of [^{3}H]choline. Light (inset) and electron microscope radioautographs show that the presynaptic calices (Ca) are loaded with labelled phospholipids transported along the preganglionic axons. The numerous silver grains found over the postsynaptic perikaryon (P) indicate the presence of label which was conveyed transsynaptically. The radioactivity is mainly related to the rough endoplasmic reticulum of the Nissl substance and to mitochondria.

active. The observed transfer of label corresponds to a passage of intact phospholipids from the axon to the myelin, probably through the Schmidt-Lantermann clefts; with time labelled phospholipids were redistributed along the myelin leaflets and finally accumulated in the paranodal regions (Fig. 6). Supplementary labelling was observed after the intracerebral injection of [³H]choline: the tracer, in part released from labelled axonal phosphotidylcholine by phospholipase or the base-exchange reaction, could leak out of the axon and be incorporated by Schwann cells into phospholipids slowly translocated to myelin as it has been observed by Gould and Dawson (1976) after an intraneural injection of [³H]choline. Thus two distinct mechanisms are involved: a direct transfer of intact phospholipids from axon to myelin and an indirect re-utilization of choline released from the axon and taken up by Schwann cells. Both intercellular exchanges of molecules contribute therefore to the maintenance of the myelin sheath (Droz et al., 1978).

(B) Transsynaptic effects exerted by pre- upon postsynaptic neurones and transneuronal transfer of molecules

In the chicken ciliary ganglion, transsynaptic influences are mainly exerted on molecules involved in cholinergic transmission. When the ganglion is denervated by section of the preganglionic nerves, the presynaptic calices degenerate within 2 days (Koenig, 1967) and ChAT and AChE located in postsynaptic

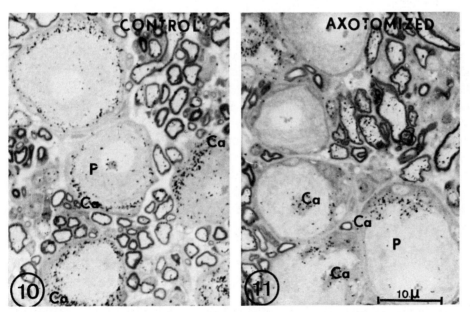

Figs. 10 and 11. Light microscope radioautographs of intact (Fig. 10) or axotomized (Fig. 11) ciliary ganglion cells 3 h after the intracerebral injection of [³H]lysine.

In control ciliary ganglion (Fig. 10), labelled proteins transported by fast axonal flow accumulate in the large presynaptic calices (Ca) which encompass the postsynaptic perikaryon (P) practically devoid of label.

Seven days after the section of the postganglionic nerves (Fig. 11), the total mount of axonally transported proteins collected in calices is reduced by 40%. Thus the postsynaptic neurone exerts a regulatory effect on presynaptic axons.

ganglion cells diminish (Chiappinelli et al., 1976; Couraud et al., unpublished). When the postganglionic nerves were sectioned and the preganglionic ones were left intact, the 7 S form of AChE was found to increase within one day; this early rise in the 7 S form could however be prevented by section of both pre- and post-ganglionic nerves. This finding indicates that presynaptic axon terminals probably exert a control upon the enzyme production in postsynaptic neurones (Koenig and Koenig, 1978). The regulation of the transsynaptic effect could be mediated by molecules transferred from one neurone to another.

After an intracerebral injection of [³H]lysine, a discrete transsynaptic passage of labelled molecules from presynaptic calices to postsynaptic ganglion cell bodies was constantly found (Fig. 10). When protein synthesis was inhibited in a ciliary ganglion by local application of puromycin, the radioactive proteins were normally conveyed to the axon terminals but the amount of label transferred to postsynaptic ganglion cells was severely impeded (Droz et al., 1973). It was therefore concluded that most of the label delivered from pre- to postsynaptic elements consists of free amino acids which are reincorporated into newly synthesized proteins by ganglion cells.

When either [2-³H]glycerol or [³H]choline was injected intracerebrally, radioactive phospholipids transported by fast axonal flow accumulated in presynaptic calices (Figs. 2 and 9, inset). However a massive transsynaptic transfer of label was observed only with [³H]choline. Since [2-³H]glycerol is not recycled once incorporated into phospholipids (Benjamins and McKhann, 1973), the absence of label in ganglion cells (Fig. 2) indicates that [³H]glycerophospholipids are not transferred through the synaptic cleft. In contrast, the intense radioautographic reaction observed over postsynaptic ganglion cells with [³H]choline (Fig. 9) would result from a transsynaptic transfer of small molecules, namely choline or its metabolites released from presynaptic phospholipids by phospholipase or the base-exchange reaction (Arienti et al., 1976; Brunetti et al., 1978); these small molecules, taken up by ganglion cells, would be reincorporated into phospholipids. Such a transsynaptic transfer of choline or metabolite contributes to the renewal of postsynaptic choline phospholipids and thereby exerts a trophic action.

(C) Transneuronal effects exerted by post- on presynaptic neurones

In embryonic ganglia, the activity of ChAT in the presynaptic axon terminals is raised when the functional innervation of the effectors in the eye is established (Chiappinelli et al., 1976). In young chickens, the interruption of the innervation of the ciliary and iris muscles by section of postganglionic nerves produces a modification of presynaptic calices: AChE activity is strongly enhanced in presynaptic calices 6 or 7 days after axotomy of ganglion cells (Taxi, 1960; Koenig, 1965; Koenig and Droz, 1971); these observations were recently supplemented by an electrophysiological investigation which reflects an alteration of synaptic transmission (Brenner and Johnson, 1976). When, at 6 or 7 days after an axotomy of ganglion cells, [³H]lysine was injected into brain, the radioautographs of the ciliary ganglion indicated that the total amount of proteins transported to the caliciform axon terminals within 3 h was reduced by 40%; this diminution of axonally transported synaptic macromolecules coincides with a remodeling of the calices (Figs. 10 and 11). These results indicate that biosynthetic activity and axonal transport in presynaptic neurones are probably influenced by signals originating from postsynaptic neurones.

(V) PERIPHERAL NEUROPATHIES AND AXONAL TRANSPORT OF SYNAPTIC MACROMOLECULES: ACRYLAMIDE-INDUCED AXONAL DYSTROPHY *

Cholinergic transmission may be impeded in various types of neurological diseases. It is therefore important to determine whether axonal transport of macromolecules towards nerve endings is affected at early stages of a pathological process and whether the alteration of axonal transport could be considered as a cause or an effect of the axonal lesion.

Acrylamide, $CH_2 = CH-CO-NH_2$, is widely used in industry and is responsible for the induction of retrograde axonal degeneration (Fullerton and Barnes, 1966; Spencer and Schaumburg, 1977). Experimental studies have shown that acrylamide intoxication produces a slight decrease of protein synthesis in the central nervous system (Schotman et al., 1977) and an impairment of the axonal transport affecting either the slow phase (Pleasure et al., 1969) or the rapid phase (Bradley and Williams, 1973). To clarify this issue, it was decided to investigate the early effects of acrylamide on the axonal transport of different molecular forms of AChE in the sciatic nerve and of radioactive proteins in the ciliary ganglion of chickens. Twenty chickens were given an intraperitoneal injection of acrylamide (100 mg per kg body weight) every two days. Ataxia was well developed after the eighth injection and the animals were studied at this time.

Alteration of fast axonal transport of molecular forms of AChE in sciatic nerve

Histological examination of the sciatic nerve in acrylamide-treated chickens showed nerve fibres, in the initial portion of nerve excised at the level of the thigh, to be unaffected or affected only slightly, whereas the distal branches were deeply altered exhibiting axonal balloons and degenerating myelin sheaths. For this reason, axonal AChE activity was measured in the initial portion of sciatic nerves and was found to be the same in control and acrylamide-treated chickens. However, the amount of the high molecular weight form, 20 S, was 5 times greater in acrylamide-treated than in control animals (Fig. 12). The axonal transport of AChE was then studied by measuring the accumulation of the various molecular forms of the enzyme in segments adjacent to a nerve transsection (section II): the fast axonal transport of the heavy forms, 11 S and 20 S, was reduced, the effect on 20 S being most marked. In contrast the slow axonal transport of the low molecular weight forms was unaffected (Fig. 13). Thus the alteration of rapid phase of axonal flow conveying the heavy forms, probably bound to membranes of the endoplasmic reticulum, appears to be an

---

* Section contributed by M. Chrétien, F. Souyri, J.Y. Couraud and L. Di Giamberardino.

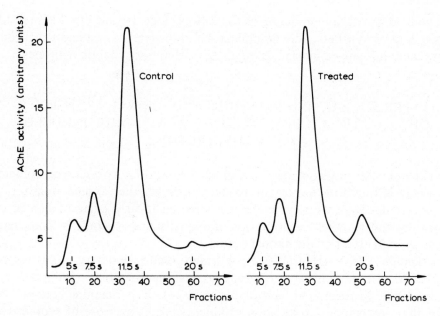

Fig. 12. Sedimentation profiles of AChE in segment of sciatic nerve of normal chickens (control) and of acrylamide-treated chickens (treated). Note the five-fold increase of the 20 S form in acrylamide-treated chickens.

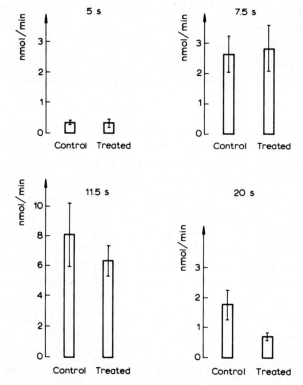

Fig. 13. Accumulation of 4 different molecular forms of AChE transported in an orthograde direction in a 2 mm segment of sciatic nerve proximal to a transsection performed 6 h before. In acrylamide-treated chickens (treated), the accumulation of the heavy molecular forms 20 S and 11 S expressed in nanomole per minute is severely reduced as compared with control chickens (control). This alteration of the movement of AChE indicates that the fast axonal transport is disturbed.

early and important effect of acrylamide intoxication; it coincides with the dramatic drop of the 20 S form in leg muscles innervated by the sciatic nerve.

Early modification of the fast axonal transport of protein in the ciliary ganglion

In contrast to the most distal portion of long axons in the sciatic nerve, the morphology of the preterminal part of the short axons of the preganglionic nerves was unaffected or little affected by the toxic agent. To determine whether or not a defect of axonal transport was concomitant with the detection of the first lesions of the nerve fibres, [³H]lysine was injected intracerebrally in control and acrylamide treated chickens; the ciliary ganglia were excised 3 h or 7 days later and processed for radioautography. The distribution of the radioactivity conveyed with rapidly and slowly transported proteins was quantitatively estimated in preganglionic axons and caliciform terminals. At 7 days, the slowly transported proteins in presynaptic calices of acrylamide-treated chickens were only slightly reduced in 40% of the animals. At 3 h, in all the acrylamide-treated chickens, the radioactivity was increased in 30% of the preganglionic axons. Several axonal profiles exhibited definite clusters of silver grains located at the periphery of the axon, frequently in the vicinity of a node of Ranvier (Figs. 14 and 15). This finding indicates that proteins transported by fast axonal flow to presynaptic calices were retained in the distal part of axons. At the ultrastructural level, the focal accumulation of silver grains was superimposed on an abnormal accumulation of profiles of smooth endoplasmic reticulum and of mitochondria (Fig. 16). The impregnation of acrylamide-treated ganglia with heavy metal salts (Thiery and Rambourg, 1976) revealed

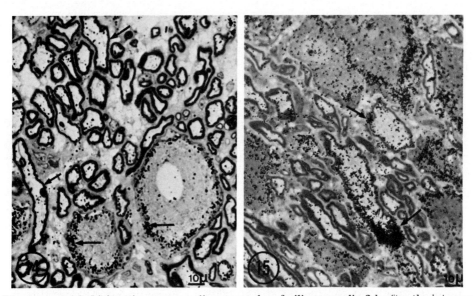

Figs. 14 and 15. Light microscope radioautographs of ciliary ganglia 3 h after the intracerebral injection of [³H]lysine to intact (Fig. 14) or to acrylamide-treated chickens (Fig. 15).

In control chickens (Fig. 14), the preganglionic axons (oblique arrows) are poorly labelled as compared to the caliciform nerve endings (horizontal arrows).

In acrylamide-treated chickens (Fig. 15), an intense labelling is found in some axonal profiles; clusters of silver grains are focally deposited at the periphery of the axon (oblique arrow).

40

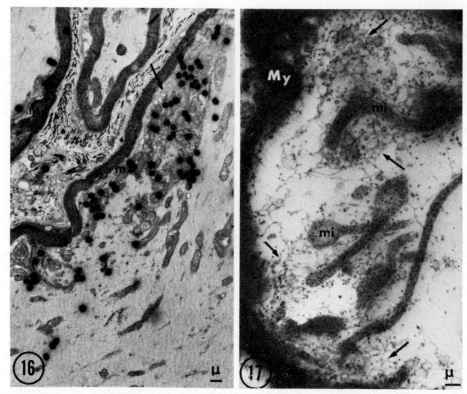

Fig. 16. Electron microscope radioautograph of a higly radioactive preganglionic axon after acrylamide treatment. By 3 h after the intracerebral injection of [³H]lysine, clumps of silver grains are located over the axoplasmic regions which are abnormally rich in tubular and vesicular profiles of axonal endoplasmic reticulum (oblique arrow). Note also the increased number of mitochondria (mi) intermingled with the endoplasmic reticulum.

Fig. 17. Preganglionic axon in a 1 μm thick section of acrylamide-treated ciliary ganglion after impregnation with heavy metal salts (Thiery and Rambourg, 1976). The axonal endoplasmic reticulum is completely disorganized: the tubules, instead of running parallel, form an irregular network (oblique arrows) of thin canaliculi and vesicles and enmesh mitochondria (mi). My: myelin sheath. (Courtesy of G. Patey.)

in 1 μm thick sections the local disorganization of the axonal endoplasmic reticulum. The loss of the longitudinal array, the development of a complex network made of large tubules and thin canaliculi, the abnormal presence of vesicles resembling synaptic vesicles and the accumulation of mitochondria intermingled with the endoplasmic reticulum network seem to represent the earliest modifications of the axonal architecture and to precede the degeneration of the axon terminal. It is therefore concluded that the abnormal growth of the axonal endoplasmic reticulum and the focal arrest of rapidly transported protein, including the 20 S form of AChE, en route to presynaptic endings, could be responsible for the defect in cholinergic transmission.

Recently Di Giamberardino et al. (1978) have investigated the axonal transport of the molecular forms of AChE in various nerves of chickens affected by an inherited muscular dystrophy. Contrary to acrylamide-induced neuropathy, the abnormal accumulation of AChE observed in muscles of dystrophic chickens does not seem to result from an alteration of axonal transport of AChE which does not differ from normal animals.

(VI) CONCLUDING REMARKS

Cholinergic synapses may be considered as dynamic structures which continuously lose and gain molecules. The biochemical events, which ensure the secretion of acetylcholine by presynaptic terminals, and the conformational changes, which convert the signals received into a postsynaptic response, take place within distinct but adjacent elements only a few nanometers apart. The small size and proximity of pre- and postsynaptic structures contrast with large size and remoteness of nerve cell bodies. The perikaryon, often located several millimeters or even decimeters from the axon terminals, is indeed the main centre manufacturing proteins, glycoproteins and phospholipids required for the maintenance of presynaptic nerve endings; the delicate problem of the distribution of macromolecules between perikaryon and axon terminals is solved by axonal transport. Thus enzymes controlling biosynthesis or breakdown of acetylcholine, the subunits which form into membrane to store or release acetylcholine and the mitochondrial components supplying energy, are selectively compartmentalized and dispatched to operative sites in presynaptic endings.

This dynamic concept of synapses echoes the views expressed some 140 years ago by Purkyně in his decisive contribution to our neurological lore. In 1837, Purkyně wrote that "cell bodies are probably central structures which relate themselves to the nerve fibres of the central and peripheral nervous system" and "they could be regarded as the collectors, the generators and the distributors of the nervous system". As emphasized by Gutmann (1971), Purkyně's concept implicitly embraces the idea that, beside the conduction of nerve impulses, "the axons also serve the equally important function of transporting essential materials, synthesized in the nerve cell soma, to the terminals and affecting other neurons or innervated tissues".

SUMMARY

The giant cholinergic axon terminal of the chicken ciliary ganglion was selected to study the renewal of presynaptic macromolecules by radioautography. In vitro incubation of the ganglion shows that the local incorporation of labelled precursors into presynaptic proteins and phospholipids is extremely low. In vivo intracerebral injection of labelled precursors of protein, glycoprotein or phospholipid allow the visualization of axonal transport of presynaptic macromolecules to cholinergic endings. Membrane constituents are transported by fast axonal flow (300 mm per day) and renewed within a few hours or weeks while the axoplasmic constituents are conveyed by slow axonal flow (1–10 mm per day). Simultaneously macromolecules are cleared from presynaptic endings by local breakdown, retrograde transport or extraneuronal release. The molecular selectivity of axonal transport was clearly shown by studying the movement of the different molecular forms of AChE in the chicken sciatic nerve. The heavy forms (20 S and 11 S) are rapidly conveyed by ortho- and retrograde transport, whereas the light forms are slowly transported in the orthograde direction. The arrangement of the endoplasmic reticulum

42

plays an important role in the distribution of fast transported membrane components to synaptic vesicles, presynaptic plasma membranes or mitochondria. In the course of the axonal transport of phospholipids, intact macromolecules are transferred to the myelin sheath; in addition, choline may be released from the axon and re-utilized by Schwann cells. A transsynaptic passage of small molecules, such as amino acids and, in particular, choline, could exert a trophic influence on the postsynaptic target cell. Reciprocally presynaptic metabolism is regulated by postsynaptic activity. A defect in the regulation of fast axonal transport coinciding with an abnormal proliferation of the endoplasmic reticulum could underlie the progressive axonal dystrophy induced by acrylamide. It is concluded that the dynamic state of presynaptic macromolecules contributes to the maintainance of neuronal interconnections and allows adaptation to various conditions.

ACKNOWLEDGEMENTS

The authors wish to express their gratitude to Mrs. J. Boyenval and R. Hassig who carried out light and electron microscope radioautography and contributed to the illustrations in this chapter.

REFERENCES

Alvarez, J. (1974) Time course and subcellular distribution of the radioactivity in a synaptic terminal after supplying the perikaryon with labelled glutamic acid. *Cell. Tiss. Res.,* 150, 11—20.

Arienti, G., Brunetti, M., Gaiti, A., Orlando, P. and Porcellati, G. (1976) The contribution of net synthesis and base-exchange reaction in phospholipid biosynthesis. In *Function and Metabolism of Phospholipids in the Central and Peripheral Nervous System,* G. Porcellati, L. Armaducci and C. Galli (Eds.), Raven Press, New York, pp. 63—78.

Benjamins, J.A. and McKhann, G.M. (1973) [2-^{3}H]glycerol as a precursor of phospholipids in rat brain: evidence for lack of recycling. *J. Neurochem.,* 20, 1111—1120.

Bennett, G., Di Giamberardino, L., Koenig, H.L. and Droz, B. (1973) Axonal migration of protein and glycoprotein to nerve endings. II. Radioautographic analysis of the renewal of glycoproteins in nerve endings of chicken ciliary ganglion after intracerebral injection of [^{3}H]fucose and [^{3}H]glucosamine. *Brain Res.,* 60, 129—146.

Boyenval, J. and Fischer, J. (1976) Dipping technique. In *Techniques in Radioautography, J. Microsc. Biol. Cell., Vol. 27,* B. Droz, M. Bouteille and D. Sandoz (Eds.), S.F.M.E., Paris, pp. 115—120.

Bradley, W.G. and Williams, M.H. (1973) Axoplasmic flow in axonal neuropathies. I. Axoplasmic flow in cats with toxic neuropathies. *Brain,* 96, 235—246.

Brenner, H.R. and Johnson, E.W. (1976) Physiological and morphological effects of postganglionic axotomy on presynaptic nerve terminals. *J. Physiol. (Lond.),* 260, 143—158.

Brunetti, M., Giuditta, A. and Porcellati, G. (1978) The synthesis of choline phosphoglycerides in the giant fibre system of the squid. *J. Neurochem.,* in press.

Chiappinelli, V., Giacobini, E., Pilar, G. and Uchimura, H. (1976) Induction of cholinergic enzymes in chick ciliary ganglion and iris muscle cells during synapse formation. *J. Physiol. (Lond.),* 257, 749—766.

Couteaux, R. (1974) Remarks on the organization of axon terminals in relation to secretory process of synapses. In *Cytopharmacology of Secretion, Advances in Cytopharmacology, Vol. 2,* B. Ceccarelli, F. Clementi and J. Meldolesi (Eds.), Raven Press, New York, pp. 369—379.

Cuénod, M. and Schonbach, J. (1971) Synaptic proteins and axonal flow in the pigeon visual pathway. *J. Neurochem.,* 18, 809—816.

Di Giamberardino, L., Bennett, G., Koenig, H.L. and Droz, B. (1973) Axonal migration of protein and glycoprotein to nerve endings. III. Cell fraction analysis of chicken ciliary ganglion after intracerebral injection of labeled precursors of proteins and glycoproteins. *Brain Res.*, 60, 147—159.

Di Giamberardino, L. and Couraud, J.Y. (1978) Rapid accumulation of high molecular weight acetylcholinesterase in transected sciatic nerves. *Nature (Lond.)*, 271, 170—172.

Di Giamberardino, L., Couraud, J.Y. and Barnard, E.A. (1979) Normal axonal transport of acetylcholinesterase forms in peripheral nerves of dystrophic chickens. Brain Res., 160, 196—202.

Droz, B. (1973) Renewal of synaptic proteins. *Brain Res.*, 62, 383—394.

Droz, B. (1975) Synthetic machinery and axoplasmic transport: maintenance of neuronal connectivity. In *The Nervous System, Vol. 1*, D.B. Tower (Ed.), Raven Press, New York, pp. 111—127.

Droz, B. and Koenig, H.L. (1971) Dynamic condition of protein in axons and axon terminals. *Acta neuropathol.*, Suppl. V, 109—118.

Droz, B., Koenig, H.L. and Di Giamberardino, L. (1973) Axonal migration of protein and glycoprotein to nerve endings. I. Radioautographic analysis of the renewal of protein in nerve endings of chicken ciliary ganglion after intracerebral injection of [^{3}H]lysine. *Brain Res.*, 60, 93—127.

Droz, B., Rambourg, A. and Koenig, H.L. (1975) The smooth endoplasmic reticulum: structure and role in the renewal of axonal membrane and synaptic vesicles by fast axonal transport. *Brain Res.*, 93, 1—13.

Droz, B., Di Giamberardino, L., Koenig, H.L., Boyenval, J. and Hassig, R. (1978) Axon-myelin transfer of phospholipid components in the course of their axonal transport as visualized by radioautography. *Brain Res.*, 153, 347—353.

Fonnum, F., Frizell, M. and Sjöstrand, J. (1973) Transport, turnover and distribution of choline acetyltransferase and acetylcholinesterase in the vagus and hypoglossal nerves of the rabbit. *J. Neurochem.*, 21, 1109—1120.

Fullerton, P.M. and Barnes, J.M. (1966) Peripheral neuropathy in rats produced by acrylamide. *Brit. J. industr. Med.*, 23, 210—221.

Gambetti, P., Autilio-Gambetti, L.A., Gonatas, N.K. and Shafer, B. (1972) Protein synthesis in synaptosomal fractions. Ultrastructural radioautographic study. *J. Cell Biol.*, 52: 526—535.

Gould, R.M. (1976) Inositol lipid synthesis localized in axons and unmyelinated fibers of peripheral nerves. *Brain Res.*, 117, 169—174.

Gould, R.M. and Dawson, R.M.C. (1976) Incorporation of newly formed lecithin into peripheral nerve myelin. *J. Cell Biol.*, 68, 480—496.

Grafstein, B. (1969) Axonal transport: communication between soma and synapse. In *Advances in Biochemical Psychopharmacology, Vol. 1*, E. Costa and P. Greengard (Eds.), Raven Press, New York, pp. 11—25.

Gutmann, E. (1971) Purkyně's concepts on nerves and some aspects of current neurobiology. In *J.E. Purkyně 1787—1869 Centenary Symposium*, V. Kruta (Ed.), Universita J.E. Purkyně, Brno, pp. 183—187.

Hendrickson, A.E. (1972) Electron microscopic distribution of axoplasmic transport. *J. compar. Neurol.*, 144: 381—397.

Koenig, H.L. (1965) Relation entre la distribution de l'activité acétylcholinestérasique et celle de l'ergastoplasme dans les neurones du ganglion ciliaire du poulet. *Arch. Anat. microc. Morphol. exp.*, 54: 937—964.

Koenig, H.L. (1967) Quelques particularités ultrastructurales des zones synaptiques dans le ganglion ciliaire du poulet. *Bull. Ass. Anat.*, 52, 711—719.

Koenig, H.L. and Droz, B. (1970) Transport et renouvellement des protéines dans les terminaisons nerveuses. *C.R. Acad. Sci. (Paris)*, 270, 2579—2582.

Koenig, H.L. and Droz, B. (1971) Effect of nerve section on protein metabolism of ganglion cells and preganglionic nerve endings. *Acta neuropathol.*, Suppl. V. 119—125.

Koenig, H.L., Di Giamberardino, L. and Bennett, G. (1973) Renewal of proteins and glycoproteins of synaptic constituents by means of axonal transport. *Brain Res.*, 62, 413—417.

Koenig, H.L., Couraud, J.Y. and Di Giamberardino, L. (1977) Increase of the 7 S molecular form of acetylcholinesterase in the ciliary ganglion of the chick after axotomy. *Proc. Sixth Intern. Meeting of Intern. Soc. Neurochem.*, Copenhagen, p. 143.

Koenig, H.L. and Koenig, J. (1978) One molecular form of AChE associated with synapses

44

in two cholinergic systems: skeletal muscle of the rat and ciliary ganglion of the chick. In *Maturation of Neurotransmission, Vol. 9*, E. Giacobini and A. Vernadak (Eds.), Karger, Basel, pp. 91–99.

Kristensson, K. and Olsson, Y. (1971) Uptake and retrograde axonal transport of peroxidase in hypoglossal neurones. Electron microscopical localization in neuronal perikaryon. *Acta neuropathol.*, 19, 1–9.

Larrabee, M.G. (1968) Transsynaptic stimulation of phosphatidylinositol metabolism in sympathetic neurons in situ. *J. Neurochem.*, 15: 803–808.

Lasek, R.J. and Hoffman, P.N. (1976) The neuronal cytoskeleton, axonal transport and axonal growth. In *Cell Motility*, R. Goldman, T. Pollard and J. Rosenbaum (Eds.), Cold Spring Harbor Laboratory, pp. 1021–1049.

Lubińska, L. and Niemierko, S. (1971) Velocity and intensity of bidirectional migration of acetylcholinesterase in transected nerves. *Brain Res.*, 27, 329–342.

Markov, D., Rambourg, A. and Droz, B. (1976) Smooth endoplasmic reticulum and fast axonal transport of glycoproteins, an electron microscope radioautographic study of thick sections after heavy metals impregnation. *J. Microsc. Biol. Cell.*, 25, 57–60.

Massoulié, J. and Rieger, F. (1969) L'acétylcholinestérase des organes électriques de Poissons (torpille et gymnote); complexes membranaires. *Europ. J. Biochem.*, 11, 441–455.

Musick, J. and Hubbard, J.I. (1972) Release of protein from mouse motor nerve terminals. *Nature (Lond.)*, 237, 279–281.

Ochs, S. (1972) Fast transport of materials in mammalian nerve fibres. *Science*, 176, 252–260.

Pleasure, D.E., Mischler, K.D. and Engel, W.K. (1969) Axonal transport of proteins in experimental neuropathies. *Science*, 166, 524–525.

Schotman, P., Gipon, L., Jennekens, F.G.I. and Gispen, W.H. (1977) Polyneuropathies and CNS protein metabolism. II. Changes in the incorporation rate of leucine during acrylamide intoxication. *Neuropathol. appl. Neurobiol.*, 3, 125–136.

Smith, B.H. and Kreutzberg, G.W. (1976) Neuron-target cell interactions. *Neurosc. Res. Progr. Bull.*, 14, 211–453.

Spencer, P.S. and Schaumburg, H.H. (1977) Central-peripheral distal axonopathy – The pathology of dying-back polyneuropathies. *Progr. Neuropathol.*, 3, 253–295.

Taxi, J. (1960) La distribution des cholinestérases dans diver ganglions du système nerveux autonome des vertébrés. *Bibl. Anat.*, 2, 73–89.

Thiéry, G. and Rambourg, A. (1976) A new staining technique for studying thick sections in the electron microscope. *J. Microsc. Biol. Cell.*, 26, 103–106.

Tuček, S. (1975) Transport of choline acetyltransferase and acetylcholinesterase in the central stump and isolated segments of a peripheral nerve. *Brain Res.*, 86, 259–270.

Uy, R. and Wold, F. (1977) Posttranslational covalent modification of proteins. *Science*, 198, 890–896.

Vigny, M., Di Giamberardino, L., Couraud, J.Y., Rieger, F. and Koenig, J. (1976) Molecular forms of chicken acetylcholinesterase: effect of denervation. *FEBS Lett.*, 69, 277–280.

Zelená, J. (1968) Bidirectional movements of mitochondria along axons of an isolated nerve segment. *Z. Zellforsch.*, 92: 186–196.

Recent Progress in the Biochemistry of Choline Acetyltransferase

D. MALTE-SØRENSSEN

Norwegian Defence Research Establishment, Division for Toxicology,
N-2007 Kjeller (Norway)

INTRODUCTION

Acetylcholine (ACh) which is generally accepted as a neurotransmitter both at peripheral (cf. Katz, 1966) and central cholinergic synapses (cf. Krnjević, 1969, 1974), is synthesized by the reversible transfer of acetyl groups from acetyl-CoA to choline catalyzed by the enzyme choline acetyltransferase (EC 2.3.1.6, ChAT). Due to the central role of ACh in the nervous transmission much attention has been paid to problems dealing with biosynthesis, storage, localization and release of ACh in nervous tissue. The answers to some of these problems hinge on a better understanding of ChAT. The basis for such understanding of ChAT is biochemical information of the enzyme itself, its catalytic function, molecular properties, intracellular localization and relation to other functional systems of the cholinergic cell and synapse.

Although a great deal of effort has been centered on the biochemical aspects of ChAT and its integration into the functioning of the cholinergic synapse, the progress in obtaining basic knowledge has been slow. The progress in the elucidation of the molecular properties of the enzyme has been hampered by the lack of sources of enzyme suitable for isolation, efficient purification methods and assay methods easy to perform. However, progress in the field was greatly accelerated by the introduction of a highly sensitive radioisotope assay for ChAT which also reduced the assay time considerably (Fonnum, 1969, 1975; McCaman and Hunt, 1965).

Thus, even though the discovery of ChAT in rabbit brain was in 1943 by Nachmansohn and Machado, it is only quite recently that reports have appeared describing the purification of the enzyme to homogeneity. The mechanism of action, the intracellular localization and its role in regulating the level and turnover of ACh is still, 35 years after its discovery, argued. The purpose of this review is to point out the main achievements obtained in the last few years and to indicate the main problems still to be solved in the biochemistry of ChAT.

PURIFICATION

Since its discovery several attempts to purify ChAT have been reported, but only quite recently has it been possible to obtain enzyme preparations with high specific activity. This slow progress has primarily been due to the lack of sources rich in ChAT and to the instability of the enzyme during purification. The latter problem has mainly been overcome by using stabilizing and protecting agents such as thiol reagents, glycerol or EDTA during the purification procedure (Chao and Wolfgram, 1973; Malthe-Sørenssen et al., 1973; Rossier, 1976a; Malthe-Sørenssen et al., 1978).

Presented in Table I are the results of the purification of ChAT from different sources. As can be seen, the specific activity of the final preparations differs in several orders of magnitude, even when the same material has been used for purification. For instance Chao and Wolfgram (1973) claimed to have purified ChAT from bovine brain to electrophoretic homogeneity although their preparation was tenfold less active than others (Ryan and McClure, 1976; Malthe-Sørenssen et al., 1978) and yet the latter preparations still had impurities present. It should be pointed out that several researchers (Rossier, J. and Mautner, H.G., personal communications; Malthe-Sørenssen, unpublished results) have found it difficult to get ChAT to move as a sharp band on electrophoresis either in an acidic, neutral or basic buffer systems. In addition it has been observed that very impure ChAT may in some systems move as a single band on electrophoresis (Mautner, 1977). It is therefore difficult to judge from electrophoresis alone if ChAT is pure. The question of deciding when an enzyme is pure, is indeed a question which troubles most biochemists dealing with purification of enzymes.

Roskoski et al. (1975) and Singh and McGeer (1974a) both claimed to have purified ChAT from human brain to homogeneity, though surprisingly to a lower specific activity than reported 10 years earlier by Morris (1966). Obviously, several of the enzyme preparations described as homogeneous contain either denatured enzyme which exhibits only a fraction of its real activity or proteins which contain ChAT only as an impurity.

TABLE I

SPECIFIC ACTIVITY OF CHOLINE ACETYLTRANSFERASE FROM DIFFERENT SPECIES

Tissue	Specific activity μmol ACh/min · mg protein	Reference
Bovine brain	1.45	Chao and Wolfgram (1973)
	28.5	Malthe-Sørenssen et al. 1978
	48	Ryan and McClure (1976)
	120	Cozzari and Hartman (1977)
Rat brain	4	Malthe-Sørenssen et al. (1973)
	20	Rossier (1976a)
	40	Ryan and McClure (1976)
Human brain	0.14	Singh and McGeer (1974a)
	0.70	Roskoski (1975)
	8.0	Hersh and Peet (1977)
Human placenta	2.3	Morris (1966)
	0.04	Roskoski (1975)
Squid head ganglia	66.7	Husain and Mautner (1973)
Drosophila melanogaster	40	Driskell et al. (1977)

TABLE II

BINDING OF CHOLINE ACETYLTRANSFERASE TO DIFFERENT AFFINITY COLUMNS

Materials	Per cent bound enzyme activity		Per cent eluted enzyme activity		
	Low ionic strength (0.1 M NaCl)	High ionic strength (0.4 M NaCl)	Low ionic strength (0.4 M NaCl)	High ionic strength (1 M NaCl)	Biospecific (0.1 M NaCl, 500 µM acetyl-CoA)
ω-Aminoalkyl-sepharose					
$-CH_2'CH_2'NH_2$	–	–	–	–	–
$-CH_2'CH_2'CH_2'NH_2$	40	–	25	40	–
$-CH_2'CH_2'CH_2'CH_2'NH_2$	100	80	–	20	–
$-CH_2'CH_2'CH_2'CH_2'CH_2'NH_2$	–	100	–	10	–
Agarose-ribosyl-acetate-CoA	100	80	20	40	80
Acrylamide-8-adenine-acetyl-CoA	20	–	20	20	–
Agarose-blue-dextran 2000	100	40	–	40	90

The enzyme preparation after DEAE-cellulose chromatography was used for affinity chromatography on ω-amino-alkylsepharose, whereas the highly purified enzyme preparation was used for chromatography on acetyl-CoA and Blue Dextran gels. The affinity material was loaded on small columns (1 × 2 cm), equilibrated with the proper salt concentration in 10 mM sodium phosphate-citrate buffer pH 7.2 with 1 mM EDTA and 7% glycerol and eluted with the same buffer containing either salt or acetyl-CoA as indicated. Enzyme activity corresponding to 0.17 units, was used in each experiment. Per cent bound and eluted enzyme activity refers to the percentage of the total enzyme activity applied to the columns. (–) mean not detectable or no binding. (Malthe-Sørenssen et al. 1978).

However, quite recently purification of ChAT from squid head ganglia, rat brain and bovine brain with high specific activity has been reported (Husain and Mautner, 1973; Rossier, 1976a; Ryan and McClure, 1976; Cozzari and Hartman, 1977; Malthe-Sørenssen et al., 1978). In one of these preparations homogeneity has been established with reasonable certainty (Cozzari and Hartman, 1977), whereas in other preparations the authors have pointed out that impurities still were present (Rossier, 1976a; Ryan and McClure, 1976; Malthe-Sørenssen et al., 1978). The recent progress in the purification of ChAT can in part be accredited to the introduction of new chromatographic techniques such as affinity chromatography and fractional precipitation with organic polymers (Malthe-Sørenssen et al., 1978). One successful approach was to introduce the thiol-binding organomercurial Sepharose that absorbed ChAT specifically. The enzyme could then be eluted by thiol reagents such as cysteine or mercaptoethanol (Husain and Mautner, 1973; Ryan and McClure, 1976; Malthe-Sørenssen et al., 1978). Another approach that resulted in impressive purification of ChAT was to introduce Sepharoses with specific ligands for ChAT either in form of an inhibitor of ChAT (Husain and Mautner, 1973), or a substrate for ChAT (Ryan and McClure, 1976). As can be seen in Table 1, ChAT from bovine brain binds to different affinity columns and can in part be specifically released from the columns. Difficulties encountered in removing ChAT from each column can be solved by introducing short spacer arms between the column material and the ligand. Long spacer arms in form of methylene groups increase the possibility of hydrophobic binding and less recovery (Table II). Similarly, the difficulties of low recovery encountered when using styrylpyridinium derivatives as ligands could be explained by hydrophobic binding.

Despite the methodological improvement in the purification of ChAT, none of the procedures published give as a final result a reasonable amount of purified homogeneous ChAT. It is important in the future to combine the present knowledge of methods for purification in a way that gives ChAT in mg amounts from a plentiful source.

HETEROGENEITY

In a series of studies Malthe-Sørenssen and Fonnum (1972), Fonnum and Malthe-Sørenssen (1973) and Malthe-Sørenssen (1976a) have demonstrated the presence of multiple molecular forms of ChAT in different species (Table III). The molecular forms differ in electric charges but not in molecular weight. The presence of multiple forms of ChAT as charge isoenzymes have recently been confirmed by different techniques (Franklin, 1976; Wenthold and Mahler, 1975; Prince and Toates, 1974; Park et al., 1977) supporting the first observations. Several other groups have also reported the presence of different molecular forms of ChAT, though most of them describe aggregates or interconvertible molecular forms of ChAT (White and Wu, 1973; Chao and Wolfgram, 1974; Banns, 1976; Polsky and Shuster, 1976a; Husain and Mautner, 1973). Contradictory results have been obtained describing the presence of aggregates of ChAT. Thus, Banns (1976) could demonstrate aggregates of ChAT from

TABLE III

MULTIPLE MOLECULAR FORMS OF ChAT FROM DIFFERENT SPECIES

Sources	Charge isoenzymes isoelectric points			Molecular weight (Daltons) of	
				ChAT	Aggregates
Rat brain	7.4	7.8	8.3(1)	67,000(2)	136,000(2)
				62,000(3)	—
Mouse brain	7.1	7.5	8.4(1)	—	—
Guinea pig brain		6.8(1)		62,000(5)	—
Rabbit brain		6.9(1)		65,000(5)	—
Human brain	5.65	6.6—7.8	7.9—8.4(4)	62,000(6)	aggregates(4)
Human placenta		—		67,000(6)	—
Bovine brain	6.8	7.8	8.6(1)	69,000(7)	60,000—1.5 · 10^6(8)
					14,000 (subunit) (9)
Squid head		—		70,000	125,000—200,000(10)
Ganglia					70,000—235,000(11)
					37,000 (subunit) (11)

(1) Malthe-Sørenssen (1976a); (2) Rossier (1976c); (3) Wenthold and Mahler 1975; (4) White and Wu (1973); (5) Bull et al. (1964); (6) Roskoski et al. (1975); (7) Malthe-Sørenssen et al. 1978; (8) Chao and Wolfgram 1974; (9) Chao 1975; (10) Husain and Mautner (1973); (11) Polsky and Shuster (1976b).

human placenta by exclusion chromatography though the aggregates only constituted a small part of the total enzyme activity, whereas Roskoski (1975) could not detect any multiple molecular forms of the same enzyme. Chao and Wolfgram (1974) also reported the presence of aggregates of ChAT from bovine brain claiming that the aggregation of the enzyme was the result of exposure to ammonium sulphate used during the purification of the enzyme (Table III). However, no such aggregation could be detected by Ryan and McClure (1976) or Malthe-Sørenssen et al. (1978) in different preparations of the same enzyme exposed to ammonium sulphate. Exclusion chromatography of ChAT from bovine brain with different specific activities gave only one molecular form with molecular weight of 65,000—70,000 Daltons (Fig. 1) which is in good agreement with values of 64,000 reported by Bull et al. (1964) for rabbit brain, 68,000 reported by Rossier (1976b) for rat brain and 67,000 reported by Roskoski et al. (1975) for human brain ChAT.

Rossier (1976b) has shown that rat brain enzyme can exist as a monomer and a dimer depending upon the presence of reducing agent which could split the dimer in monomer. This could partially explain the presence of aggregates in some preparations. Despite the apparent discrepancies there seems to be general agreement on the existence of multiple molecular forms of ChAT in several mammals and certain invertebrates. The notion that there should exist subunits of ChAT has been suggested for the bovine brain enzyme (Chao, 1975) and squid enzyme (Polsky and Shuster, 1976b). However, this has not been confirmed by others.

So far any attempt to discern any physiological importance of the multiple molecular forms of ChAT has been unsuccessful. The subcellular distribution of the enzyme did not reveal any preferential apportionment between different subcellular compartments. The charge isoenzymes of rat brain ChAT could all be identified in axons, synaptosomes and perikarya (Fonnum and Malthe-

50

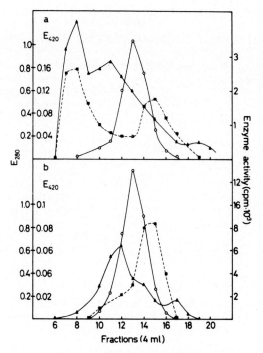

Fig. 1. Exclusion chromatography of choline acetyltransferase on Ultrogel AC-34. Enzyme preparations after ammonium sulphate fractionation were used. 1 ml of each preparation was applied to the column (1.5—45 cm) and eluted at a flow rate of 9 ml/h. Fractions of 4 ml were collected. (a) 0—50% saturated ammonium sulphate fraction, 25 mg protein/ml; (b) 55—65% saturated ammonium sulphate fraction, 18 mg protein/ml. o———o, ChAT; ▲-----▲, E_{280}; ■-----■, E_{420}. (Malthe-Sørenssen et al., 1978).

Sørenssen, 1973). Since the different charge isoenzymes have different membrane affinity one possible function in vivo could be that the isoenzyme, with the highest membrane affinity is linked either to the outside of synaptic vesicles close to one of the compartments of ACh (Fonnum and Malthe-Sørenssen, 1972), or to the presynaptic membrane at the site for the high affinity uptake of choline. Rossier et al. (1977) and Barker and Mittag (1976) have proposed a coupling between the membrane complex for high affinity uptake of choline and ChAT, but further investigations are necessary to establish such a link.

REACTION MECHANISM

Several kinetic mechanisms have been proposed for the enzymatic reaction of ChAT, where the catalysis involves a two substrate reaction with the reversible transfer of an acetyl group from acetyl-CoA to choline. Different kinetic parameters have been measured for ChAT from different species. Apart from their intrinsic interest, kinetic parameters such as K_m and V_{max} provide information about the concentration of the substrates likely to be required for synthesis of ACh in nervous tissue, and product-inhibition studies may reveal regulatory mechanisms by which the activity of cholinergic neurones are self-limiting.

Of the kinetic mechanisms which have been suggested most recent studies favour the Theorell-Chance mechanism, where the substrates bind to the enzyme in a sequential order and any steady state concentration of the ternary complex is very low. The sequence generally assumed is illustrated below.

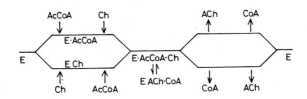

In this scheme acetyl-CoA is assumed to be the leading substrate and CoA the last product to be detached. The model is based on measurement of initial velocities and product inhibition patterns. The order of substrate binding and product release cannot be decided from kinetic data alone. Very recent evidence based on the formation of dead end inhibitor-enzyme complexes obtained by Hersh and Peet (1977) and proposed earlier by Prince and Hide (1971), suggests a rapid equilibrium random mechanism illustrated below, where the interconversion of the ternary complex is partially rate determining.

Rapid equilibrium, random bi-bi

However, since both CoA and acetyl-CoA bind more tightly to the enzyme than choline and ACh (Potter et al., 1968; Rama Sastry and Henderson, 1972; Emson et al., 1974), it seems likely that there is an ordering of substrate binding. Morris et al. (1971), Prince and Hide (1971) and Rossier et al. (1977) have all noted that the binding of the substrates to ChAT is dependent upon the salt concentration in the medium and that differences in ion concentration may affect the reaction mechanism. Rossier et al. (1977) have suggested that the concentration of chloride ion regulates the activity of the enzyme by changing the V_{max} and K_m. If this holds true it introduces an interesting aspect in the regulation of the ACh synthesis coupled to the activity of the cholinergic neurone. However, most of the evidence is in favour of a general salt effect and not an ion-specific effect.

More detailed knowledge of the active site has recently become available. As mentioned previously there is strong evidence for a hydrophobic "pocket" on the enzyme. ChAT binds strongly to long alkyl chains (Table II) (Malthe-Sørenssen et al., 1978 and Rossier, 1977) and to hydrophobic molecules such as styrylpyridinium derivatives. In addition mixed disulphides of CoA such as methyl-, ethyl- and propyl-CoA disulphides were better inhibitors of ChAT than CoA (Currier and Mautner, 1976), which strengthened the evidence for a

52

hydrophobic "pocket" on the enzyme. Blue dextran with its chromophore Reactive Blue has been shown to be a potent competitive inhibitor of ChAT with respect to acetyl-CoA (Roskoski et al., 1975; Rossier, 1977). Since Blue dextran is thought to be a marker for a dinucleotide fold in enzymes (Thompson et al., 1975), it is reasonable to postulate that ChAT contains a dinucleotide fold in the active site. Acetyl-CoA binds in the fold at least partially through some hydrophobic interaction with the adenine moiety of acetyl-CoA. Recent results have shown that a free adenine group is of importance in the binding (Table II; Malthe-Sørenssen et al., 1978).

It has been known for a long time that ChAT contains reactive thiol groups important for the activity of the enzyme. Whether a thiol group is part of the active site of ChAT participating in the reaction mechanism has been a controversial question. Roskoski (1973) claimed to have isolated an acetyl-thio derivative of bovine brain ChAT as an intermediate in the acetyl transfer reaction by incubating ChAT with radioactive acetyl-CoA. The labelled acetylated enzyme could be isolated by gel filtration and the acetyl group could then be transferred to choline (Fig. 2a). The enzyme preparation used in these experiments was of low specific activity (<0.1 μmol ACh formed/min · mg protein). The formation of the acetyl-thio enzyme intermediate was abolished in the presence of agents reacting with thiol groups. Attempts to isolate an acetyl-thioenzyme

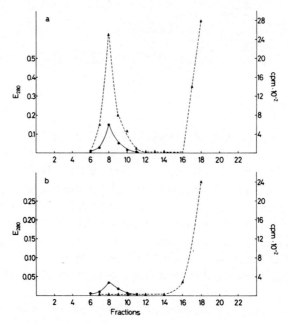

Fig. 2. Gel filtration of different ChAT preparations incubated with [$1-^{14}$C]acetyl-CoA, on Sephadex G-50 fine. The enzyme was incubated in 200 μl of a medium which contained: 5 mM potassium phosphate, 50 mM NaCl, 100 μM physostigmine, 60 μM [$1-^{14}$C]acetyl-CoA (59.2 μCi/μmol) and enzyme activity of 0.43 μmol ACh/min at pH 7.4. After incubation (10 min at 37°C) the incubation solution was subjected to gel filtration. Elution buffer was 5 mM potassium phosphate, pH 5.9. Fractions of 1 ml were collected and their radioactivity and protein content (E_{280}) were measured. (a) Gel filtration of ChAT with low specific activity (0.075 μmol/min · mg protein); (b) Gel filtration of highly purified ChAT (specific activity of 2.5 μmol ACh/min · mg protein). ▲- - - - -▲, [$1-^{14}$C]radioactivity; ●————●, E_{280}. (Malthe-Sørenssen, 1976b).

53

derivative using a highly purified enzyme preparation from bovine brain with high specific activity (>8 μmol ACh formed/min · mg protein) were unsuccessful (Fig. 2b), although the preparation could be inhibited by thiol reagents. Recently, Malthe-Sørenssen (1976b) presented evidence that a reactive thiol group is present in the vicinity of the active site of bovine brain ChAT. When the bovine brain enzyme was treated with thiol reagents with bulky residues such as 5,5'-dithiobis(2-nitrobenzoate) the enzyme was completely inhibited. When the thionitrobenzoate derivative was transformed to a thiocyanate derivative, the enzyme activity was restored. The discrepancies observed in the protection of ChAT by substrates against various thiol reagents, and the degree of inhibition produced by these reagents, could then be explained by differences in the distance between the thiol group and the active site in ChAT from different species.

Several lines of evidence have been accumulated which suggest the participation of a histidine residue in the acetyl transfer reaction of ChAT. White and Cavallito (1970) postulated on the basis of inhibition experiments with styrylpyridiniums derivatives and cupric ion the presence of a catalytically important imidazole residue of histidine. In model studies Burt and Silver (1973) and Hebb et al. (1975) demonstrated that imidazole itself could catalyze the formation of the ACh from acetyl-CoA and choline, which is evidence in support of the concept that histidine takes part in the acetyl transfer reaction. That a histidine residue was essential for catalytic activity of ChAT has been verified in several studies. Roskoski (1974) and Malthe-Sørenssen (1976b) using ethoxyformic anhydride could inhibit the enzyme activity and reactivate the enzyme by treatment with hydroxylamine. The inhibition was followed by a concomitant increase in the absorbance at 242 nm due to acylation of histidine, which disappeared after reactivation (Malthe-Sørenssen, 1976b). Currier and Mautner (1974) were able to demonstrate the photo-oxidation of a histidine residue by rose bengal and the simultaneous inhibition of the enzyme.

Currier and Mautner (1974) have concluded that a general base catalysis takes place, basing this on the observation of a D_2O isotope effect seen in the transacetylation reaction. Malthe-Sørenssen (1976b) has proposed a nucleophilic catalysis for the transacetylation (Fig. 3) in accordance with the noncatalytic reaction. However, caution should be used in interpreting isotope effects, especially in enzymatic reactions where the majority of the isotope effects studied so far are "abnormal" (Jenks, 1969). An isotope effect of 1.8 measured by Currier and Mautner (1974) is difficult to interpret in terms of mechanism since imidazole catalyzed reactions of both the general base and nucleophilic

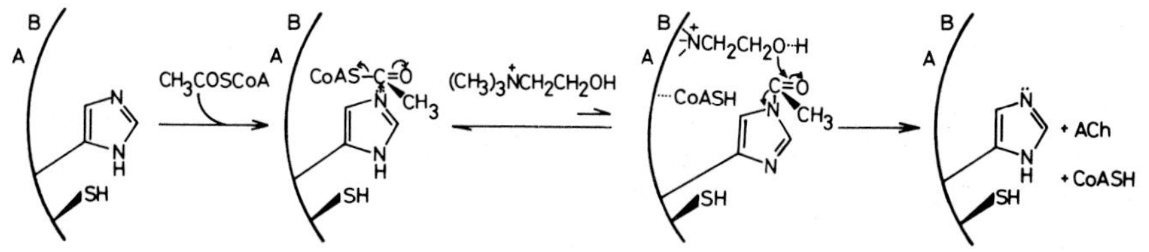

Fig. 3. Proposed participation of a histidine residue in the nucleophilic catalyzed reaction of choline acetyltransferase. A and B represent the binding site for acetyl-CoA and choline respectively.

type give isotope effects in this range (Johnson, 1967). It is perhaps pertinent at this point to call to mind that whereas experimental observations are facts which, if correct, are invariable with time, reaction mechanism and other derived concepts can change as new facts are observed.

IMMUNOLOGICAL PROPERTIES

In recent years one of the aims of the effort to purify ChAT has been to obtain an antibody against the enzyme which could be used for immunochemical detection. This technique would provide information on two difficult subjects. First, the intracellular localization of the enzyme and the position it occupies in the nerve terminal, and second, the identification of cholinergic neurones. Antibodies against ChAT would also make it possible to study structural feature of ChAT, its turnover, axonal transport and phylogeny. Since the immunohistochemical detection of ChAT will be dealt with in detail in another chapter I will concentrate on the immunological properties of ChAT and problems concerned with the preparation of the antibodies and their specificity.

Several problems have been encountered in obtaining antibodies against ChAT. The first attempts reported by Rossier et al. (1973) and Shuster and O'Tool (1974) were hampered by a low antigeneity of rat brain and mouse brain ChAT. Repeated immunizations of rabbits with purified enzyme preparations of ChAT over several months were necessary to obtain antisera active against ChAT. The antisera obtained had a high titre and cross reacted with a variety of species and inactivated ChAT almost 90% (Table IV). Although the preparation of rat brain ChAT used for immunization had a high specific activity (>20 μmol ACh/min · mg protein), the antibody preparation obtained was not monospecific. Several components could be detected by immunodiffusion (Rossier, 1976c). Differences in the immunogenic properties of ChAT exist. Eng et al. (1974), Malthe-Sørenssen (1975) and McGeer et al., (1974) obtained

TABLE IV

RELATIVE IMMUNOLOGICAL ACTIVITY OF DIFFERENT ANTIBODIES AGAINST CHOLINE ACETYLTRANSFERASE FROM DIFFERENT SPECIES

Species	% Inhibition of ChAT			
	Bovine ChAT IgG(1)	Human ChAT IgG(2)	Rat ChAT IgG(3)	Mouse ChAT IgG(4)
Bovine	98	–	48	68
Human	–	56	61	–
Cat	75	–	65	–
Rat	82	51	91	80
Mouse	80	36	85	82
Guinea pig	82	59	60	–
Pigeon	18	–	28	–
Chicken	26	0	–	79

The data are obtained from: (1) Malthe-Sørenssen et al., 1978; (2) Singh and McGeer (1974b); (3) Shuster and O'Toole (1974); (4) Rossier et al. (1973). The highest inactivation of ChAT with the different antibodies obtained is presented.

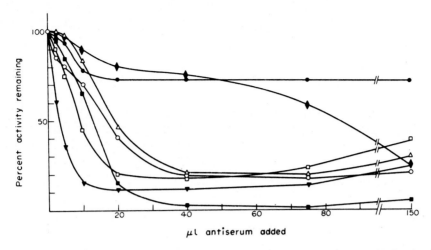

Fig. 4. Inactivation of ChAT from different species by antibody against the bovine brain enzyme. The same amount of enzyme activity (0.004 units) from the different species was used in each experiment and the enzyme activity tested after centrifugation. ChAT from brains of: (■), cow; (○), rat; (▼), cat; (△), mouse; (□), guinea-pig; (♦), chicken and (●), pigeon. Identical precipitation patterns were observed with the following preparation of bovine brain enzyme: homogenate, ammonium sulphate precipitate and highly purified enzyme (28 μmol ACh/min · mg protein). (Malthe-Sørenssen et al., 1978).

antibodies against ChAT from bovine brain and human brain respectively only two to three weeks after the first immunization. Only one or two booster injections of ChAT were necessary. The antisera obtained differed greatly in their capability to inactivate ChAT and in cross reactivity. While the antibody preparation of Eng et al. (1974) and Singh and McGeer (1974b) inactivated ChAT only to a certain extent, maximum 60%, the antiserum against bovine brain ChAT obtained by Malthe-Sørenssen et al. (1978) inactivated ChAT from different species almost completely by immunoprecipitation (Fig. 4). The differences in reactivity of the antisera could be due to differences in the titre of the sera. However, quite recently Cozzari and Hartman (1977) in agreement with our observation reported the preparation of an antiserum against bovine brain ChAT capable of the immunoprecipitation of 95% of ChAT activity. It seems more likely, to the author, that the differences in the reactivity of the antisera is due to the fact that Eng et al. (1974) and Singh and McGeer (1974b) used enzyme preparation with low specific activity while Cozzari and Hartman (1977) and Malthe-Sørenssen et al. (1978) used preparation with high specific activity.

Surprisingly Eng et al. (1974) and Singh and McGeer (1974a) and Singh and McGeer (1974a) claimed to have obtained monospecific antibodies with antigen preparations of low specific activities whereas others (Rossier, 1976b; Shuster and O'Toole, 1974; Polsky and Shuster, 1976b; Malthe-Sørenssen et al., 1978; Cozzari and Hartman, 1977) were all careful to point out that the antisera obtained by them were not monospecific. Although, the antigen preparations used for the immunization by the latter were far higher in specific activity. It is difficult to give an explanation for the differences in specificity, and it can only be observed that striking differences exist in the specificity of antibodies obtained using enzyme preparations with quite different specific activity for immunization.

It should be pointed out that an absolute prerequisite for ultrastructural localization of ChAT by immunohistochemical techniques is a monospecific antibody. Quite recently Cozzari and Hartman (1977) have reported the preparation of antibodies against ChAT using enzyme with high specific activity (128 μmol ACh/min mg protein) for immunization. The antiserum was not monospecific, but it was possible to absorb the non-active contaminant and thus they were able to produce a monospecific antibody. However, at present no detailed report has appeared on this study.

SUMMARY

In the recent years great progress has been achieved in the purification of the ChAT. The progress is mainly due to improved techniques in column chromatography, fractionation techniques, affinity chromatography and in stabilizing the purified enzyme. Several laboratories have prepared enzyme preparations with specific activity higher than 20 μmol/min · mg protein. Different molecular forms of ChAT exist both as charge isoenzymes and as aggregates. However, the presence of aggregates seems to be dependent upon experimental conditions.

Various reaction mechanisms have been proposed for ChAT. Strong evidence has been presented that a histidine residue in the active site participates in the acetyl transfer reaction.

Antibodies raised against bovine brain ChAT inactivate ChAT almost completely in immunoprecipitation experiments. The antibodies crossreact with several species which is of importance in terms of immunohistochemical localization of ChAT.

However, the specificity of several of the antibody preparations is still questioned, although encouraging results have recently been obtained in preparing monospecific antibodies against bovine brain ChAT of high specific activity.

REFERENCES

Banns, H.E. (1976) A study of the molecular sizes and stability of choline acetyltransferase. *J. Neurochem.*, 26, 967–971.

Barker, L.A. and Mittag, T.W. (1976) Synaptosomal transport and acetylation of 3-trimethyl-aminopropan-1-ol. *Biochem. Pharmacol.*, 25, 1931–1933.

Burt, A.M. and Silver, A. (1973) Non-enzymatic imidazole-catalysed acyl transfer reaction and acetylcholine synthesis. *Nature New Biol. (Lond.)*, 243, 157–159.

Bull, G., Feinstein, A. and Morris, D. (1964) Sedimentation behaviour and molecular weight of choline acetyltransferase. *Nature (Lond.)*, 201, 1326.

Chao, L.P. (1975) Subunits of choline acetyltransferase. *J. Neurochem.*, 25, 261–266.

Chao, L.P. and Wolfgram, F. (1973) Purification and some properties of choline acetyltransferase (EC 2.3.1.6) from bovine brain. *J. Neurochem.*, 20, 1075–1081.

Chao, L.P. and Wolfgram, F. (1974) Activation, inhibition and aggregation of choline acetyltransferase. *J. Neurochem.*, 23, 697–701.

Cozzari, C. and Hartman, B.K. (1977) Purification and preparation of antibodies to choline acetyltransferase from beef caudate nucleus. *Proc. int. Soc. Neurochem.*, Vol. 6, Copenhagen, Denmark, p. 140.

Currier, S.F. and Mautner, H.G. (1974) On the mechanism of choline acetyltransferase. *Proc. nat. Acad. Sci. (Wash.)*, 71, 3355–3358.

Currier, S.F. and Mautner, H.G. (1976) Evidence for a thiol reagent inhibiting choline acetyltransferase by reacting with the thiol group of coenzyme A forming a potent inhibitor. *Biochem. Biophys. Res. Commun.*, 69, 431–436.

Driskell, J., Weber, J. and Roberts, E. (1977) Purification of choline acetyltransferase of Drosophila Melanogaster. *Proc. Eighth Ann. Meet. Am. Soc. Neurochem.*, Denver, Col., p. 68.

Emson, P.C., Malthe-Sørenssen, D. and Fonnum, F. (1974) Purification and properties of choline acetyltransferase from the nervous system of different invertebrates. *J. Neurochem.*, 22, 1089–1098.

Eng, L.T., Uyeda, C.T., Chao, L.P. and Wolfgram, F. (1974) Antibody to bovine choline acetyltransferase and immunofluorescent localization of the enzyme in neurons. *Nature (Lond.)*, 250, 243–245.

Fonnum, F. (1969) Radiochemical micro assay for the determination of choline acetyltransferase and acetylcholinesterase activities. *Biochem. J.*, 115, 465–472.

Fonnum, F. (1975) A rapid radiochemical method for the determination of choline acetyltransferase. *J. Neurochem.*, 24, 407–409.

Fonnum, F. and Malthe-Sørenssen, D. (1972) Molecular properties of choline acetyltransferase and their importance for the compartmentation of acetylcholine. In *Biochemical and Pharmacological Mechanisms Underlying Behaviour*, P.B. Bradley and R.W. Brimblecombe (Eds.), *Progr. Brain Res., Vol. 36*, Elsevier, Amsterdam, pp. 13–27.

Fonnum, F. and Malthe-Sørenssen, D. (1973) Membrane affinities and subcellular distribution of the different molecular forms of choline acetyltransferase from rat. *J. Neurochem.*, 20, 1351–1359.

Franklin, G.I. (1976) A stain for detection of choline acetyltransferase after electrophoresis. *J. Neurochem.*, 26, 639–642.

Hebb, C., Mann, S.P. and Mead, J. (1975) Measurement and activation of choline acetyltransferase. *Biochem. Pharmacol.*, 24, 1007–1011.

Hersh, L.B. and Peet, M. (1977) Re-evaluation of the kinetic mechanism of the choline acetyltransferase reaction. *J. Biol. Chem.*, 252, 4796–4802.

Husain, S.S. and Mautner, H.G. (1973) The purification of choline acetyltransferase of squid head ganglia. *Proc. nat. Acad. Sci. (Wash.)*, 70, 3749–3753.

Jencks, W.P. (1969) in *Catalysis in Chemistry and Enzymology*, McGraw Hill, New York, pp. 243–281.

Johnson, S.L. (1967) in *Advances in Physical Organic Chemistry, Vol. 5*, V. Gold (Ed.), Academic Press, New York, pp. 237–330.

Katz, B. (1966) in *Nerve, Muscle and Synapse*, McGraw Hill, New York.

Krnjević, K. (1969) Central cholinergic pathways. *Fed. Proc.*, 28, 113–120.

Krnjević, K. (1974) Chemical nature of synaptic transmission in vertebrates. *Physiol. Rev.*, 54, 418–540.

Malthe-Sørenssen, D. (1975) Heterogeneity of choline acetyltransferase in cholinergic mechanism. In *Cholinergic Mechanisms*, P.G. Waser (Ed.), Raven Press, New York, pp. 257–262.

Malthe-Sørenssen, D. (1976a) Molecular properties of choline acetyltransferase from different species investigated by isoelectric focusing and ion exchange absorption. *J. Neurochem.*, 26, 861–865.

Malthe-Sørenssen, D. (1976) Choline acetyltransferase, evidence for acetyl transfer by a histidine residue. *J. Neurochem.*, 27, 873–881.

Malthe-Sørenssen, D., Eskeland, T. and Fonnum, F. (1973) Purification of rat brain choline acetyltransferase, some immunochemical properties of a highly purified preparation. *Brain Res.*, 62, 517–522.

Malthe-Sørenssen, D. and Fonnum, F. (1972) Multiple forms of choline acetyltransferase in several species demonstrated by isoelectric focusing. *Biochem. J.*, 127, 229–236.

Malthe-Sørenssen, D., Lea, T., Eskeland, T. and Fonnum, F. (1978) Molecular characterization of choline acetyltransferase from bovine brain caudate nucleus and some immunological properties of the highly purified enzyme. *J. Neurochem.*, 30, 35–46.

Mautner, H.G. (1977) Choline acetyltransferase. In *CRC critical reviews in biochemistry*. New York, pp. 341–370.

McCaman, R.E. and Hunt, J.M. (1965) Microdetermination of choline acetylase in nervous tissue. *J. Neurochem.*, 12, 253–259.

McGeer, P.L., McGeer, E.G., Singh, U.K. and Chase, W.H. (1974) Choline acetyltransferase

58

localization in the central nervous system by immunohistochemistry. *Brain Res.*, 81, 373–379.

Morris, D. (1966) The choline acetyltransferase of human placenta. *Biochem. J.*, 98, 754–762.

Morris, D., Maneckjee, A. and Hebb, C. (1971) The kinetic properties of human placental choline acetyltransferase. *Biochem. J.*, 125, 857–863.

Nachmansohn, D. and Machado, A.L. (1943) The formation of acetylcholine. A new enzyme: choline acetylase. *J. Neurophysiol.*, 6, 397–403.

Park, P.H., Joh, T.H. and Reis, D.J. (1977) Choline acetyltransferase in rat brain exists as charge isozymes. *Proc. Eighth Ann. Meet. Amer. Soc. Neurochem.*, Denver, Col., p. 169.

Polsky, R. and Shuster, L. (1976a) Preparation and characterization of two isozymes of choline acetyltransferase from squid head ganglia. I. Purification and properties. *Biochim. Biophys. Acta*, 445, 25–42.

Polsky, R. and Shuster, L. (1976b) Preparation and characterization of two isozymes of choline acetyltransferase from squid head ganglia. II. Self association, molecular weight determinations, and studies with inactivating antisera. *Biochim. Biophys. Acta*, 445, 43–46.

Potter, L.T., Saelens, J.K. and Glover, V.A.S. (1968) Choline acetyltransferase from rat brain. *J. Biol. Chem.*, 243, 3864–3870.

Prince, A.L. and Hide, E.G. (1971) Activation of choline acetyltransferase by salts. *Nature New Biol., (Lond.)*, 234, 222–223.

Prince, E.K. and Toates, P. (1974) The heterogeneity of goldfish brain and muscle choline acetyltransferase. *J. Brit. Pharmacol.*, 50, 447(P).

Rama-Sastry, B.V. and Henderson, G.I. (1972) Kinetic mechanism of human placental choline acetyltransferase. *Biochem. Pharmacol.*, 21, 787–802.

Roskoski, R. (1973) Choline acetyltransferase. Evidence for an acetyl-enzyme reaction intermediate. *Biochemistry*, 12, 3709–3714.

Roskoski, R. (1974) Choline acetyltransferase and acetylcholinesterase. Evidence for essential histidine residues. *Biochemistry*, 13, 5141–5144.

Roskoski, R., Lin, C.T. and Roskoski, L.M. (1975) Human brain and placental choline acetyltransferase; purification and properties. *Biochemistry*, 14, 5105–5110.

Rossier, J. (1976a) Purification of rat brain choline acetyltransferase. *J. Neurochem.*, 26, 543–548.

Rossier, J. (1976b) Immunological properties of rat brain choline acetyltransferase. *J. Neurochem.*, 26, 549–553.

Rossier, J. (1976c) Biophysical properties of rat brain choline acetyltransferase. *J. Neurochem.*, 26, 555–559.

Rossier, J. (1977) Acetyl-Coenzyme A and Coenzyme A analogues. *Biochem. J.*, 165, 321–326.

Rossier, J., Bauman, A. and Benda, P., Antibodies to rat brain choline acetyltransferase: species and organ specificity. *FEBS Lett.*, 36, 43–48.

Rossier, J., Spantidakis, Y. and Benda, P. (1977) The effect of Cl^- on choline acetyltransferase kinetic parameters and a proposed role of Cl^- in the regulation of acetylcholine synthesis. *J. Neurochem.*, 29, 1007–1012.

Ryan, R.L. and McClure, W.O. (1976) Purification and some properties of choline acetylase from bovine caudate nucleus. *Proc. Fed. Am. Soc. Exp. Biol.*, 35, 1647.

Singh, V.K. and McGeer, P.L. (1974a) Antibody production to choline acetyltransferase purified from human brain. *Life Sci.*, 15, 901–913.

Singh, V.K. and McGeer, P.L. (1974b) Crossimmunity of antibodies to choline acetyltransferase in various vertebrate species. *Brain Res.*, 82, 356–359.

Shuster, L. and O'Toole, C. (1974) Inactivation of choline acetyltransferase by an antiserum to the enzyme from mouse brain. *Life Sci.*, 15, 645–656.

Thompson, S.T., Cass, K.H. and Stellwangen, E. (1975) Blue dextran-Sepharose. An affinity column for the dinucleotide fold in proteins. *Proc. nat. Acad. Sci. (Wash.)*, 72, 669–679.

Wenthold, R.J. and Mahler, H.R. (1975) Purification of rat brain choline acetyltransferase and an estimation of its half life. *J. Neurochem.*, 24, 963–967.

White, H.L. and Cavallito, C.J. (1970) Choline acetyltransferase. Enzyme mechanism and mode of inhibition by a styrylpyridine analogue. *Biochim. Biophys. Acta*, 205, 343–358.

White, H.L. and Wu, J.C. (1973) Separation of apparent multiple forms of human brain choline acetyltransferase by isoelectric focusing. *J. Neurochem.*, 21, 939–947.

Immunohistochemistry of Choline Acetyltransferase

E.G. McGEER, P.L. McGEER, T. HATTORI and V.K. SINGH

*Kinsmen Laboratory of Neurological Research, Department of Psychiatry,
University of British Columbia, Vancouver, B.C., V6T 1W5 (Canada)*

INTRODUCTION

Acetylcholine was the first neurotransmitter to be discovered. Despite a long history of investigation, little is yet known about what cells in brain tissue contain this material. Histochemical methods for acetylcholinesterase have been used to gain information on the possible localization of cholinergic neurons but this hydrolytic enzyme is present in cholinoceptive structures as well as cholinergic ones. The synthetic enzyme choline acetyltransferase (ChAT) is a definitive marker for cholinergic neurons and their processes. In recent years, the techniques of immunohistochemistry have been applied with encouraging results to the precise localization of various transmitter synthetic enzymes in the central nervous system. Thus antibodies have been made and immunohistochemical localizations carried out for tyrosine hydroxylase, dopadecarboxylase, dopamine-β-hydroxylase and phenylethanolamine-N-methyltransferase in catecholaminergic systems, for tryptophan hydroxylase in serotoninergic systems, and for glutamic acid decarboxylase in GABAergic.

The development of an immunohistochemical method for any new system depends on the preparation of the pure enzymic antigen and the production of monospecific antibodies of sufficiently high titer to allow a good reaction with the antigenic enzyme in tissue which has been fixed sufficiently for structural preservation. In this chapter we describe some of the problems and successes so far encountered in our work on the immunohistochemistry of ChAT-containing systems in brain and spinal cord.

ENZYME PURIFICATION AND PROPERTIES

The purification of ChAT from human striatal tissue and the preparation of monospecific antibodies to this ChAT in rabbits have been described in detail elsewhere (Singh and McGeer, 1974a). In the initial preparation much of the catalytic activity of the enzyme was lost and, in subsequent work, Dr. F. Peng in our laboratory has modified the purification steps somewhat and produced a material with a specific activity of 12,694 nmol/mg of protein per 30 min, some 3-fold more active than the initially purified protein. Although the

Fig. 1. Immunodiffusion plate showing reaction of purified, active ChAT (center well) with antibodies raised against itself (outer well 5) or the previously purified, somewhat denatured protein (outer well 6). Both bottom wells (c) contain control rabbit serum.

catalytic activities differ, their behavior on SDS gel electrophoresis are identical and the purified protein with high catalytic activity yields single lines on double diffusion plate with antibodies prepared against either itself or the previously purified, somewhat denatured protein (Fig. 1). Such results, as well as the removal of ChAT activity from crude homogenates with the antibody (Table 1), should dispel the doubts which have been raised in the literature as to the purity and identity of our enzymic antigen and the antibodies raised thereto. This skepticism seems to have been largely engendered because of the difficulty in obtaining electrophoretically pure enzyme from rat brain (Rossier, 1976), but ChAT has not only been purified from human tissue by us and by others (Roskoski et al., 1975) following our procedure but also from bovine brain (Chao and Wolfgram, 1973), squid head ganglia (Husain and Mautner, (1973), and beef caudate nucleus (Cozzari and Hartman, 1977). Monospecific antibodies to ChAT have also been raised by Cozzari and Hartman (1977) and by Kan et al. (1977).

TABLE 1

PRECIPITATION OF CHOLINE ACETYLTRANSFERASE (ChAT) ACTIVITY BY ANTIBODIES

	ChAT activity (%)
Pure ChAT	100
Pure ChAT and rabbit anti-ChAT serum	52
Pure ChAT + rabbit anti-ChAT serum and goat anti-rabbit IgG	9.5
Pure ChAT and normal rabbit serum and goat anti-rabbit IgG	96
Striatal homogenate + rabbit anti-ChAT serum + goat anti-rabbit IgG	3
Striatal homogenate + normal rabbit serum + goat anti-rabbit IgG	98

0.1 ml pure ChAT (or striatal homogenate) and 0.1 ml anti-ChAT serum (or normal rabbit serum) + 0.1 ml goat anti-rabbit IgG + 0.1 ml H_2O were kept at 4°C for 2 days, centrifuged at $29000 \times g$ for 1 h ChAT activity was then measured in the supernate. Enzyme assays were run on equivalent dilutions.

In the process of purification of ChAT from human striatal tissue, two separate proteins with ChAT activity were separated on phosphocellulose columns. The major and more stable component was highly antigenic and produced the specific anti-ChAT antibodies in rabbit. The minor and less stable component appeared to be devoid of any antigenicity when injected into rabbit and did not react with the antibodies produced against the major component (Singh et al., 1975).

The problem of antigenic activity of various forms of ChAT is an intriguing one and has practical importance in the application of the antibodies to ChAT immunohistochemistry. The antibodies produced against the major component of human brain ChAT cross-reacted with ChAT from a variety of mammalian species but showed no sign of reaction with ChAT obtained from birds or invertebrates (Singh and McGeer, 1974b). Furthermore, the reaction was clearly more intense with ChAT from human or guinea-pig tissue than with that from rat or mouse. This has dictated our choice of the guinea-pig as the species to be used for most of the immunohistochemical work to be reported here.

The molecular weight of this predominant enzyme component used in antibody production was about 67,000 daltons as measured by molecular filtration through sephadex G-100. The purified enzyme has a K_m value of 8 μM for acetyl-coenzyme A and 250 μM for choline. Enzyme activity was enhanced by potassium chloride, sodium chloride, ammonium sulfate and chelating agents such as ethylenediaminetetraacetic acid (Singh and McGeer, 1977).

General immunohistochemical procedures

In order to minimize background staining, we found it necessary to shake the sera with liver powder or to purify them partially by means of DEAE-cellulose and sephadex G-200 column chromatography. The control and anti-ChAT serum fractions were adjusted to identical protein concentrations with 0.05 M phosphate-saline buffer, pH 6.8. Goat anti-rabbit immunoglobulin (IgG) (as such and labeled with fluorescein) and rabbit peroxidase—antiperoxidase complex (PAP) were obtained from Cappel Laboratories. Initial studies at the light microscopic level using the indirect immunohistochemical methods with fluorescein-labeled antibodies (McGeer et al., 1974) indicated that good staining could be obtained if the fresh frozen tissue was sectioned on a cryostat and then exposed to cold ether for 30 min. Staining was not obtained without the exposure to ether or in sections from fixed tissue. Subsequently, using light and electron-microscopic techniques, it was found that staining could be achieved by the PAP method in guinea-pig tissue fixed by perfusion with 4% paraformaldehyde. Use of glutaraldehyde or prolonged postfixation in formaldehyde seemed to abolish the reaction. Vibrotome rather than other sectioning techniques also seemed to be essential to successful staining.

Similar technical problems have plagued the application of immunohistochemical techniques to other central nervous system enzymes and a number of problems, such as lack of penetration of the antibodies, have not been solved as will be evident in subsequent paragraphs.

Nevertheless, with these techniques we have identified cholinergic neurons in the spinal cord, cerebral cortex, basal ganglia and habenula by light microscopy.

We have applied the electron microscopic method to the detection of neuronal processes in cell bodies in the caudate nucleus and, by means of double labeling with 6-hydroxydopamine, have shown a dopaminergic-cholinergic linkup. We have also demonstrated cholinergic nerve endings in the interpeduncular nucleus at the electron microscopic level. A wealth of ChAT-containing structures has been observed in preliminary observations of other areas of brain, consistent with what would be anticipated from the biochemical distribution studies that have already been performed on acetylcholine and ChAT. Others (Cozzari and Hartman, 1977; Kan et al., 1977) have reported specific immunofluorescence in the anterior horn cells of the spinal cord and Kan et al. have reported ChAT staining in cell bodies in the cerebral cortex and in mossy fibers of the cerebellum.

Examples of immunohistochemical results

(A) *Spinal cord.* The spinal cord has been selected as a test system by ourselves and others (Kan et al., 1977; Cozzari and Hartman, 1977) because it was already clear from other types of work which structures should stain and which should not. It is therefore reassuring that all three laboratories which have worked at the problem have found intense and specific staining in the anterior horn cells of the spinal cord (Fig. 2) which are known to be cholinergic. There was also intense staining of the axon bundles of anterior horn cells passing through anterior white matter on their way to form the ventral roots. Dorsal roots showed no positive staining and laminae 1–3, which do stain for acetylcholinesterase, showed very little staining for ChAT (McGeer et al., 1974).

(B) *Cerebral cortex.* The presence of cholinoceptive cells in the cerebral cortex has long been established but lesion experiments gave equivocal results as to the existence of long cholinergic neurons originating or terminating in the cortex. Early immunohistochemical observations (McGeer et al., 1974) estab-

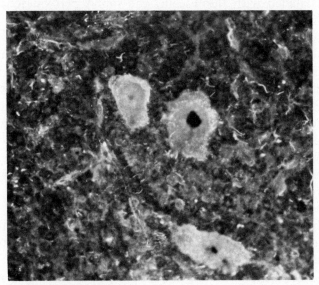

Fig. 2. Positively stained anterior horn cells of beef spinal cord. ×350.

lished the presence of cholinergic nerve cell bodies in the cortex; they have been confirmed by Kan et al., (1977) using similar immunohistochemical techniques. More recent results (McGeer et al., 1977) suggest that the majority of these cholinergic cells are interneurons but more definitive evidence is needed and will hopefully be obtained by more detailed immunohistochemical work.

(C) *Habenular-interpeduncular tract.* The cholinergic nature of many neurons in the habenular-interpeduncular tract was initially indicated by the sharp fall of ChAT activity in the interpeduncular nucleus following lesions of this tract (Kataoka et al., 1973) and subsequently confirmed by measurements of choline uptake (Sorimachi and Kataoka, 1974) and other biochemical

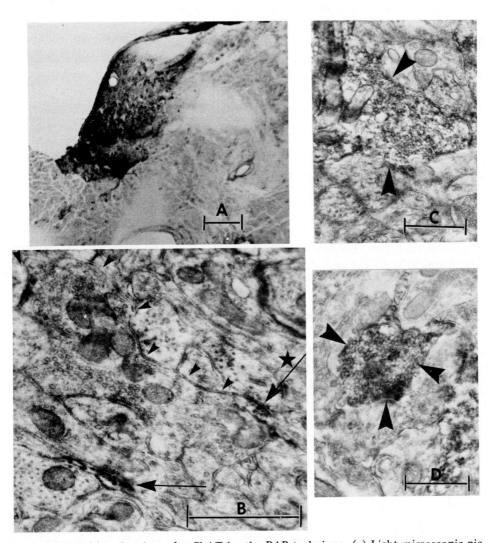

Fig. 3. Immunohistochemistry for ChAT by the PAP technique. (a) Light microscopic picture of the positive staining in the medial habenula. Bar indicates 100 μm. (b) Electron micrograph of two heavily stained unmyelinated axons (arrows) in the interpeduncular nucleus, one of which (starred) can be traced to the large type I bouton (arrow heads). Bar indicates 1 μm. (c, d) Weakly (c) and heavily (d) stained boutons in the interpeduncular nucleus. The outer membranes of the vesicles and mitochondria were usually more intensely stained than their interior. Bars indicate 1 μm.

cholinergic markers (Leranth et al., 1975) in the interpeduncular nucleus of lesioned animals.

In light microscopic studies the cell somata of the guinea pig medial habenula could be seen strongly and specifically stained for ChAT by the immunohistochemical (PAP) technique, with more diffuse and faint staining observed in the lateral habenula (Fig. 3a). Electron microscopic studies of the interpeduncular nucleus indicated that the staining for ChAT was heavy in some unmyelinated axons (Fig. 3b). It was clear that not all cholinergic boutons were stained since it was often possible to trace axons from a strongly stained portion to an unstained larger bouton which typically made several asymmetrical synaptic contacts with dendritic elements. Although the boutons were quite irregularly stained (Figs. 3c, d), the positive boutons always contained moderately packed, large round vesicles. The outer membranes of the vesicles and mitochondria were usually more intensely stained than their interior. The morphology of the stained boutons and of the boutons which originated from intensely stained axons was always identical and was very similar to that of hippocampal boutons which are preferentially labelled by radioactive proteins axonally transported from the medial septal area in the cholinergic septal-hippocampal tract. Boutons of this type degenerate after unilateral lesions to the habenula and are preferentially labelled by radioactive proteins transported from the medial habenula. These observations (Hattori et al., 1977) not only provided new information as to the morphological characteristics of the cholinergic nerve endings in the interpeduncular nucleus but, since only one type of boutons was involved, offered further evidence for the specificity of the ChAT immunohistochemical technique.

(D) *Neostriatum.* ChAT-containing structures of the neostriatum have been visualized at the light and electron microscopic level by immunohistochemistry using the peroxidase-antiperoxidase (PAP) multisandwich technique. Positively staining cell bodies, dendrites and nerve endings were observed in striatal slices from guinea pig brain (Fig. 4a). The cell bodies were seven to fifteen micrometers in diameter, with the smooth nuclei and few Nissl bodies characteristic of striatal interneurons (Kemp and Powell, 1971). Dendrites extending from these cells were stained, particularly over microtubules. The staining extended into dendritic spines (Fig. 4b). Axons and nerve endings were more sparsely stained. The stained nerve endings invariably made asymmetric synapses with non-staining spines (Fig. c). The boutons of the cholinergic interneurons of the striatum are much smaller in size (0.2–0.3 μm) and contain somewhat smaller round vesicles (25–40 nm in diameter) than the cholinergic boutons of the long axons from the habenulo-interpeduncular and septo-hippocampal tracts.

Intraventricular administration of 6-hydroxydopamine was used to bring about degeneration of dopaminergic nerve endings in the neostriatum. The degenerating nerve endings were seen to make synaptic contacts with major dendrites or dendritic spines positively staining for ChAT (Fig. 4d), providing the first direct evidence of the long hypothesized dopaminergic-cholinergic link (McGeer et al., 1961). Not all degenerating nerve endings were in contact with dendrites positively staining for ChAT, nor were all structures positively staining for ChAT seen to be innervated with degenerating dopaminergic nerve endings. It has not yet been possible to obtain definite estimates of the percent-

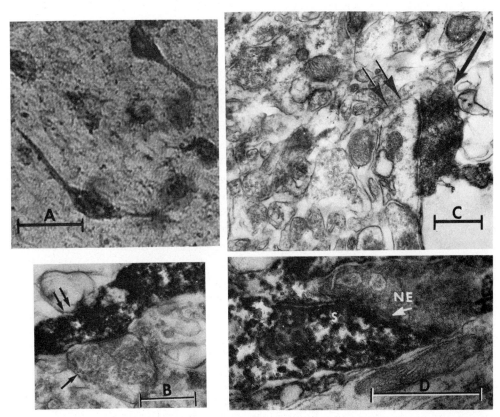

Fig. 4. Staining for ChAT in the guinea pig neostriatum by PAP technique. (a) Light microscopic micrograph. Note the staining of dendritic processes. Bar = 25 μm. (b) Positively
stained dendrite (⇉) receiving asymmetrical synapse from non-staining nerve ending. Bar
indicates 1 μm. (c) Damaged, positively stained nerve endings (→) making asymmetrical
synapse with non-staining dendritic spine (⇉). Bar indicates 0.5 μm. (d) Degenerating nerve
ending (NE) following 6-OHDA administration. Asymmetrical synaptic contact (→) is being
made with dendritic spine (S) staining positively for ChAT. Bar indicates 0.5 μm.

age of dopaminergic nerve endings in the striatum that do innervate cholinergic
interneurons; not all dopaminergic nerve endings may be seen as degenerating
at any selected time of sacrifice, and the immunohistochemical technique may
not be able to stain all ChAT-containing structures due to poor penetration of
antibodies into the tissue. Despite these difficulties, it is already clear that there
is heavy innervation of cholinergic neurons by dopaminergic terminals (Hattori
et al., 1976).

Pharmacological studies from a number of laboratories agree that agents that
enhance dopaminergic activity cause a decreased release of acetylcholine in the
striatum and a rise in its level, since synthesis seems to continue. Agents that
decrease dopaminergic activity, on the other hand, increase the release of
acetylcholine and decrease its level (McGeer et al., 1976). Hence the direct
innervation of cholinergic neurons by dopaminergic nerve endings seen in
Fig. 4d is not unexpected on pharmacological grounds. The existence of
cholinergic interneurons in the striatum is also expected since lesioning the
known pathways to and from the striatum failed to cause any significant loss of
ChAT or acetylcholinesterase (McGeer et al., 1971; Butcher and Butcher,
1974).

DISCUSSION

Immunohistochemical techniques involve a complex series of poorly understood chemical reactions with obvious opportunities for artifacts to appear. Precise conditions must be found which permit detection of a specific immunohistochemical reaction not masked by the non-specific background staining which is frequently present, particularly with the PAP technique.

The antibodies must be specific for the protein being localized. In the case of the antibody fraction used in this study, the rabbit serum was obtained from animals immunized against ChAT, the homogeneity of which had been established by demonstration of a single band on polyacrylamide gel electrophoresis at two different pH's. The crude serum, when reacted against impure brain extracts from several species, including guinea pig, always gave a single precipitin band on double diffusion plates, indicating that the antibodies were monospecific. The serum globulin was further purified to increase the intensity of the reaction. Therefore, it seems highly unlikely that the results could have been distorted by specific reactions of the immune serum with proteins other than ChAT. This supposition is further supported by the morphological specificity of the staining we have seen.

Other possible difficulties must also be considered. Since the tissue must first be partially fixed in order to preserve subcellular structure, the catalytic activity, solubility, and antigenic properties of the substance being localized may be changed by the fixative. Because of its large size the serum antibody may not reach the antigen in adequate amounts, or the antigen-antibody complex may be bound to subcellular structures not reflecting the original in vivo antigenic sites. The PAP complex itself, which contains the peroxidase marker, has a very high molecular weight which could well retard its ability to penetrate small structures even if the original antigen-antibody reactions were satisfactory. Finally, small amounts of oxidized diaminobenzidine, the product of the peroxidase reaction, may be generated by residual endogenous peroxidase in tissues or may diffuse from sites of the specific PAP reaction. These potential artifacts must be kept in mind in interpreting immunohistochemical results. The validity of the ChAT localizations indicated in the figures is supported, however, not only by the use of control non-antigenic serum but also by the fact that the localizations indicated by the immunohistochemical technique are in accord with the biochemical, physiological and pharmacological data.

The relative degree of staining of certain subcellular structures deserves special comment. There was only a sparse staining of nerve endings which is contrary to what would be anticipated from the known subcellular localization of ChAT from biochemical studies. The PAP technique, under the conditions that we have used, is therefore not labelling all of the existing ChAT. The fact that damaged nerve endings are more frequently seen to be stained then intact ones suggests that the problem is one of penetration. In other studies involving PAP localization of tyrosine hydroxylase, staining of nerve endings was also incomplete (Pickel et al., 1975).

The staining of the nuclei in ChAT-containing cells in the striatum was also unexpected. While light, it was nevertheless present. However, this too has been found for tyrosine hydroxylase in catecholaminergic cells (Pickel et al., 1975).

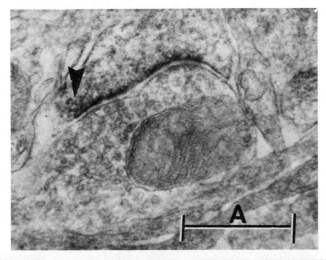

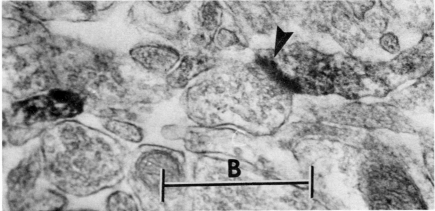

Fig. 5. (a, b) Guinea pig neostriatum stained for choline acetyltransferase. Note densities in the dendritic spine particularly near the postsynaptic membrane. Bars = 1 μm.

In subcellular biochemical studies, small amounts of enzymes such as ChAT and tyrosine hydroxylase can be found in the nuclear fraction (McGeer et al., 1965), but thorough washing removes nearly all the activity, suggesting that this is weak, probably non-specific binding. Such a form of binding might occur as well under the conditions used for immunohistochemistry.

A more interesting result was the staining of dendrites and dendritic spines which seems to be both intense and specific. Again this phenomenon has been noted for tyrosine hydroxylase. The association with microtubules suggests that the enzymes are not present in the dendrites accidentally. Further observations have suggested that both ChAT (Fig. 5) and tyrosine hydroxylase (McGeer et al., unpublished observations) are in fact associated with structures which are in close proximity to the dendritic postsynaptic membrane. These morphological observations, coupled with chemical evidence such as the presence of dopamine-sensitive adenylate cyclase in afferents to the substantia nigra (Phillipson et al., 1977; Spano et al., 1977), the release of dopamine from nigral dendrites (Nieoullon et al., 1977) and the effect of cholinergic agents on dopamine release in the striatum (Giorguieff et al., 1976), have led us to

68

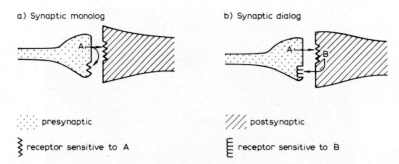

a) Synaptic monolog

b) Synaptic dialog

::::: presynaptic

//// postsynaptic

receptor sensitive to A

receptor sensitive to B

Fig. 6. Hypothesized monolog (a) and dialog (b) across a synapse involving (a) presynaptic auto receptors sensitive to the transmitter of neuron A and (b) presynaptic heterologous receptors sensitive to the dendritically released transmitter from neuron B.

hypothesize that there may be a dialog rather than a monolog across the synapse. According to this hypothesis there are, instead of or in addition to the homologous presynaptic receptors suggested by Carlsson (Fig. 6a), heterologous presynaptic receptors responsive to neurotransmitter released from the dendrites (Fig. 6b). Only further work involving many techniques in addition to the immunohistochemical ones can tell us whether such synaptic dialogs do in fact occur; if they do, it will clearly revolutionize all our concepts regarding the intercellular communication in the central nervous system.

SUMMARY

ChAT was purified from human striatum and used to raise monospecific antibodies in rabbit. These have been used for immunohistochemistry at the light and electron microscopic level to demonstrate cholinergic cells in spinal cord and cortex, and to study the morphology of cholinergic boutons in the neostriatum and interpeduncular nucleus. Direct evidence for dopaminergic innervation of cholinergic dendrites in the neostriatum has been obtained. Some of the problems in immunohistochemistry are discussed. On the basis of the heavy staining for ChAT in dendrites, its association with structures near the postsynaptic membrane, and of other evidence in the literature, the hypothesis is advanced that communication across the synapse involves a dialog rather than a monolog.

ACKNOWLEDGEMENT

This work is supported by the Medical Research Council of Canada.

REFERENCES

Butcher, S.G. and Butcher, L.L. (1974) Origin and modulation of acetylcholine activity in the neostriatum. *Brain Res.*, 71, 161–171.

Chao, L.O. and Wolfgram, F. (1973) Purification and some properties of choline acetyltransferase from bovine brain. *J. Neurochem.*, 20, 1075–1081.

Cozzari, C. and Hartman, B.K. (1977) Purification and preparation of antibodies to choline acetyltransferase from beef caudate nucleus. *Proc. int. Soc. Neurochem.*, 6, 112.

Giorguieff, M.F., Le Floc'h, M.L., Westfall, T.C., Glowinski, J. and Besson, M.J. (1976) Nicotinic effect of acetylcholine on the release of newly synthesized [³H]dopamine in rat striatal slices and cat caudate nucleus. *Brain Res.*, 106, 117–131.

Hattori, T., McGeer, E.G., Singh, V.K. and McGeer, P.L. (1977) Cholinergic synapse of the interpeduncular nucleus. *Exp. Neurol.*, 55, 666–679.

Hattori, T., Singh, V.K., McGeer, E.G. and McGeer, P.L. (1976) Immunohistochemical localization of choline acetyltransferase containing neostriatal neurons and their relationship with dopaminergic synapse. *Brain Res.*, 102, 164–173.

Husain, S.S. and Mautner, H.G. (1973) The purification of choline acetyltransferase of squid head ganglion. *Proc. nat. Acad. Sci. (Wash.)*, 70, 3749–3753.

Kan, K.-S., Chao, L.-P., Wolfgram, F. and Eng, L.F. (1977) Immunohistochemical localization of choline acetyltransferase in rabbit CNS. *Proc. int. Soc. Neurochem.*, 6, 115.

Kataoka, K., Nakamura, Y. and Hassler, R. (1973) Habenulo-interpeduncular tract: a possible cholinergic neuron in rat brain. *Brain Res.*, 62, 264–267.

Kemp, J.M. and Powell, T.P.S. (1971) The site of termination of afferent fibres in the caudate nucleus. *Phil. Trans. roy. Soc. B. (Lond.)*, 262, 383–401.

Leranth, C.S., Brownstein, M.J., Zaborsky, L., Jaranyi, Z.S. and Palkovits, M. (1975) Morphological and biochemical changes in rat interpeduncular nucleus following the transsection of the habenulo-interpeduncular tract. *Brain Res.*, 99, 124–128.

McGeer, P.L., Bagchi, S.P. and McGeer, E.G. (1965) Subcellular localization of tyrosine hydroxylase in beef caudate nucleus. *Life Sci.*, 4, 1859–1867.

McGeer, P.L., Boulding, J.E., Gibson, W.C. and Foulkes, R.G. (1961) Drug induced extra pyramidal reactions. *J. Amer. med. Ass.*, 177, 665–670.

McGeer, P.L., Hattori, T., Singh, V.K. and McGeer, E.G. (1976) Cholinergic systems in extrapyramidal function. In *The Basal Ganglia*, M.D. Yahr (Ed.), Raven Press, New York, pp. 213–226.

McGeer, P.L. and McGeer, E.G. (1976) Enzymes associated with the metabolism of catecholamines, acetylcholine, and GABA in human controls and patients with Parkinson's disease and Huntington's chorea. *J. Neurochem.*, 26, 65–76.

McGeer, P.L., McGeer, E.G., Fibiger, H.C. and Wickson, V. (1971) Neostriatal choline acetylase and acetylcholinesterase following selective brain lesions. *Brain Res.*, 35, 308–314.

McGeer, P.L., McGeer, E.G., Scherer, U. and Singh, K. (1977) A glutamatergic corticostriatal path? *Brain Res.*, 128, 369–373.

McGeer, P.L., McGeer, E.G., Singh, V.K. and Chase, W.H. (1974) Choline acetyltransferase localization in the central nervous system by immunohistochemistry. *Brain Res.*, 81, 273–379.

Nieoullon, A., Cheramy, A. and Glowinski, J. (1977) Release of dopamine in vivo from cat substantia nigra. *Nature (Lond.)*, 266, 375–377.

Phillipson, O.T., Emson, P.C., Horn, A.S. and Jessell, T. (1977) Evidence concerning the anatomical localization of the dopamine stimulated adenylate cyclase in the substantia nigra. *Brain Res.*, 136, 45–58.

Pickel, V.M., Tong, H.J. and Reis, D.J. (1975) Ultrastructural localization of tyrosine hydroxylase in noradrenergic neurons of brain. *Proc. nat. Acad. Sci. (Wash.)*, 72, 659–663.

Roskoski, R., Lim, C.T. and Roskoski, L.M. (1975) Human brain and placental choline acetyltransferase. *Biochem.*, 14, 5105–5110.

Rossier, J. (1976) Immunological properties of rat brain choline acetyltransferase. *J. Neurochem.*, 26, 549–553.

Singh, V.K. and McGeer, P.L. (1974a) Antibody production to choline acetyltransferase purified from human brain. *Life Sci.*, 15, 901–913.

Singh, V.K. and McGeer, P.L. (1974b) Cross immunity of antibodies to choline acetyltransferase in various vertebrate species. *Brain Res.*, 82, 356–359.

Singh, V.K. and McGeer, P.L. (1977) Studies on choline acetyltransferase isolated from human brain. *Neurochem. Res.*, 2, 281–291.

Singh, V.K., McGeer, E.G. and McGeer, P.L. (1975) Two immunologically different choline acetyltransferases in human neostriatum. *Brain Res.*, 96, 187–191.

Sorimachi, M. amd Kataoka, K. (1974) Choline uptake by nerve terminals: a sensitive and a specific marker of cholinergic innervation. *Brain Res.*, 72, 350–353.

Spano, P.F., Trabucchi, M. and DiChiara, G. (1977) Localization of nigral dopamine sensitive adenylate cyclase on neurons originating from the corpus striatum. *Science*, 196, 1343–1345.

Sodium-Dependent High Affinity Choline Uptake

MICHAEL J. KUHAR

*Departments of Pharmacology and Experimental Therapeutics and Psychiatry
and the Behavioral Sciences, The Johns Hopkins University School
of Medicine, Baltimore, MD 21205 (U.S.A.)*

INTRODUCTION

The biochemical components involved in the synthesis of acetylcholine (ACh) have been the subject of intense investigation over many years. Various aspects of this question have been reviewed several times (Hebb, 1972; Whittaker et al., 1972; Fonnum, 1973, 1975; Browning, 1976; Freeman and Jenden, 1976; Haubrich and Chippendale, 1977; Kuhar and Murrin, 1978). An especially critical component in the control of ACh synthesis seems to be the sodium-dependent, high affinity choline uptake (SDHACU). Since this topic has been recently reviewed (Kuhar and Murrin, 1978) this chapter will summarize the main features of the SDHACU, discuss the latest literature reports and their significance, stress its role in the regulation of ACh synthesis and discuss the physiological factors that control SDHACU.

CHARACTERISTICS OF SDHACU

The following is a brief summary of the characteristics of SDHACU. For detailed discussions and references see Kuhar and Murrin (1978).

The choline transport system appears uniquely localized to cholinergic nerve terminals. This follows from a variety of denervation studies which show that depletion of cholinergic nerve terminals in a given tissue results in a selective loss of SDHACU. Such a "strategic" localization implies an important role in ACh synthesis.

SDHACU and choline acetyltransferase are closely related. This was evident even in the earliest experiments since the bulk of the choline taken up by the transport system is converted to ACh. Thus, the view has evolved that SDHACU and choline acetyltransferase are coupled in some manner although the precise nature of this coupling is the subject of ongoing debate.

It has been shown that the radiolabeled ACh derived from SDHACU is released by a calcium-dependent and magnesium-inhibited mechanism. Increasing depolarization results in increased release. Thus, the releasable ACh derived from the SDHACU has the characteristics long associated with neurotransmitter release.

One of the most interesting properties of the SDHACU system is that its capacity appears to be coupled to neuronal activity. When the activity of cholinergic neurons is increased, one observes increased uptake by the SDHACU. The opposite results are found when the activity of cholinergic neurons is decreased. The finding that the capacity of the uptake system changes with neuronal activity identifies it as a likely site for the control of ACh synthesis.

Structure-activity studies reveal that SDHACU has strict molecular requirements. These requirements include a quaternary nitrogen and a free hydroxyl group or a group that is iso-electronic with a free hydroxyl group. These studies further show that the carrier is optimally designed for choline itself rather than for some other closely related molecule.

A number of other studies have shown that the SDHACU has a marked ionic and energy dependence. The transport system shows a fairly strict requirement for sodium, potassium and chloride. The sodium-dependence implies a cotransport of sodium and choline by the carrier. The energy dependence has also been studied. Introducing various metabolic inhibitors and reducing the incubation temperature reveals some involvement of energy. While the mechanism of transport is not understood and is discussed further below, the ionic, energy and temperature dependencies, the saturability, the structural specificity and the apparent ability to take up choline against the concentration gradient satisfy most classical requirements for SDHACU being an active, carrier-mediated transport system.

SDHACU AND THE REGULATION OF ACh SYNTHESIS

The synthesis of ACh may be controlled by several biochemical components: the activity of choline acetyltransferase, the supply of acetyl-CoA, and the supply of choline by either a low affinity system or by SDHACU. Which of these is the rate-limiting step in the synthesis of ACh? The bulk of available evidence suggests that, of these components, SDHACU is the most critical or rate-limiting step in synthesis. Some of the evidence for this can be summarized as follows. Most substances depleting tissues of their ACh content by preventing its synthesis are inhibitors of SDHACU. Moreover, their relative potency for depleting ACh is proportional to their relative potency for inhibiting SDHACU (Simon et al., 1975; Guyenet et al., 1973; Holden et al., 1975). A recent paper by Somogyi et al. (1977) shows that 1 μM hemicholinium-3 is sufficient to block the stimulus-induced release of ACh from the guinea pig ileum. Szerb et al. (1970) had shown earlier similar results with brain tissue. The concentration of hemicholinium-3 used is sufficient to inhibit SDHACU almost completely, but not sufficient to inhibit the low affinity choline transport or the enzyme choline acetyltransferase (Mann and Hebb, 1975; Yamamura and Snyder, 1973). Polak et al. (1977) have recently shown that the capacity of SDHACU is sufficient to account for the quantity of ACh released by potassiumdepolarized brain slices. Thus, the evidence is steadily accumulating that SDHACU is a critical and rate-limiting step in the formation of releasable ACh. What about choline acetyltransferase? It had been thought for many years that it was the major factor in the control of the synthesis of ACh. Present evidence suggests that the enzyme is present in excess and is not rate-limiting in comparison to SDHACU. It has been noted that the in vivo synthesis of ACh is considerably lower than that obtainable in vitro using soluble preparations of the enzyme (Hebb, 1972). Also, an important experiment by Krell and Goldberg (1975) showed that the administration of an inhibitor of choline acetyltransferase causes partial inhibition of brain enzyme but does not lower brain levels of ACh. It is not yet possible to decide whether the availability of acetyl-CoA has any role in regulation of synthesis, but it may. Low affinity choline transport may also supply some choline for ACh synthesis, especially in cases where serum choline is greatly elevated (Kuntscherová, 1972; Haubrich et al., 1974; Cohen and Wurtman, 1976).

It has been known for some time that the synthesis of ACh is in some way controlled by impulse-flow. When cholinergic neurons increase their firing the synthesis of ACh is accelerated and when the neurons are inactive, the synthesis rate is reduced. It has been thought that choline acetyltransferase could be involved in this (and it may be), but the finding that the capacity of the uptake system changes with neuronal activity (Simon et al., 1976; Murrin and Kuhar, 1976), identifies it as a likely site for the control of ACh synthesis during changes in neuronal activity.

THE PHYSIOLOGICAL MECHANISM OF SDHACU

What is the molecular mechanism of SDHACU and of its control? At this point in time, we do not have any model that is solidly supported by data. The following discussion is based on work on transport systems in general (Schulz and Curran, 1970).

As described above, the properties of SDHACU are compatible with the notion that is is an active, carrier-mediated transport. The sodium dependence suggests that the sodium gradient is the driving force and that sodium is co-transported with choline. The chloride dependence may reflect a co-transport of anion for charge neutralization to maintain the sodium electrochemical gradient. The energy dependence could reflect the necessity of the sodium-potassium ATPase for maintaining the sodium gradient which directly provides energy for the transport.

The sodium electrochemical gradient appears to have sufficient energy to transport choline against a concentration gradient of about 100- to 1000-fold. The transport against the gradient would continue until the choline concentration builds up inside the nerve terminal. The observation that the K_m of choline acetyltransferase for choline is 100-fold higher than the K_T of SDHACU for choline is compatible with the notion that the choline inside the nerve terminal, or in the "synthetic compartment", is higher than it is outside the nerve terminal. The intrasynaptosomal choline would be the immediate source of substrate for the enzyme and production of releasable ACh (see Fig. 1).

How do changes in neuronal activity result in changes in the capacity of SDHACU? Again it is not possible to answer this question with any degree of certainty. One reasonable explanation is that the intrasynaptosomal pool of

74

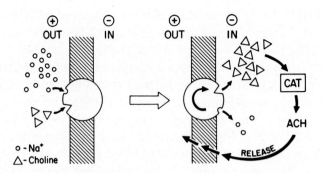

Fig. 1. Role of the sodium-dependent choline carrier in the synthesis of ACh. Both sodium and choline bind to the carrier, which is depicted as a transmembranous rotating macromolecule in the membrane (shaded). The sodium electrochemical gradient provides the driving force and seems capable of establishing a choline concentration gradient of 100—1000-fold. When the maximal gradient is established, choline transport stops. The pool of transported choline would be the immediate supply for the enzyme, choline acetyltransferase (CAT). The enzyme CAT is in equilibrium with substrates and products. When acetylcholine (ACh) is released, CAT uses more choline to replace ACh, the intrasynaptosomal pool of choline falls, and the sodium gradient then drives more choline into the neuron. In this model, intrasynaptosomal choline concentrations directly control the activity of sodium-dependent choline transport; the transport system, enzyme, choline and ACh are in an equilibrium which is controlled by impulse-flow and ACh release.

choline is depleted by the increased synthesis of ACh which was depleted by increased release. When this intrasynaptosomal pool is depleted, the sodium electrochemical gradient will drive additional choline into the terminal until an equilibrium is re-established. If this model were correct, then the intrasynaptosomal pool of choline would be directly affecting the rate of choline transport and there would be an equilibrium between SDHACU, choline acetyltransferase and releasable ACh (Figure 1). While this model is reasonable and perhaps likely, other models have been proposed. The observed inverse relationship between ACh levels and choline uptake has suggested that intrasynaptosomal ACh levels somehow regulate the transport system (Jenden et al., 1976). Other studies showing that the uptake activation requires the presence of divalent cations suggest that some divalent cation-mediated metabolic event could control the rate of transport (Murrin et al., 1977). If the energy for the transport is provided by the sodium gradient, then alterations in the local sodium gradient could be responsible for the coupling between the impulse flow and the transport. Still other possibilities exist, but all remain speculative until the transport system can be understood more clearly in molecular terms.

SUMMARY

SDHACU has the characteristics of an active, carrier-mediated transport system. Available data point to it as the rate-limiting step in the synthesis of ACh. While the transport is not understood in molecular terms, available data suggest, by analogy with other more studied transport systems, some possible mechanisms.

ACKNOWLEDGEMENTS

The author acknowledges the support of USPHS Grants MH 25951 and MH 00053, the contributions of his colleagues, including H. Rommelspacher, S.F. Atweh, J.R. Simon and L.C. Murrin and the clerical assistance of Carol Kenyon and Victoria Rhodes.

REFERENCES

Browning, E.T. (1976) Acetylcholine synthesis: substrate availability and the synthetic reaction. In *Biology of Cholinergic Function*, A.M. Goldberg and I. Hanin (Eds.), Raven Press, New York, pp. 187–202.

Cohen, E.L. and Wurtman, R.J. (1976) Brain acetylcholine: control by dietary choline. *Science*, 191, 561–562.

Fonnum, F. (1973) Recent developments in biochemical investigations of cholinergic transmission. *Brain Res.*, 62, 497–507.

Fonnum, F. (1975) Review of recent progress in the synthesis, storage and release of acetylcholine. In *Cholinergic Mechanisms*, P.G. Waser (Ed.), Raven Press, New York, pp. 145–160.

Freeman, J.J. and Jenden, D.J. (1976) The sources of choline for acetylcholine synthesis in brain. *Life Sci.*, 19, 949–962.

Guyenet, P., Lefresne, P., Rossier, J., Beaujouan, J.C. and Glowinski, J. (1973) Inhibition by hemicholinium-3 of [^{14}C]acetylcholine synthesis and [^{3}H]choline high affinity uptake in rat striatal synaptosomes. *Molec. Pharmacol.*, 9, 630–639.

Haubrich, D.R. and Chippendale, T.J. (1977) Regulation of acetylcholine synthesis in nervous tissue. *Life Sci.*, 20, 1465–1478.

Haubrich, D.R., Wedeking, P.W. and Wang, P.F.L. (1974) Increase in tissue concentration of acetylcholine in vivo induced by administration of choline. *Life Sci.*, 14, 921–927.

Hebb, C. (1972) Biosynthesis of acetylcholine in nervous tissue. *Physiol. Rev.*, 52, 918–957.

Holden, J.T., Rossier, J., Beaujouan, J.C., Guyenet, P. and Glowinski, J. (1975) Inhibition of high affinity choline transport in rat striatal synaptosomes by alkyl bisquaternary ammonium compounds. *Molec. Pharmacol.*, 11, 19–28.

Jenden, D.J., Jope, R.S. and Weiler, M.H. (1976) Regulation of acetylcholine synthesis: does cytoplasmic acetylcholine control high affinity choline uptake. *Science*, 194, 635–637.

Krell, R.D. and Goldberg, A.M. (1975) Effects of choline acetyltransferase inhibitors on mouse and guinea pig brain choline and acetylcholine. *Biochem. Pharmacol.*, 24, 391–396.

Kuhar, M.J. and Murrin, L.C. (1978) Sodium-dependent high affinity choline uptake. *J. Neurochem.*, 30, 15–21.

Kuntscherová, J. (1972) Effect of short-term starvation and choline on the acetylcholine content of organs of Albino rats. *Physiol. bohemoslovaca*, 21, 655–660.

Mann, S.P. and Hebb, C. (1975) Inhibition of choline acetyltransferase by quaternary ammonium analogues of choline. *Biochem. Pharmacol.*, 24, 1013–1017.

Murrin, L.C., DeHaven, R.N. and Kuhar, M.J. (1977) On the relationship between [^{3}H]choline uptake activation and [^{3}H]acetylcholine release. *J. Neurochem.*, 29, 681–687.

Murrin, L.C. and Kuhar, M.J. (1976) In vitro activation of high affinity choline uptake by depolarizing agents. *Molec. Pharmacol.*, 12, 1082–1090.

Polak, R.L., Molenaar, P.C. and VanGelden, M. (1977) Acetylcholine metabolism and choline uptake in cortical slices. *J. Neurochem.*, 29, 477–485.

Schultz, S.G. and Curran, P.F. (1970) Coupled transport of sodium and organic solutes. *Physiol. Rev.*, 50, 637–718.

Simon, J.R., Atweh, S. and Kuhar, M.J. (1976) Sodium-dependent high affinity choline uptake: a regulatory step in the synthesis of acetylcholine. *J. Neurochem.*, 26, 909–922.

Simon, J.R., Mittag, T. and Kuhar, M.J. (1975) Inhibition of synaptosomal uptake of choline by various choline analogs. *Biochem. Pharmacol.*, 24, 1139–1142.

Somogyi, G.T., Vizi, E.S. and Knoll, J. (1977) Effect of hemicholinium-3 on the release and net synthesis of acetylcholine in Auerbach's plexus of guinea pig ileum. *Neurosci.*, 2, 791–796.

76

Szerb, J.C., Malik, H. and Hunter, E.J. (1970) Relationship between acetylcholine content and release in cat cortex. *Can. J. Physiol. Pharmacol.,* 48, 780—790.

Whittaker, V.P., Dowdall, M.J. and Boyne, A.F. (1972) The storage and release of acetylcholine by cholinergic nerve terminals: recent results with non-mammalian preparations. *Biochem. Soc. Symp.,* 36, 49—68.

Yamamura, H.I. and Snyder, S.H. (1973) High affinity transport of choline into synaptosomes of rat brain. *J. Neurochem.,* 21, 1355—1374.

Relationship of Choline Uptake to Acetylcholine Synthesis and Release

R.M. MARCHBANKS and S. WONNACOTT

*Department of Biochemistry, Institute of Psychiatry, De Crespigny Park,
London SE5 8AF (United Kingdom)*

INTRODUCTION

The depletion of neurotransmitter stores by release from the terminal is compensated for by an increase in synthesis. This has been widely demonstrated for acetylcholine (ACh) in a variety of preparations both in vivo and in vitro. (Mann et al., 1938; Birks and MacIntosh, 1961; Polak and Meeuws, 1966; Richter and Marchbanks, 1971). The compensatory mechanism sometimes overshoots slightly so that the amount of ACh in the terminals can be greater after release has been evoked. These effects are demonstrable in isolated nerve terminals, as shown in Fig. 1.

The first question to ask in considering control mechanisms is to identify the rate limiting step in the overall process of synthesis and release. A comparison

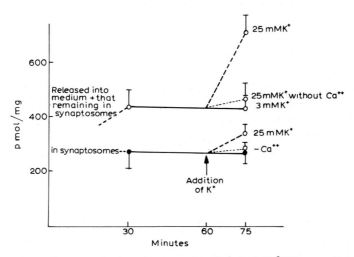

Fig. 1. Synaptosomes from cerebral cortex were incubated at 37°C in a medium containing 180 mM NaCl, KCl as indicated, 2 mM $CaCl_2$, 2 mM $MgCl_2$ 10 mM glucose, 10 mM sodium phosphate buffer pH 7.2 and 10 μM eserine. At the point marked by the arrow the synaptosomes were washed twice by centrifugation (10,000 $\times$ g for 10 s) in a medium without calcium and then resuspended as indicated. Note that in 3 mM KCl there is little release of ACh between 30 min and 75 min. The initial release of ACh in this medium is probably related to recovery from the preparative process.

78

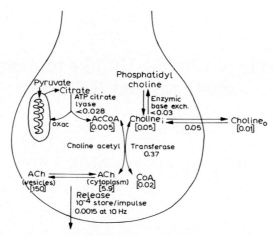

Fig. 2. Concentrations (µmol/g denoted thus [ ]) and *maximal* rates µmol/min · g of components of the cholinergic system in cerebral cortex synapses. Choline transport (this paper Fig. 8), choline (Shea and Aprison, 1977), enzymic base exchange (Saito et al., 1975), ATP citrate lyase (Tuček, 1967) acetyl-CoA and CoA (Guynn, 1976) have been assumed to be distributed equally throughout the cells and their processes in the cerebral cortex. Choline acetyltransferase (Tuček, 1967) and ACh (Shea and Aprison, 1977) have been assumed to be present only in cholinergic terminals which constitute about 10% of a total synaptosomal volume of 30 µl/g of cerebral cortex (Marchbanks, 1967). The equilibrium constant for the formation of ACh has been estimated (Potter et al., 1968) to be 500. If the figures above represented equilibrium concentrations the constant would be 472.

of the different rates is given in Fig. 2 for a "cholinergic" nerve terminal in the cerebral cortex. Certain assumptions (described in the legend) have to be made in deriving these figures, so that they should be regarded as crude approximations only. Nonetheless they will serve as a guide and it becomes apparent that the maximal rate of synthesis by choline acetyltransferase (ChAT) far exceeds both the rate of release and the rate of supply of precursors. There are situations in which the rate of synthesis does not exceed the rate of release by such a disproportionate factor (Marchbanks, 1977) but they are unusual in mammalian tissues.

Since ChAT is in excess it follows that the substrates and products in the synthetic reaction are likely to be at thermodynamic equilibrium. The ratio of these in the cholinergic terminal approximates to the equilibrium constant as would be expected if this proposal is true. Experimental support is also found in a comparison of the rate of choline (Ch) uptake with the rate of its acetylation (Fig. 3) in cerebral cortex synaptosomes. It can be seen that a more or less constant proportion of radioactive Ch is acetylated independently of the extent of its uptake into the terminal.

ChAT is a fully reversible enzyme and this fact taken into consideration along with the equilibrium position of the reactants has certain consequences for the interpretation of isotopic conversion data. In the first place, measurements of the metabolic turnover of ACh compartments are likely to reflect isotopic exchange of the Ch moiety into the ACh pool rather than release of ACh balanced by synthesis. Secondly, the widely observed decrease in isotopic conversion as the Ch concentration increases (Yamamura and Snyder, 1973; Haga and Noda, 1973) is readily explicable (as follows) without recourse to assump-

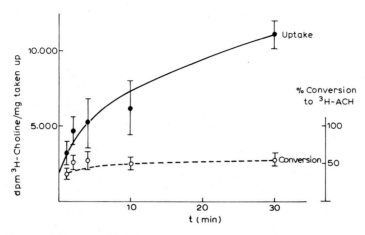

Fig. 3. Synaptosomes from cerebral cortex were incubated as described in Fig. 1 with [³H]-choline 2 μM, 0.1 μCi/ml. At the times indicated samples were withdrawn, washed by centrifugation and the total radioactivity taken up estimated. Radioactive [³H]ACh and [³H]Ch were extracted with kalignost in vinyl acetonitrile and separated by TLC (Marchbanks and Israël, 1971).

tions about the existence and nature of a "high affinity" Ch uptake coupled to ACh synthesis.

If the system is at equilibrium isotopic Ch that is taken up will exchange fully with both the ACh and Ch pools. The ratio of isotopic ACh to Ch will therefore be equal to the ratio of the pool sizes. The relative pool sizes will be given by the mass action equation so that:

(radioactive ACh/radioactive Ch) = K_{eq}[acetyl-CoA]/[CoA]

As the net Ch content of the synaptosomes increases during uptake of radioactive Ch some will be converted to ACh and this will require acetyl-CoA and produce CoA. The right hand side of the equation will therefore decrease as more Ch enters the synaptosomes unless the supply of acetyl-CoA can match the influx of choline. As the Ch flux rises at higher external Ch concentration, the isotopic conversion to ACh will decrease as the external Ch concentration rises.

Returning to the rate controlling stage of ACh synthesis, the data in Fig. 2 suggest that the Ch supply either by carrier mediated transport (Marchbanks, 1968) or by breakdown of phospholipids might be rate limiting. Acetyl-CoA in the cytoplasm depends on the activity of ATP citrate lyase and this will become rate controlling at high Ch influx rates. However we will confine our attention to the role of transport. The stimulation of Ch uptake by procedures that cause the release of ACh has been reported by a number of workers (Kuhar et al., 1973; Carrol and Goldberg, 1975; Simon et al., 1976; Polak et al., 1977; Collier and Ilson, 1977). The effect can be seen in isolated nerve terminals. A prior depolarization by 25 mM potassium concentrations causes release of ACh; if synaptosomes treated in this way are returned to non-depolarizing medium and Ch uptake measured an enhanced transport can be observed. Fig. 4 shows these effects and it is so to be noted that transport is dependent on Ca²⁺ being present during the depolarization phase. If Ca²⁺ is not

80

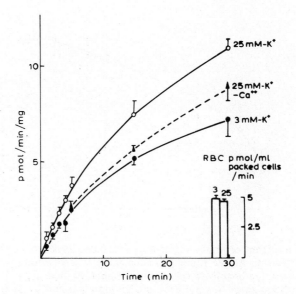

Fig. 4. Synaptosomes were pre-incubated for 30 min in depolarizing media (25 mM K⁺, control 3 mM K⁺) with and without calcium, after this they were returned to a medium containing 3 mM-K⁺ and the time course of [³H]choline (2 μM, 0.1 Ci/ml) uptake investigated. The inset shows the uptake of choline at 30 min into erythrocytes after similar treatment.

present or 18 mM Mg^{2+} is present during depolarization then subsequent enhancement of transport is not observed suggesting that the effect is related to ACh release rather than depolarization per se. This conclusion is strengthened by some experiments on Ch transport in erythrocytes, the results of which are shown in the inset to Fig. 4. Erythrocytes do not contain ACh and Ch trans-

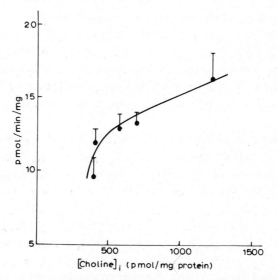

Fig. 5. Synaptosomes were preloaded for 30 min with varying concentrations of [¹⁴C]choline. The internal concentration of choline ([Choline]$_i$) was determined from the uptake and the contribution of endogenous choline (400 pmol/mg protein). After 2 rapid washes the uptake of [³H]choline at a concentration of 2 μM was determined 4 min after the addition of isotope. Dilution of the isotope was determined and corrected for by measuring the efflux of the preloaded [¹⁴C]choline. The rates of uptake have been corrected for passive influx which was usually less than 10% of the total.

port was not enhanced by prior treatment of the cells with high concentrations of potassium. We concluded that the stimulation of transport resulted directly from the release of ACh rather than from the depolarization that causes release.

Since transport of Ch is one of the rate controlling steps of ACh production it seemed possible that this represented a physiological control mechanism and it was investigated further. Any concentration changes resulting from ACh release will be inside the terminal so that attention was directed towards processes that might affect Ch influx from inside (i.e. *trans* to the direction of transport). Synaptosomes were preloaded with Ch and the rate of influx was measured as a function of the extent of preloading. The results are shown in Fig. 5. As can be seen preloading with Ch causes a considerable increase in influx. *Trans*-activation is rather a characteristic feature of carrier mediated transport and has been demonstrated for Ch transport in erythrocytes (Martin, 1968) though not previously reported in nervous tissue. By preloading with [^{14}C]Ch it is possible to measure the efflux of Ch from the terminals. As shown in Fig. 6 efflux also shows saturation kinetics with regard to the concentration of Ch inside the terminal and is also activated by Ch outside.

The existence of pronounced *trans*-activation for both influx and efflux has certain consequences which are important in the design of transport experiments and the conclusions that can be drawn from them. It is evident that results are not going to be reproducible unless the concentration of choline inside the nerve terminals is kept constant. This is quite difficult because the breakdown of phospholipids is constantly contributing to the internal concentration of Ch so that the uptake of Ch will depend considerably on the previous handling of the synaptosomes. Since there is no easy way of control-

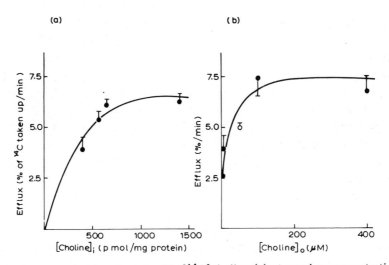

Fig. 6. Synaptosomes were preloaded with [^{14}C]choline (a) at varying concentrations, and the efflux measured at an external choline concentration of 2 μM; and (b) at a concentration of 2 μM while efflux was measured with the outside choline concentration as shown on the abcissa. Efflux was measured by separating the synaptosomes from their incubation media by a brief centrifugation (10,000 $\times$ g for 10 s) and determination of the radioactivity in the supernatant fluid. It was assumed that the radioactivity taken up in the preloading phase had mixed completely with the endogenous choline.

82

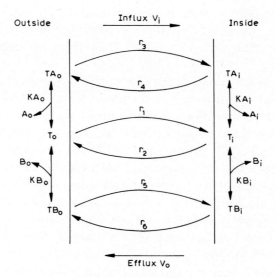

Fig. 7. Processes involved in carrier mediated transport. A and B are two possible substrates at concentration A_0, etc binding to the carrier (T) with dissociation constants K_{A0} etc. $r_1 ... r_6$ are the rate constants with which the various species T_0, TA_0 ... etc traverse the membrane.

ling the breakdown of phospholipids within the terminal their contribution to Ch stores was measured and found to be about 400 pmol/mg protein. This amount was taken into account when the Ch concentration inside the terminal was manipulated by pre-incubation.

It is also clear that the application of simple Michaelis-Menten kinetics to transport phenomena is quite inappropriate. The essential feature necessary to explain *trans*-activation is that the carrier must be returned to the outside for influx to proceed, and to the inside for efflux to continue. This is embodied in the model shown in Fig. 7 which is about the simplest that will explain the various phenomena. For generality two possible substrates A and B are considered; the carrier molecule, denoted T, is shown as traversing the membrane but it could equally and without changing the formal properties of this model be stationary but having inward or outward facing conformations.

In generating the flux equations from this model the following assumptions have been made:

(i) The binding reactions of A, B with T described by the dissociation constants K_{A0} etc are very fast compared with the rate constants ($r_1 ... r_6$) for the traverse of the membrane by various species of T.

(ii) The overall concentration of all species of the carrier T remains constant (conservation principle).

(iii) Over the period of measurement of the initial rate the concentration of the various species of T on one side of the membrane equals those on the other side (steady-state assumption). The detailed algebra necessary to develop the flux equation is simple but rather long winded so it will not be repeated here. Numerous equivalent forms can be generated but perhaps the most useful for present purposes expresses influx (v_i) as a function of efflux (v_e);

$$V_i = \frac{A_0(V_{max}(r_3 + r_4) + V_e(r_4 - r_2))}{K_{A_0}r_4(r_1 + r_2) + A_0(r_2r_4 + r_3r_4)}$$

The dependency of influx on efflux is immediately apparent. A plot of v_i against concentration (A_0) will be a rectangular hyperbola only if v_e is constant. The constants are complicated and not at all equivalent to the K_m and V_m of enzyme activity. The maximal velocity of transport calls for some comment. Influx will only be maximal when the *trans*-concentration of substrate is saturating, similarly for efflux so that the maximal velocity in both directions will be the same and it can be readily shown that:

$$V_{max} = \frac{T_t r_3 r_4}{r_3 + r_4}$$

where T_t is the total amount of carrier.

That this corresponds to experiment can be seen from Fig. 8 where reciprocal flux is plotted against reciprocal *cis*-concentration at saturating *trans* concentrations. As predicted both lines intercept at the common maximal velocity.

The claim that there is a "high affinity" Ch transport system specifically associated with cholinergic cells is based largely on the application of Michaelis-Menten kinetics to measurement of Ch influx at various outside Ch concentrations (Yamamura and Snyder, 1973; Guyenet et al., 1973; Dowdall and Simon, 1973; Suskiw and Pilar, 1976). From the data given above it would seem that this method of analysis is inadequate and likely to lead to erroneous conclusions about the existance of a separate "high affinity" binding. The analysis of our own data using the expression relating influx to efflux does not suggest that two uptake systems exist in cerebral cortex synaptosomes.

The effect of a competing substrate for transport (B in Fig. 7) can be rather complicated depending on the affinity of B for the carrier relative to A, and on the rate of transport of B relative to A. A quantitative treatment is possible but rapidly leads to rather opaque equations. In Table 1 are set out a few simple

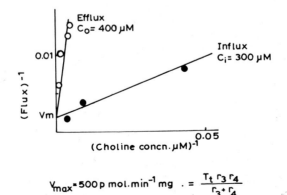

Fig. 8. Choline flux (pmol · min^{-1} · mg^{-1}) plotted as its reciprocal against the reciprocal of the *cis* choline concentration at saturating *trans* concentrations of choline.

TABLE 1

EFFECT OF A COMPETING SUBSTRATE ON FLUX

Characteristics			Effect on flux
Affinity	Transport rate	Position relative to origin of flux	
High	High	*cis*	inhibition
		trans	activation
Low	High	*cis*	slight inhibition
		trans	activation
High	Low	*cis*	inhibition
		trans	inhibition
Low	Low	*cis*	slight inhibition
		trans	slight inhibition

rules of thumb of the effects of competitors on flux in relation to their characteristics and position.

ACh is known to compete for the site of Ch transport (Potter, 1968; Martin, 1969; Marchbanks, 1969) and it appears to be transported itself but only slowly. It would seem possible therefore that the increased Ch uptake after depolarization could be the result of the reduction in the cytoplasmic concentration of ACh consequent upon its release. This reduction would result in reduced inhibition of Ch influx. If so, the effect of depolarization in increasing Ch influx should be abolished by increasing the internal ACh concentration. The effect of preloading with ACh and Ch on the activation of choline uptake is shown in Fig. 9. As is to be expected, preloading with Ch reduces the activation because the higher concentration of Ch displaces ACh from the carrier thus nullifying its inhibitory effect. Preloading with ACh also reduces the

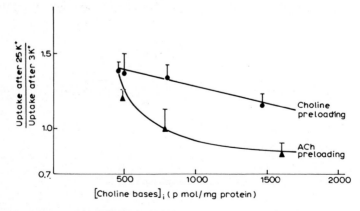

Fig. 9. Synaptosomes were preloaded with either choline (●) or ACh (▲) to give the total concentration of choline bases shown on the abscissa. At the same time they were either depolarised (25 mM K^+) or kept in normal medium (3 mM K^+). Subsequently they were all returned to a medium containing 3 mM K^+ and the rate of uptake of [^{3}H]choline measured as described in the legend to Fig. 5.

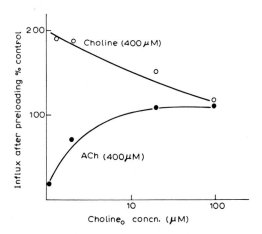

Fig. 10. Synaptosomes were preloaded (as described in legend to Fig. 5) with either 400 μM choline (○) or, 400 μM ACh (●). After washing, influx was measured at various choline concentrations (shown for convenience logarithmically on the abscissa). Influx is shown as a percentage of the control in which there were no choline bases in the preloading medium.

activation but more strongly and at lower intra-terminal concentrations suggesting that the decrease in cytoplasmic ACh is the primary reason for the activation.

A more direct demonstration of this effect can be made by comparing the Ch influx after preloading with either ACh or Ch. As is shown in Fig. 10 preloading with ACh decrease Ch influx, while preloading with Ch, as would be expected increases influx. The effect of ACh is that expected of a competing substrate with low affinity and transport rate.

These experiments indicate that in synaptosomes the internal ACh concentration can modulate the influx of its precursor Ch and that this occurs by the relief of *trans*-inhibition of the Ch carrier. A full quantitative description can only be given when the various rate constants have been determined but qualitatively the effect seems clear. Our results agree with those of Collier and Ilson (1977) on the superior cervical ganglion except that they were unable to demonstrate that high Mg^{2+} depressed the activation of transport. However, as also in the study by Murrin et al. (1977), the transport of Ch and its analogues was measured during the period when release was modified by high Mg^{2+} concentrations so it remains possible that there is a direct effect of the cation itself on Ch transport which masked the effect mediated through ACh release.

Although many features of synaptosome transmitter metabolism mimic the in vivo situation they are artefactual particles and it is germane to consider how far the effects demonstrated here are physiologically relevant. One of the ways in which Ch metabolism of synaptosomes is artificial is in the rapid breakdown of phospholipids and consequently much enhanced intra-terminal Ch concentration (probably about 5-fold). A higher intra-terminal Ch concentration will tend to nullify the effect of ACh in inhibiting transport so that it would be expected that at the lower tissue Ch concentrations found in vivo the activation resulting from ACh release would be much more dramatic. On a qualitative basis, it seems likely that the enhanced supply of Ch achieved by this regulatory mechanism would be important in maintaining supplies of ACh in the face of its depletion by stimulated release.

This mechanism could also explain the overshoot effect if there was a slight delay in the conversion of the increased intra-terminal Ch to ACh. This is likely because the acetyl-CoA supply is rate limiting, and it would allow the accumulation of more Ch (and thus ACh) before its transport was inhibited by ACh.

Over the past few years the effect of ACh release on Ch transport has led to the concept that there is an obligatory coupling between the process of Ch transport and ChAT (Barker and Mittag, 1974; Simon et al., 1976). This is usually linked with the proposal that cholinergic terminals specifically contain the so called "high affinity" Ch transport system with which the enzyme of synthesis is closely associated (Kuhar et al., 1973; Haga and Noda, 1973). We would not wish to comment on the existence or otherwise of a separate "high affinity" system except to note that our results do not require its postulation and indeed, suggest that the analogy with enzyme kinetics that led to its proposal is false.

However, the question of an obligatory coupling with ChAT is more important since there is considerable evidence that at physiological ionic strengths the enzyme is freely soluble and not bound to membranes (Fonnum, 1968). The explanation that we have given of the way in which ACh levels control Ch transport does not require re-evaluation of this finding or the postulation of a molecular association for which there is no direct experimental evidence.

SUMMARY

(1) The mechanism of the increase in choline (Ch) transport induced by acetylcholine (ACh) release has been investigated. The esterification of Ch is not rate controlling.

(2) Ch influx is enhanced after a period of ACh release from synaptosomes induced by depolarization with potassium. The ACh release is dependent on Ca^{2+} and inhibited by high Mg^{2+} concentrations, the subsequent activation of Ch influx behaves likewise.

(3) Ch influx is activated by high concentrations of Ch inside the synaptic terminal, and its efflux activated by high concentrations outside. A model of this process is presented and the relationship between influx and efflux derived.

(4) ACh binds to the Ch carrier but is transported itself only slowly. It could be predicted therefore that a decrease in cytoplasmic ACh would result in the decrease in the immobilisation of the carrier on the inside ot the synaptic membrane and an increase in Ch influx.

(5) Ch influx is inhibited by raising the intra-synaptosomal ACh concentration.

(6) Raising the intra-synaptosomal ACh concentration by pre-loading also abolishes the stimulatory effect of depolarization.

(7) It is concluded that a decrease in the cytoplasmic ACh concentration can stimulate the influx of its precursor Ch by relief of *trans*-inhibition of the Ch carrier.

ACKNOWLEDGEMENTS

These studies were made possible by Grants (to RMM) G974/907/N and G976/298/N from the Medical Research Council of the United Kingdom.

REFERENCES

Barker, L.A. and Mittag, T.W. (1974) Comparative studies of substrates and inhibitors of choline transport and choline acetyltransferase. *J. Pharmacol. exp. Ther.*, 192, 86–94.

Birks, R. and MacIntosh, F.C. (1961) Acetylcholine metabolism of a sympathetic ganglion. *Canad. J. Biochem. Physiol.*, 39, 788–827.

Carroll, P.T. and Goldberg, A.M. (1975) Relative importance of choline transport to spontaneous and potassium depolarized release of ACh. *J. Neurochem.*, 25, 523–527.

Collier, B. and Ilson, D. (1977) The effect of preganglionic nerve stimulation on the accumulation of certain analogues of choline by a sympathetic ganglion. *J. Physiol. (Lond.)*, 264, 489–509.

Dowdall, M.J. and Simon, E.J. (1973) Comparative studies on synaptosomes: uptake of [*N-Me*-^{3}H]choline by synaptosomes from squid optic lobes. *J. Neurochem.*, 21, 969–982.

Fonnum, F. (1968) Choline acetyltransferase binding to and release from membranes. *Biochem. J.*, 109, 389–397.

Guyenet, P., Lefresne, P., Rossier, J., Beaujouan, J.C. and Glowinski, J. (1973) Inhibition by hemicholinium-3 of [^{14}C]ACh synthesis and [^{3}H]choline high affinity uptake in rat striatal synaptosomes. *Mol. Pharmacol.*, 9, 630–639.

Guynn, R.W. (1976) Effect of ethanol on brain CoA and acetyl-CoA. *J. Neurochem.*, 27, 303–304.

Haga, T. and Noda, H. (1973) Choline uptake systems of rat brain synaptosomes. *Biochim. Biophys. Acta*, 291, 564–575.

Kuhar, M.J., Sethy, V.H., Roth, R.M. and Aghajanian, G.K. (1973) Choline: selective accumulation by central cholinergic neurons. *J. Neurochem.*, 20, 581–593.

Mann, P.J.G., Tennenbaum, M. and Quastel, J.H. (1939) Acetylcholine metabolism in the central nervous system. The effects of potassium and other cations on acetylcholine liberation. *Biochem. J.*, 33, 822–835.

Marchbanks, R.M. (1967) The osmotically sensitive potassium and sodium compartments of synaptosomes. *Biochem. J.*, 104, 148–157.

Marchbanks, R.M. (1968) The uptake of [^{14}C]choline in synaptosomes in vitro. *Biochem. J.*, 110, 533–541.

Marchbanks, R.M. (1969) Biochemical organization of cholinergic nerve terminals in the cerebral cortex. In *Cellular dynamics of the neuron*, S.H. Barondes (Ed.), Academic Press, New York, pp. 115–135.

Marchbanks, R.M. (1977) Turnover and release of acetylcholine. In *Synapses*, G.A. Cottrell and P.N.R. Usherwood (Eds.), Blackie, Glasgow, pp. 81–101.

Marchbanks, R.M. and Israël, M. (1971) Aspects of acetylcholine metabolism in the electric organ of Torpedo marmorata. *J. Neurochem.*, 18, 439–448.

Martin, K. (1968) Concentrative accumulation of choline by human erythrocytes. *J. Gen. Physiol.*, 51, 497–516.

Martin, K. (1969) Effects of quaternary ammonium compounds on choline transport in red cells. *Brit. J. Pharmacol.*, 36, 458–469.

Murrin, L.C., Dehaven, R.N. and Kuhar, M.J. (1977) On the relationship between [^{3}H]choline uptake activation and [^{3}H]acetylcholine release. *J. Neurochem.*, 29, 681–687.

Polak, R.L. and Meeuws, M.M. (1966) The influence of atropine on the release and uptake of acetylcholine by the isolated cerebral cortex of the rat. *Biochem. Pharmacol.*, 15, 989–992.

Polak, R.L., Molenaar, P.C. and van Gelder, M. (1977) Acetylcholine metabolism and choline uptake in cortical slices. *J. Neurochem.*, 29, 477–485.

Potter, L.T. (1968) Uptake of choline by nerve endings isolated from the rat cerebral cortex. In *The interaction of drugs and subcellular components on animal cells*, P.N. Campbell (Ed.), Churchill, London, pp. 293–304.

88

Potter, L.T., Glover, V.A.S. and Saelens, J.K. (1968) Choline acetyltransferase from rat brain. *J. Biol. Chem.*, 243, 3864–3870.

Richter, J.A. and Marchbanks, R.M. (1971) Synthesis of radioactive acetylcholine from [^{3}H]choline and its release from cerebral cortex slices in vitro. *J. Neurochem.*, 18, 691–703.

Saito, M., Bourque, E. and Kanfer, J. (1975) Studies on base-exchange reaction of phospholipids in rat brain particles and a "solubilized" system. *Arch. Biochem. Biophys.*, 169, 304–317.

Shea, P.A. and Aprison, M.H. (1977) The distribution of acetyl-CoA in specific areas of the CNS of the rat as measured by a modification of a radio-enzymatic assay for acetylcholine and choline. *J. Neurochem.*, 28, 51–58.

Simon, J.R., Atweh, S. and Kuhar, M.J. (1976) Sodium-dependent high affinity choline uptake: a regulatory step in the synthesis of acetylcholine. *J. Neurochem.*, 26, 909–922.

Suskiw, J.B. and Pilar, G. (1976) Selective localization of a high affinity choline uptake system and its role in ACh formation in cholinergic nerve terminals. *J. Neurochem.*, 26, 1133–1138.

Tuček, S. (1967) Subcellular distribution of acetyl-CoA synthetase, ATP citrate lyase, citrate synthase, choline acetyltransferase, fumarate hydratase and lactate dehydrogenase in mammalian brain tissue. *J. Neurochem.*, 14, 531–545.

Yamamura, H.I. and Snyder, S.H. (1973) High affinity transport of choline into synaptosomes of rat brain. *J. Neurochem.*, 21, 1355–1374.

Choline Uptake and Metabolism by Nerve Cell Cultures

R. MASSARELLI, TUEN YEE WONG, CHANTAL FROISSART and J. ROBERT

*Centre de Neurochimie du CNRS and Institute of Biological Chemistry,
Faculty of Medicine, Université Louis Pasteur, 11 Rue Humann,
67085 Strasbourg (France)*

INTRODUCTION

The transport of choline (Ch) into synaptosomes has been suggested to be mediated by two mechanisms which differ in their affinity towards the substrate. Initial studies showed only one Km (about 10^{-5} M) for the uptake of choline into synaptosomes (Diamond and Milfay, 1972; Hemsworth et al., 1971; Marchbanks, 1968; Diamond and Kennedy, 1969). However, on lowering Ch concentrations and using a larger range (from 0.5 to 100 μM), it was shown by a Lineweaver-Burk plot that the experimental points did not fit a straight line, but could be resolved into two straight lines which correspond to the high and low affinity mechanisms (Yamamura and Snyder, 1973).

A correlation was shown between cholinergic markers and high affinity uptake (HAU) of Ch in various areas of rat brain, and HAU of Ch was lost after degeneration of cholinergic terminals (Kuhar et al., 1975a, b); thus, it was suggested that HAU was specific for cholinergic neurons, and there is considerable evidence supporting the concept that Ch transported by this mechanism is necessary for, and may even limit, ACh synthesis (see Kuhar and Murrin, 1978). However, using nerve cell cultures as a model of the nervous system, several laboratories have shown that HAU of Ch can be demonstrated in non-cholinergic neurons as well as in glial cells (Haber and Hutchison, 1976; Massarelli and Mandel, 1976).

Why should a non-cholinergic cell show a mechanism which has so clearly been demonstrated as characteristic of and necessary for the synthesis of ACh? We have previously shown (Massarelli, 1978) that the incubation of nerve cell cultures in Krebs-Ringer phosphate containing no Ch can create a non-steady-state in the endogenous concentration of Ch; the steady-state appeared only to be maintained by incubating cells with the normal Ch concentration of the growth medium (30 μM). Thus incubation with Ch at concentrations necessary to measure high affinity uptake (considerably less than 30 μM) appears to disturb the equilibrium of the endogenous Ch pool. We suggest that, under these conditions, the cell may regulate the transport of choline depending on its needs. The high and low affinities shown in a Lineweaver Burk plot may just be the graphical representation of conformational changes brought to a single transport mechanism by varying the exo/endocellular concentrations of choline.

The present experiments were made in the hope that they might provide evidence for or against this postulate.

MATERIALS AND METHODS

Culture

Cultures from mouse neuroblastoma C1300, clone M_1 cells, were grown in Falcon plastic Petri dishes and used at confluency. The cells did not show any choline acetyltransferase (EC 2.3.1.6, ChAT) activity. The growth medium was Dulbecco's modified Eagle's containing 10% foetal calf serum.

Uptake experiments

Cells attached at the bottom of Petri dishes were washed 3 times with 0.147 M NaCl at 37°C and the preincubation was performed in Krebs-Ringer phosphate pH 7.4 (137 mM NaCl, 2.6 mM KCl, 0.7 mM $CaCl_2$, 0.5 mM $MgCl_2$, 3.2 mM Na_2HPO_4, 1.4 mM KH_2PO_4 and 10 mM glucose). Various concentrations of [^{14}C]choline (Amersham), kept at the same specific radio activity, were added for various periods of incubation, after which the Petri dishes were washed 3 times with 0.147 mM NaCl, and the cells were digested with concentrated formic acid. Aliquots of the acid extract were used for the determination of total radioactivity; this represented only 2–3% of medium radioactivity at the longest time of incubation.

In the studies that measured Ch and its metabolites, a mixture of ice-cold 1 N formic acid/acetone (15/85) was added after washing the cells, the cells were then scraped off, and homogenized. After centrifugation at 3000 × g for 20 min, the supernatant was collected, frozen with liquid N_2, and lyophilized. 50 μl of 0.04 N HCl containing 10 mM Ch-chloride and phosphorylcholine (PCh) chloride were added to the dry material, 10 μl were counted (acid total extract), and 10 μl were spotted on TLC cellulose plates. Chromatography used the system of Marchbanks and Israël (1971) (Butanol/ethanol/acetic acid/water, 100 : 20 : 17 : 33); the areas containing Ch and PCh (identified from the migration of standards) were scraped and the radioactivity determined.

Lipids from the pellet remaining after acid homogenization were extracted in chloroform/methanol 1 : 1 and aliquots were taken for scintillation counting. The tissue residue was dissolved in NaOH (1 M) and aliquots were taken for assay of protein by the method of Lowry et al. (1951).

RESULTS AND DISCUSSION

Uptake and metabolism

Initial experiments characterized the uptake and metabolism of Ch by the M_1 cells. The accumulation of radioactivity into the total acid extract of M_1 cells incubated with varying concentrations of [^{14}C]Ch was approximately linear

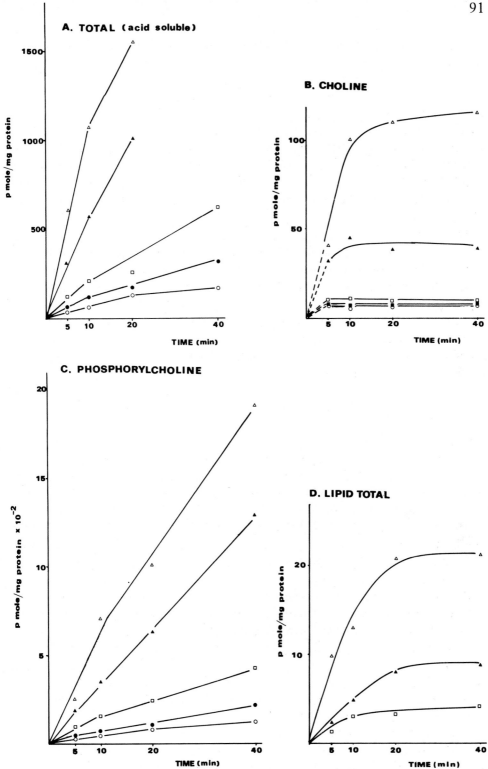

Fig. 1. Distribution of radioactivity originating from [^{14}C]choline in the incubation medium. Cells were incubated with various concentrations of [^{14}C]Ch kept at the same specific activity and the various compartments extracted and separated as described in the Materials and Methods section. The data are expressed as pmoles of [^{14}C]Ch incorporated into each compartment/mg protein. Each point represents the average of two experiments. ○, 0.38 μM; ●, 0.64 μM; □, 1.28 μM; ▲, 6.41 μM; △, 19.23 μM.

with time (Fig. 1A). However, when the free Ch compartment was measured, it appeared that a saturation plateau was already present after 5 min of incubation (Fig. 1B). The incorporation of Ch into PCh increased linearly with time (Fig. 1C), so that the ratio PCh/Ch varied between 5 and 50. Thus, choline enters these cells, rapidly saturates the free Ch compartment, and is essentially directed towards the synthesis of PCh; the lipid incorporated much less radioactivity, as is shown in Fig. 1D. These experimental results are interpreted as supporting the finding (Massarelli, 1978) that endogenous Ch is in a non steady-state when cells are exposed to low exogenous Ch; The radioactivity in the free Ch compartment reaches different saturation plateaus at different exogenous Ch concentrations (Fig. 1B).

The kinetic parameters of the uptake process could be calculated, but we do not consider them to be valid. When tissue is incubated with labelled Ch, the amount of total tissue radioactivity can only be analysed as Ch *transport* if it is clearly demonstrated that uptake of Ch into its free endocellular compartment is linear with time, and that the production of metabolites (i.e. the activity of the enzymes involved in Ch metabolism) is unimportant. If the latter requirement is not fulfilled, a Michaelis-Menten treatment of total uptake of radioactivity would represent several enzymatic activities and not uptake alone. This was clearly so for the uptake of choline in our cell cultures.

Effect of temperature on choline uptake

One possible way to characterize the uptake mechanism is to study its behaviour at different temperatures. Such a study made on cultured cells incubated with 1 μM Ch is shown in Fig. 2. The accumulation of radioactivity in the total acid extract showed a break in the Arrhenius plot at 17°C. The two resulting energies of activation would suggest that the uptake may change its energy requirement with temperature and that, at physiological temperature, this requirement is rather low. Similar results have already been published (Massarelli et al., 1976) and it was suggested that, in nerve cell cultures, the HAU of Ch may be a facilitated diffusion mechanism. However, when the effect of temperature was studied on the incorporation of Ch into the endocellular free Ch compartment (representing thus the transport of Ch) a rather strange Arrhenius plot was obtained (Fig. 2B). One possible explanation for this bell shaped curve is that, when cells are incubated with 1 μM [^{14}C]Ch, there are great differences in the metabolism of the endogenous Ch pool at different temperatures. An anomalous Arrhenius plot was also observed for Ch incorporation into the lipid fraction (Fig. 2D), while PCh synthesis gave only one energy component (Fig. 2C).

These data may reflect the non steady-state conditions referred to earlier, and it is possible that different metabolic pathways intervene at different temperatures. If this were so, at a concentration of external Ch that would not change the endogenous size of the pool, the temperature dependency should not show anomalies. When cells were incubated with 30 μM choline (normal concentrations in the medium) the Arrhenius plots showed only one energy of activation for all parameters measured (Fig. 3).

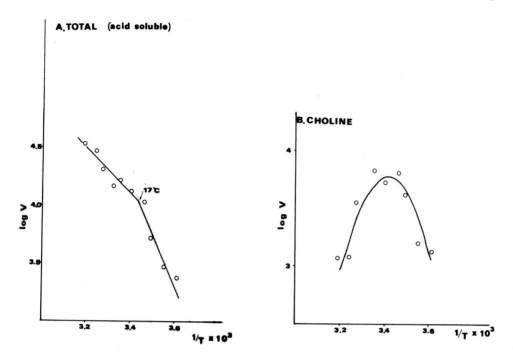

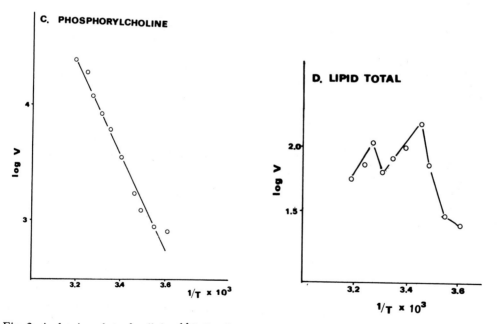

Fig. 2. Arrhenius plot of cellular ^{14}C-distribution after incubation with 1 μM [^{14}C]Ch. (A) The values of E_a are: 8.6 kcal $\cdot$ mol^{-1} above 17°C and 19.1 kcal $\cdot$ mol^{-1} below 17°C. (C) ^{14}C-incorporation in PCh showed an E_a of 18.1 kcal $\cdot$ mol^{-1}. Each point is the average of three independent determinations. The points were fitted by regression analysis.

94

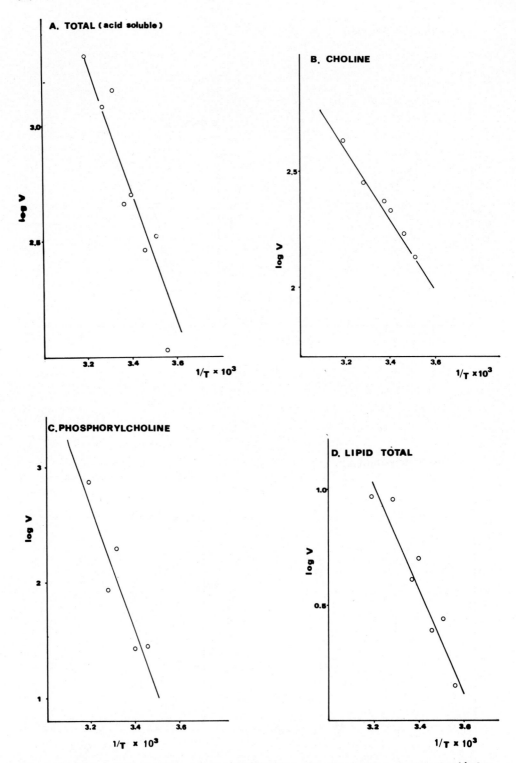

Fig. 3. Arrhenius plot of cellular choline distribution after incubation with 30 μM [^{14}C]Ch.
(A) E_a = 12.7 kcal · mol^{-1}; (B) E_a = 6.7 kcal · mol^{-1}; (C) E_a = 29.0 kcal · mol^{-1}; (D) E_a = 20.2 kcal · mol^{-1}. Each point is the average of three independent determinations.

CONCLUSIONS

These results and other arguments which have been developed elsewhere (Massarelli, 1978) suggest that we can not exclude the possibility that a single uptake mechanism exists; it is energy dependent (not as much, though, as the other processes involved in Ch metabolism as can be judged by their energies of activation (Fig. 3)), and it may change in conformation depending on the concentration of external Ch and/or on endocellular steady-state conditions. It should be emphasized that these suggestions are based upon experiments on cell cultures, and it would be inappropriate to extrapolate the suggestions to intact tissues or to synaptosomes. It would, however be interesting to test whether synaptosomes under HAU conditions can maintain steady-state concentrations of Ch, and if these may influence Ch uptake as we have shown in nerve cell structures.

SUMMARY

Uptake and metabolic distribution of Me[^{14}C]choline was studied in clone M1 of mouse neuroblastoma 1300. Total incorporation of ^{14}C and labelling of phosphorylcholine were linear with time, while the free Ch compartment was rapidly saturated. Incubation of cells at different temperature, and under high affinity uptake conditions (1 μM of exogenous Ch), resulted in abnormal Arrhenius plots. This anomaly was not shown when cells were incubated with 30 μM choline (concentration used normally in the growth medium). The results support previously published data suggesting that conditions used to study high affinity Ch uptake produce a non steady-state in the endocellular pool of choline. It is postulated that the uptake of Ch may be regulated by the exo-endocellular concentrations of choline, and that the high and low affinity uptake systems may represent a single mechanism which changes in conformation with different concentrations of substrate.

REFERENCES

Diamond, I. and Kennedy, E.P. (1969) Carrier-mediated transport of choline into synaptic nerve endings. *J. Biol. Chem.*, 244, 3258–3263.

Diamond, I. and Milfay, D. (1972) Uptake of [^{3}H-*methyl*]choline by microsomal, synaptosomal, mitochondrial and synaptic vesicle fractions of rat brain. *J. Neurochem.*, 19, 1899–1909.

Haber, B. and Hutchison, H.T. (1976) Uptake of neurotransmitters and precursors by clonal cell lines of neural origin. In *Transport Phenomena in the Nervous System*, G. Levi, L. Battistin and A. Lajtha (Eds.), Plenum Press, New York, pp. 179–198.

Hemsworth, B.A., Darmer, K.I. and Bosmann, H.B. (1971) The incorporation of choline into isolated synaptosomal and synaptic vesicles fractions in the presence of quaternary ammonium compounds. *Neuropharmacol.*, 10, 109–120.

Kuhar, M.J., Dehaven, R.M., Yamamura, H.I., Rommelspacher, H. and Simon, J.R. (1975a) Further evidence for cholinergic habenulo-inter-penduncular neurons: pharmacologic and functional characteristics. *Brain Res.*, 97, 265–275.

Kuhar, M.J., Sethy, V.H., Roth, R.H. and Aghajanian, G.K. (1975b) Choline: selective accumulation by central cholinergic neurons. *J. Neurochem.*, 20, 581–593.

Kuhar, M.J. and Murrin, L.C. (1978) Sodium-dependent, high affinity choline uptake. *J. Neurochem.*, 30, 15–21.

Lowry, O.H., Rosebrough, N.J., Farr, A.L. and Randall, R.J. (1951) Protein measurement with the Folin Phenol Reagent. *J. Biol. Chem.*, 193, 265–275.

Marchbanks, R.M. (1968) The uptake of [^{14}C]choline into synaptosomes in vitro. *Biochem. J.*, 110, 533–541.

Marchbanks, R.M. and Israël, M. (1971) Aspects of acetylcholine metabolism in the electric organ of Torpedo marmorata. *J. Neurochem.*, 18, 439–448.

Massarelli, R. (1978) Uptake of choline in nerve cell cultures: correlation with the endogenous pool of choline. In *Cholinergic Mechanisms and Psychopharmacology*, D.J. Jenden (Ed.), Plenum Press, New York, pp. 539–550.

Massarelli, R. and Mandel, P. (1976) On the uptake mechanism of choline in nerve cell cultures. In *Transport phenomena in the nervous system* (Levi, G., Battistin, L. and Lajtha, A., eds.), pp. 199–209, Plenum Press, New York.

Massarelli, R., Stefanovic, V. and Mandel, P. (1976) Cholinesterase activity and choline uptake in intact nerve cell cultures. *Brain Res.*, 112, 103–112.

Yamamura, H.I. and Snyder, S.H. (1973) High affinity transport of choline into synaptosomes of rat brain. *J. Neurochem.*, 21, 1355–1374.

Regulation of Acetylcholine Synthesis in Cholinergic Nerve Terminals

GUILLERMO PILAR and KEN VACA

Physiology Section, Biological Sciences Group, University of Connecticut, Storrs, CT 06268 (U.S.A.)

INTRODUCTION

Initial studies on acetylcholine (ACh) metabolism in cat superior cervical ganglion (SCG) led to the conclusion that cholinergic nerve endings possess a high capacity for neurotransmitter synthesis (Brown and Feldberg, 1936). However, until recently, the mechanisms regulating the rate of ACh synthesis have remained elusive. A great deal of information is available on the properties of choline acetyltransferase (ChAT) which converts acetyl-CoA and choline (Ch) to ACh (see Mautner, 1977, for review); this enzyme tends to be present in considerable excess in cholinergic neurons and there is no convincing evidence for physiological regulation of its kinetic parameters. Na^+-coupled transport of organic substances was first observed by Quastel and co-workers (see Quastel, 1965) who also did much of the pioneering work on ACh metabolism using brain slices. The many excellent studies on ACh metabolism in brain slices will not be discussed in this brief survey. The present paper will only deal with ACh synthesis in presynaptic endings with emphasis on our own work in the ciliary nerve terminals of the chick iris.

The use of synaptosomes in rapid tracer flux studies was undoubtedly helpful in the resolution of a Na^+-dependent, high affinity choline (Ch) uptake system (SDHACU) (Haga and Noda, 1973; Yamamura and Snyder, 1973). It was immediately evident that the rate of synthesis in neurons corresponded much more closely to the rate of Ch uptake than the very much higher soluble ChAT activity, suggesting that availability of precursors, particularly Ch, might control the rate of ACh synthesis. Yet, most of the work on the physiology of ACh as a neurotransmitter has been done on more intact preparations, especially the neuromuscular junction (see Katz, 1969). It seemed desirable to find a preparation where both physiological and metabolic parameters could be explored. Although high affinity Ch uptake could not be detected in rat diaphragm (Chang and Lee, 1970; Potter, 1970) SDHACU was kinetically resolved in the wafer-thin, doughnut shaped striated muscle of the chick iris (Suszkiw and Pilar, 1976) which allows relatively fast (~1 min) equilibration with extracellular markers (Beach et al., 1978). This muscle is also multiply innervated; see Fig. 1 for a typical picture of its neuromuscular junction. The electrophysiology of synaptic transmission has been well characterized in this preparation

98

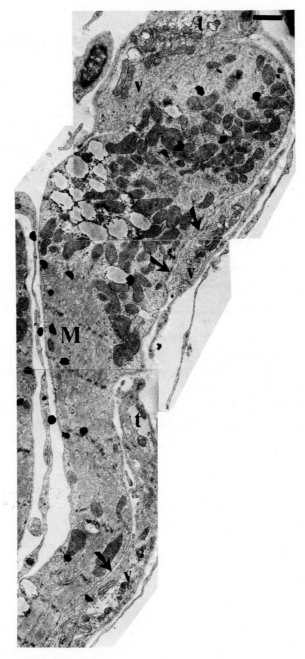

Fig. 1. Electron micrographic montage of a striated constrictor muscle fibre from the chick iris. The section is along the longitudinal axis of the fibre. Three regions of contact between nerve terminals (t) and a single muscle cell (M) are observed in the illustration. Terminals, filled with clear and round shaped synaptic vesicles (V) are separated from the muscle by gaps of approximately 100 nm. Few postsynaptic specializations (arrows) are seen. Large accumulations of mitochondria are present in the muscle cytoplasm in the vicinity of the synaptic area. Calibration bar: 1 μM.

(Pilar and Vaughan, 1969); short-term synaptic plasticity is currently being investigated (Vaca and Pilar, work in progress).

When cat SCG is stimulated electrically, the rates of Ch uptake and ACh synthesis increase (Collier and MacIntosh, 1969) sufficiently to sustain rapid rates of ACh turnover although release gradually declines (Birks and MacIntosh, 1961). Following in vivo alterations of electrical activity, corresponding changes of SDHACU and ACh synthesis were noted in synaptosomes with higher levels of uptake and synthesis following elevated activity (Simon et al., 1976). Furthermore, conditions which prevent high affinity uptake, such as hemicholinium-3 or absence of Na^+, also prevent the replenishment of ACh during heightened activity (Birks and MacIntosh, 1961; Birks, 1963). Guyenet et al. (1973) found a good correlation between the inhibition of SDHACU by hemicholinium-3 and inhibition of ACh synthesis using either Ch, glucose or pyruvate as radiolabelled precursors in synaptosomes. Log-log plots of Na^+ concentration versus SDHACU and versus ACh synthesis in the nerve terminals of the iris have almost identical slopes, close to 1 (Vaca and Pilar, 1979).

ENERGETICS OF CHOLINE TRANSPORT

Because the high affinity Ch transport appears to be obligatory for ACh synthesis, it would be appropriate to first consider the energetics of SDHACU prior to examining its coupling to ACh synthesis. Studies with a wide variety of metabolic inhibitors in both synaptosomal and neuromuscular preparations indicate that SDHACU is partially dependent on ongoing metabolism (summarized in Table I). Inhibition of energy metabolism brought a reduction of Ch

TABLE I

SHORT TERM ENERGY REQUIREMENTS OF SDHACU IN SYNAPTOSOMES AND MOTOR NERVE TERMINALS

Tissue preparation	Q_{10}	Maximum inhibition by metabolic inhibitors	Maximum inhibition as a result of blocking Na, K-ATPase	Reference
Rat brain synaptosomes	2.5 (20–30°C)	72%	45%	Yamamura and Snyder 1973
		<5%	26%	Haga and Noda 1973
	2.7 (27–37°C)	48%	79%	Simon and Kuhar 1976
Guinea-pig myenteric plexus		53%	68%	Pert and Snyder 1974
Chick iris	3.6 (20–30°C) 1.3 (30–40°C)	83%	84%	Beach et al. 1978

The term "maximum inhibition" refers to the maximum effect observed after a short exposure (usually 4 or 8 min) to one of the several inhibitors used in a given study. The original references should be consulted for details of the experimental protocol.

uptake in 10 min or less, with respiratory inhibitors and uncouplers of oxidative phosphorylation generally being most effective. In the temperature range 20–30°C, Q_{10} values close to 3 have been observed, although in nerve terminals of the iris, there is a transition above 30° to a lower Q_{10} (approximately 1.3) typical of passive ion movements. Comparison of the effect of inhibiting energy metabolism with the effect of inhibiting Na, K-ATPase on SDHACU (Table I) suggests that the apparently active nature of Ch transport is mostly, perhaps completely, due to its dependence on continued active pumping of Na^+ and K^+. Inhibition of Na,K-ATPase causes a loss of asymmetric ion distribution, notably the Na^+ gradient, and a reduction in membrane potential. Indirect evidence suggests that the electrogenic aspects of the Na^+ pump makes a contribution to SDHACU (Vaca and Pilar, 1979).

Depolarization can reduce SDHACU by as much as 90% (Murrin and Kuhar, 1976; Beach et al., 1978); both high extracellular $[K^+]$ and veratridine are effective. The effect of depolarization does not depend on ACh release since it does not require the presence of Ca^{2+}; furthermore, because the effect is on initial velocity of Ch uptake, it represents a reduction of a unidirectional flux (Vaca and Pilar, 1979) – homoexchange of Ch is not detectable (Simon et al., 1976; Vaca, unpublished observations). The linear relationship between Ch transport and log $[K^+]_{out}$ suggests that SDHACU is linearly related to membrane potential (Vaca and Pilar, 1979). High affinity Ch transport into nerve terminals thus bears very many similarities to Na^+-coupled transport of organic solvents in non-neural tissues (Schultz and Curran, 1970). The term "secondary active transport" has been used to describe transport which is not directly coupled to a hydrolytic reaction (cf. Geck and Heinz, 1976). SDHACU appears to be "secondary active transport". As will be discussed later, the increased availability of Ch for ACh synthesis during electrical activity may be explicable solely in terms of alterations in the combined electrochemical gradients for Na^+ and Ch.

RELATIONSHIP BETWEEN CHOLINE TRANSPORT AND ACETYLCHOLINE SYNTHESIS IN THE RESTING STATE

In the absence of synaptic activity in motor nerve terminals, synthesis of new ACh from radiolabelled Ch proceeds quite slowly (Potter, 1970; Vaca and Pilar, 1978). Similarly, in the resting state, a relatively small proportion of Ch transport occurs via the high affinity system in the iris (Vaca and Pilar, 1978). However, under a variety of experimental conditions, the percentage of Ch taken up by the high affinity system which is then converted to ACh is quite constant, typically 35–45% (Suszkiw and Pilar, 1976; Beach et al., 1978; Vaca and Pilar, 1979), provided that not much ACh release is evoked. In synaptosomes, percent conversion of Ch to ACh is also relatively constant, typically 60–70% (Haga and Noda, 1973; Yamamura and Snyder, 1973; Simon et al., 1976). All these studies suggest a close connection between SDHACU and ACh synthesis. In fact, if cytoplasmic cholinesterases are inactivated by preincubation with paraoxon, approximately 96% of Na^+-dependent Ch transport results in conversion to ACh (Table II). Using $K_{eq} = 12$ for ChAT in a physiological

TABLE II

Na⁺-DEPENDENT Ch UPTAKE AND ACh SYNTHESIS AFTER ACETYLCHOLIN-
ESTERASE INHIBITION

Preincu- bation	Incuba- tion	Total Ch uptake (pmol/iris/8 min)	P	ACh Synthesis (pmol/iris/8 min)	P
Tyrode (control)	Tyrode	1.771 ± 0.148 (15)		0.391 ± 0.067 (15)	
Tyrode	Na⁺-free Tyrode	0.854 ± 0.136 (11)	<0.001	0.061 ± 0.011 (11)	<0.001
Paraoxon	Tyrode	1.797 ± 0.177 (14)	NS	0.982 ± 0.155 (14)	<0.01
Paraoxon	Na⁺-free	0.855 ± 0.099 (11)	<0.001	0.073 ± 0.011 (11)	<0.001

Irises were dissected out from ten-day-old chicks and preincubated 20 min at 20°C in Tyrode solution with and without 200 μM paraoxon. They were rinsed, equilibrated to 37°C in the presence or absence of Na⁺, and then incubated in [^{3}H]Ch containing solutions. [Ch⁺] = 0.7 μM. For the Na⁺-free experiments, Na⁺ was replaced by Li⁺ (n = 4) or isoosmotic sucrose (n = 7). The uptake was terminated by rinsing the tissues with ice cold Tyrode solution and after homogenization, the [^{3}H]Ch⁺ taken up and [^{3}H]ACh synthesized were determined by liquid scintilation spectrometry and high-voltage electrophoresis. Results are expressed as mean $\pm$ S.E. Numbers of observations are indicated in parentheses. P-values determined by two-tailed student's t test indicate the significance of difference from control values.

salt solution (Pieklik and Guynn, 1975), one expects that at equilibrium, in the absence of cholinesterase activity, most of the Ch free in the cytosol of the nerve terminal will be converted to ACh. That equilibration of the newly taken up Ch with ACh is rapidly attained is not surprising in view of the vast excess of ChAT activity, which is close to 50 nmol/h/iris when solubilized. Several important conclusions follow from these findings: (1) In the resting state, much of the newly transported Ch remains in the nerve terminal cytosol where it is available for conversion to ACh. (2) Cytoplasmic Ch and ACh are maintained in a dynamic equilibrium, as determined by mass action considerations and by opposing ChAT and AChE activities.

CHANGES OF Ch TRANSPORT AND ACh SYNTHESIS ACCOMPANYING EVOKED TRANSMITTER RELEASE

Replacement of endogenous ACh with newly synthesized ACh formed from radiolabelled Ch is greatly accelerated by electrical stimulation of cholinergic neurons (Collier and MacIntosh, 1969; Potter, 1970; Vaca and Pilar, 1979). Increased SDHACU by synaptosomes in vitro following in vivo manipulations of electrical activity led to the suggestion that regulation of the rate of ACh synthesis was a direct consequence of a coupling of SDHACU to impulse flow (Simon et al., 1976). In ciliary nerve terminals of the iris, SDHACU and ACh synthesis are roughly proportional to the frequency of electrical stimulation, increasing at least 5-fold over the range 0 to 30 Hz (Vaca and Pilar, 1979). If the accelerated Ch transport is blocked by hemicholinium-3 or Na⁺-free medium, terminal stores of ACh are largely depleted by stimulation (Birks and MacIntosh, 1961; Birks, 1963; Potter, 1970). The increase in SDHACU and

TABLE III
ACh CONTENT IN CHICK IRIS DURING DEPOLARIZATION

Incubation	ACh pmoles	P
Normal tyrode	26.2 ± 1.8	
55 mM K^+, 3 mM Ca^{++}, 1 mM Mg^{2+}	15.0 ± 2.1	<0.01
55 mM K^+O Ca^{2+}, 10 mM Mg^{2+}	24.1 ± 2.8	N.S.

Irises were dissected out and after 10 minutes incubation in the indicated solutions, they were homogenized and ACh was determined following the method of Shea and Aprison (1973). Membrane potential depolarization was elicited by raising the $[K^+]$ concentration in the incubation medium from 3 mM (normal Tyrode) to 55 mM.

ACh synthesis in the iris (Vaca and Pilar, 1979) and in the uptake and acetylation of Ch analogues in cat SCG (Collier and Ilson, 1977) requires the presence of Ca^{2+} during electrical stimulation.

A conditioning depolarization with elevated K^+ or with veratridine is also effective in accelerating SDHACU in nerve terminals which are allowed to repolarize (Murrin and Kuhar, 1976; Vaca and Pilar, 1979). The effect of the conditioning depolarization is to increase the V_{max} for SDHACU without significantly affecting the K_M for Ch. Indeed, changes in K_M would not be very effective in the chick, where the plasma [Ch] = 5 μM and the K_M for SDHACU is 1–2 μM; even if the K_M for Ch were to become vanishingly small, transport could increase only 20–40%. However, transport increases in direct proportion to the V_{max}, approximately doubling after a conditioning depolarization.

A 10 min depolarization with 55 mM K^+ Tyrode's solution decreases endogenous ACh levels in the iris by more than 40% (Table III). This depleting effect of depolarization is prevented by removing Ca^{2+} and raising $[Mg^+]$ (Table III), conditions which decrease the rate of depolarization-induced release of preformed $[^3H]$ACh approximately 50-fold (Vaca, unpublished observations). The effect of Ca^{2+} and Mg^{2+} concentrations during the conditioning depolarization on the subsequent activation of Ch transport and ACh synthesis is illustrated in Fig. 2. It is quite clear that conditions which inhibit transmitter release also substantially reduce the activation of uptake and synthesis. The simplest explanation for these phenomena is that the reduction of nerve terminal ACh drives ACh synthesis by mass action. This would in turn reduce the chemical gradient for Ch and thus increase the velocity of SDHACU. If much of the released ACh originates from a vesicular pool, as is widely believed, the newly synthesized ACh would be expected to replenish the recently emptied vesicles, where it would be protected from cytoplasmic cholinesterases. As a result, one would expect an increase in the percentage of recently transported Ch converted to ACh — this is in fact observed. Following a conditioning depolarization, the V_{max} for ACh synthesis from $[^3H]$Ch increases more than 5-fold, compared with 2-fold for SDHACU (Vaca and Pilar, 1979). Recent evidence from *Torpedo* electromotor synapses is consistent with the sequence of events suggested above; after exhaustive stimulation, recovery of the vesicular population is paralleled, with a slight delay, by recovery of ACh — the newly synthesized ACh is preferentially sequestered in freshly reformed vesicles (Suszkiw and Whittaker, this volume).

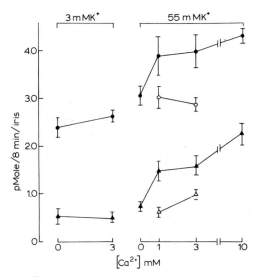

Fig. 2. Effect of preincubation conditions on subsequent [³H]Ch uptake and [³H]ACh synthesis. Irises were preincubated at 37°C in Tyrode solution containing normal, 3 mM [K⁺] (left) or elevated, 55 mM [K⁺] (right) at various concentrations of divalent cations for 6 min. They were then washed for 1 min in normal Tyrode's solution (with 3 mM Ca²⁺) and transferred to normal Tyrode's solution containing 0.7 μM [³H]Ch for 8 min. Finally they were homogenized and ACh and total uptake determined as in Table II. In the graph it can be seen that preincubation with a depolarizing solution (55 mK⁺) greatly increases the Ch uptake and ACh synthesis, and that the presence of Ca²⁺ at physiological levels is needed for this effect to occur. Large concentration of Ca²⁺ (10 mM) further increase the synthesis of ACh. The effects of Ca²⁺ are almost totally blocked by high Mg²⁺ concentration, which is reported to inhibit Ca²⁺ influx into nerve terminals (Blaustein, 1975).
●, [³H]Ch uptake; ▲, [³H]ACh synthesis following preincubation in Tyrode solution containing 1 mM Mg²⁺. ○, [³H]Ch uptake; △, [³H]ACh synthesis following preincubation in Tyrode solution containing 20 mM Mg²⁺. Vertical bars, ±S.E.M.

Although, in the ciliary nerve terminals, ACh release appears to be the major prerequisite for acceleration of SDHACU and ACh synthesis, other factors may be involved. A small proportion of the activation of SDHACU is dependent on the influx of Na⁺ through a TTX-sensitive channel during the conditioning depolarization (Vaca and Pilar, 1979). Elevated [Mg²⁺] allows an activation of SDHACU in synaptosomes (Murrin et al., 1977) and uptake and acetylation of Ch analogues in cat SCG (Collier and Ilson, 1977) even at concentrations which greatly reduce ACh release. These findings can be explained, based on energetic considerations, if an after-hyperpolarization of the nerve terminal membrane

TABLE IV

ACh CONTENT IN CHICK IRIS AFTER DEPOLARIZATION

Preincubation	Incubation	ACh pmoles	P
Normal Tyrode	Tyrode	28.0 ± 4.7	
55 mM K⁺	Tyrode	26.9 ± 2.8	NS

Irises were dissected out and preincubated in the indicated solutions for 10 min at 37°C. Then they were transferred to normal Tyrode solution with 5 μM Ch and incubated for an additional 10 min. Afterwards their content of ACh was determined as in Table III.

104

can be demonstrated. Either a Na⁺-activated electrogenic pump (Thomas, 1972) or a divalent cation-activated K⁺ conductance (Meech, 1976) would increase SDHACU by increasing the driving force for Na:Ch co-transport. Post-tetanic hyperpolarization has been observed in cultured ciliary neurons (J.B. Tuttle, personal communication).

One might expect that these exquisitely coordinated mechanisms for increasing the rate of ACh synthesis might be quite efficient in rapidly replenishing ACh during repetitive activity. This turns out to be true. After more than 40% of the endogenous ACh has been depleted by depolarization with high [K⁺], return to normal medium containing physiological levels of Ch results in a rapid return of ACh to control levels (Table IV). Similarly, Potter (1970) estimated that during nerve stimulation, synthesis rates were sufficient to replenish 35% of the store size in 5 min.

SUMMARY

It is possible to incorporate these observations into a relatively simple model of the regulation of ACh synthesis (Fig. 3). The diagram shows that Ca²⁺ influx

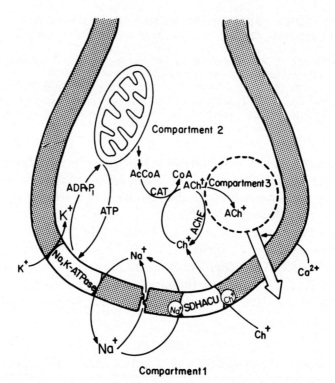

Fig. 3. Regulation of ACh synthesis in nerve terminals during electrical stimulation (after Fonnum, 1973): Compartment 1, extracellular space; compartment 2, cytoplasm; compartment 3, protected ACh fraction (probably vesicular). Na, K-ATPase, energized by ATP, maintains a low intracellular Na⁺ concentration against its electrochemical gradient. The Na⁺ gradient and the membrane potential drive a Na⁺ coupled Ch influx represented by the Na⁺ dependent high affinity Ch uptake system (SDHACU). After the action potential Ca²⁺ influx results in release of ACh from compartment 3, and Na⁺ entry through a TTX sensitive channel, activates Na, K-ATPase.

accompanying depolarization elicits release of ACh from compartment 3 (probably vesicular) thus driving translocation of cytoplasmic ACh (compartment 2) into the vesicles. This in turn stimulates ACh synthesis by mass action, with a consequent reduction in cytoplasmic Ch. Therefore, the Ch chemical gradient is reduced across the nerve membrane, and this brings about a further increase in its uptake. The Na^+ entry results in an activation of Na, K-ATPase, with a consequent Na^+ extrusion and probably a membrane hyperpolarization. Thus, the Na^+ gradients is maintained and the potential difference across the terminal membrane may be increased, with an increased driving force for the coupled influx of Ch. Consequently, more intracellular Ch becomes available for ACh synthesis, and ChAT converts Ch to ACh, limited by AChE activity in compartment 2.

The critical role of SDHACU in the function of cholinergic neurons is well established. Beyond its essential role in maintaining ACh levels, it may be of deeper significance. Liley and North (1953) suggested, in mathematically explicit terms, that the availability of precursor might not only limit the size of the releasable pool of transmitter but also account for the elevated release of transmitter during post-tetanic potentiation. We are currently investigating this possibility.

ACKNOWLEDGEMENTS

The authors wish to thank their collaborators in this study, Dr. J. Suszkiw and Dr. R. Beach for their discussions and suggestions during the different stages of this work, and Ms. Pat Vaillancourt and Ms. Donna Woolam for their unfailing secretarial help. This review was based on research supported by NIH Grant NS-10338 and The University of Connecticut Research Foundation.

REFERENCES

Beach, R., Vaca, K. and Pilar, G. (1978) Metabolic requirements for high affinity choline uptake and acetylcholine synthesis in nerve terminals at the neuromuscular junction, submitted for publication.

Birks, R.I. (1963) The role of sodium ions in the metabolism of acetylcholine. *Canad. J. Biochem. Physiol.*, 41, 2573–2597.

Birks, R.I. and MacIntosh, F.C. (1961) Acetylcholine metabolism of a sympathetic ganglion. *Canad. J. Biochem. Physiol.*, 39, 787–827.

Blaustein, M.P. (1975) Effects of potassium, veratridine and scorpion venom on calcium accumulation and transmitter release by nerve terminals in vitro. *J. Physiol. (Lond.)*, 247, 617–655.

Brown, G.L. and Feldberg, W. (1936) The acetylcholine metabolism of a sympathetic ganglion. *J. Physiol. (Lond.)*, 88, 265–283.

Chang, C.C. and Lee, C. (1970) Studies on the [^{3}H]choline uptake in rat phrenic nerve diaphragm preparations. *Neuropharmacology*, 9, 223–233.

Collier, B. and Ilson, D. (1977) The effect of preganglionic nerve stimulation on the accumulation of certain analogues of choline by a sympathetic ganglion. *J. Physiol. (Lond.)*, 264, 489–509.

Collier, B. and MacIntosh, F.C. (1969) The source of choline for acetylcholine synthesis in a sympathetic ganglion. *Canad. J. Physiol. Pharmacol.*, 47, 127–136.

106

Fonnum, F. (1973) Recent developments in biochemical investigations of cholinergic transmission. *Brain Res.* 62, 497–507.

Geck, P. and Heinz, E. (1976) Coupling in secondary transport. Effect of electrical potentials on the kinetics of ion linked co-transport. *Biochim. Biophys. Acta,* 443, 49–63.

Guyenet, P., Lefresne, P., Rossier, J., Beaujouan, J.C. and Glowinski, J. (1973) *Molec. Pharmacol.,* 9, 630–639.

Haga, T. and Noda, H. (1973) Choline uptake systems of rat brain synaptosomes. *Biochim. Biophys. Acta,* 291, 564–575.

Katz, B. (1969) The Release of Neural Transmitter Substances. C.C. Thomas, Springfield, pp. 1–55.

Liley, A.W. and North, K.A.K. (1953) An electrical investigation of effects of repetitive stimulation on mammalian neuromuscular junction. *J. Neurophysiol.,* 16, 509–526.

Mautner, H.G. (1977) Choline acetyltransferase. *CRC Critical Rev. Biochem.,* 5, 341–370.

Meech, R.W. (1976) Intracellular calcium and the control of membrane permeability. *Symp. Soc. exp. Biol.,* 30, 161–191.

Murrin, L.C., DeHaven, R.N. and Kuhar, M.J. (1977) On the relationship between [^{3}H]choline uptake activation and [^{3}H]acetylcholine release. *J. Neurochem.,* 29, 681–687.

Murrin, L.C. and Kuhar, M.J. (1976) Activation of high-affinity choline uptake in vitro by depolarizing agents. *Molec. Pharmacol.,* 12, 1082–1090.

Quastel, J.H. (1965) Molecular transport at cell membranes. *Proc. roy. Soc. B.,* 163, 169–196.

Pert, C.B. and Snyder, S.H. (1974) High affinity transport of choline into the myenteric plexus of guinea-pig intestine. *J. Pharmacol. exp. Ther.,* 191, 102–108.

Pieklik, J.R. and Guynn, R.W. (1975) Equilibrium constants of the reactions of choline acetyltransferase, carnitine acetyltransferase, and acetylcholinesterase under physiological conditions. *J. biol. Chem.,* 250, 4445–4450.

Pilar, G. and Vaughan, P. (1969) Electrophysiological investigations of the pigeon iris neuromuscular junctions. *Comp. Biochem. Physiol.,* 29, 51–72.

Potter, L.T. (1970) Synthesis, storage and release of [^{14}C]acetylcholine in isolated rat diaphragm muscles. *J. Physiol. (Lond.),* 206, 145–166.

Schultz, S.G. and Curran, P.F. (1970) Coupled transport of sodium and organic solutes. *Physiol. Rev.,* 50, 637–718.

Shea, P.A. and Aprison, M.H. (1973) An enzymatic method for measuring picomole quantities of acetylcholine and choline in CNS tissue. *Anal. Biochem.,* 56, 165–177.

Simon, J.R. and Kuhar, M.J. (1976) High affinity choline uptake: ionic and energy requirements. *J. Neurochem.,* 27, 93–99.

Simon, J.R., Atweh, S. and Kuhar, M.J. (1976) Sodium-dependent high affinity choline uptake: a regulatory step in the synthesis of acetylcholine. *J. Neurochem.,* 26, 909–952.

Suszkiw, J.B. and Pilar, G. (1976) Selective localization of a high affinity choline uptake system and its role in ACh formation in cholinergic nerve terminals. *J. Neurochem.,* 26, 1133–1138.

Suszkiw, J. and Whittaker, V. (1979) Role of vesicle recycling in vesicular storage and release of ACh in Torpedo electroplaque synapses. In *The Cholinergic Synapse,* S. Tuček (Ed.), *Progr. Brain Res., Vol. 49,* Elsevier/North-Holland Biomedical Press, Amsterdam, pp. 153–162.

Thomas, R.C. (1972) Electrogenic sodium pump in nerve and muscle cells. *Physiol. Rev.,* 52, 563–594.

Vaca, K. and Pilar, G. (1979) Mechanism controlling choline transport and acetylcholine synthesis in motor nerve terminals during electrical stimulation, *J. gen. Physiol.,* in press.

Yamamura, H.I. and Snyder, S.H. (1973) High affinity transport of choline into synaptosomes of rat brain. *J. Neurochem.,* 21, 1355–1374.

Cholinergic False Transmitters

B. COLLIER, P. BOKSA and S. LOVAT

*Department of Pharmacology and Therapeutics, McGill University,
Montreal, Quebec (Canada)*

INTRODUCTION

A false neurotransmitter is a substance which is not normally present in a nerve ending, but which can accumulate in the sites that are usually occupied by the physiological neurotransmitter, and can be released by stimuli that normally release the physiological transmitter. The criteria that are considered essential to the identification of a compound as a false transmitter are usually accepted to be those defined for the adrenergic nervous system by Kopin (1968): presence, release by a process requiring calcium ions, and storage in those subcellular site or sites that contain the physiological transmitter. It is usual, but not essential, that a false transmitter has a quantitatively different, but a qualitatively similar, pharmacological action to the normal transmitter; most false transmitters are less active than is the physiological transmitter, and this is so for all cholinergic false transmitters that have yet been identified. If the false "transmitter" is close to inactive it appears still to be considered a false transmitter, but as it is unlikely to transmit information, it maybe requires some other name such as a transmitter diluent.

The idea that false transmitters could be formed in cholinergic nerves is far from new, and was clearly expounded more than 30 years ago (Keston and Wortis, 1946) for 2-hydroxyethyl triethylammonium (triethylcholine), the triethyl-analogue of choline. The suggestion of Keston and Wortis was based upon the observation that triethylcholine prevented the contractile effect of choline, but not of acetylcholine (ACh), on the frog's rectus abdominis preparation; their interpretation was that choline had to be acetylated to be active and that triethylcholine interfered with' this. This experiment, of course, could not distinguish, e.g. between an action of triethylcholine to block choline uptake from one involving false transmitter formation. Indeed, it was not known at that time that triethylcholine could act as a substrate for choline acetyltransferase; this observation was made in 1956 by Burgen et al., who demonstrated the enzymic acetylation of a number of choline analogues including mono-, di-, and triethylcholine.

The first clear demonstration that a choline analogue could form a false neurotransmitter now appears to have been that of Reitzel and Long (1959) showing that choline or its monoethyl analogue, monoethylcholine, could antagonize a hemicholinium-induced block of neuromuscular transmission; monoethylcholine was one sixth as effective as was a similar concentration of choline. Although the authors did not make this conclusion, the most likely explanation of their results is that monoethylcholine was synthesized to a false transmitter, acetylmonoethylcholine, and that this was released as a weak neurotransmitter; acetylmonoethylcholine is about one fifth as active as is ACh on nicotinic receptors (Holton and Ing, 1949). Despite these early indications that false cholinergic transmitters could be formed, direct demonstration that a choline analogue could be acetylated and released in a cholinergic synapse appears not to have been made until it was shown for triethylcholine much later (Ilson and Collier, 1975).

FORMATION OF FALSE TRANSMITTERS

In the adrenergic nervous system, false neurotransmitters can be introduced into nerve terminals in at least 3 ways (reviewed by Muscholl, 1972): synthesis of the false transmitter within the nerve ending from a precursor; uptake of the false transmitter itself by the neuronal membrane amine uptake mechanism; and accumulation of a false transmitter following inhibition of enzymes involved in the normal biosynthesis and metabolism of noradrenaline. In the cholinergic nervous system, the only false transmitters that have so far been identified are synthesized from a false precursor, and all of the false precursors have been analogues of choline. The possibilities of creating false transmitters from analogues of acetyl-CoA have not yet been explored. Similarly, there is no example of the direct formation of a false transmitter as the result of administering an ACh-like compound, and this is unlikely to occur, because ACh itself is not accumulated by cholinergic nerve endings, and ACh re-use appears not to occur even when its hydrolysis is prevented (Katz et al., 1973; Kuhar and Simon, 1974).

In order for a choline analogue to form a false transmitter, the analogue has to gain entry to the nerve terminal, a process that presumably operates most efficiently via choline's high affinity transport mechanism. It must then be acetylated by choline acetyltransferase, which is considered to be a cytoplasmic enzyme (see Fonnum, 1975) and, if releasable transmitter is stored in synaptic vesicles, the acetylated analogue must be accumulated from cytosol by vesicles. Finally, the compound must be released along with or instead of acetylcholine. Each of the above steps that are involved in transmitter synthesis, storage and release, presumably have some specificity, and a false transmitter can only be formed if that specificity is relative rather than absolute.

RATIONALE FOR STUDYING FALSE NEUROTRANSMITTERS

There are at least two objectives in studies of false neurotransmitter turnover. (1) There is the possibility that such compounds, or their precursors, can provide experimental tools with which to dissect the sequential mechanisms associated with ACh turnover. For example, a substrate for the choline transport mechanism that is a poor substrate for choline acetyltransferase can dissociate transport and acetylation, processes that are normally quite tightly coupled (e.g. Barker and Mittag, 1975; Lefresne et al., 1975; Suszkiw and Pilar, 1976); this sort of approach allows study of some of the factors that control choline uptake by cholinergic nerve endings (e.g. Collier and Ilson, 1977). Alternatively, advantage can be taken of different rates of metabolism of choline and its analogues to differentially label subfractions of transmitter stores (e.g. von Schwarzenfeld, 1978). (2) There is the possibility that false transmitters can be used pharmacologically to alter synaptic function; this can be particularly useful in the central nervous system (see Glick et al., 1975 for an example) where the functional significance of cholinergic activity is still not at all clear.

SPECIFICITY OF CHOLINE UPTAKE

The first barrier in an intact nerve terminal to the formation of a cholinergic false transmitter is the imerpeability of the cell membrane to charged molecules, a problem that is circumvented for choline by the high affinity uptake mechanism. The process is not entirely specific for choline and other compounds can serve as substrates. Compounds (Table I) that are now known to be transported into cholinergic nerve terminals by the choline uptake mechanism are: sulfocholine (high affinity not tested), monoethylcholine, diethylcholine, triethylcholine, pyrrolidinecholine, and homocholine (Frankenberg et al., 1973; Barker and Mittag, 1975; Collier et al., 1977). Compounds that are now known to inhibit choline transport without being transported themselves include hemicholinium (Collier, 1973; Slater and Stonier, 1973) and N-methyl-3-quinucleodol (Dowdall, 1978). ACh (Katz et al., 1973; Kuhar and Simon, 1974) and 4-hydroxybutyl trimethylammonium (Collier, unpublished data) are not transported instead of choline by nerve terminals. In general, the choline carrier appears to require the quaternary group of choline; there can be considerable N-alkyl substitution as well as replacement of N by S, without much loss of affinity; the hydroxyl group of choline appears to be essential

TABLE I

UPTAKE OF CHOLINE OR OF CHOLINE ANALOGUES BY MAMMALIAN BRAIN SYNAPTOSOMES

Compound	"High affinity uptake" K_m (μM)	"Low affinity uptake" K_m (μM)
Choline	1–3	50–120
Monoethylcholine	2.7	
Diethylcholine	2.5	
Triethylcholine	5.0	15
Pyrrolidinecholine	5.5	
Homocholine	3.0	14.5
Sulfocholine		160

(but see Barker (1978) for a possible exception); and there can be some, but not much, increase of the spatial separation of the -N and -OH groups (c.f. homocholine (3-hydroxypropyltrimethylammonium) and 4-hydroxybutyl-trimethylammonium).

SPECIFICITY OF CHOLINE ACETYLATION AND THE SYNTHESIS OF FALSE TRANSMITTERS

It has been clear for a long time that choline acetyltransferase can acetylate compounds other than choline (Burgen et al., 1956; Dauterman and Mehrotra, 1963; Hemsworth and Smith, 1970). Of the compounds identified above (Table I) as being substrates for the choline uptake mechanism, all except homocholine appear to be substrates for choline acetyltransferase, although some are better than others (Table II). However, there is some question about whether the in situ specificity of choline acetyltransferase is the same as that determined by in vitro tests (Barker and Mittag, 1975). Certainly homocholine, which has never been shown to be acetylated in vitro (Burgen et al., 1956; Dauterman and Mehrotra, 1963; Currier and Mautner, 1974; Collier et al., 1977) is acetylated in vivo (Collier et al., 1977; Dowdall, 1978). The significance of this apparent difference between in vitro and in situ acetylation is not clear, but homocholine must be added to monoethyl-, diethyl-, triethyl-, pyrrolidine-, and sulfo-cholines as likely precursors to false transmitters.

The synthesis of ACh is well known to be accelerated by stimuli that induce transmitter release (e.g. Mann et al., 1938; Birks and MacIntosh, 1961; Collier and MacIntosh, 1969; Potter, 1970; Grewaal and Quastel, 1973), as if turnover at rest is depressed by some control mechanism that is relieved during synaptic activity so that transmitter synthesis can replenish transmitter release; the mechanism of this control is uncertain. The in situ acetylation of choline analogues is similarly increased during synaptic activity. This has been clearly shown for triethylcholine and homocholine in ganglion (Ilson et al., 1977;

TABLE II

ACETYLATION OF CHOLINE ANALOGUES BY CHOLINE ACETYLTRANSFERASE IN VITRO

Compound	Acetylation relative to choline		References *
	K_m (choline = 1)	V_{max} (choline = 100)	
Monoethylcholine	2–7	67–100	1,3,6
Diethylcholine	40–60	70	3,6
Triethylcholine	150–170	70	6,7
Pyrrolidinecholine	40	53	6
Homocholine	–	–	1,2,5,8
Sulfocholine	7	88	4

* References: (1) Burgen et al. (1956); (2) Dauterman and Mehrotra (1963); (3) Hemsworth and Smith (1970); (4) Frankenberg et al. (1973); (5) Currier and Mautner (1974); (6) Barker and Mittag (1975); (7) Mann and Hebb (1975); (8) Collier et al. (1977).

Collier et al., 1977) and for pyrrolidinecholine in Torpedo (Zimmermann and Dowdall, 1977) and in myenteric plexus of guinea-pig gut (Kilbinger, 1977). The magnitude of the increased synthesis of acetylated analogues is difficult to compare to that of acetylcholine in the different tissues because experimental conditions have not always been maintained constant, but it appears unlikely that there are large quantitative differences between the analogues and choline in any one tissue. This is clearly so for pyrrolidinecholine in Torpedo, where tissue was simultaneously exposed to [^{14}C]choline and [^{3}H]pyrrolidecholine: stimulation increased ACh synthesis about 5-fold and increased acetylpyrrolide-choline synthesis about 6-fold (Zimmermann and Dowdall, 1977). Similarly, in ganglia, the 6-fold increase in acetyltriethylcholine synthesis induced by 20 Hz stimulation for 60 min in the presence of 10^{-5} M triethylcholine (Ilson et al., 1977) compares well with the 5-fold increase in ACh synthesized from choline under very similar conditions (Collier and MacIntosh, 1969). The more modest (50%) increase of acetylpyrrolidinecholine synthesis induced by stimulation in myenteric plexus (Kilbinger, 1977) is comparable to the small increase of ACh formation induced by stimulating that tissue (Szerb, 1975).

Thus, as far as one can yet analyse this question, the factors that control ACh synthesis appear to exert a similar controlling influence upon the synthesis of acetylated analogues of choline. This aspect of false transmitter turnover might be significant to studies of the control of neurotransmitter synthesis, for the different choline analogue appear to differ from choline in their affinity for choline acetyltransferase, and certain choline analogues affect endogenous ACh levels. For example, the similar increase of triethylcholine and choline acetylation during stimulation argues against product inhibition as being significant to the control of choline acetyltransferase activity (whether directly (Kaita and Goldberg, 1969), or involving isoenzymes (Singh et al., 1975) or anions (Rossier et al., 1977)), because stimulation in the presence of triethylcholine induces a considerable depletion of ACh stores, which are not nearly replaced by acetyltriethylcholine (Ilson et al., 1977), whereas ACh stores during stimulation in choline's presence are maintained (Birks and MacIntosh, 1961).

The in situ acetylation of homocholine is unexpected considering its apparent lack of acetylation by choline acetyltransferase in vitro. There are several possible explanations for this, and the following are offered as examples. Firstly, in situ acetylation might occur by the action of a membrane-bound choline acetyltransferase (which likely exists (Fonnum, 1975) with substrate specificity different from non membrane-bound enzyme; although there is no evidence that the two forms of enzyme differ with respect to choline acetylation (Fonnum, 1968; Kuczenski et al., 1975), homocholine has not been tested. Secondly, the in situ acetylation might occur by the action of a choline acetyltransferase isoenzyme that does not survive solubilization; isoenzymes have been clearly demonstrated (Malthe-Sørenssen and Fonnum, 1972), but their affinity for homocholine has not been tested, although they do not differ in their affinity for choline. Thirdly, a substrate presented to choline acetyltransferase via the choline transport mechanism might be preferentially acetylated if transport and acetylation are linked in some way as suggested by Barker and Mittag (1975). It is clear that further experimentation is necessary to approach these important problems.

112

RELEASE OF FALSE TRANSMITTERS

Of the six compounds identified above as likely to form false transmitters, all except sulfocholine have been tested to determine whether release can be induced. The result of a typical experiment designed to test this for diethylcholine using the cat's superior cervical ganglion is illustrated by Fig. 1. In this experiment the ganglion was first perfused for 60 min with medium containing eserine and [^{3}H]diethylcholine (10^{-6} M), during which time the preganglionic trunk was stimulated (20 Hz) to increase transmitter turnover. Following this, perfusion was continued with eserine-Krebs solution (no diethylcholine), the ganglion was at rest, and after 20 min the preganglionic nerve was again stimulated. Clearly, stimulation evoked the release of radioactive material, and assay of this material showed that the extra radioactivity released during stimulation was acetyldiethylcholine, whereas in the absence of nerve stimulation the resting efflux of radioactivity was unchanged diethylcholine. This kind of result has been obtained in similar experiments using monoethylcholine, triethylcholine, pyrrolidinecholine, and homocholine (Collier et al., 1976; Ilson et al., 1977; Collier et al., 1977). With all compounds, nerve stimulation releases acetylated product and the efflux of unchanged analogue is never clearly altered by stimulation. Thus, the transmitter release mechanism, at least in ganglia, is specific for acetylated choline analogues and does not accommodate the amino alcohols. The amount of acetylated choline analogues accumulated and released by ganglia varies somewhat between analogues and considerably between different experimental protocols used in the studies. However, expressing release as a proportion of stored false transmitter normalizes the difference and should allow releasability to be compared. This comparison (Fig. 2) shows that all of the compounds so far studied are as available for release as is

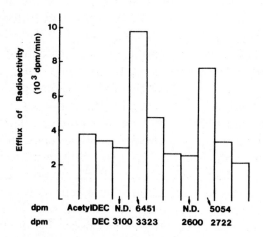

Fig. 1. The release of acetyldiethylcholine (acetylDEC) from cat superior cervical ganglion. The ganglion had been exposed to [^{3}H]diethylcholine (DEC) for 60 min during preganglionic nerve stimulation and then washed by perfusion (no diethylcholine, no stimulation) for 20 min. After this, the perfusate was collected in 1-min periods. Each vertical column represents radioactivity collected in a 1-min collection period and during the 4th and the 8th of these the preganglionic nerve was stimulated. The values below indicate the content of DEC and acetylDEC in the 3rd, 4th, 7th, and 8th samples.

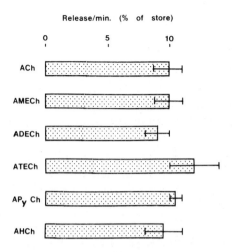

Fig. 2. Relative releasability of true and false transmitters from cat superior cervical ganglion. Release is expressed as the % of the stored compound released per min during preganglionic nerve stimulation (20 Hz). ACh is acetylcholine; AMECh is acetylmonoethylcholine; ADECh is acetyldiethylcholine; ATECh is acetyltriethylcholine; APyCh is acetylpyrrolidinecholine; and AHCh is acetylhomocholine. For details of the experiments that generated many of these values see Collier et al. (1976, 1977) and Ilson et al. (1977).

ACh. Thus, as long as the compound is acetylated, the release mechanism shows no structural specificity.

The calcium-dependence of the release of all the compounds included in Fig. 2 has been tested in ganglia, and in the absence of ionized calcium, nerve stimulation does not evoke the release of these compounds. Thus, in terms of synthesis and release, acetylmonoethyl-, acetyldiethyl-, acetyltriethyl-, acetyl-pyrrolidine-, and acetylhomo-choline fulfill the requirements of a false trans-mitter.

The synthesis and release of some of these false transmitters has also been shown in other tissues. Thus pyrrolidinecholine is acetylated and released by cholinergic nerve endings in Torpedo electric organ (Zimmermann and Dowdall, 1977) and myenteric plexus of guinea-pig ileum (Kilbinger, 1977); and brain slices have been shown to synthesize and release acetyldiethylcholine (Boksa, unpublished) and acetyltriethylcholine (Ilson et al., 1977). In these tissues, as in the experiments on ganglia, only acetylated choline analogue, and not un-changed analogue is released, and, where test has been made, false transmitter and true transmitter appear equally available for release (Zimmermann and Dowdall, 1977).

Thus, in cholinergic nerve terminals, the transmitter release mechanism appears to be specific for acetylated compounds, but to be non-specific with respect to the chemical structure of the acetylated choline analogues. The now classical concept that releasable transmitter is sequestered within synaptic vesicles and is released therefrom by exocytosis lacks definitive neurochemical evidence, and has not escaped unchallenged (see discussion in reviews by Mac-Intosh and Collier, 1976; MacIntosh, 1978). The principle alternative to the vesicle hypothesis is that transmitter escapes the nerve terminal by membrane gates, the opening of which for a predetermined time allows quantal ACh release to occur from the nerve ending's cytosol. Such a membrane gate or pore might be expected to have a degree of chemical specificity, and the evidence that release shows little such specificity is at least circumstantial evidence against the hypothesis.

However, if the equal releasability of false neurotransmitters is to be used to favour the vesicle hypothesis, one has to postulate that the mechanism by which newly synthesized acetylated compounds is sequestered into releasable stores shows little structural specificity.

Little is known about the accumulation of ACh from the presumed site of synthesis in the cytosol to its presumed site of storage in synaptic vesicles, but if it involves an active uptake process it might be expected to show chemical selectivity. If the transfer of ACh from the cytoplasm to the synaptic vesicles is electrogenic, as has recently been suggested (Carpenter and Parsons, 1978), the process would show little chemical specificity, but this mechanism would not distinguish between acetylated and non-acetylated choline analogues, unless the synthesis and vesicular uptake of ACh were coupled in some way as has been suggested from other evidence (e.g. Collier and Katz, 1971; Hebb, 1972; Aquilonius et al., 1973).

Some of these problems can be approached by studying the subcellular distribution of the false neurotransmitters, and the next section summarizes the information available so far on this point.

SUBCELLULAR LOCALIZATION OF FALSE TRANSMITTERS

The last section summarized the evidence suggesting that, on the basis of synthesis and release, the following might be false neurotransmitters: acetyl-monoethyl-, acetyldiethyl-, acetyltriethyl-, acetylpyrrolidine-, and acetyl-homo-choline. There is some information about the subcellular localization of all except acetylmonoethylcholine.

The most complete studies have used pyrrolidinecholine. Zimmermann and Dowdall (1977) demonstrated unequivocally that Torpedo electric organ exposed to pyrrolidinecholine synthesizes acetylpyrrolidinecholine and that this false transmitter is stored in synaptic vesicles. However, they did not measure acetylpyrrolidinecholine in the cytoplasmic fraction prepared from this tissue, so it is unclear whether or not the proportion of false transmitter in cytosol and vesicle fractions is different from that of ACh, but it appears unlikely that there was much difference for the ratio of false to true transmitter in vesicles and in total tissue was similar.

Von Schwarzenfeld (1978) injected [^{14}C]choline and [^{3}H]pyrrolidinecholine intraventicularly to guinea pig, and measured the ratio of acetylated products in subcellular fractions prepared from cerebral cortex taken from animals killed at varying times following the injection. She found the false transmitter in all subfractions that contained ACh, but the ratio of [^{14}C]- to [^{3}H]acetyl products was different in cytoplasm, monodispersed synaptic vesicles (D vesicles by terminology of Whittaker and Barker (1972)) and in disrupted synaptosomes (H fraction). The meaning of these results is not yet entirely clear but they do indicate that the relationship between intracellular compartments is likely more complicated than can be explained by considering only two transmitter pools, cytoplasmic and vesicular. The results of von Schwarzenfeld clearly show that metabolic heterogeneity exists for synaptic vesicles, an idea that was suggested earlier by Barker et al., (1972) and that has recently been supported by experiments on Torpedo (Zimmermann and Denston, 1977). The ratio of true to false transmitter achieved in the different sub-fractions in these experiments by von Schwarzenfeld was highest in fraction D, intermediate in fraction O (cytoplasmic), and lowest in fraction H; if fraction H contains the metabolically most active vesicles, as postulated by Barker et al. (1972), this result presumably indicates either the preferential uptake of acetylpyrrolidinecholine or the preferential release of ACh by these vesicles. The evidence discussed in the last section concluded that the false transmitters were equally available for release in ganglia, and if this were so in brain, the result of von Schwarzenfeld would indicate better uptake of the false transmitter than of ACh by vesicles of fraction H, a somewhat surprising result. Obviously, much more work is necessary to clarify this point, but whatever the interpretation, the comparison of the ratio of false and true transmitter in subcellular fractions to that of released transmitters probably offers a new biochemical approach to the problem of studying ACh release mechanisms.

Our own attempts to determine the subcellular distribution of false transmitters were unsophisticated when compared to the studies referred to above. They have been confined to measuring the proportion of acetylated choline or analogue of choline present in cytoplasmic and vesicle fractions prepared from slices of brain exposed to the appropriate precursor. The results obtained so far are summarized in Table III, which shows that the acetylated choline analogues can be detected in both vesicle and cytosol fractions; the relative distribution of false transmitters and ACh is not dramatically different, although acetylhomocholine appears to be relatively less well accumulated in vesicles than is ACh.

Thus, the available evidence suggests that the false transmitters have a similar subcellular distribution to the normal transmitter. There is as yet no ACh analogue that uniquely localizes to any one of the subcellular fractions containing ACh, but there may well be more subtle differences in the relative

TABLE III

SUBCELLULAR DISTRIBUTION OF ACETYLCHOLINE AND ANALOGUES IN RAT BRAIN EXPOSED TO CHOLINE OR CHOLINE ANALOGUE

Substrate	Acetylated compound in soluble fraction ('free') [*] / Acetylated compound in particulate fractions ('bound')
Choline [**]	3.1 ± 0.61
Triethylcholine [**]	2.9 ± 0.47
Homocholine [**]	4.8 ± 0.83
Triethylcholine [***]	2.3 ± 0.46
Diethylcholine [***]	2.0 ± 0.35

[*] Results are mean ± S.E.M. of 4 experiments.
[**] Tissue not stimulated.
[***] Tissue stimulated.

TABLE IV

UNCHAGED CHOLINE OR CHOLINE ANALOGUE IN NERVE-ENDING BOUND FRAC-
TION PREPARED FROM LYSED SYNAPTOSOMES MADE FROM RAT BRAIN EX-
POSED TO THESE COMPOUNDS

Substrate	% Unmetabolized compound
Choline	69 ± 4.3
Triethylcholine	76 ± 2.3
Homocholine	69 ± 2.3

distribution of false and true transmitters between compartments. These differences might well be explored to provide important information about transmitter turnover and release.

In most of these studies on the subcellular storage of false transmitters that
have been accumulated in tissue following its exposure to a radioactive
precursor, only part of the radioactivity is recovered as acetylated compound
(e.g. Table IV). The association of appreciable unchanged choline analogue
with the vesicle fraction (see also von Schwarzenfeld, 1978) is particularly
interesting in view of the apparent unreleasability of non-acetylated analogues,
but the nature of the binding of unchanged analogues to vesicles has yet to be
properly studied.

Although the biochemical studies of false transmitter subcellular distribution
clearly show the presence of ACh and ACh analogues in synaptic vesicles, they
do not indicate whether the true and the false transmitter co-exist in a single
vesicle, or whether some vesicles contain only ACh and others only ACh
analogue. This question is being approached by the elegant biophysical analysis
of Large and Rang (1978; see also this volume) making use of their observation
(Colquhoun et al., 1977) that the action of the false transmitter, acetylmonoethylcholine, at the neuromuscular junction can be distinguished from that of
ACh by analysis of the time constant of decay of end-plate currents induced by
the two drugs. After partial replacement of ACh by acetylmonoethylcholine,
miniature end-plate potentials were recorded and shown to result from the
release of *both* true and false transmitter. In other words a quantum of transmitter, and in terms of the classical hypothesis, a vesicle, must contain not
either ACh or acetylmonoethylcholine, but both together. Thus, there is
either rapid mixing of true and false transmitter between vesicles, or quanta are
formed upon demand from a store of pre-formed transmitters. A similar conclusion was made previously by Elmqvist and Quastel (1965) from their analysis of hemicholinium's effect on quantum size.

DISPOSITION OF FALSE TRANSMITTERS

Except for acetylhomocholine, there is no evidence to suggest that the disposition of cholinergic false transmitters differs significantly from that of ACh
after their release from nerve terminals. All of the compounds identified in the
preceding sections as false transmitters are substrates for acetylcholinesterase,
although the rate of hydrolysis of some of them is slower than that of ACh and

TABLE V

HYDROLYSIS OF ACETYLCHOLINE ANALOGUES BY ACETYLCHOLINESTERASES

Compound	Relative rate (ACh = 100)	References [*]
Acetylmonoethylcholine	100	1,2
Acetyldiethylcholine	96	1
Acetyltriethylcholine	94	1
Acetylpyrrolidinecholine	70	2
Acetylhomocholine	4	3

[*] References: (1) Holton and Ing (1949); (2) Collier et al. (1976); (3) Collier et al. (1977).

acetylhomocholine is a poor substrate (Table V). Acetylhomocholine differs from ACh also in that there appears to be no inhibition of acetylcholinesterase by excess substrate (Mehrota and Dauterman, 1963), and this could possibly be used to assess the physiological significance, if any, of the phenomenon. The catalytic potential of acetylcholinesterase is such (see Silver, 1974) that even the slow rate of hydrolysis of acetylhomocholine may well be enough to dispose of released compound, but this has not been properly tested. If released acetylhomocholine is not fully hydrolysed by acetylcholinesterase, it likely escapes from the synaptic space by diffusion. The possibility that false transmitters might be disposed of by uptake mechanisms has not been tested but is not likely considering that ACh uptake appears not to function to remove released transmitter even when acetylcholinesterase is inactivated (Katz et al., 1973).

PHARMACOLOGICAL ACTIONS OF FALSE TRANSMITTERS

The ACh-like effects of the compounds identified above as cholinergic false transmitters are summarized in Table VI. All are less potent than ACh on both muscarinic and nicotinic receptors, but the ratios of the two effects differ for the different compounds, as would be expected. The action of acetylmono-

TABLE VI

PHARMOCOLOGICAL EFFECTS OF CHOLINERGIC FALSE TRANSMITTERS

Compound	Equipotent molar ratio (ACh = 1)		References [*]
	Muscarinic	Nicotinic	
Acetylmonoethylcholine	2–3	5	1,2
Acetyldiethylcholine	400–700	300	1,2
Acetyltriethylcholine	1700–2000	5000	1,2
Acetylpyrrolidinecholine	3–4	15–30	5,6,7
Acetylhomocholine	40–50	5–8	

[*] References: (1) Holton and Ing (1949); (2) Barlow et al. (1963); (3) Curtis and Ryall (1966); (4) Barrass et al. (1970); (5) Cho et al. (1972); (6) Collier et al. (1976); (7) von Schwarzenfeld and Whittaker (1977).

ethylcholine has been analysed in some detail at the neuromuscular junction (Colquhoun et al., 1977) to show that this false transmitter has not only a lower affinity for ACh receptors than has ACh but also activates ionic channels with a shorter lifetime than those activated by ACh; these two differences account for the different potency of the compound.

It is not yet clear whether the main pharmacological effects of all choline analogues upon synaptic transmission or upon behavior (e.g. Glick et al., 1975) in intact animals are due to the incorporation of the compounds into false transmitters or are due to some other action of the drugs. For monoethylcholine, the formation and release of the false transmitter appears to dominate the pharmacological effect on isolated nerve-muscle preparation (Colquhoun et al., 1977); the weaker effect of acetylmonoethylcholine compared to ACh reduced the margin of safety for junctional transmission. A similar reduction of the safety for transmission at nicotinic synapses might be expected for homocholine, but this has not yet been tested. A more significant depressant effect of pyrrolidinecholine might be expected from the lesser potency of its acetyl derivative, but where this has been tested the effect of pyrrolidinecholine was not particularly dramatic (Kilbinger et al., 1976) or was due to a postsynaptic not a presynaptic action (von Schwarzenfeld and Whittaker, 1977). The neuromuscular blocking effect of diethylcholine (Chiou, 1974) and triethylcholine (Bowman and Rand, 1961) appear to be consistent with one of two presynaptic mechanisms: the release of poorly effective false transmitters, or the depletion of endogenous ACh stores; at least in ganglia (Ilson et al., 1977), the effects of triethylcholine appear more likely the result of drug-induced inhibition of ACh synthesis than the result of false transmitter release, although the latter can hardly be expected to assist transmission.

SUMMARY

All putative cholinergic false transmitters so far studied are formed from choline analogues. The criteria of presence, subcellular localization, and release have been demonstrated for acetyldiethylcholine, acetyltriethylcholine, acetylpyrrolidinecholine, and acetylhomocholine. The criteria of presence and release have been demonstrated for acetylmonethylcholine, and the criterion of presence for acetylsulfocholine. All of these presumed false transmitters are less active than is ACh at nicotinic and muscarinic receptor sites, and all can be hydrolysed by acetylcholinesterase. Studies on the uptake, acetylation, storage, release and turnover of choline analogues and the false transmitters formed therefrom are likely to be of some use in clarifying presynaptic cholinergic mechanisms.

ACKNOWLEDGEMENTS

The work of the authors is supported by the Medical Research Council of Canada; P. Boksa holds an MRC Studentship.

REFERENCES

Aquilonius, S.-M., Flentge, F., Schuberth, J., Sparf, B. and Sundwall, A. (1973) Synthesis of acetylcholine in different compartments of brain nerve terminals in vivo as studied by the incorporation of choline from plasma and the effect of pentobarbital on this process. *J. Neurochem.*, 20, 1509–1521.

Barker, L.A. (1978) Synaptosomal uptake of [^{3}H]-N-Me]-N,N,N,-trimethyl-N-propynyl-ammonium. *Fed. Proc.*, 37, 649.

Barker, L.A. and Mittag, T.W. (1975) Comparative studies of substrates and inhibitors of choline transport and choline acetyltransferase. *J. Pharmacol. exp. Ther.*, 192, 86–94.

Barker, L.A., Dowdall, M.J. and Whittaker, V.P. (1972) Choline metabolism in the cerebral cortex of guinea pigs. *Biochem. J.*, 130, 1063–1080.

Barlow, R.B., Scott, K.A. and Stephenson, R.P. (1963) An attempt to study the effects of chemical structure on the affinity and efficacy of compounds related to acetylcholine. *Brit. J. Pharmacol.*, 21, 509–522.

Barrass, B.C., Brimblecomb, R.W., Rich, P. and Taylor, J.V. (1970) Pharmacology of some acetylcholine homologues. *Brit. J. Pharmacol.*, 39, 40–48.

Birks, R. and MacIntosh, F.C. (1961) Acetylcholine metabolism of a sympathetic ganglion. *Canad. J. Biochem. Physiol.*, 39, 787–827.

Bowman, W.C. and Rand, M.J. (1961) Actions of triethylcholine on neuromuscular transmission. *Brit. J. Pharmacol.*, 17, 176–195.

Burgen, A.S.V., Burke, G. and Desbarats-Schonbaum, M.-L. (1956) The specificity of brain choline acetylase. *Brit. J. Pharmacol.*, 11, 308–312.

Carpenter, R.S. and Parsons, S.M. (1978) Electrogenic behavior of synaptic vesicles from Torpedo californica. *J. Biol. Chem.*, 253, 326–329.

Chiou, C.Y. (1974) Studies on action mechanisms of a possible false cholinergic transmitter, (2-hydroxyethyl) methyldiethylammonium. *Life Sci.*, 14, 1721–1733.

Cho, A.K., Jenden, D.J. and Lamb, S.I. (1972) Rates of alkaline hydrolysis and muscarinic activity of some aminoacetates and their quaternary ammonium analogs. *J. Med. Chem.*, 15, 391–394.

Collier, B. (1973) The accumulation of hemicholinium by tissues that transport choline. *Canad. J. Physiol. Pharmacol.*, 51, 491–495.

Collier, B. and Ilson, D. (1977) The effect of preganglionic nerve stimulation on the accumulation of certain analogues of choline by a sympathetic ganglion. *J. Physiol. (Lond.)*, 264, 489–509.

Collier, B. and Katz, H.S. (1971) The synthesis, turnover and release of surplus acetylcholine in a sympathetic ganglion. *J. Physiol. (Lond.)*, 214, 537–552.

Collier, B. and MacIntosh, F.C. (1969) The source of choline for acetylcholine synthesis in a sympathetic ganglion. *Canad. J. Physiol. Pharmacol.*, 47, 127–135.

Collier, B., Barker, L.A. and Mittag, T.W. (1976) The release of acetylated choline analogues by a sympathetic ganglion. *Molec. Pharmacol.*, 12, 340–344.

Collier, B., Lovat, S., Ilson, D., Barker, L.A. and Mittag, T.W. (1977) The uptake, metabolism and release of homocholine: studies with rat brain synaptosomes and cat superior cervical ganglion. *J. Neurochem.*, 28, 331–339.

Colquhoun, D., Large, W.A. and Rang, H.P. (1977) An analysis of the action of a false transmitter at the neuromuscular junction. *J. Physiol. (Lond.)*, 266, 361–395.

Currier, S.F. and Mautner, H.G. (1974) On the mechanism of action of choline acetyltransferase. *Proc. nat. Acad. Sci. (Wash.)*, 71, 3355–3358.

Curtis, D.R. and Ryall, R.W. (1966) Pharmacological studies upon spinal presynaptic fibres. *Exp. Brain Res.*, 1, 195–204.

Dauterman, W.C. and Mehrotra, K.N. (1963) The N-alkyl group specificity of choline acetylase from brain. *J. Neurochem.*, 10, 113–117.

Dowdall, M.J. (1978) Nerve terminal sacs from Torpedo electric organ. In *Cholinergic Mechanisms and Psychopharmacology*, D.J. Jenden (Ed.), Plenum Press, New York, pp. 359–375.

Elmqvist, D. and Quastel, D.M.J. (1965) Presynaptic action of hemicholinium at the neuromuscular junction. *J. Physiol. (Lond.)*, 177, 463–482.

Fonnum, F. (1968) Choline acetyltransferase: binding to and release from membranes. *Biochem. J.*, 109, 389–398.

Fonnum, F. (1975) Recent progress in the synthesis, storage, and release of acetylcholine. In *Cholinergic Mechanisms*, P.G. Waser (Ed.), Raven Press, New York, pp. 145–159.

Frankenberg, L., Heimburger, G., Nilsson, C. and Sörbo, B. (1973) Biochemical and pharmacological studies on the sulfonium analogues of choline and acetylcholine. *Europ. J. Pharmacol.*, 23, 37–46.

Glick, S.D., Crane, A.M., Barker, L.A. and Mittag, T.W. (1975) Effects of pyrrolcholine on behaviour in rats. *Neuropharmacol.*, 14, 561–564.

Grewaal, D.S. and Quastel, J.H. (1973) Control of synthesis and release of radioactive acetylcholine in brain slices from the rat. *Biochem. J.* 132, 1–14.

Hebb, C. (1972) Biosynthesis of acetylcholine in nervous tissue. *Physiol. Rev.*, 52, 918–957.

Hemsworth, B.A. and Smith, J.C. (1970) The enzymic acetylation of choline analogues. *J. Neurochem.*, 17, 171–177.

Holton, P. and Ing, H.R. (1949) The specificity of the trimethylammonium group in acetylcholine. *Brit. J. Pharmacol.*, 4, 190–196.

Ilson, D. and Collier, B. (1975) Triethylcholine as a precursor to a cholinergic false transmitter. *Nature (Lond.)*, 254, 618–620.

Ilson, D., Collier, B. and Boksa, P. (1977) Acetyltriethylcholine: a cholinergic false transmitter in cat superior servical ganglion and rat cerebral cortex. *J. Neurochem.*, 28, 371–381.

Kaita, A.A. and Goldberg, A.M. (1969) Control of acetylcholine synthesis – the inhibition of choline acetyltransferase by acetylcholine. *J. Neurochem.*, 16, 1185–1191.

Katz, H.S., Salehmoghaddam, S. and Collier, B. (1973) The accumulation of radioactive acetylcholine: failure to label endogenous stores. *J. Neurochem.*, 20, 569–579.

Keston, A.S. and Wortis, S.B. (1946) Antagonistic action of choline and its triethyl analogue. *Proc. Soc. exp. Biol. Med.*, 61, 439–440.

Kilbinger, H. (1977) Formation and release of acetylpyrrolidinecholine as a false cholinergic transmitter in the myenteric plexus of the guinea-pig small intestine. *Naunyn-Schmideberg's Arch. Pharmacol.*, 296, 153–158.

Kilbinger, H., Wagner, A. and Zerban, R. (1976) Some pharmacological properties of the false cholinergic transmitter acetylpyrrolidinecholine and its precursor. *Naunyn-Schmiedeberg's Arch. Pharmacol.*, 295, 81–87.

Kopin, I.J. (1968) False adrenergic transmitters. *Ann. Rev. Pharmacol.*, 8, 377–394.

Kuczenski, R., Segal, D.S. and Mandell, A.J. (1975) Regional and subcellular distribution and kinetic properties of rat brain choline acetyltransferase. *J. Neurochem.*, 24, 39–45.

Kuhar, M.J. and Simon, J.R. (1974) Acetylcholine uptake: lack of association with cholinergic neurons. *J. Neurochem.*, 22, 1135–1137.

Large, W.A. and Rang, H.P. (1978) Incorporation of acetylmonoethylcholine into the transmitter pool at the mammalian neuromuscular junction. *J. Physiol. (Lond.)*, 275, 61P–62P.

Lefresne, P., Guyenet, P., Beaujouan, J.C. and Glowinski, J. (1975) The subcellular localization of ACh synthesis in rat striatal synaptosomes investigated with the use of Triton X-100. *J. Neurochem.*, 25, 415–422.

MacIntosh, F.C. (1978) The present status of the vesicle hypothesis. In *Cholinergic Mechanisms and Psychopharmacology*, D.J. Jenden (Ed.), Plenum Press, New York, pp. 297–322.

MacIntosh, F.C. and Collier, B. (1976) Neurochemistry of cholinergic terminals. *Handb. exp. Pharmacol.*, 42, 99–228.

Malthe-Sørenssen, D. and Fonnum, F. (1972) Multiple forms of choline acetyltransferase in several species demonstrated by isoelectric focusing. *Biochem. J.*, 127, 229–236.

Mann, S.P. and Hebb, C. (1975) Inhibition of choline acetyltransferase by quaternary ammonium analogues of choline. *Biochem. Pharmacol.*, 24, 1013–1017.

Mann, P.J.G., Tennenbaum, M. and Quastel, J.H. (1938) On the mechanism of acetylcholine formation in brain in vitro. *Biochem. J.*, 32, 243–261.

Mehrotra, K.N. and Dauterman, W.C. (1963) The specificity of rat brain acetylcholinesterase for N-alkyl analogues of acetylcholine. *J. Neurochem.*, 10, 119–123.

Muscholl, E. (1972) Adrenergic false transmitters. *Handb. exp. Pharmacol.*, 33, 618–660.

Potter, L.T. (1970) Synthesis, storage, and release of [^{14}C]acetylcholine in isolated rat diaphragm muscles. *J. Physiol. (Lond.)*, 206, 145–166.

Reitzel, N.L. and Long, J.P. (1959) Hemicholinium antagonism by choline analogues. *J. Pharmacol. exp. Ther.*, 127, 15–21.

Rossier, J., Spantidakis, Y. and Benda, P. (1977) The effect of Cl$^-$ on choline acetyltransferase kinetic parameters and a proposed role for Cl$^-$ in the regulation of acetylcholine synthesis. *J. Neurochem.*, 29, 1007–1012.

von Schwarzenfeld, I. (1978) The uptake of acetylpyrrolidinecholine – a false cholinergic transmitter — into mammalian cerebral cortical synaptic vesicles. In *Cholinergic Mechanisms and Psychopharmacology*, D.J. Jenden (Ed.), Plenum Press, New York, pp. 657—672.

von Schwarzenfeld, I. and Whittaker, V.P. (1977) The pharmacological properties of the cholinergic false transmitter, *N*-2-acetoxyethyl-*N*-methylpyrrolidinium and its precursor. *Brit. J. Pharmacol.*, 59, 69—74.

Silver, A. (1974) *The Biology of Cholinesterases.* North Holland, Amsterdam.

Singh, V.K., McGeer, E.G. and McGeer, P.L. (1975) Multiple forms of choline acetyltransferase. *Abstr. 5th Ann. Meet. Soc. Neurosci.*, New York, p. 384.

Slater, P. and Stonier, P.D. (1973) The uptake of hemicholinium-3 by rat brain cortex slices. *J. Neurochem.*, 20, 637—639.

Suszkiw, J.B. and Pilar, G. (1976) Selective localization of a high affinity choline uptake system and its role in ACh formation in cholinergic nerve terminals. *J. Neurochem.*, 26, 1133—1138.

Szerb, J.C. (1975) Endogenous acetylcholine release and labelled acetylcholine formation from [^{3}H]choline in the myenteric plexus of the guinea-pig ileum. *Canad. J. Physiol. Pharmacol.*, 53, 566—574.

Whittaker, V.P. and Barker, L.A. (1972) The subcellular fractionation of brain tissue. *Methods Neurochem.*, 2, 1—52.

Zimmermann, H. and Denston, C.R. (1977) Separation of synaptic vesicles of different functional states from the cholinergic synapses of the Torpedo electric organ. *Neuroscience*, 2, 715—730.

Zimmermann, H. and Dowdall, M.J. (1977) Vesicular storage and release of a false cholinergic transmitter (acetylpyrrolcholine) in the Torpedo electric organ. *Neuroscience*, 2, 731—739.

PART II

STORAGE AND RELEASE OF ACETYLCHOLINE

On the Mechanism of Acetylcholine Release

M. ISRAËL and Y. DUNANT

*Département de Neurochemie, Laboratoire de Neurobiologie Cellulaire du C.N.R.S.,
91190 Gif-sur-Yvette (France) and
Département de Pharmacologie, CH-1211 Genève 4 (Switzerland)*

ACETYLCHOLINE, VESICLES AND SYNAPTOSOMES IN THE ELECTRIC ORGAN OF *TORPEDO*

In spite of the powerful and specific methods now available, it has been difficult to study the biochemistry of cholinergic transmission at the neuromuscular junction. The reason for this is that only a minute fraction of the tissue volume is actually occupied by the neuromuscular synapse. In contrast, the electric organ of the marine fish *Torpedo* offers a more favourable material for such studies since the presynaptic nerve terminals in this tissue represent as much as 2 to 3‰ of the total volume.

The cholinergic nature of transmission in the electric organ of *Torpedo* was discovered before the 2nd World War by Feldberg, Fessard and Nachmansohn in the "Station de Biologie Marine" of Arcachon, France (Fig. 1). They demonstrated that the electric organ contains high amounts of acetylcholine (ACh) and the enzyme acetylcholinesterase; tissue ACh was shown to be released by stimulating the nerves in the presence of eserine; moreover, an intra-arterial injection of exogenous ACh generated a strong electrical response (Feldberg, Fessard and Nachmansohn, 1940; Feldberg and Fessard, 1942; for a pleasing account of these historical experiments, see Feldberg, 1977).

For many years, biochemists gave little attention to the electric organ of *Torpedo,* in spite of the homology of this organ with the neuromuscular junction of the voluntary muscles. The discovery of choline acetyltransferase (ChAT) by Nachmansohn and Machado (1943) was performed on the *Electrophorus* electric organ and various other tissues. Later, the isolation of synaptosomes from the brain of mammals (Whittaker, 1959; de Robertis et al., 1961) attracted the interest of many biochemists. However, it is well known that the fractions of synaptosomes and synaptic vesicles isolated from the brain are very heterogenous with respect to the transmitters they contain.

Purely cholinergic synaptic vesicles were isolated from the electric organ of *Torpedo* in 1968 (Israël, Gautron and Lesbats, 1968, 1970); as much as 40% of the ACh present in the homogenate was recovered in the purified vesicles. It was later shown that all the ACh of the homogenate was bound and associated with the vesicles (Israël et al., 1977). During the numerous steps of the isola-

126

Fig. 1. A copy of a photograph taken in Arcachon, where Feldberg, Fessard and Nachmansohn started their historical work on Torpedo electric organ. Blaschko was also there. Many scientists came to the Marine Station that year. Discussions must have been interesting and vivid. For a moment they had to interrupt their images of the future and smile at the photographer.

tion procedure, the vesicular ACh is not destroyed since it is protected by the vesicular membrane from the tissue cholinesterase. It was also shown in the same experiments that these pure cholinergic vesicles do not contain the enzyme ChAT (Israël, Gautron and Lesbats, 1970). This confirmed an earlier demonstration by Fonnum (1966) that, in various tissues, ChAT is a cytoplasmic enzyme.

In contrast to the work on mammalian brain, it was more difficult to isolate intact nerve terminals (synaptosomes) than vesicles from the electric organ. Even by gentle homogenization procedures, the *Torpedo* nerve terminals are easily disrupted. Only recently, using a completely different method, has it been possible to isolate them in a satisfactory physiological state (Israël et al., 1976; Morel et al., 1977; see Morel et al., p. 191 of this volume). The synaptosomes from *Torpedo* are completely free from postsynaptic structures. They provide a direct demonstration that nearly all the ACh of the tissue is contained in the presynaptic nerve terminals.

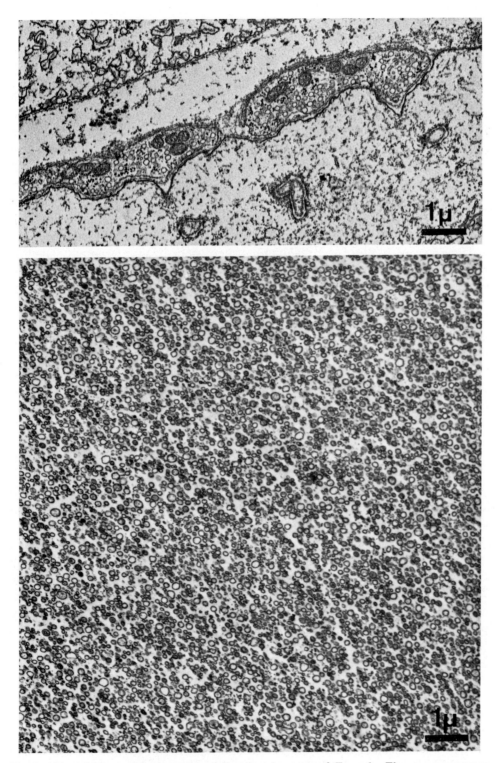

Fig. 2. (Above) Electron micrograph of the electric organ of *Torpedo*. The numerous nerve terminals show many synaptic vesicles. (Bottom) Synaptic vesicles from the electric organ of *Torpedo*. They were isolated in 1968 and found to contain an important amount of acetyl-choline.

TURNOVER OF CYTOPLASMIC AND VESICULAR ACh AT REST

In experiments with synaptosomes as with intact tissue, only a fraction of the transmitter store was found to be associated with the synaptic vesicles. As mentioned earlier, this latter fraction is described as the "bound ACh" which survives homogenization. On the other hand, 40–50% of the total ACh is extra-vesicular, most probably cytoplasmic. This pool is hydrolysed in the first stage of the isolation procedure when the membranes of the nerves terminals are opened by homogenization. The cytoplasmic ACh then diffuses out and is rapidly destroyed by cholinesterases. The same phenomenon occurs with the isolated synaptosomes. They loose their cytoplasmic ACh, but not their vesicular ACh after one exposure to freezing and thawing. It is this labile pool, representing about half of the total ACh, which has been called "free ACh".

The "free ACh" pool is a real compartment since it is preferentially labelled when radioactive precursors are supplied to the tissue. This is in fact not surprising, since ChAT, the enzyme responsible for the synthesis of ACh, is cytoplasmic. In the electric organ of *Torpedo,* external choline and acetate are both utilized and incorporated into ACh with an equal efficiency (Marchbanks and Israël, 1971; Israël and Tuček, 1974). At rest, the turnover of ACh is slower in the electric organ than in most of mammalian preparations (see MacIntosh and Collier, 1976). The spontaneous ACh release is balanced by resynthesis so that the ACh in the two compartments remains constant. In contrast, nerve stimulations unbalances this equilibrium and causes dramatic changes in the intra-terminal level of transmitter.

DYNAMICS OF ACh COMPARTMENTS DURING SYNAPTIC ACTIVITY

The ACh compartments were analysed after stimulation of the nerves in vivo and in vitro which supply the electric organ. Attention was at first focused on the behaviour of vesicular ACh, since the idea was prevalent that the immediately available transmitter was stored as a large population of preformed quanta, possibly in synaptic vesicles. This view was of course strengthened by the demonstration that synaptic vesicles, especially those prepared from the *Torpedo* electric organ, do indeed contain ACh. Therefore in the initial experiments we were surprised to find that the vesicular ACh was depleted only by prolonged stimulation, a long period after the exhaustion of transmission (Dunant et al., 1972, 1974). This finding was later criticized by Zimmermann and Whittaker (1974) who reported earlier decreases of vesicular content during stimulation. This apparent discrepancy is now clearly resolved; the early loss of vesicular ACh occurs when the fish is stimulated out of sea-water. Under this condition, the electrical energy cannot be dissipated as naturally as in the external medium and the animal drives, by its own discharge, a large leakage current back through the tissue. This causes a profound alteration of the nerve terminals (Dunant et al., 1976). In contrast, when the tissue is stimulated in vivo or in vitro under more physiological conditions, the vesicular ACh remains stable in the couse of activity until the synaptic transmission is exhausted and even later.

The question now arises as to what is the immediate source of the released transmitter. It is interesting that in all of the synapses investigated, the most recently synthesized transmitter is preferentially released (Collier, 1969; Potter, 1970; see review by MacIntosh and Collier, 1976). Knowing the cytoplasmic localization of ChAT, one can therefore suspect that cytoplasmic ACh is more directly involved in transmission than was thought previously. This was investigated with the electric organ of *Torpedo* where the cytoplasmic ACh can be measured immediately after the stimulation period as free ACh, i.e. as the difference between the total ACh and the vesicle bound ACh. The result was clear. After several minutes of stimulation in the presence of a labelled external precursor, the specific radioactivity of free ACh but not that of bound ACh, increased (Dunant et al., 1972, 1974). This showed that it was in fact the cytoplasmic ACh which was used and renewed during activity.

In later experiments tissue samples were taken and analysed at shorter time intervals in the course of stimulation. Surprisingly, the level of free ACh, but not that of bound ACh, showed dramatic fluctuations. These were a rapid oscillation superposed on a wave of slower time course (Fig. 3). The rapid oscillation is undamped and has a period of 5 sec. It is probably related to the complex regulation of the enzymatic reactions responsible for the synthesis of

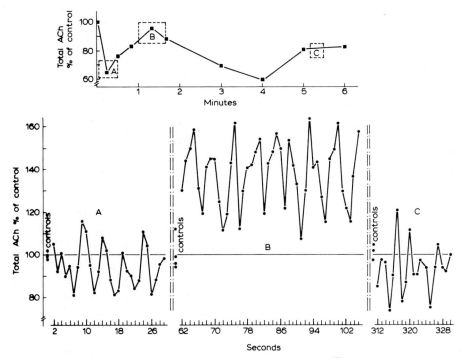

Fig. 3. Slow wave and fast oscillation of total ACh in the course of stimulation at 5/sec. Inset shows a curve describing the slow ACh changes of the tissue which appeared when the sample was performed with a low time resolution. This curve with its successive peaks has been called "slow wave". Regions A, B and C of the slow wave have been analyzed with a high time resolution of sampling (every sec) in the lower graphs. The rapid oscillations became apparent. They occurred at different mean levels corresponding to three phases of the slow wave. The rapid oscillations were undamped and involved about 35% of total ACh. Their period was approximately 5 sec. Lower graphs A, B and C are from separate experiments. The results are expressed as percentage of unstimulated samples.

transmitter (Dunant et al., 1977). Important in this connection is the fact that, under the same conditions, the tissue ATP was also shown to oscillate, in phase and in a ratio 1 to 1 with ACh (Israël et al., 1977). The rapid oscillation of ACh must be an intracellular phenomenon which is apparently not reflected in corresponding fluctuations of the transmitter release. This is suggested by the fact that neither the amplitude nor the shape of the electrical discharge oscillate within a period of 5 sec.

In contrast to the rapid oscillation, the slow wave of free ACh shows some correlation with electrophysiological changes. The electrical discharge of the *Torpedo* electric organ is the sum of the individual endplate potentials generated by ACh on each electroplaque. When the nerves receive repetitive stimulation, the amplitude of the successive electrical responses declines in the course of activity. Its decay curve shows characteristic phases which correspond in time with inflexions points of the slow wave of free ACh. Moreover, changes in temperature affect the kinetics of the discharge and that of free ACh (Dunant et al., 1974, 1977) in a similar way.

Fig. 4 shows the amplitude changes of the physiological responses and the slow wave of ACh during repetitive stimulation at 5 Hz for 5 min. Within the first 15 to 30 sec of stimulation, about half of the free ACh is lost while the electroplaque discharge loses 50 to 70% of its initial amplitude. The mean level of free ACh then increases for about one minute and attains its initial value; during this phase the electroplaque discharge remains constant. After this increase, free ACh will decline again to approximatively 30% of its initial level, while the discharge decays in a couple of minutes to a few percent of the initial response.

QUANTITATIVE DESCRIPTION OF THE CHANGES OCCURRING DURING STIMULATION

The agreement between the electrophysiological and biochemical results in these experiments permits a quantitative description of the changes occurring during synaptic activity. The cumulative amount of ACh released at any time R(t) during the experiment shown in Fig. 4a was calculated as follows. The potentials of the successive discharges shown were averaged over 10-sec intervals and added; each average represents 50 discharges. During the first 15-sec period of stimulation 450 nmol/g of ACh was lost and the cumulative potential was 337 mV. On the assumption that no synthesis of ACh occurred during the initial period these figures can be used to relate the cumulative potential to the amount of ACh released that caused it. This is shown on the graduated curve of Fig. 5a. The initial amount of ACh (I) was 770 nmol/g and about 200 nmol/g remained at the end of the experiment. It can be seen from the asymptote of R(t) in Fig. 5a that a further 1800 nmol/g of ACh(S) was synthesized during the 4 min of the experiment. The function R(t) can be expressed as the sum of an exponential function of the initial store and a sigmoid function of the total amount synthesized;

$$R(t) = I(1 - \exp^{-k_2 t}) + S(1 - \exp^{-k_1 t})^{n_1}$$

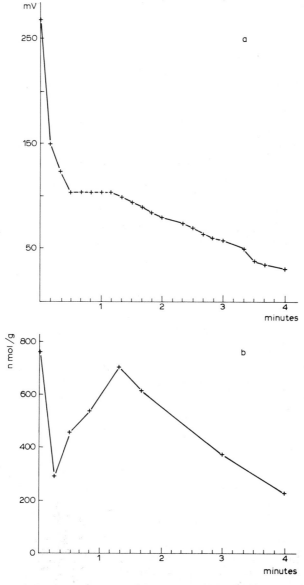

Fig. 4. Stimulation of the electric organ of *Torpedo* at 5 Hz for 5 min. Decay of the electrical response of the tissue in the course of stimulation. Changes in the mean level of free ACh in the same stimulation. In this experiment, the sampling intervals are too large to reveal the rapid oscillation. Only the slow wave of free ACh is seen. Its inflexion points correspond in time with the characteristic phases of the physiological response.

Values of k_1 k_2 and n that fit the data are given in the legend to Fig. 5a.

The amount of ACh synthesized as a function of time can also be arrived at by adding R(t) to the free ACh. If added to the initial amount of ACh the total amount of ACh present and synthesized at any given time A(t) can be calculated. The experimental figures are shown (crosses) in the upper part of Fig. 5a.

A(t) can also be expressed as a sigmoid function of S;

$$A(t) = I + S(1 - exp^{-k_3 t})^{n_3}$$

132

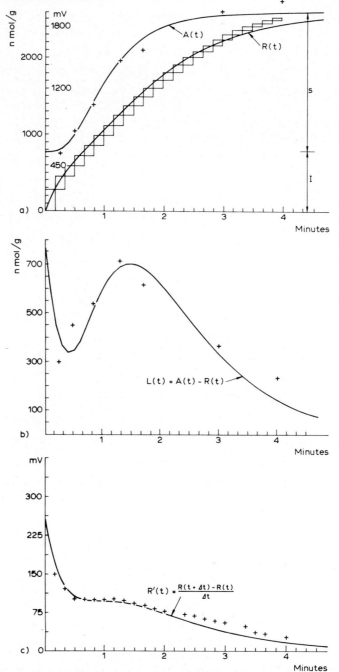

Fig. 5. Quantitative description of the changes occurring during a repetitive stimulation at 5 Hz for 5 min. Experimental data from the example illustrated in Fig. 4.

(a) Lower curve: Cumulated outputs obtained by adding up the mean electroplaque discharge every 10 sec. Ordinate calibrated in mV and nmol ACh/g tissue. I = 770 nmol/g is the initial amount of free ACh; S is the amount synthesized. S = 1800 nmol/g. R(t) (continuous line) fits well the cumulated experimental curve. Upper curve: By adding to the cumulated outputs the amount of ACh left in the nerve terminal as free ACh, we get a description of cumulated inputs of synthesis (+). The sigmoid A(t) (continuous lines) fits well the data. A(t) cumulates a maximum amount S equal to the maximum released. See the equations and further explanation in the text. The values of the constants used were $k_1 = 0.9$, $k_2 = 2.8$, $k_3 = 1.8$, $n_1 = 3.3$ and $n_3 = 4$.

(b) Plot of the free ACh equation L(t) = A(t) − R(t) describing the variation of free ACh in the course of a 5 Hz stimulation. Theoretical curve (continuous line) is close to the experimental points (+).

(c) Plot of the derivative of cumulated outputs. Since R(t) is an equation describing cumulated outputs, its derivative R'(t) should describe the output per impulse. The plot of R'(t) (continuous line) fits well the experimental discharge decay curve (+). The time interval Δt is the time between two successive discharges.

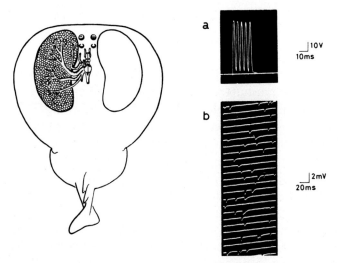

Fig. 6. (Left) Drawing of a *Torpedo marmorata*. The two electric organs are situated on both sides of the central nervous system. Each of them is innervated by four electric nerves, which originate from the electric lobe, a large nucleus made of modified motoneurones.

(a) (Right) Reflex discharge of a *Torpedo* recorded from a whole electric organ in open circuit. When a *Torpedo* discharges normally in sea-water, the voltage is much lower (10–15 V) since a large current is driven outside, in the well conducting environment.

(b) Miniature end-plate potentials recorded from the electric organ of *Torpedo* using an extracellular microelectrode. These miniature potentials are very similar to the much described potentials of the neuromuscular junctions. They are abolished by curare. Their time course is prolonged by anticholinesterase drugs.

and is shown as the continuous trace in the upper part of Fig. 5a using the constants given in the legend.

The amount of free ACh present at any time in the terminal is described by $L(t) = A(t) - R(t)$. This theoretical function is seen in Fig. 5b, in good agreement with the actual measurements of free ACh. In turn the derivative of $R(t)$ will describe the amount released at each moment and should correspond to the kinetics of the physiological response. This is shown in Fig. 5c, with time intervals equal to the intervals between the successive discharges. Since $R(t) = A(t) - L(t)$, the changes in the electrophysiological response appear as the difference between the speed of the synthesis and the speed of the free ACh variation.

This quantitative analysis provides a useful description of the mean ACh variations occurring during stimulation of electric organ. However it cannot give any information about the very mechanism by which the transmitter is released. Furthermore, it does not take into account the quantal phenomena which are observed mainly at neuromuscular junctions (see Katz, 1969) but which certainly occur at most synapses (see Fig. 6 for the electric organ of *Torpedo*). A general hypothesis for this mechanism will be presented in the subsequent chapters.

HOW THE TRANSMITTER METABOLISM CAN ENSURE QUANTAL RELEASE?

The biochemical approach for cholinergic transmission has demonstrated that a large store of transmitter is present in synaptic vesicles, but this store is not immediately available for release. By contrast, the cytoplasmic pool of ACh is continually used and renewed during nerve activity. In most synapses, the ACh stores are not depleted by prolonged activation at rather high frequencies provided that stimulation is applied under the most physiological conditions (see MacIntosh and Collier, 1976). This implies that the enzymatic machinery responsible for ACh metabolism must operate at the high rates observed by the biophysicists in the phenomena of transmission. Therefore, it is to be expected that the successive steps of ACh metabolism are ensured by structurally organized proteins, perhaps a multi-enzyme complex, located at, or close to the presynaptic membrane (for a similar view, see Lefresne, 1974).

According to the results of the physiological experiments, the transmitter is released in the form of multimolecular packets, or quanta (see Katz, 1969). This has led to the idea that the released transmitter originates from a very large population of preformed quanta stored as such inside the nerve terminal, possibly in the vesicles. The dynamics of release, on this view, would not be closely related to the reactions responsible for ACh resynthesis; it would be explained just by the movements and interactions of the preformed quanta with their releasing sites.

Therefore an apparent discrepancy exists between the conclusions of the biochemical and the physiological experiments. This can be solved only by a mechanism which would be able to utilize the newly synthesized cytoplasmic ACh and to release it in a quantal manner.

A WORKING HYPOTHESIS

Let us suppose that a definite number of *operative units* are plugged into the presynaptic membrane (Fig. 7). These operators would be able to bind axoplasmic ACh in a reversible and saturable manner. The task of the operators would be to deliver their ACh into the synaptic cleft when triggered by calcium entry. The geometry of the synapse is such that the carriers for precursors, the enzymes responsible for the synthesis of ACh, and the operator would work as a sort of multi-enzyme complex. In this way, during activity, the operators will be promptly reloaded with the latest synthesized transmitter.

The amount of transmitter released by a given nerve impulse will thus be given by

$$L' = npq = np \frac{q_{max} A}{K_s + A}$$

where n is the number of operative units, p their mean probability of individual discharge (the parameter used in the poissonnian or binomial analysis of the release), q the number of ACh molecules loaded on one unit, A the axoplasmic

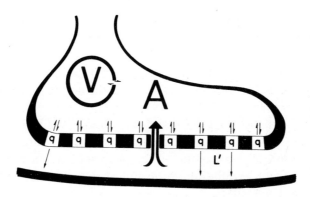

Fig. 7. Hypothesis on the quantal release of axoplasmic acetylcholine (ACh). A, axoplasmic ACh; v, vesicular ACh; q, the quantity of transmitter loaded on one of the n operators; and L' the amount of ACh liberated by a given nerve impulse. See further explanations in the text.

ACh concentration and K_s a constant corresponding to the saturation kinetics of the operators.

When confronted with experimental data, this rather simple hypothesis may clarify many apparent discrepancies: (i) the newly-synthesized, axoplasmic transmitter will be preferentially released in comparison with the more stable vesicular ACh. During prolonged activity, however, vesicular ACh can act as a sort of reserve, compensating for losses of the continually recycling pool of transmitter. (ii) As long as the amount of cytoplasmic ACh remains above a critical level, the operators will be satured. A spontaneous individual discharge will give rise to one quantum (q_{max}) and an evoked discharge to the release of an integral number of quanta (npq_{max}). Within limits, the amount liberated will then be independent of the ACh concentration. (iii) In most of synapses the speed of transmitter resynthesis is not rate limiting under physiological conditions. Thus the operators are expected to become recharged very rapidly after an impulse and the quantum size is to remain constant. However a desaturation will occur when the ACh concentration decreases below the critical value. This is consistent with the fact that, after inhibition of ACh resynthesis, the amplitude of miniature end-plate potentials declines after the utilization of a given amount of the stored transmitter (Elmqvist and Quastel, 1965).

PROPORTION OF THE AVAILABLE ACh STORED AS PREFORMED QUANTA

Since the operators postulated in the present hypothesis are closely associated with the presynaptic membrane, their number (n) is expected to be a function of the surface of the membrane. This is in accordance with the described correlation of end-plate size with the number of released quanta: under optimal conditions the cholinergic endings have been shown to release about one quantum per μm^2 of presynaptic membrane (Nishi et al., 1967; Kuno et al., 1971; Hubbard, 1973). This value is of the same order as the number of the active zones, characterized in electron micrographs by densely staining structures in the membrane, associated with a number of synaptic vesicles (Couteaux and Pécot-Déchavassine, 1973). It is also in good correspondence with the number of "releasing sites" computed from the statistical analysis of quantal release (about 400 sites for the phrenic-diaphragm junction of the rat, see Hubbard, 1973).

An obvious consequence of the present hypothesis is that only a small fraction of the available ACh would be stored as preformed quanta. During repetitive stimulation, most of the quanta would be formed at the membrane in the course of activity from axoplasmic ACh and also, very efficiently, from the latest resynthesized transmitter.

Let us calculate this for a rat neuromuscular junction. Careful estimations have recently shown that the number of molecules in a quantum is approximately 6000 (Kuffler and Yoshikami, 1975). Thus, taking q_{max} = 6000 and n = 400 for a mammalian neuromuscular junction, the amount of transmitter existing at a given moment as preformed quanta would be $nq_{max} = 2.4 \times 10^6$ molecules. Now, the axoplasmic ACh, which represents the available store in the present hypothesis, can be estimated to about A = 2.43×10^9 molecules (on the basis of a "free" volume of 150 μm^3 (see Potter, 1970) and of the concentration of 27 mM, reported for other synapses (Dunant et al., 1974; Katz and Miledi, 1977). Thus, if the present hypothesis is correct, only 0.1 percent of the available store would be present as preformed quanta.

Moreover the same calculation shows that axoplasmic ACh in this synapse will be sufficient to ensure the release of an amount equivalent to A/q_{max} = 4×10^5 quanta in the absence of resynthesis. This value compares well with the estimation made by Elmqvist and Quastel (1965) on the basis of electrophysiological experiments. These authors showed that, after inhibition of ACh synthesis, a rat neuromuscular junction is able to deliver a finite amount of transmitter, corresponding to about 3×10^5 quanta of "normal" size.

THE CYTOLOGICAL REALITY OF THE OPERATORS

The hypothesis proposed here is a description of the properties one should expect for the process of ACh release. In this case, the question of what these operators are in the terminal remains completely open. Three principal propositions can be made: first the operators units could correspond to "gates" which open for a short time in the presynaptic membrane. These gates would have to be specific for ACh and to behave quite differently from those which are believed to cause passive transfer of sodium and/or potassium ions during action potentials or postsynaptic potentials. They would have to be saturable from inside since the release of ACh appears, within limits, independent of its electrochemical gradient. Secondly, the operators could correspond to a minute fraction of synaptic vesicles inserted in the membrane, for example the double row of vesicles shown by electron microscopy in the active zones of motor nerve terminals. This interpretation seems attractive since it takes into account the increased turnover of the vesicle population following stimulation (Heuser and Reese, 1973). However, the particular vesicles operating at the membrane would have to gain new properties in this position, especially the capacity to become saturated with cytoplasmic ACh within a few millisec. There is no evidence that this occurs with the main vesicle population nor with subfractions as those recently reported by Zimmermann and co-workers (Zimmermann and Whittaker, 1977; Zimmermann and Denston, 1977).

The question then is raised as to the role of the main vesicular population. It certainly contains a reserve of ACh which remains rather stable in the first stages of activity but can be mobilized upon calcium accumulation (Babel-Guérin, 1974; Israël et al., 1977; Gautron, 1978). On the other hand, the vesicle population most probably undergoes cycles of endo- and exocytosis, but this can involve substances other than ACh. An attractive hypothesis would be that, after having delivered their ACh into the cytoplasm, the vesicles would serve to extrude the calcium by exocytosis that had entered during nerve activity. However, even though it was confirmed by radio-autography that calcium accumulates in nerve terminals during stimulation, most of this extra calcium was found to be associated with the presynaptic membrane rather than with the main vesicle population (Babel-Guérin et al., 1977).

Finally it is quite possible that operators are groups of membrane proteins, able to bind ACh and, under the action of calcium, deliver it into the synaptic cleft. In this view the appearance of subminiature potentials (Kriebel et al., 1976) could well correspond to separate discharges from subunit of the groups.

It may be, that when considered at the molecular level these three possibilities are not very different to each other since, in the present hypothesis, it is the saturability of the operators which is the most important property. It will allow the synapse to deliver constant pulses even if the available store indergoes rather large fluctuations. Such remarkable constancy is of course expected for the transmission of reliable messages through the synapses.

SUMMARY

(1) The electric organ of *Torpedo* is a purely cholinergic tissue, from which synaptic vesicles and synaptosomes can be isolated and purified. Most of the tissue acetylcholine (ACh) is localized in nerve terminals, but only a part of it is found to be associated with synaptic vesicles. The remaining ACh, about half of the total store, is extravesicular most probably axoplasmic. This pool is preferentially labelled when external precursors are supplied to the tissue. This is not surprising since choline acetyltransferase is a cytoplasmic enzyme.

(2) Upon nerve stimulation, the axoplasmic pool of ACh is preferentially used and renewed from external precursors. Its level is found to undergo dramatic oscillations during synaptic activity. In contrast, the vesicular ACh remains stable during the first stages of stimulation, provided that the experiment is performed under physiological conditions. Therefore, the axoplasmic ACh appears as the compartment of transmitter available for release during synaptic activity. A quantitative description is proposed of the changes occurring during repetitive stimulation of the electric organ.

(3) A general hypothesis is proposed for the release of ACh. In this hypothesis the quantal release of axoplasmic ACh would be provided by saturable

"operators" situated at the presynaptic membrane. These operators would be closely related to the enzymatic machinery responsible for ACh synthesis. This hypothesis is then discussed in the cytological context of the cholinergic synapse.

ACKNOWLEDGEMENTS

We are grateful to our colleagues J. Corthay, A.-F. Diebler, L. Eder, P. Jirounek, A. Kato, B. Lesbats, F. Loctin, R. Manaranche, P. Mastour-Frachon, F. Meunier, N. Morel and Y. Morot-Gaudry for valuable help and discussion. We also thank Dr. A. Kato and Mrs. N. Collet for help with the manuscript.

REFERENCES

Babel-Guérin, E. (1974) Métabolisme du calcium et libération de l'acétylcholine dans l'organe électrique de la Torpille. *J. Neurochem.*, 23, 525–532.

Babel-Guérin, E., Boyenval, J., Droz, B., Dunant, Y. and Hassig, R. (1977) Accumulation of calcium in cholinergic axon terminals after nerve activity. Localization by electron microscope radio-autography at the nerve-electroplaque junction of *Torpedo*. *Brain Res.*, 121, 348–352.

Collier, B. (1969) The preferential release of newly synthesized transmitter by a sympathetic ganglion. *J. Physiol. (Lond.(,* 205, 341–352.

Couteaux, R. and Pécot-Déchavassine, M. (1973) Données ultrastructurales et cytochimiques sur le mécanisme de libération de l'acétylcholine dans la transmission synaptique. *Arch. ital. Biol.*, III, 231–262.

De Robertis, E.D.P., Pellegrino de Iraldi, A., Rodriguez de Lorez Arnaiz, G. and Gomez, J. (1961) On the isolation of nerve endings and synaptic vesicles. *J. Biophys. Biochem. Cytol.*, 9, 229–235.

Dunant, Y., Gautron, J., Israël, M., Lesbats, B. and Manaranche, R. (1972) Les comparti-ments d'acétylcholine de l'organe électrique de la Torpille et leurs modifications par la stimulation. *J. Neurochem.*, 19, 1987–2002.

Dunant, Y., Gautron, J., Israël, B. and Manaranche, R. (1974) Evolution de la décharge de l'organe électrique de la Torpille et variations simultanées de l'acétylcholine au cours de la stimulation. *J. Neurochem.*, 23, 635–643.

Dunant, Y., Israël, M., Lesbats, B. and Manaranche, R. (1976) Loss of vesicular acetyl-choline in *Torpedo* electric organ on discharge against high external resistance. *J. Neurochem.*, 27, 975–977.

Dunant, Y., Israël, M., Lesbats, B. and Manaranche, R. (1977) Oscillation of acetylcholine during nerve activity in the *Torpedo* electric organ. *Brain Res.*, 125, 123–140.

Elmqvist, D. and Quastel, D.M.J. (1965) Presynaptic action of hemicholinium at the neuro-muscular junction. *J. Physiol. (Lond.)*, 177, 463–382.

Feldberg, W. (1977) The early history of synaptic and neuromuscular transmission by acetyl-choline. Reminiscence of an eye witness. In *The Pursuit of Nature*, A.L. Hodgkin (Ed.), Cambridge University Press, Cambridge, pp. 65–83.

Feldberg, W. and Fessard, A. (1942) The cholinergic nature of the nerves of the electric organ of the *Torpedo (Torpedo marmorata)*. *J. Physiol. (Lond.)*, 101, 200–215.

Feldberg, W., Fessard, A. and Nachmansohn, D. (1940) The cholinergic nature of the nervous supply of the electric organ of the *Torpedo (Torpedo marmorata)*. *J. Physiol. (Lond.)*, 97, 3P.

Fonnum, F. (1966) Is choline acetyltransferase present in synaptic vesicles? *Biochem. Pharmacol.*, 15, 1641–1643.

Gautron, J. (1978) Effets du calcium et de la stimulation sur les terminaisons nerveuses des jonctions nerf-électroplaque de la Torpille. *Biologie Cell.*, 31, 33–44.

Heuser, J.E. and Reese, T.S. (1973) Evidence for the recycling of synaptic vesicle membrane during transmitter release at the frog neuromuscular junction. *J. Cell Biol.*, 57, 315–344.

Hubbard, J. (1973) Microphysiology of vertebrate neuromuscular transmission. *Physiol. Rev.*, 53, 674—723.

Israël, M., Gautron, J., Lesbats, B. (1968) Isolement des vésicules synaptiques de l'organe électrique de la Torpille et localisation de l'acétylcholine à leur niveau. *C.R. Acad. Sci. (Paris)*, 266, 273—275.

Israël, M., Gautron, J. and Lesbats, B. (1970) Fractionnement de l'organe électrique de la Torpille: Localisation subcellulaire de l'acétylcholine. *J. Neurochem.*, 17, 1441—1450.

Israël, M., Lesbats, B., Manaranche, R., Marsal, J., Mastour-Frachon, P. and Meunier, F.M. (1977) Related changes in amounts of ACh and ATP in resting and active *Torpedo* nerve-electroplaque synapses. *J. Neurochem.*, 28, 1259—1267.

Israël, M., Manaranche, R., Mastour-Frachon, P. and Morel, N. (1976) Isolation of pure cholinergic nerve endings from the electric organ of *Torpedo marmorata*. *Biochem. J.*, 160, 113—115.

Israël, M. and Tuček, S. (1974) Utilizationof acetate for the synthesis of acetylcholine in the electric organ of *Torpedo. J. Neurochem.*, 22, 487—491.

Katz, B. (1969) *The Release of Neurotransmitter Substances.* Liverpool University Press, Liverpool.

Katz, B. and Miledi, R. (1977) Transmitter leakage from motor nerve endings. *Proc. roy. Soc. (Lond.), B.* 196, 59—72.

Kriebel, M.E., Llados, F. and Matteson, D.R. (1976) Spontaneous subminiature endplate potentials in mouse diaphragm muscle. Evidence for synchronous release. *J. Physiol. (Lond.)*, 262, 553—581.

Kuffler, S.W. and Yoshikami, D. (1975) The number of transmitter molecules in a quantum: An estimate from iontophoretique application of acetylcholine at the neuromuscular synapse. *J. Physiol. (Lond.)*, 251, 465—482.

Kuno, M., Turkanis, S.A. and Weakly, J.N. (1971) Correlation between nerve terminal size and transmitter release at the neuromuscular junction of the frog. *J. Physiol. (Lond.)*, 213, 545—556.

Lefresne, P. (1974) *Contribution à l'étude du métabolisme de l'acétylcholine dans les terminaisons nerveuses du néostriatum chez le rat.* Thèse Université de Paris VI. C.N.R.S., Paris.

MacIntosh, F.C. and Collier, B. (1976) Neurochemistry of nerve terminals. In *Neuromuscular Junction,* E. Zaimis (Ed.), *Handb. exp. Pharm., Vol. 42,* Springer Verlag, pp. 99—228.

Marchbanks, R.M. and Israël, M. (1971) Aspects of acetylcholine metabolism in the electric organ of *Torpedo marmorata. J. Neurochem.*, 18, 439—448.

Morel, N., Israël, M., Manaranche, R. and Mastour-Frachon, P. (1977) Isolation of pure cholinergic nerve endings from *Torpedo* electric organ. Evaluation of their metabolic properties. *J. Cell Biol.*, 75, 43—55.

Nachmansohn, D. and Machado, A.L. (1943) The formation of acetylcholine. A new enzyme, "choline acetylase". *J. Neurophysiol.*, 6, 397—403.

Nishi, S., Soeda, H. and Koketsu, K. (1967) Release of acetylcholine from sympathetic preganglionic nerve terminals. *J. Neurophysiol.*, 30, 114—134.

Potter, L.T. (1970) Synthesis, storage and release of [^{14}C]acetylcholine in isolated rat diaphragm muscles. *J. Physiol. (Lond.)*, 206, 145—166.

Whittaker, V.P. (1959) The isolation and characterization of acetylcholine containing particles from brain. *Biochem J.*, 72, 694—706.

Zimmermann, H. and Denston, C.R. (1977) Separation of synaptic vesicles of different functional states from cholinergic synapses of the *Torpedo* electric organ. *Neuroscience*, 2, 715—730.

Zimmermann, H. and Whittaker, V.P. (1974) Effect of electrical stimulation on the yield and composition of synaptic vesicles of the electric organ of *Torpedo*: A combined biochemical, electrophysiological and morphological study. *J. Neurochem.*, 22, 435—450.

Zimmermann, H. and Whittaker, V.P. (1977) Morphological and biochemical heterogeneity of cholinergic synaptic vesicles. *Nature (Lond.)*, 267, 633—635.

Vesicular Heterogeneity and Turnover of Acetylcholine and ATP in Cholinergic Synaptic Vesicles

H. ZIMMERMANN

Abteilung Neurochemie, Max-Planck-Institut für biophysikalische Chemie, Postfach 968, 3400 Göttingen (F.R.G.)

INTRODUCTION

A major obstacle in the understanding of the molecular processes underlying acetylcholine (ACh) release is the consistent finding in various systems, that there is preferential release of newly synthesized transmitter whilst the synaptic vesicle fraction does not contain newly synthesized transmitter to the same extent (for literature see MacIntosh and Collier, 1976). Generally, the metabolism of ACh in such studies was investigated by application of radiolabelled precursors. The specific radioactivity of the transmitter was then measured in intact tissue, or in various subcellular fractions obtained, and, upon stimulation, in the tissue perfusate. However, this methodological approach is hampered by two major problems:

(1) The only compartments of transmitter which can be directly investigated are total tissue, synaptosomal and vesicular contents. Synaptosomes contain, in addition to vesicle-bound ACh, a pool of transmitter that appears not to be associated with synaptic vesicles and this pool has been called names like labile, free or cytoplasmic ACh; it is assumed to correspond to an independent subcellular compartment. However, subcellular compartments which are inferred from our subcellular fractionation methods do not necessarily correspond to physiological compartments in the cell.

(2) Calculations on the dynamics of subcellular transmitter compartments were usually based on the assumption that the transmitter is distributed in these compartments in a homogeneous way (e.g. Marchbanks, 1977). However, as far as synaptic vesicles are concerned one might expect that, like other cell organelles, they can exist in different functional states within the same cell. Also a possible free compartment of cytoplasmic ACh might have a heterogeneous distribution within the nerve terminal. Only if the isolated synaptic vesicle fraction is homogeneous and contains only particles of identical metabolic state, can the average value for the specific radioactivity (SRA) of ACh obtained from the vesicle fraction be taken as a measure of vesicle function as such.

Biochemical evidence for vesicle heterogeneity was presented by Barker et al. (1972). A pool of highly labelled vesicles isolated from the guinea pig cortex sedimented together with membranes at a density corresponding to 1.0 M sucrose, (H-fraction), rather than at 0.4 M sucrose (D-fraction) where the bulk of synaptic vesicles sedimented. This result was interpreted as indicating that synaptic vesicles still attached to their releasing site, the preterminal membrane, were harvested in the H-fraction.

Using the *Torpedo* electric organ system, we approached the problem of vesicular storage and release of acetylcholine in the following ways:

In a series of morphological experiments, we perfused blocks of electric tissue with an extracellular marker (dextran) in order to study its uptake into synaptic vesicles. By using low frequency stimulation (0.1 Hz), we choose experimental conditions which do not cause any sign of serious exhaustion of nerve terminal resources like mitochondrial swelling, infoldings of the pre-synaptic membrane or loss of synaptic vesicles (Zimmermann and Denston, 1977a, b).

In order to study the turnover of the vesicle constituents, ACh and ATP, radiolabelled precursors were applied to the perfusate and vesicular uptake was investigated with and without low frequency stimulation. Vesicles were isolated on zonal sucrose gradients which had a high density resolution in the range of synaptic vesicle bouyant density (Zimmermann and Whittaker, 1977; Zimmermann and Denston, 1977b; Zimmermann, 1978).

A third approach uses the double labelling of synaptic vesicles with the real and a false transmitter. In experiments with [^{3}H]pyrrolcholine as a false precursor together with [^{14}C]choline we could show that acetylpyrrolcholine and acetylcholine are released from cholinergic nerve terminals in the same ratio as they are lost from synaptic vesicles on stimulation (Zimmermann and Dowdall, 1977). This approach will be explored further in this volume (von Schwarzenfeld et al., p. 163).

EFFECT OF LOW FREQUENCY STIMULATION ON SYNAPTIC FINE STRUCTURE AND UPTAKE OF EXTRACELLULAR MARKER

Stimulation of the electric organ at 5 Hz in vivo causes substantial alterations in the fine structure of the nerve terminals, such as loss of synaptic vesicles and infoldings of the external presynaptic membrane (Zimmermann and Whittaker, 1974a, b). By contrast, low frequency stimulation (0.1 Hz) does not lead to significant alterations in the general appearance of the nerve terminals in vivo or in vitro. Counts of vesicle numbers revealed that there is a small increase (about 35%) in vesicle number per nerve terminal profile. There is, however, the stimulus-dependent appearance of a population of smaller synaptic vesicles (25% smaller in diameter) (Fig. 1a, Fig. 2, inset). Clusters of these small vesicles can regularly be observed in close apposition to the presynaptic membrane. In order to evaluate the functional role of the two vesicle populations observed in the nerve terminal, dextran (mol. wt. 10,000, 6–10%) was added to the perfusate. If blocks were not stimulated, dextran was not taken up into synaptic vesicles nor did the small vesicle population appear. On low frequency stimulation the small vesicle population takes up extracellular marker, whilst the larger vesicles contain dextran only to a small extent (Fig. 1b). After 720–1800 impulses (4 experiments), 80% of all vesicles could be shown to belong to

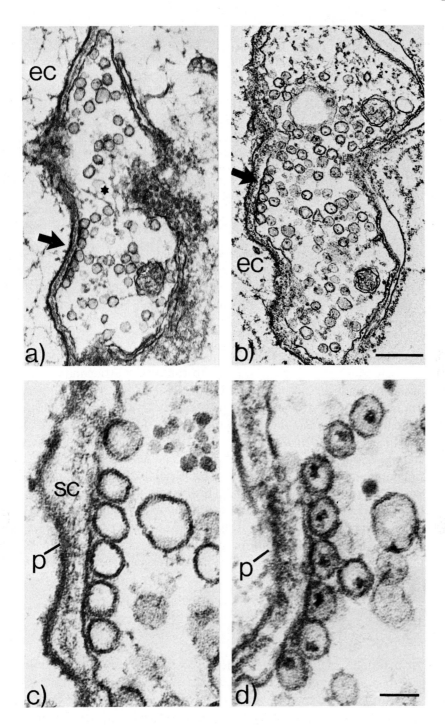

Fig. 1. Effect of low frequency stimulation (1800 impulses, 0.1 Hz) on synaptic fine structure and uptake of dextran into synaptic vesicles. (a) Stimulation in vivo via the electric lobe; most of the vesicles belong to the small diameter population. (b–d) Stimulation of perfused tissue blocks via the attached nerve. (b) Stimulation in the presence of dextran in the perfusate (MW 10,000, 10%). (c) Cluster of vesicles (small vesicle population) in close apposition to the presynaptic membrane after stimulation in the absence of dextran and (d) in the presence of dextran (10%). Arrows: clusters of vesicles; *, cisternae of axoplasmic reticulum; ec, electrocyte; p, postsynaptic membrane; sc, synaptic cleft. Bars indicate 400 nm (a,b) and 100 nm (c,d).

TABLE I

PROPERTIES OF CLUSTERS OF SYNAPTIC VESICLES IN CLOSE APPOSITION TO
THE PRESYNAPTIC MEMBRANE IN STIMULATED NERVE ENDINGS

Cluster diameter		Distance between clusters		Clusters per profile		Vesicles in cluster as % of total vesicles	
(μm)	(t)	(μm)	(t)		(t)		(t)
0.38 ± 0.02	88	1.17 ± 0.39	14	0.70 ± 0.11	88	10.67 ± 1.51	88

Values are means ± S.E.M. of 4 experiments; t, number of terminals from which values are derived; 1440–1800 impulses (0.1 Hz) were applied to perfused tissue blocks.

the small vesicle population, 91% of which contained electron dense particles. The electron dense particles usually form a polymorphous structure in the center of the vesicle core and are different from "calcium spots" (Bohan et al., 1973) which have an eccentric position on the inner vesicle membrane and a homogeneous electron dense appearance.

Fig. 1c,d shows a detailed picture of clusters of synaptic vesicles in close apposition to the presynaptic membrane with and without application of dextran. These clusters might correspond to the vesicle attachment sites of the neuromuscular junction observed by Couteaux (1974). In our experiments, there was no indication of a postsynaptic counterpart to the clusters of vesicles lined up on the presynaptic membrane. As shown in Table I these clusters, on average, have a diameter of about 400 nm which would correspond to about 5 adjacent small synaptic vesicles in a cross section. As shown by tangential sections the vesicles form patches of close apposition rather than single rows. In four experiments, we found about 2 clusters for every 3 terminal profiles analyzed; in only 14 terminal profiles could we observe more than one cluster. Therefore no graph on the distribution of distances between clusters can be given to summarize the distribution pattern of vesicle clusters on the presynaptic membrane. About 10% of all synaptic vesicles were found apposed to the presynaptic membrane.

DISCUSSION

The experiments suggest that synaptic vesicles become morphologically heterogeneous on stimulation. The new population of smaller vesicles can be shown to take up dextran from the extracellular medium. Thus these vesicles must have had a porous contact with the extracellular space or have gone through one or more cycles of exocytosis and endocytosis. Vesicular uptake of extracellular marker has been shown for peripheral as well as central synapses (see Holtzman, 1977). From the present results it cannot be decided whether a vesicle fuses completely on exocytosis and is retrieved from the external membrane by an additional process, or whether exocytosis is transient (partial) thus sustaining the integrity of the particle for a number of release cycles. The apposition of vesicles to the presynaptic membrane in the form of clusters could indicate that vesicle recycling and transmitter turnover take place mainly in these areas very close to the presynaptic membrane.

BIOCHEMICAL HETEROGENEITY OF SYNAPTIC VESICLES AND TURNOVER OF ACh

Can synaptic vesicles which have already gone through one or more cycles of exo- and endocytosis, and, therefore, may contain extracellular marker, still be

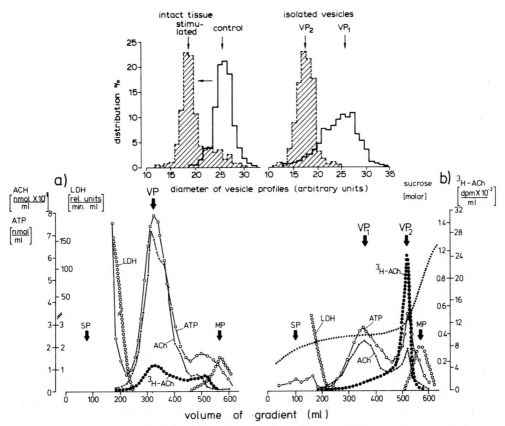

Fig. 2. Isolation of synaptic vesicles after previous perfusion of tissue blocks with [^{3}H]-acetate (1.5 μCi/ml perfusate, 5 μM). (a) 8-h perfusion without stimulation; [^{3}H]ACh shows one peak coincident with that of vesicular ACh and ATP (fraction VP) and a second smaller one in the denser sucrose at 520 ml. (b) Block perfused in parallel but stimulated (1800 impulses, 0.1 Hz) after 3 h of preperfusion with radiolabel. There is a loss of ACh and ATP in the "typical" vesicle peak (fraction VP$_1$) and the appearance of a second peak of coinciding activities of ACh and ATP in the denser sucrose (fraction VP$_2$). This fraction contains the majority of [^{3}H]ACh and is separated from the membrane fraction (MP) as indicated by the activity of lactate dehydrogenase (LDH). The amount of parent fraction loaded on the gradient corresponds to 40 g of wet wt. tissue in both cases. SP, soluble peak (supernatant); dotted line, sucrose gradient. Inset: Effect of stimulation (0.1 Hz) on synaptic vesicle profile diameter (normalized distributions). Intact tissue: Blocks were perfused in parallel. One served as unstimulated control and the other one was stimulated (1800 impulses). Isolated vesicles: Vesicle diameters of peak function VP$_1$ and VP$_2$ isolated from perfused blocks after stimulation with 1800 impulses. Betwen 282 and 702 vesicle profiles were analyzed for each distribution (Data from Zimmermann and Denston, 1977a,b).

regarded as functional for sustaining transmission? Can such vesicles become reloaded with ACh and ATP and can they release it again?

By application of radiolabelled precursor to perfused blocks of electric tissue, vesicular actylcholine can be labelled. We isolated labelled synaptic vesicles from perfused and stimulated tissue in blocks in order to observe the vesicular heterogeneity which was apparent from the fine structural experiments.

On zonal centrifugation with sucrose gradients, synaptic vesicles can be isolated from electric tissue as a defined peak showing coincident activities of ACh and ATP (for methods see Whittaker et al., 1972; Zimmermann and Denston, 1977b). If vesicles are isolated from perfused tissue, they sediment at a density of 0.4 M sucrose which is identical for vesicles isolated from intact tissue. Fig. 2a shows a zonal separation of vesicles prepared from a tissue block that had been perfused for 8 h in the presence of [^{3}H]acetate, which is a good precursor of vesicular ACh (Israël and Tuček, 1974). ACh and ATP, as vesicle markers, show a coincident peak (VP). Newly synthesized ACh ([^{3}H]ACh) is mainly incorporated in this fraction. A second tissue block was perfused in parallel but after 3 h preperfusion with radiolabel, it was stimulated at 0.1 Hz via the attached nerve (1800 impulses). Stimulation alters drastically the distribution of the vesicle markers on the gradient whilst the distribution of lactate dehydrogenase activity is unchanged (Fig. 2b). There is a loss of both ACh and ATP in the region of 0.4 M sucrose (fraction VP$_1$), and a new peak containing both ACh and ATP appears at a density of about 0.5 M sucrose (fraction VP$_2$). Most of the newly synthesized ACh is found in this second peak of ACh and ATP activity. The specific radioactivity of ACh in fraction VP$_2$ is about 12 times higher than that in fraction VP$_1$.

Ultrastructural analysis of the fractions showed that both fraction VP$_1$ and VP$_2$ contain mainly synaptic vesicles. Fraction VP$_2$ however is contaminated by membranes to a larger extent than is fraction VP$_1$, since it overlaps the rising portion of the membrane peak (Fig. 2b). The vesicles isolated in fraction VP$_2$ are about 25% smaller in diameter than those in fraction VP$_1$. The inset in Fig. 2 compares the change in diameter distribution of vesicles on stimulation as observed in intact tissue to that of peak fractions VP$_1$ and VP$_2$. The diameter distributions suggest that the vesicles which become smaller on stimulation (and which can be shown to take up extracellular marker) sediment mainly in fraction VP$_2$, whilst vesicles in fraction VP$_1$ are identical in diameter to those from unstimulated tissue. The possibility that the ACh recovered in fraction VP$_2$ is not vesicular, but represents ACh of a cytoplasmic pool entrapped in synaptosome-like particles, could be ruled out. The fine structural analysis of fraction VP$_2$ did not provide evidence for the presence of such particles. Furthermore, parent fractions prepared for the isolation of synaptic vesicles and labelled with [^{3}H]ACh were mixed with parent fractions for the isolation of nerve terminals (Dowdall and Zimmermann, 1977) and labelled with [^{14}C]-ACh. The [^{14}C]ACh-containing particles clearly separated from the [^{3}H]ACh-labelled particles in fraction VP$_2$ and coincided with the activity of ChAT in the membrane peak (Zimmermann and Denston, 1977b).

On average, stimulation (1080—1800 impulses) caused about 50% loss of vesicular ACh. In five experiments, the vesicles isolated in fraction VP$_2$ contained about 25% of all vesicular ACh, 30% of vesicular ATP, and most of the newly synthesized ACh recovered on the gradient. The specific radioactivity of ACh in peak fraction VP_2 was about 16 times higher than that of vesicles sedimenting at the original density and 9 times higher than the average of all vesicles isolated.

DISCUSSION

Stimulation makes apparent metabolic and morphological heterogeneity of cholinergic synaptic vesicles. By use of very flat continuous sucrose gradients in zonals rotors, the new population of smaller vesicles, which contains the bulk of newly synthesized ACh, can by physically separated from previously passive vesicles. At present the reason for the increase in density of the new vesicle population is not completely clear. If the new vesicles have the same density of membrane and core matrix as the original ones, they would be expected to be denser because of their smaller radius.

If all smaller vesicles were isolated in fraction VP_2, it can be calculated that they would, on average, contain only 13% of the ACh content of vesicles sedimenting in fraction VP_2 or vesicles isolated from unstimulated tissue. This might explain their high capacity for uptake of newly synthesized ACh. It is likely that the population of vesicles isolated in fraction VP_2 is more heterogeneous than the vesicles in fraction VP_1. Some vesicles might have gone through several cycles of release, whilst others might have only discharged once, or might have recovered already most of their ACh content. Not all of these vesicles might be immediately involved in renewed release.

In a further study, the specific radio-activity (SRA) of ACh released from perfused tissue blocks on stimulation was compared to that of vesicle fractions VP_1 and VP_2 (Suszkiw et al., 1978; Suszkiw and Whittaker, p. 153). These studies revealed that the SRA of released ACh is not significantly different from that of ACh in fraction VP_2, whilst the SRA of ACh in vesicle fraction VP_1 and total tissue ACh was about 10 times lower than that of released ACh.

The data, thus, suggest that vesicles which have released transmitter by going through at least one cycle of exo- and endocytosis have the highest capacity for uptake of newly synthesized ACh, and also the highest probability for releasing the transmitter again. The morphologically described population of vesicles in close apposition to the presynaptic membrane (which is a subpopulation of the new smaller vesicle population) might by functional in achieving a rapid cycle of reuptake and renewed release. The notion of vesicular heterogeneity, thus, resolves the paradox of vesicular storage and release of transmitter derived from studies where only average values for subcellular transmitter pools could be obtained.

If synaptic vesicles are the counterparts of ACh quanta, our data would imply that the original view (Del Castillo and Katz, 1974) assuming equal probability of release for each quantum present in the nerve terminal, would have to be revised. Our data suggest that under all experimental conditions there will always be vesicles with a higher turnover and, therefore, higher probability of release than others. Using electrophysiological methods, this

148

might only become apparent under experimental conditions when refilling of vesicles with transmitter is slower than transmitter release.

VESICULAR TURNOVER OF ADENINE NUCLEOTIDES

Like other amine storage particles, cholinergic synaptic vesicles isolated from the *Torpedo* electric organ contain ATP. The molecular ratio of ACh to ATP was found to be about 5 : 1 (Dowdall et al., 1974). The function of vesicular ATP is still unknown. We investigated whether synaptic vesicles show a similar turnover for ATP as they do for ACh.

Experiments were performed on tissue blocks perfused with [³H]adenosine. If [³H]adenosine was applied to resting tissue, a peak of [³H]-labelled material was superimposed to the synaptic vesicle peak; 83% of the radiolabel was in the form of ATP. If blocks were stimulated at low frequency most of the newly synthesized ATP was found in fraction VP_2. Fig. 3a shows an experiment in which [³H]adenosine was applied together with [¹⁴C]choline. The distribution of the [¹⁴C] and [³H]label indicates that the newly synthesized ACh and ATP are preferentially contained in the vesicle population sedimenting in the denser sucrose (fraction VP_2). In a parallel experiment (Fig. 3b), a tissue block was stimulated at low frequency (0.1 Hz) and then washed in the presence of hemicholinium-3 and dipyridamole to block reuptake of choline and adenosine

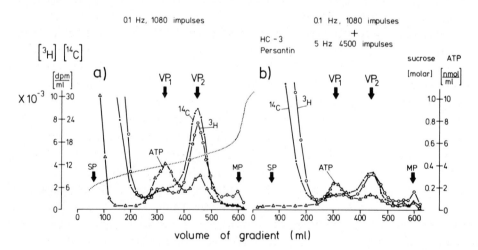

Fig. 3. Isolation of synaptic vesicles from stimulated tissue blocks after previous perfusion with [³H]adenosine (3 μCi/ml perfusate, 0.13 μM) and [¹⁴C]choline (1 μCi/ml perfusate, 20 μM). Blocks (a) and (b) were perfused in parallel. (a) After 3 h of perfusion with radiolabel, 1080 impulses (0.1 Hz) were applied. Both [¹⁴C]- and [³H]-labelled products are preferentially incorporated into fraction VP_2. (b) After stimulation (1080 impulses, 0.1 Hz) the block was washed (3 h) in the presence of hemicholinium-3 (100 μM) and dipyridamole (persantine, 5 μM). It was then restimulated at 5 Hz (4500 impulses). There is loss of total ATP as well as [¹⁴C] and [³H] label from fraction VP_2. The amount of parent fraction loaded on the gradient corresponds to 45 g wet wt. for (a) and to 43.5 g wet wt. for (b). SP, soluble (supernatant) fraction; VP_1, first vesicle peak; VP_2, second vesicle peak; MP, membrane peak; dotted line, sucrose gradient.

respectively. Restimulation at 5 Hz caused a loss of both newly synthesized ACh and ATP from fraction VP_2.

An analysis of vesicular contents of adenine nucleotides yielded the following results: besides ATP, synaptic vesicles contained about 15% ADP and traces of AMP. Under all experimental conditions, and for all types of vesicles isolated, ATP is the predominant vesicular adenine nucleodite. On the other hand, there is an overall loss of about 50% of vesicular ATP content on stimulation (similar to the loss of ACh). Since we did not find a corresponding increase in metabolic products of ATP in synaptic vesicles, we conclude that ATP is lost from synaptic vesicles on stimulation as is ACh.

DISCUSSION

The data suggest that the new population of synaptic vesicles appearing on stimulation has a high turnover rate for both ACh and ATP. Since there is a major component of postsynaptic ATP release on stimulation (Meunier et al., 1975), it is difficult to establish whether ATP is released from nerve terminals together with ACh. There is, however, recent evidence (Meunier, 1977) that ATP can be released by K^+ from nerve terminals isolated from electric tissue. Thus both ACh and ATP might be released by and exocytotic mechanism.

Since synaptic vesicles were found to contain actin (Tashiro and Stadler, 1977) and, after stimulation, also a myosin-like protein (Stadler et al., 1977), the possibility of the involvement of an ATP-activated actomyosin system in vesicle exocytosis has to be considered.

If there is release of vesicular nucleotides from the nerve terminal, our data would suggest that the nerve terminals possess systems for rapid degradation, reuptake, resynthesis and vesicular reuptake of adenine nucleotides that are about as effective as those for the transmitter. The presence of a Na^+, K^+-ATP-ase and 5′-nucleotidase on the outer membrane of the cholinergic nerve ending, as well as an uptake system for adenosine with a K_m-value similar to the high affinity choline uptake system ($K_m \approx 1\ \mu M$) (Dowdall, 1977), is a first hint in this direction.

SUMMARY

Experiments were performed on perfused blocks of innervated *Torpedo* electric tissue. Low frequency stimulation leads to the appearance of a population of vesicles about 25% smaller in diameter than normal. If dextran (mol. wt. 10,000) is added to the perfusate, these smaller vesicles become labelled during stimulation.

Synaptic vesicles isolated from perfused and stimulated blocks of electric tissue are heterogeneous. The smaller vesicles have a higher density than those of original size, and can be separated on zonal sucrose density gradients. By application of [³H]acetate or [³H]adenosine to the perfusate, it can be shown that the denser population of synaptic vesicles has a high turnover rate for both newly synthesized ACh and ATP. The data suggest that synaptic vesicles can be reloaded with ACh and ATP and that vesicles which have gone through at least one cycle of exo-endocytosis have a higher probability of reuptake and release than have previously resting vesicles.

REFERENCES

Couteaux, R. (1974) Remarks on the organization of axon terminals in relation to secretory processes at synapses. In *Advanc. Cytopharmacol.*, Vol. 2, B. Ceccarelli, F. Clementi and J. Meldolesi (Eds.), Raven Press, New York, pp. 369–379.

Barker L.A., Dowdall, M.J. and Whittaker, V.P. (1972) Choline metabolism in the cerebral cortex of guinea pig. *Biochem. J.*, 130, 1063–1080.

Bohan, T.P., Boyne, A.F., Guth, P.S., Narayanan, Y. and Williams, T.H. (1973) Electron dense particles in cholinergic synaptic vesicles. *Nature (Lond.)*, 244, 32–34.

Del Castillo, J. and Katz, B. (1954) Quantal components of the endplate potential. *J. Physiol. (Lond.)*, 124, 560–573.

Dowdall, M.J. (1977) Adenine nucleotides in cholinergic transmission: presynaptic aspects. *Abstr. Nucleotides and Neurotransmission. Conferences on Neurobiologie de Gif, 1977*, p. 7–8.

Dowdall, M.J. and Zimmermann, H. (1977) The isolation of pure cholinergic nerve terminals sacs (T-sacs) from the electric organ of juvenile *Torpedo. Neuroscience*, 2, 401–421.

Dowdall, M.J., Boyne, A.F. and Whittaker, V.P. (1974) Adenosine triphosphate, a constituent of cholinergic synaptic vesicles. *Biochem. J.*, 140, 1–12.

Holtzman, E. (1977) The origin and fate of secretory packages, especially synaptic vesicles. *Neuroscience*, 2, 327–355.

Israël, M. and Tuček, S. (1974) Utilization of acetate and pyruvate for the synthesis of "total", "bound" and "free" acetylcholine in the electric organ of *Torpedo. J. Neurochem.*, 22, 487–493.

MacIntosh, F.C. and Collier, B. (1976) Neurochemistry of cholinergic nerve terminals. In *Handbook of Experimental Pharmacology*, Vol. 42, pp. 99–228.

Marchbanks, R.M. (1977) Turnover and release of acetylcholine. In *Synapses, Proc. Int. Symp. Scottisch electronphysiol. Soc.*, G.A. Cottrell and P.N.R. Usherwood (Eds.), Blackie, Glasgow–London, pp. 81–101.

Meunier, F.M. (1977) Effect de la dépolarisation sur la libération d'ATP pre-et post-synaptique. *Abstr. Nucleotides and Neurotransmission. Conferences de Neurobiologie de Gif, 1977*, p. 15.

Meunier, F.M., Israël, M. and Lesbats, B. (1975) Release of ATP from stimulated nerve electroplaque junctions. *Nature (Lond.)*, 257, 407–408.

Stadler, H., Tashiro, T. and Zimmerman, H. (1977) Changes in the protein composition of synaptic vesicles during stimulation. *Proc. Int. Soc. Neurochem.*, Vol. 6, Copenhagen, p. 167.

Suszkiw, J.B., Zimmermann, H. and Whittaker, V.P. (1978) Vesicular storage and release of acetylcholine in *Torpedo electroplaque* synapses. *J. Neurochem.*, in press.

Tashiro, T. and Stadler, H. (1977) Proteins of cholinergic synaptic vesicles. *Proc. Int. Neurochem.*, Vol. 6, Copenhagen, p. 169.

Whittaker, V.P., Essman, W.B. and Dowe, G.H.C. (1972) The isolation of pure synaptic vesicles from the electric organs of elasmobranch fish of the family *Torpedinidae. Biochem. J.*, 128, 833–846.

Zimmermann, H. (1978) Turnover of adenine nucleotides in cholinergic synaptic vesicles of the *Torpedo* electric organ. *Neuroscience*, 3, 827–836.

Zimmermann, H. and Whittaker, V.P. (1974a) Effect of electrical stimulation on the yield and composition of synaptic vesicles from the cholinergic synapses of the electric organ of *Torpedo*: a combined biochemical, electrophysiological and morphological study. *J. Neurochem.*, 22, 435–450.

Zimmermann, H. and Whittaker, V.P. (1974b) Different recovery rates of the electrophysiological, biochemical and morphological parameters in the cholinergic synapses of the *Torpedo* electric organ after stimulation. *J. Neurochem.*, 22, 1109–1114.

Zimmermann, H. and Denston, C.R. (1977b) Separation of vesicles of different functional states from the cholinergic synapses of the *Torpedo* electric organ. *Neuroscience*, 2, 715–730.

Zimmermann, H. and Denston, C.R. (1977a) Recycling of synaptic vesicles in the cholinergic synapses of the *Torpedo* electric organ during induced transmitter release. *Neuroscience, 2,* 695–714.

Zimmermann, H. and Dowdall, M.J. (1977) Vesicular storage and release of a false cholinergic transmitter (acetylpyrrolcholine) in the *Torpedo* electric organ. *Neuroscience, 2,* 731–739.

Zimmermann, H. and Whittaker, V.P. (1977) Morphological and biochemical heterogeneity of cholinergic synaptic vesicles. *Nature (Lond.), 267,* 633–635.

Role of Vesicle Recycling in Vesicular Storage and Release of Acetylcholine in *Torpedo* Electroplaque Synapses

J.B. SUSZKIW * and V.P. WHITTAKER

Abteilung Neurochemie, Max-Planck-Institut für biophysikalische Chemie, Postfach 968, 3400 Göttingen (F.R.G.)

INTRODUCTION

Recent work has made it clear that the vesicle population in cholinergic nerve terminals is functionally and metabolically heterogeneous (for brain cortex see Barker et al., 1972 and for *Torpedo* electromotor synapses see Zimmermann and Whittaker, 1977 and Zimmermann and Denston, 1977a,b). In *Torpedo*, the subject of this contribution, stimulation, at frequencies (0.1 Hz) low enough to conserve vesicle numbers, generates a second population of vesicles, smaller and denser than those present initially; as stimulation continues, this almost entirely replaces the initial population. The two populations can be isolated as separate fractions, designated VP_1 and VP_2. Experiments with blocks of tissue perfused with dextran and/or radioactive ACh precursors (Zimmermann and Whittaker, 1977; Zimmermann and Denston, 1977a,b) show (1) that the smaller, denser vesicles (VP_2) take up dextran and therefore (since dextran cannot penetrate through membranes) must have undergone one or more cycles of exocytosis and reformation, and (2) that they preferentially acquire newly synthesized transmitter. By contrast, higher frequency stimulation (5–10 Hz) at least in vivo causes extensive transmitter and vesicle depletion with invagination of the presynaptic plasma membrane (changes which, however, are readily reversible in a few hours of rest). The vesicles remaining in depleted terminals are, again, of smaller diameter than the original population (Zimmermann and Whittaker, 1974a,b; Boyne et al., 1975, results with *Narcine*).

We now describe work with perfused blocks of electric tissue which contrasts the morphological changes taking place under different conditions of stimulation and shows that in experiments in which the ACh stores have been labelled with radioactive precursors the ACh released on stimulation has, independently of the conditions of labelling, a specific radioactivity (SRA) indistinguishable from that in the highly-labelled vesicle subpopulation. Part of this work is in course of publication (Suszkiw et al., 1978).

---

* Permanent address: Physiology Section, Biological Sciences Group U-42, University of Connecticut, Storrs, Conn. 06268 (U.S.A.).

METHODS

In experiments designed to create maximum vesicle heterogeneity with minimum depletion of vesicle numbers, blocks of electric tissue were perfused in a closed circuit and vesicles isolated essentially as described by Zimmermann and Denston (1977b) except that the stimulus frequency was 1 Hz instead of 0.1 Hz and blocks were allowed to recover for 3 h after stimulation before analysis or restimulation. In experiments designed to cause maximum vesicle depletion and thus, after recovery, minimum vesicle heterogeneity, blocks were perfused with a choline-free medium in open circuit; stimulation was again at 1 Hz.

To label the ACh pools, [Me-^{3}H]acetate was added to the perfusates during the stimulus and subsequent recovery period. In release studies labelled blocks were submitted to a second ("release") stimulus at 1 Hz (closed perfusion experiments) or 10 Hz (open perfusion), the perfusate collected and analysed for total and radioactive ACh. In the first case stimulation was via the nerve, and in the second case, since tissue was subdivided in small slices of approximately 4 g each, for determination of the time course of recovery, a randomly selected slice was field stimulated. An anticholinesterase (paraoxon, 100 μM or neostigmine, 50 μM as indicated) and a blocker of choline uptake (hemicholinium-3, 50 μM) were added to the perfusate before the release stimulus to stabilize released ACh and to block reutilization of choline.

RESULTS

Experiments with closed circuit perfusion

The results essentially confirmed and extended earlier work (Zimmermann and Denston, 1977a,b). When up to 1080 stimuli were applied at 1 Hz, vesicle numbers were not significantly reduced, but vesicular ACh was (Table I, column 2, Suszkiw et al., 1978). Vesicles isolated from the tissue and fractionated on sucrose density gradients in a zonal rotor after a recovery period of 3 h separated into the expected two peaks (Fig. 1, VP$_1$ and VP$_2$) and [^{3}H]-acetate added to the perfusate was preferentially incorporated into VP$_2$ as

TABLE I

RELATIONSHIP BETWEEN TOTAL VESICULAR ACh AND [^{3}H]ACh IN THE VP$_2$ SUB-POPULATION OF VESICLES, UNDER CONDITION OF CONSERVATION OF VESICLE NUMBERS

No of impulses (1 Hz)	Total vesicular ACh (nmoles/g)	Calculated fraction of empty vesicles	[^{3}H]ACh in VP$_2$ (d.p.m./g)
0	97.4 ± 13.2	0	2934 ± 1598
120 *	78.7 ± 2.2	0.19 ± 0.03	20075 ± 7196
360	63.3 ± 8.8	0.35 ± 0.10	27990 ± 8158
1080	53.9 ± 13.2	0.45 ± 0.17	27270 ± 3075

Results are means ± S.D., n = 3; * mean ± range, n = 2. Fraction of empty vesicles calculated from 1 − (Vesicular ACh after n impulses)/(Vesicular ACh in control, unstimulated preparation). Experimental conditions are the same as described in Fig. 1.

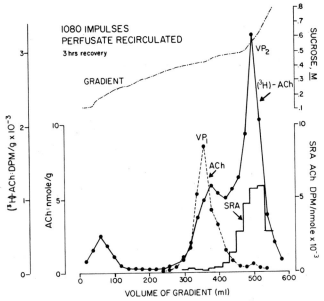

Fig. 1. Zonal sucrose density profile of vesicular ACh and [³H]ACh isolated from tissue stimulated without reduction in vesicle numbers and rested for 3 h post-stimulation. The tissue, perfused by recirculating *Torpedo* Ringer containing 1.25 μCi of [³H]acetate, was stimulated for 18 min at 1 Hz and was allowed to recover for 3 h at rest. Results are expressed per g of tissue.

[³H]ACh. As seen in Table I, column 4, the amount of [³H]ACh incorporated is relatively constant and independent of total vesicular ACh depletion. This suggests that only a proportion of the vesicles retain their capacity for storage, the remainder becoming at least for a time nonfunctional or "demobilized". On the assumption that vesicles do not partially refill, the proportion of empty vesicles can be calculated (column 3).

The source of ACh released from blocks whose ACh stores have been previously labelled by perfusing [³H]acetate and stimulating at 1 Hz would appear to be the vesicles of fraction VP_2 as judged by the correspondence between the SRA's of the released ACh and that of fraction VP_2 (Fig. 2).

Experiments with open circuit perfusion

When stimuli were applied at 1 Hz to blocks submitted to open circuit perfusion, their effect on the morphology of the presynaptic nerve terminals was considerable. Fig. 3 shows the sequence of changes observed when blocks were perfused in open circuit for 2 h with choline-free *Torpedo* Ringer solution and then stimulated at 1 Hz till the response fell to 15% of the initial, at which point mean vesicle numbers had decreased by 51% and vesicular ACh had fallen from 100 ± 10 to 29 ± 4 nmol/g of tissue (mean of 5 experiments ±S.E.M.).

Open perfusion alone (Fig. 3A) produced no detectable alteration in synaptic morphology. By contrast, tissues fixed immediately after stimulation (Figs. 3B,C) showed pronounced ultrastructural changes. Vesicle numbers were obviously depleted and arrays of parallel membranes, cisternae and vacuoles are prevalent. After 60 min at rest in Ringer solution supplemented with 0.1 mM choline chloride, infoldings, cisternae and vacuoles are less frequently seen and small vesicles and tubular sacs are more obvious (Fig. 3d). Segmentation of

156

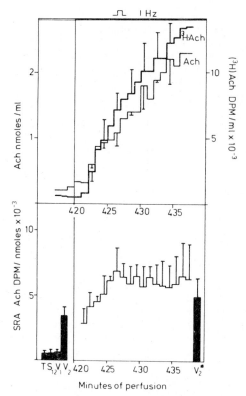

Fig. 2. Release of ACh from tissue preloaded with [³H]ACh after brief loading stimulation (120 stimuli at 1 Hz). (Top) Total released ACh and [³H]ACh per ml per 1 min collection period. (Bottom) SRA of the released ACh is compared with that of the total tissue ACh (T), total vesicular extract (S_{12}), peak of VP_1 vesicles (V_1), and peak of the VP_2 vesicles (V_2), all derived from a control, unstimulated tissue block, equivalent to tissue conditions just before the release stimulation. V_2^* is the peak SRA of VP_2 vesicles isolated from the restimulated tissue. Results are means of three experiments ±S.D. After the first few collection periods, the SRA of the relased ACh is not significantly different from that of V_2^*.
For further experimental details see Suszkiw et al. (1978).

tubules (arrows) into lengths approximately equal to the diameter of synaptic vesicles appears to be taking place. After 3 h at rest, vesicle numbers returned to 80 ± 8% of controls and after 5 h to 97 ± 8%; at that time, the morphology is almost back to its prestimulus state (Fig. 3E).

In contrast to the time course of ultrastructural recovery, that of tissue and vesicular ACh is much slower (Fig. 4); only after 20 h has tissue ACh returned to near (91 ± 10% of) the control level and vesicular ACh is at that time still depressed (69 ± 8% of control). These results closely parallel those of Zimmer-

Fig. 3. Ultrastructure of control and stimulated terminals at various stages of recovery. (A) Unstimulated preparation. Synaptic vesicles are relatively uniform in size. (B,C) Two terminals illustrating the range of ultrastructural changes seen immediately after stimulation. An array of parallel membranes enclosing cytoplasm (*) is typical. Cisternae (c), vacuoles (v) and blebs or micropits (arrow) can be observed. (D) After 1 h at rest. The terminal has been selected to illustrate the possible reformation of vesicles from tubular membrane (arrows). (E) Preparation after 5 h at rest. Morphologically uniform vesicles appear to be smaller in size from those in the control. Parallel membranes are absent but cisternae (c) can occasionally be observed. Osmium fixation. Magnification: X26,250 (A, B, C, and E) and X30,000 (D).

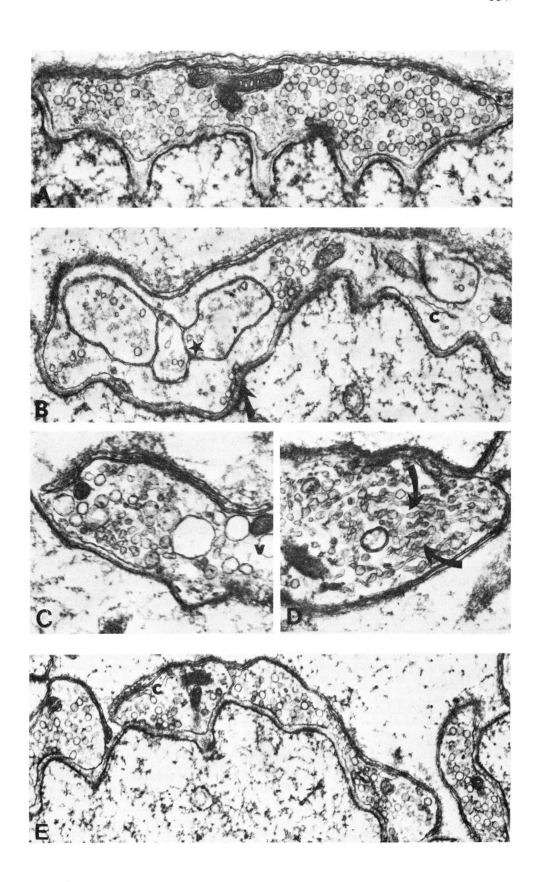

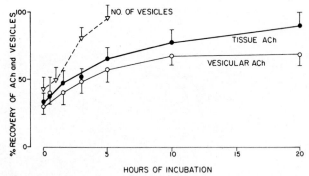

Fig. 4. Recovery of vesicles and of ACh in tissue depleted of both by electrical stimulation. Tissue was depleted of ACh and of vesicles as described in the text. Slices of control and stimulated tissue were incubated for the indicated periods of time in a large volume of *Torpedo* Ringer solution containing 100 μM choline, pH 7.4, at 18°. At the indicated times random fragments were frozen in liquid N_2; total tissue and vesicular ACh were determined as described elsewhere (Suszkiw et al., 1978). Results are means of 5 or more experiments ±S.E.M. Recovery of vesicle numbers was determined from counts in 56 terminals (2190 vesicles counted) from unstimulated tissue, 131 terminals (2859 vesicles) immediately after stimulation, 84 terminals (1704 vesicles) after 1 h, 92 terminals (3672 vesicles) after 3 h, and 120 terminals (4176 vesicles) after 5 h at rest.

mann and Whittaker (1974a,b) on tissue stimulated in vivo through the electric lobe, in which 2-3 days was needed for complete recovery of vesicular ACh, and are also somewhat analogous to those seen in frog neuromuscular junctions by Ceccarelli et al. (1972, 1973). It was also observed in the earlier work (see also Whittaker et al., 1975) that whereas the response of the organ to single shocks recovered pari passu with vesicle number, fatiguability (defined as the reciprocal of the number of impulses required to reduce the response of the organ to 1% of initial) did not return to normal until vesicular ACh did so.

More sophisticated ultrastructural studies than the preliminary ones presented here will be needed to elucidate the precise significance of the changes in synaptic morphology induced by stimulation under conditions of open circuit perfusion (and possible choline depletion). However, the changes might most simply be explained by postulating that under the conditions selected vesicle reformation (endocytosis) is inhibited (perhaps by excessive mobilization of intraterminal Ca^{2+}) and multiple "piggy back" exocytosis occurs, with the formation of tunnel-like invaginations of the presynaptic terminal plasma membrane composed of fused vesicle membrane and readily reverting by a pinching-off process to reconstituted vesicles.

Vesicles extracted from tissue initially depleted of vesicles and then allowed to recover for 10 h in the presence of [^{3}H]acetate and 0.1 mM choline no longer separated into two distinct subpopulations in sucrose density gradients (Fig. 5, upper diagram). The reformed vesicles labelled with [^{3}H]ACh sediment more like the VP_1 than like the VP_2 vesicles isolated from tissue under conditions of stimulation not accompanied by vesicle depletion. However the SRA peak of the vesicular ACh does not correspond to the peak of endogenous ACh. This indicates that not all of the labelled vesicles have recovered completely the characteristics of the virgin, VP_1 population, and suggests that within the population of recovering vesicles a subpopulation still exists which resemble the activated vesicles isolated in the VP_2 fraction from tissue stimulated under conditions which conserve vesicle numbers.

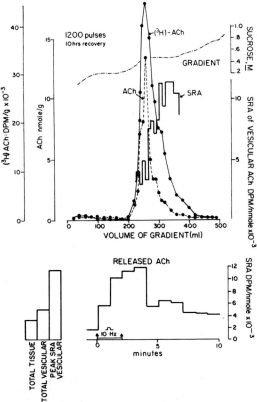

Fig. 5. (Top) Distribution of vesicular ACh and [³H]ACh in zonal sucrose density gradients: Tissue initially depleted of vesicles and ACh. The tissue was washed with choline-free Ringer solution and was stimulated at 1 Hz for 20 min. After stimulation, slices of tissue were incubated for 10 h in the presence of 100 μM choline and [³H]acetate (1.25 μCi/ml) before extraction of vesicles. For details of zonal separation of vesicles, see Whittaker et al. (1972). Results are expressed per g of tissue. (Bottom) Comparison of SRA of released ACh with that of total tissue, total vesicular and peak SRA fraction. A fragment of the same tissue was washed with radioactivity-free Ringer containing 50 μM neostigmine bromide and was field-stimulated for 2 min at 10 Hz. The perfusate was collected at 1-min intervals, and ACh and [³H]ACh were determined. The SRA of the total vesicular ACh refers to the SRA of the vesicles (S₁₂ fraction), before separation on the gradient. The peak SRA is that of the gradient fraction with the highest SRA.

This subpopulation of highly radioactive vesicles, like the VP₂ vesicles, appears to be the preferred source of released transmitter. As shown in Fig. 5 (lower diagram), the SRA of ACh released from the (partially) recovered blocks is close to the maximum SRA of the isolated vesicles.

DISCUSSION

Our results confirm and extend earlier conclusions that the pattern of vesicle exocytosis and thus the morphology of the presynaptic terminal can be profoundly modified by the conditions of stimulation in a manner consistent with eventual recovery. Under conservative conditions, vesicle reformation is immediate and total vesicle numbers are not affected though the fact that vesicles have gone through at least one cycle of exo- and endocytosis is revealed by their ability to sequestrate extracellular dextran and their reduction in diam-

160

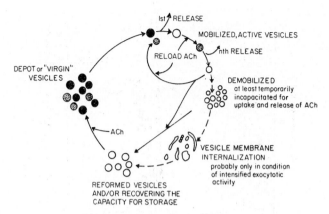

Fig. 6. Scheme of vesicle membrane cycling in relation to ACh storage and release.

eter (Zimmermann and Denston, 1977a,b). Under more vigorous conditions, multiple or "piggy back" exocytosis appears to take place, whereby vesicles fuse with developing or ingrowing indentations of the plasma membrane formed by vesicles that have previously exocytosed and have not been reformed. During recovery, which must be assumed to be an ongoing process even when the steady state configuration is pushed strongly in the direction of exocytosis, segmentation or pinching off of these multiple vesicle fusions occurs with the internalizing of vesicle membrane and the reformation of vesicles. Further morphological studies will be required to distinguish the details of this process and to decide, for example, whether all membranous cisternae or vacuoles are in continuity with the extracellular space or represent a phase in vesicle reformation. It seems clear however that vesicle membrane recycling, at least in *Torpedo*, cannot under all conditions of stimulation be represented by a single cycle but requires a more complex scheme such as that shown in Fig. 6.

The observation that the SRA of ACh released from blocks whose ACh stores had been prelabelled with radioactive acetate did not differ significantly from that of the highly labelled subpopulation of synaptic vesicles resolves a paradox the existence of which has often been cited as one of the main objections to the vesicle theory, namely that the SRA of released ACh may frequently be markedly different from that of the vesicle population as a whole.

SUMMARY

(1) Vesicle turnover and acetylcholine (ACh) metabolism have been compared in isolated blocks of *Torpedo* electric tissue under conditions selected to produce (a) maximum conservation of synaptic vesicle numbers and maximum metabolic heterogeneity (b) maximum depletion of synaptic vesicles and maximum metabolic homogeneity. The former involved crossed circuit perfusion, stimulation at 1 Hz and a 3 h recovery period, the latter open circuit perfusion, stimulation at 1 Hz and up to 20 h recovery.

(2) The expected conservation of vesicle number and appearance of two

separable vesicle populations, designated VP_1 and VP_2 under condition (a) was confirmed. When ACh stores were labelled with [^{3}H]acetate and the blocks restimulated, the specific radioactivity (SRA) of the released ACh did not differ significantly from that of the ACh in fraction VP_2, though it was much higher than that of the whole tissue and VP_1.

(3) Under condition (b) vesicle numbers were severely depleted and invaginations of the presynaptic plasma membrane, vacuoles and cisternae made their appearance. These changes reversed themselves and vesicle numbers were restored to normal after 5 h rest, but vesicular ACh, also severely depleted, had not completely recovered at 20 h. When ACh stores were labelled with [^{3}H]-acetate and the blocks restimulated, only one peak of vesicular ACh was separated, but evidence was again obtained for the presence of a subpopulation of highly labelled vesicles which were the source of the ACh released on stimulation.

(4) The demonstration that the SRA of released ACh does not significantly differ from that of a subpopulation of vesicles active in taking up newly synthesized transmitter under two very different conditions removes an often cited objection to the vesicle theory of transmitter storage and release.

(5) Vesicle cycling follows different, though partially overlapping routes depending on the stimulation conditions. One of these is thought to involve transient exo- and endocytosis, the other multiple or "piggyback" exocytosis and segmental pinching-off of invaginations with the reformation of vesicles. Both types of reconstituted vesicles appear to be capable of renewed transmitter storage.

ACKNOWLEDGEMENTS

Appreciation is expressed to Dr. Lamia H. Khairallah, University of Connecticut EM Service, for her assistance with electron microscopy. J.B.S. was supported by a Max-Planck Fellowship and in part by a travel grant from the Research Foundation of the University of Connecticut, Storrs, Conn., U.S.A.

REFERENCES

Barker, L.A., Dowdall, M.J. and Whittaker, V.P. (1972) Choline metabolism in the cerebral cortex of guinea pigs. Stable-bound acetylcholine. *Biochem. J.*, 130, 1063–1080.

Boyne, A.F., Bohan, T.P. and Williams, T.H. (1975) Changes in cholinergic synaptic vesicle populations and the ultrastructure of the nerve terminal membranes of *Narcine brasiliensis* electric organ stimulated to fatigue in vivo. *J. Cell Biol.*, 67, 814–825.

Ceccarelli, B., Hurlbut, W.P. and Mauro, A. (1972) Depletion of vesicles from frog neuromuscular junctions by prolonged tetanic stimulation. *J. Cell Biol.*, 54, 30–38.

Ceccarelli, B., Hurlbut, W.P. and Mauro, A. (1973) Turnover of transmitter and synaptic vesicles at the frog neuromuscular junction. *J. Cell Biol.*, 57, 499–524.

Suszkiw, J.B., Zimmermann, H. and Whittaker, V.P. (1978) Vesicular storage and release of acetylcholine in *Torpedo* electroplaque synapses. *J. Neurochem.*, 30, 1269–1280.

Whittaker, V.P., Essman, W.B. and Dowe, G.H.C. (1972) The isolation of pure cholinergic synaptic vesicles from the electric organs of elasmobranch fish of the family Torpedinidae. *Biochem. J.*, 128, 833–845.

Whittaker, V.P., Zimmermann, H. and Dowdall, M.J. (1975) The biochemistry of cholinergic

162

synapses as exemplified by the electric organ of *Torpedo*. *J. neural Trans.*, Suppl. XII, 39–60.

Zimmermann, H. and Whittaker, V.P. (1974a) Effect of electrical stimulation on the yield and composition of synaptic vesicles from the cholinergic synapses of the electric organ of *Torpedo*; a combined biochemical, electrophysiological and morphological study. *J. Neurochem.*, 22, 435–450.

Zimmermann, H. and Whittaker, V.P. (1974b) Different recovery rates of the electrophysiological, biochemical and morphological parameters in the cholinergic synapses of the *Torpedo* electric organ after stimulation. *J. Neurochem.*, 22, 1109–1114.

Zimmermann, H. and Denston, C.R. (1977a) Turnover of synaptic vesicles in the cholinergic synapses of the *Torpedo* electric organ during induced transmitter release. *Neuroscience*, 2, 695–714.

Zimmermann, H. and Denston, C.R. (1977b) Separation of synaptic vesicles of different functional states from cholinergic synapses of the *Torpedo* electric organ. *Neuroscience*, 2, 715–730.

Zimmermann, H. and Whittaker, V.P. (1977) Morphological and biochemical heterogeneity of cholinergic synaptic vesicles. *Nature (Lond.)*, 267, 633–635.

Vesicular Storage and Release of Cholinergic False Transmitters

I. VON SCHWARZENFELD [a], G. SUDLOW [b] and V.P. WHITTAKER [b]

Pharmakologisches Institut der Universität Mainz [a] and Abteilung Neurochemie,
Max-Planck-Institut für biophysikalische Chemie, Postfach 986,
3400 Göttingen [b] (F.R.G.)

INTRODUCTION

The vesicular theory of transmitter storage and release views neurotransmitter release as a modified form of cell secretion, in which the synaptic vesicles are regarded as miniature storage granules, their small size being accounted for by the exigencies of transport from their site of formation in the Golgi membranes of the neuronal perikarya to their site of utilization in the nerve terminal, by way of the axon. The specialized geometry of the nerve cell imposes a second modification: the synaptic vesicles do not, apparently, undergo total exocytosis, but are reformed in a functional state and can be recharged with transmitter freshly synthesized or freshly recaptured from the synaptic cleft. In spite of much recent criticism (MacIntosh and Collier, 1976; cf. Rang, 1978) vesicular storage and release remains an attractive theory, since it links neurotransmitter release to a general cellular mechanism, that of secretion, and also neatly explains the generation of miniature postsynaptic potentials by resting nerve terminals. The magnitude of these clearly requires that transmitter must be released in packets or "quanta" of several, perhaps many, thousands of molecules at a time; the vesicle provides a simple morphological basis for this quantization. However, it turns out to be technically difficult to measure, on the same tissue, the size of the quantum and the transmitter content of a vesicle with the accuracy needed to provide a critical test of the theory. Thus other kinds of evidence are needed.

Vesicular storage and release of transmitter seems to be accepted without question in the adrenergic system (for review see Smith, 1972). Important supporting evidence has come from the study of "false transmitters" (see Kopin, 1968, Collier, this volume, p. 107). Noradrenergic terminals and the dense-cored vesicles within them readily take up a large number of phenylethylamines; however, the specificity of the uptake system (permease) in the external

---

* This article was originally planned as a joint contribution of the three authors named. The untimely death of Irene von Schwarzenfeld on April 6, 1978 in her 37th year has deprived the surviving authors of her judgement and approval of the final manuscript, and her country of a neurochemist of great promise. A full account of her work was submitted for publication to Neuroscience just before she died and is now in press.

164

presynaptic membrane is different from that in the vesicles, with the consequence that some phenylethylamines enter only the cytoplasm while others are taken up into vesicles. *On stimulation, only those phenylethylamines that penetrate into vesicles are released.*

The cholinergic system is more complicated: when ACh is released, it is broken down to choline and acetate and the former is recaptured by the nerve terminal. The choline is reacetylated in the cytoplasm (the relevant enzyme, choline acetyltransferase which utilizes acetylcoenzyme A as its other substrate is located in the cytoplasm) and the ACh so formed is taken up into vesicles. Thus there are three steps instead of two, each with a different specificity, which an analogue of choline must negotiate before it can enter the vesicle compartment. However, in principle, it should be possible to utilize the same approach in the cholinergic system as in the adrenergic and some analogues of choline have indeed been shown to be false transmitter precursors (Fig. 1; see legend for references).

In the work to be described here two well characterized types of cholinergic terminal have been investigated, those of the guinea-pig cortex (von Schwarzenfeld, 1978a,b) and the *Torpedo* electric organ (G. Sudlow, preliminary results). Earlier studies with *Torpedo* electric organ have shown that the incorporation of newly synthesized ACh into vesicular pools is greatly increased by stimulation (Zimmermann and Whittaker, 1977; Zimmermann and Denston, 1977a,b; Suszkiw, Zimmermann and Whittaker, 1978; Suszkiw and Whittaker, 1979). Thus in the present work the administration of putative radiolabelled false transmitter precursors has been accompanied by low frequency stimulation which, at any rate in *Torpedo* using closed circuit perfusion, does not significantly reduce the vesicle population or seriously deplete transmitter stores. This "loading" stimulus is followed, after a suitable "washout" period to remove extracellular precursors, by a higher frequency "release" stimulus, which causes release of transmitter and false transmitter from the subcellular stores. The ratio in which natural and false transmitters are released is then compared with the ratio in which they are present in the various subcellular pools, as separated by subcellular fractionation applied to the tissue immedi-

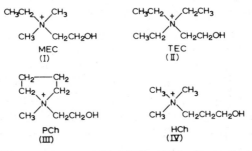

Fig. 1. False transmitter precursors: (I) MEC, *N*-2-hydroxyethyl-*N*-ethyl-*N,N*-dimethyl-ammonium; (II) TEC, *N*-2-hydroxyethyl-*N,N,N*-triethylammonium. (III) PCh, *N*-methyl-*N*-2-hydroxyethylpyrrolidinium; (IV) HCh, *N*-3-hydroxypropyl-*N,N,N*-trimethylammonium (homocholine). The acetyl esters of compounds I–IV have been shown to be false transmitters in the cat superior cervical ganglion (Ilson and Collier, 1976; Collier, Barker, and Mittag, 1976; Collier, Lovat, Ilson, Barker and Mittag, 1977) and the acetyl ester of compound III in *Torpedo* electric organ (Zimmermann and Dowdall, 1977) and the myenteric plexus of the guinea-pig ileum (Kilbinger, 1977).

ately upon termination of the release stimulus. Provided the two compounds distribute themselves sufficiently differently among the pools, this should enable the subcellular origin of the released transmitter to be identified.

Our subcellular fractionation studies have confirmed the concept of vesicular heterogeneity and the presence within the total vesicle population of a sub-population of metabolically active vesicles which preferentially take up newly synthesized transmitter (Barker, Dowdall and Whittaker, 1972; Dowdall and Zimmermann, 1974; Zimmermann and Whittaker, 1977; Zimmermann and Denston, 1977a,b; von Schwarzenfeld 1978a,b; Zimmermann, 1979; Suszkiw and Whittaker, 1979). For cortex, this is fraction H of the fractionation scheme of Whittaker, Michaelson and Kirkland (1964), a fraction of synaptic vesicles adhering to fragments of the external terminal membrane; for electric organ, it is the denser vesicle fraction (VP_2) of Zimmermann and colleagues (Zimmermann and Whittaker, 1977; Zimmermann and Denston, 1977b). HCh (Fig. 1, IV) has proved to be a particularly useful precursor since its acetyl ester tends to accumulate in the cytoplasm, causing marked differences in the composition of the vesicular and cytoplasmic pools. This has provided a sensitive test of the vesicle theory.

METHODS

Guinea-pig cortex

The methods have been fully described elsewhere (von Schwarzenfeld, 1978b) and will only be briefly summarized here. To study release, the "Oborin cup" technique (Mitchell, 1963) was used. Animals were anaesthetized with urethane (1.5 g/kg) and maintained at 37°C. The skin was removed over the parietal areas of the skull, a hole 9.1 mm in diameter was drilled through each parietal bone and the dura mater removed. A Perspex cylinder, bevelled at the lower edge (height 10 mm, internal diameter 9 mm) was inserted snugly into each hole until it rested on the cortex and was sealed into position with acrylyl cyanide. The animals were then transferred to a stereotaxic frame and the cups formed by the cylinders filled with warm (37°C) Locke solution containing 100 μM paraoxon (0.5 ml). For stimulation, stainless steel electrodes were inserted into each cortex through the cup area to a depth of 1.5 mm; the indifferent electrode was placed in the neck muscle. After 30 min the Locke solution was replaced and radioactive choline and choline analogue (11 μCi of [^{14}C]choline, 50 mCi/mmol + 72 μCi of [^{3}H]PCh, 2.2 Ci/mmol or 12 μCi of [^{14}C]choline + 86 μCi of [^{3}H]HC, 2.2 Ci/mmol) was injected into the cortex below the cup at a depth of 1.5 mm. Low frequency electrical stimulation (0.1 Hz, 1.5 ms, 10 mA) was commenced and continued for 40 min to enhance vesicular uptake of the transmitters. Radioactivity was allowed to diffuse for 20 min, the cups were then rinsed 6 times with 0.5-ml Locke solution/cup at 5 min intervals to reduce radioactivity due to unincorporated precursors. This rinsing was followed by a series of up to five 15-min collection periods. During the second, third or fifth collection period, high frequency stimulation (30 Hz, 1.5 ms, 10 mA) was applied. The protocol is summarized in Fig. 2a. At the end

166

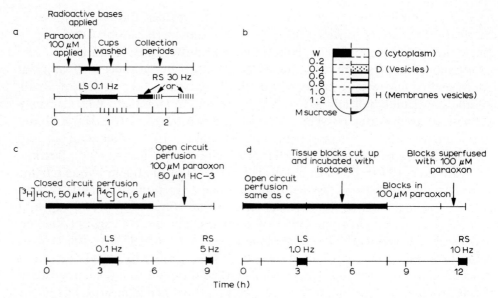

Fig. 2. Summary of experimental procedures using (a) guinea-pig cortex, (c,d) *Torpedo* electric organ. Note that the incorporation of radiolabelled precursors is promoted by low frequency (0.1 Hz a,c; 1 Hz, d) stimulation (LS) and release of labelled stored is achieved by higher (RS) frequencies [5 (c), 10 (d) or 30 (a) Hz]. Paraoxon (100 μM) was used to block esteratic hydrolysis of released esters and HC-3 (c) to block reutilization of precursors during release. (b) Subfractionation of guinea-pig cortical synaptosomal-mitochondrial fraction after disruption by osmotic shock.

of stimulation the underlying cortical tissue ($\sim$300 mg) was excised and used to prepare synaptic vesicles and soluble terminal cytoplasm essentially as described by Whittaker and Barker (1972), but with the following modifications (Fig. 2b). The synaptosomal-mitochondrial fraction P_2 was suspended in water containing 100 μM eserine (3 ml/g of tissue) to preserve cytoplasmic ACh and its analogues, and the whole fraction (W) (approx. 0.9 ml) placed on a discontinuous gradient into which a layer of 0.2 M sucrose (4 ml/tube) had been inserted between the sample and the layers (1.5–2.0 ml/tube) of 0.4 to 1.2 M sucrose differing by 0.2 M normally used. This modification reduced cytoplasmic contamination of the fraction of monodispersed synaptic vesicles (fraction D). The top layer (diluted cytoplasm, fraction O), the 0.4 M sucrose layer (fraction D) and the 1.0–1.2 M interface (synaptic vesicles in association with external cytoplasmic membranes, fraction H) were collected after centrifuging.

Extraction and separation of radiolabelled compounds

Radioactive compounds were extracted from 0.8 ml of the combined cup solutions and 1.0–1.5 ml of the subcellular fractions using essentially the method of Barker, Dowdall and Whittaker (1972). Residues obtained after liquid ion-exchange extraction into tetraphenylboride in allyl cyanide (10 mg/ml) and removal of tetraphenylboride ion on Dowex 1 resin (Cl$^-$ form) were taken up in methanol and separated on cellulose thin layer chromatography sheets using butanol : methanol : glacial acetic acid : H_2O (8 : 2 : 1 : 3) (choline, PCh) or propanol : formic acid : H_2O (8 : 1 : 1) (choline, HCh) (Barker and Mittag, 1975). Areas corresponding to the separated esters (detected by Dragendorff's reagent) were scraped into scintillation vials and the radioactivity counted. APCh and AHCh were corrected for 1—1.5% contamination by the unacetylated bases, ACh for enzymic hydrolysis in the cups (5%) and all three esters for losses during extraction and separation (15%). Results for fraction D are corrected for contamination from fraction O (2.3—4.8% depending on the compound).

Torpedo electric organ

Innervated blocks of electric tissue comprising the territory of one electromotor nerve and its accompanying blood vessels were dissected out, perfused and stimulated as described by Zimmermann and Denston (1977b); the protocol is summarized in Fig. 2c. A second procedure (Suszkiw and Whittaker, 1979) was also used and gave clearer results due to the lower background of unutilized radiolabelled compounds before "release" stimulation. In this procedure (Fig. 2d) tissue blocks were initially dissected, perfused and stimulated as before except that the perfusion medium was not recirculated and the blocks were stimulated at 1 Hz for the "loading" stimulus. The blocks were then cut into smaller pieces, incubated and washed and then reperfused with wash-out medium. Field stimulation at 10 Hz was employed for release. In all experiments, control blocks were treated in an identical manner except for the final "release" stimulus.

At the end of the "release" stimulus, the control and stimulated tissue blocks were frozen in liquid nitrogen and vesicles pelletted (Zimmermann and Dowdall, 1977) or separated by zonal centrifuging (Zimmermann and Denston, 1977b). Vesicles were identified by their content of ATP which was assayed by the luciferin—luciferase method of Stanley and Williams (1969) as modified by Dowdall, Boyne and Whittaker (1974).

Extraction and separation of radiolabelled compounds. This was achieved essentially as described for the guinea-pig cortex except that the compounds were separated by high-voltage electrophoresis (Brooker and Harkiss, 1974) using 8% aqueous formic acid as solvent.

RESULTS

Guinea-pig cortex

Experiments with PCh. Earlier experiments (von Schwarzenfeld, 1978a) showed that after intraventricular injection of [^{14}C]choline and [^{3}H]PCh, both bases are rapidly taken up into brain tissue and acetylated; the acetyl esters are incorporated into vesicles. The ratios in which the esters are present in the fractions differ significantly at any one time from one fraction to another (Fig. 3). Fraction H took up more label than fraction D and also more APCh relative to ACh than fraction D, which is consistent with earlier conclusions (Barker, Dowdall and Whittaker, 1972) that it more avidly than fraction D takes up newly synthesized ester. Low frequency electrical stimulation after intra-

168

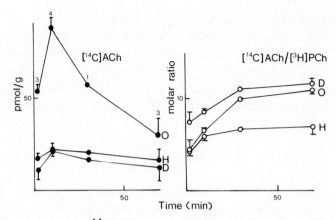

Fig. 3. (a) Incorporation of [^{14}C]ACh into brain fractions O, H and D following intraventricular injection of [^{14}C]choline and [^{3}H]PCh. (b) Incorporation of [^{3}H]APCh in the same experiments, expressed as the molar ratio of [^{14}C]ACh to [^{3}H]APCh. Points are the means and bars the S.E.M.s of the number of experiments indicated.

cortical injection of the precursors greatly enhanced uptake of esters into fraction H (by 330% for ACh and 280% for APCh) but had less effect on fraction D (197% increase for ACh and none for APCh). By contrast, high frequency stimulation decreased the amount of transmitters in fraction H, but had little effect on fraction D (Fig. 4).

Fig. 5a shows the release of radioactive ACh and APCh into the cups after loading the cortex with the radioactive precursors. The application of high frequency stimulation approximately doubled the release and also increased the

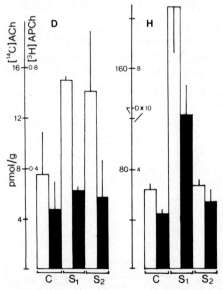

Fig. 4. Effect of stimulation on the incorporation of [^{14}C]ACh (white blocks) and [^{3}H]-APCh (black blocks) into brain fractions. Blocks are mean values, and bars are S.E.M.s; the number of experiments was 3 (control, 0.1 Hz stimulation) or 4 (30 Hz stimulation). For further details see text. C, control; S_1, 0.1 Hz; S_2, 30 Hz. Note that the scale of H is is 10 times that of D.

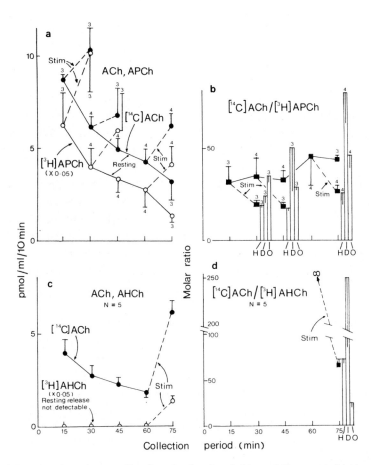

Fig. 5. (a,c) Resting (continuous lines) and stimulated (dotted lines, stim) release of [^{14}C]-ACh (filled circles) and either [^{3}H]APCh (a) or [^{3}H]AHCh (c) (open circles) into cups placed on the surface of the guinea-pig cortex. For the procedure see text and Fig. 2a. (b,d) Ratios of acetyl esters (ACh/APCh, b; ACh/AHCh, d) in the cortical superfusate (squares) at rest (continuous lines) or after stimulation (dotted lines). Bars give ratios in fractions containing terminal soluble cytoplasm (O), monodispersed synaptic vesicles (D) or synaptic vesicles associated with external presynaptic membranes (H) prepared at the end of stimulation. For preparation of fractions see text and Fig. 2b. Note close agreement between the ratio in which the natural and false transmitters are released on stimulation and that in which they are found in fraction H. All points and blocks are mean values of the number of experiments indicated; bars are S.E.M.s. The ratios for stimulated release have been corrected by extrapolation for the contribution made to total release by resting release.

output of endogenous ACh (not shown) irrespective of the collection period in which it was applied. Fig. 5b shows that the effect of the release stimulus was to lower significantly the [^{14}C]ACh/[^{3}H]APCh ratio. When fractionation was carried out on the tissue subjacent to the cup, the ratio of the two radioactive esters in fraction H was found to be significantly different from those in fractions O and D but *not significantly different from that of the released esters*. This finding was again independent of the collection period during which the release stimulus was applied. By contrast, the ratio in which the transmitters were released under resting conditions was close to that found in the cytoplasmic fraction O.

Experiments with HCh. Fig. 5c shows the results of experiments similar to those in Fig. 5a, but with [³H]HCh in place of PCh. Under resting conditions [¹⁴C]ACh was released, in this series of experiments, essentially as before, but after appropriate corrections had been made for background material, no release of [³H]AHCh could be detected. On high frequency stimulation, however, not only was the output of [¹⁴C]ACh increased, but clearly detectable amounts of [³H]AHCh made their appearance. The finite ratio in which the two transmitters were released during this collection period was not significantly different from that in which they were present in fraction H though quite different from those of fractions D and O (Fig. 5d).

A comparison of the ratio [¹⁴C]ACh/[³H]AHCh with the corresponding ratio for APCh in fractions O and H shows that AHCh is a much poorer false transmitter than APCh. HCh is not only less efficiently taken up and acetylated than PCh; the acetylated product is much less efficiently taken up into vesicles and tends to accumulate in the cytoplasm. Thus the ratio for fraction O is much smaller than that for fraction H or than the corresponding ratio for fraction O in the PCh experiments. It is thus clear that the ratio in which two transmitters are released from cortical cholinergic endings on stimulation is determined by their ratio in the active vesicles associated with the presynaptic plasma membrane (fraction H) whatever their ratio may be in the cytoplasm (fraction O) or the pool of inactive vesicles (fraction D).

Release of endogenous ACh. Endogenous ACh was also measured in those experiments in which stimulation was applied in the 60–75-min collection period. In both series of experiments (those with PCh and those with HCh), there was excellent agreement between the specific radioactivity of the released ACh and that in fraction H; both were significantly different from those of fractions O and D.

Some control experiments. Electrical stimulation does not produce non-specific liberation of material from either the intracellular or extracellular compartments of the brain. These were labelled by injecting [2-³H]deoxyglucose and [¹⁴C]inulin respectively. Electrical stimulation did not evoke any release of these substances additional to that from unstimulated tissue. Other experiments showed that fraction H contained no significant amounts of soluble cytoplasmic markers or entrained cytoplasm.

Torpedo electric organ

Fig. 6a shows the release of ³H and ¹⁴C from small tissue blocks prepared according to the protocol shown in Fig. 2d as a result of field stimulation at 10 Hz after previous loading with [³H]HCh and [¹⁴C]choline. There is a low background radioactivity; the effect of stimulation is to cause a marked fall in the ratio, due to the preferential release of [¹⁴C]ACh (Fig. 6b). In Fig. 6b the ratio of the two isotopes in the perfusing solution is compared with that in the total tissue (TT) and the vesicle fraction (V) derived from the control (left) and stimulated (right) blocks. It will be seen that the ³H/¹⁴C ratio in the tissue is several times higher than that in the vesicles. Electrophoresis showed that the vesicles contained ¹⁴C mainly as ACh and ³H as a mixture of HCh + AHCh. Thus HCh evidently passed into the vesicles. The radioactivity of the tissue extracts was similarly constituted; however, as indicated by their higher ³H/¹⁴C

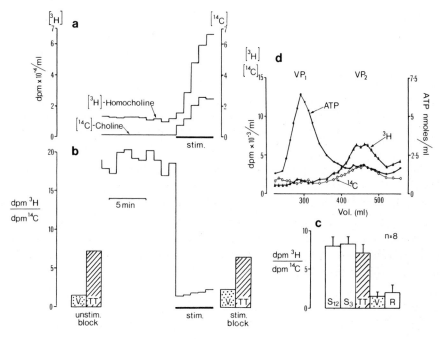

Fig. 6. Release of [³H] and [¹⁴C] as a result of high frequency stimulation (stim.) from a block of electric tissue preloaded with [³H]HCh and [¹⁴C]choline, according to the procedure outlined in Fig. 2d. (b) Isotope ratio in the perfusate and in extracts of the total tissue (TT) and a crude vesicle pellet (V) from blocks that either did (stim.) or did not (unstim.) receive the "release" stimulus. The vesicle pellet was sedimented from a cytoplasmic extract in iso-osmotic glycine of crushed and frozen tissue at $100,000 \times g$ for 60 min after removal of coarse- and medium-sized particles at $10,000 \times g$ for 30 min. Other experiments show that the additional radioactivity is due to ACh [¹⁴C] or ACH + HCh [³H]. Note that on stimulation the isotope ratio falls rapidly to that of the vesicles and is well below that of the whole tissue. (c) Pooled results of eight experiments showing mean ratios (blocks) + S.E.M. (bars) after a release stimulus in the perfusate (R), vesicles (V), total tissue (TT), vesicle supernatant (S_3) and cytoplasmic extract (S_{12}) of the frozen and crushed organ. The perfusate ratios have been corrected for background radioactivity. Note that there is no significant difference between R and V suggesting that the esters are released from the vesicular compartment. (d) Zonal separation of vesicles isolated from tissue blocks prepared according to the protocol shown in Fig. 2c. Note that both radioactive components are preferentially incorporated in the metabolically active fraction (VP_2) of synaptic vesicles. Thus the radioactivity found in the crude vesicle pellets can be largely ascribed to the presence of the VP_2 population.

ratio there was, relative to [¹⁴C]ACh, far more AHCh and HCh than in vesicles. Thus AHCh (and HCh) are poor substrates for the vesicles permease (though better than choline).

This conclusion is confirmed by the results shown in Fig. 6c in which the ratios found in two cytoplasm-rich fractions, the cytoplasmic extract (S_{12}) from which the vesicles are derived, and the supernatant (S_3) left after sedimenting the vesicles are also seen to be high. The ratio in the perfusate during high frequency stimulation (Figs. 6b,c) did not differ significantly from that of the vesicle fraction, but was far below that of whole tissue or cytoplasm-rich fractions.

Fig. 6d shows, as expected, that newly formed false transmitter is preferentially incorporated into the same VP_2 fraction of vesicles as is known preferentially to accumulate newly synthesized ACh (Zimmermann and Denston, 1977b) and ATP (Zimmermann, 1979). This fraction can thus be regarded as the *Torpedo* counterpart to fraction H of guinea-pig brain. In the experiment shown in Fig. 6d, the isotope ratio in VP_1 (equivalent to fraction D) was different from that in VP_2 but this was not a constant finding.

DISCUSSION AND CONCLUSIONS

Earlier work with false transmitters (Ilson and Collier, 1976; Collier, Barker and Mittag, 1976; Collier, Lovat, Ilson, Barker and Mittag, 1977; Kilbinger, 1977) simply demonstrated that these compounds were released on stimulation: their subcellular distribution was not investigated. Zimmermann and Dowdall (1977) were the first to show, in the *Torpedo* electric organ, that one of these compounds, APCh, was incorporated into vesicles as well as being released on stimulation and that the ratio in which natural and false transmitters were released was consistent with a vesicular origin of the released transmitters. However, the distribution of APCh in the subcellular pools did not differ sufficiently from that of ACh to provide a sensitive test of the vesicle theory.

The experiments presented here show that in both mammalian cortex and *Torpedo* electric organ, natural and false transmitters are released in a ratio not significantly different from that in which they are found in a metabolically active subpopulation of vesicles *irrespective of the ratio in which they occur in the cytoplasm or the tissue as a whole*. This must be regarded as strong evidence in favour of the vesicular theory of transmitter storage and release.

SUMMARY

Several analogues of choline are known which are taken up by cholinergic nerve terminals and released in the acetylated form on stimulation. Such compounds are thus precursors of cholinergic false transmitters. We have now studied two such compounds, N-methyl-N-2-hydroxyethylpyrrolidinium (PCh) and homocholine (HCh), using guinea-pig cortex and the electric organ of *Torpedo marmorata*. In the experiments with cortex, the false transmitter precursors were injected into the tissue and released transmitters collected in Locke solution applied to the surface of the cortex in plastic sleeves. Free vesicles and those attached to external membranes were isolated from tissue subjacent to the cup. Blocks of electric tissue corresponding to the territory of one nerve were perfused through the accompanying venous sinus and vesicles isolated by the technique currently employed in our laboratory. In both cases uptake of [^{14}C]choline and the tritiated analogue were promoted by electrical stimulation. Further stimulation caused the release of true and false transmitters in the same ratio as they were found in a metabolically active subpopulation of vesicles even when this differed greatly from that found in cytoplasm (brain) or whole tissue (*Torpedo*).

ACKNOWLEDGEMENTS

I. von Schwarzenfeld's work was supported by the Deutsche Forschungs-gemeinschaft; Miss A. Mergler gave valuable technical help in the later stages of the work. G. Sudlow is grateful for an Auslandsstipendium from the Max-Planck-Gesellschaft. We are grateful to Dr. M.J. Dowdall for much helpful advice.

REFERENCES

Barker, L.A. and Mittag, T.W. (1975) Comparative studies of substrates and inhibitors of choline transport and choline acetyltransferase. *J. Pharmacol. exp. Ther.*, 192, 86–94.

Barker, L.A., Dowdall, M.J. and Whittaker, V.P. (1972) Choline metabolism in the cerebral cortex of guinea-pigs: stable bound acetylcholine. *Biochem. J.*, 130, 1063–1080.

Brooker, S.E. and Harkiss, K.J. (1974) High voltage electrophoresis of choline and its esters. *J. Chromatog.*, 89, 96–98.

Collier, B., Barker, L.A. and Mittag, T.W. (1976) The release of acetylated choline analogues by a sympathetic ganglion. *Molec. Pharmacol.*, 12, 340–344.

Collier, B., Lovat, S., Ilson, D., Barker, L.A. and Mittag, T.W. (1977) The uptake, metabolism and release of homocholine: studies with rat brain synaptosomes and cat superior cervical ganglion. *J. Neurochem.*, 28, 331–339.

Dowdall, M.J. and Zimmermann, H. (1974) Evidence for heterogeneous pools of acetylcholine in isolated cholinergic synaptic vesicles. *Brain Res.*, 71, 160–166.

Dowdall, M.J., Boyne, A.F. and Whittaker, V.P. (1974) Adenosine triphosphate, a constituent of cholinergic synaptic vesicles. *Biochem. J.*, 140, 1–12.

Ilson, D. and Collier, B. (1976) Triethylcholine as a precursor to a cholinergic false transmitter. *Nature (Lond.)*, 254, 618–620.

Kilbinger, H. (1977) Formation and release of acetylpyrrolidinecholine (*N*-methyl-*N*-acetoxyethylpyrrolidinium) as a false cholinergic transmitter in the myenteric plexus of the guinea pig small intestine. *Naunyn-Schmiedeberg's Arch., Pharmacol.*, 296, 153–158.

Kopin, I.J. (1968) False adrenergic transmitters. Ann. Rev. Pharmacol., 8, 377–394.

MacIntosh, F.C. and Collier, B. (1976) Neurochemistry of cholinergic terminals. In *Neuromuscular Junction*, E. Zaimis (Ed.), *Handb. exp. Pharmacol., New Ser., Vol. 42*, Springer Verlag, Berlin, pp. 99–228.

Mitchell, J.F. (1963) The spontaneous and evoked release of acetylcholine from the cerebral cortex. *J. Physiol. (Lond.)*, 165, 98–116.

Rang, H.P. (1978) Chemically transmitting synapses [review of G.A. Cottrell and P.N.R. Usherwood (Eds.) Synapses, Blackie, Glasgow, 1977] *Nature (Lond.)*, 271, 191.

Schwarzenfeld, I. von (1978a) The uptake of acetylpyrrolidinecholine – a false cholinergic transmitter – into mammalian cerebral cortical synaptic vesicles. In *Cholinergic Mechanisms and Psychopharmacology*, D.J. Jenden (Ed.), Plenum Press, New York, pp. 657–672.

Schwarzenfeld, I. von (1978b) Origin of transmitters released by electrical stimulation from a small, metabolically very active vesicular pool of cholinergic synapses in guinea-pig cerebral cortex. *Neuroscience*, in press.

Smith, A.D. (1972) Cellular control of the uptake, storage and release of noradrenaline in sympathetic nerves. *Biochem. Soc. Symp.*, 36, 103–131.

Stanley, P.E. and Williams, S.G. (1969) Use of liquid scintillation spectrometer for determining adenosine triphosphate by the luciferase enzyme. *Analyt. Biochem.*, 29, 381–392.

Suszkiw, J.B. and Whittaker, V.P. (1979) Role of vesicle recycling in vesicular storage and release of ACh in *Torpedo* electroplaque synapses. In *The Cholinergic Synapse, S. Tuček (Ed.), Progr. Brain Res., Vol. 49*, Elsevier, Amsterdam, pp. 153–162 (this volume).

Suszkiw, J.B., Zimmermann, H. and Whittaker, V.P. (1978) Vesicular storage and release of acetylcholine in *Torpedo* electroplaque synapses. *J. Neurochem.*, 30, 1269–1280.

Whittaker, V.P. and Barker, L.A. (1972) The subcellular fractionation of brain tissue with special reference to the preparation of synaptosomes and their component organelles. *Methods Neurochem., 2,* 1–52.

Whittaker, V.P., Michaelson, I.A. and Kirkland, R.J.A. (1964) The separation of synaptic vesicles from nerve-ending particles ("synaptosomes"). *Biochem. J., 90,* 293–303.

Zimmermann, H. (1979) Vesicular heterogeneity and turnover of acetylcholine and ATP in cholinergic synaptic vesicles. In *The Cholinergic Synapse, S. Tuček (Ed.), Progr. Brain Res., Vol. 49,* Elsevier, Amsterdam, pp. 141–151 (this volume).

Zimmermann, H.and Denston, C.R. (1977a) Recycling of synaptic vesicles in the cholinergic synapses of the *Torpedo* electric organ during induced transmitter release. *Neuroscience, 2,* 695–714.

Zimmermann, H. and Denston, C.R. (1977b) Separation of vesicles of different functional states from the cholinergic synapses of the *Torpedo* electric organ. *Neuroscience, 2,* 715–730.

Zimmermann, H. and Dowdall, M.J. (1977) Vesicular storage and release of a false cholinergic transmitter (acetylpyrrolcholine) in the *Torpedo* electric organ. *Neuroscience, 2,* 731–739.

Zimmermann, H. and Whittaker, V.P. (1977) Morphological and biochemical heterogeneity of cholinergic synaptic vesicles. *Nature (Lond.), 267,* 633–635.

Facilitation, Augmentation, and Potentiation of Transmitter Release

K.L. MAGLEBY

Department of Physiology and Biophysics, University of Miami School of Medicine, Miami, FL 331 (U.S.A.)

INTRODUCTION

When a synapse is stimulated repetitively, the amount of transmitter released by each nerve impulse is not constant, but is a function of the previous frequency and duration of stimulation (Feng, 1941; Liley, 1956). This chapter presents a brief summary of some of the progress that has been made in recent years towards understanding the dynamic properties of transmitter release. The experiments to be described are mainly concerned with the neuromuscular junction, and the topics to be discussed are those that reflect personal interests. Recent reviews (Hubbard, 1973; Barrett and Magleby, 1976) and the cited papers may be consulted for more complete details.

It is now generally accepted that evoked transmitter release occurs in multi-molecular packets (quanta) (del Castillo and Katz, 1954a), and that it is the entry of Ca^{2+} into the nerve terminal at the time of the action potential that leads to the release of the quanta (Katz and Miledi, 1969). Evidence is also accumulating which suggests that one quantum of transmitter is released when a synaptic vesicle discharges its contents of transmitter into the synaptic cleft by exocytosis (Heuser, this volume, p. 478), and that synaptic vesicles are discharged at distinct release sites or active zones in the nerve terminal (Peper et al., 1974).

It is not known how Ca^{2+} acts to cause transmitter release, but the observation that the number of quanta released by a nerve impulse is related to the fourth power of Ca^{2+} in the bathing solution has led to the suggestion that four Ca^{2+} ions may act together to cause the release of a quantum of transmitter (Dodge and Rahamimoff, 1967).

From the above observations it appears that changes in transmitter release could result from changes in Ca^{2+} entry or sequestering, changes in the number or positions of synaptic vesicles at the release sites, changes in the number of activated release sites, or changes in the factors that lead to fusion of the synaptic vesicle membrane with the nerve terminal membrane. There are undoubtedly also other factors not considered above which affect transmitter release.

It is obvious that it will be necessary to have a complete understanding of all the factors involved in transmitter release before the mechanism of transmitter release is thoroughly understood. The experimental method we have used to study transmitter release is an extension of the approach used by Feng (1941) and formalized by Mallart and Martin (1967). The effect of repetitive stimulation on transmitter release has been studied to look for and characterize the processes in the nerve terminal that affect transmitter release.

PROCESSES THAT AFFECT TRANSMITTER RELEASE

The processes which affect transmitter release can be separated into two classes: those that act to increase release (facilitation, augmentation, and potentiation) and those that act to decrease release (depression). Under conditions of normal levels of transmitter release, repetitive stimulation first leads to an increase, and then following a number of impulses, to a decrease (depression) in the amount of transmitter released by each nerve impulse (Magleby, 1973b). The mechanism of depression may involve a depletion of transmitter available for release (Thies, 1965; and see Bennett and Fisher, 1977).

When transmitter release is reduced to low levels by decreasing the amount of Ca^{2+} that enters the nerve terminal (the concentration of the Ca^{2+} in the bathing solution is reduced and the concentration of Mg^{2+} is increased), repetitive stimulations leads to a progressive increase in transmitter release, and depression is no longer apparent (Magleby, 1973b). Under these conditions it is thus possible to study the processes that increase transmitter release without the complication of depression.

ESTIMATING TRANSMITTER RELEASE

Under the conditions of low quantal content, changes in end-plate potential (EPP) amplitudes during and following repetitive stimulation result from changes in the number of quanta released from the nerve terminal and not from any changes in post-synaptic sensitivity (del Castillo and Katz, 1954; Magleby and Zengel, 1976a). Thus, changes in transmitter release can be estimated from changes in EPP amplitudes if appropriate precautions are taken (Martin, 1955).

PROCESSES THAT INCREASE TRANSMITTER RELEASE

The processes which increase transmitter release have traditionally been distinguished from one another by differences in their time courses of decay. On this basis at least four processes have been described: first and second components of facilitation (Mallart and Martin, 1967; Younkin, 1974; Magleby, 1973a); augmentation (Magleby and Zengel, 1976a,b); and potentiation, also called posttetanic potentiation or PTP (Liley, 1956; Hubbard, 1963; Rosenthal, 1969; Magleby, 1973b; Magleby and Zengel, 1975a,b). The processes and their approximate time constants of decay are listed in Table I. These four processes are all expressed in a similar manner, as an increase in transmitter release (observed experimentally as an increase in EPP ampli-

TABLE I

PROCESSES WHICH INCREASE TRANSMITTER RELEASE AT THE FROG NEURO-
MUSCULAR JUNCTION AND THEIR APPROXIMATE TIME CONSTANTS OF DECAY
AT 20°C

Process	Time constant
Facilitation	
First component	50 msec
Second component	300 msec
Augmentation	7 sec
Potentiation	30 sec to min *

* The time constant of decay of potentiation increases with the duration of stimulation.

tudes), and their formal definitions are similar; each one is defined as the fractional increase of a test EPP amplitude over a control in the absence of the
other three processes. Experimentally, however, it is not always possible to
measure one process in the absence of the other three. Consequently, the magnitude and time course of the individual processes are derived mathematically
from the fractional change in EPP amplitude. The experimentally determined
estimation of the contribution that facilitation, augmentation and potentiation
make to increasing transmitter release during repetitive stimulation relies on the
fact that these different processes have distinct and non-overlapping time constants which characterize their decays. Following a conditioning train the first
component of facilitation decays to insignificant levels in about 200 msec, the
second component decays in about 2 s, and augmentation decays in about 20 s.
Potentiation can then be measured in the absence of these processes from the
data points collected after facilitation and augmentation have decayed. The
fractional increases in the testing EPP amplitudes following the conditioning
trains are plotted semilogarithmically against time and potentiation is estimated
by a least squares fit to the data points beyond a time when the faster processes
would have decayed. The projection of the decay of potentiation to intercept
the axis at the end of the conditioning train gives the initial magnitude of
potentiation. After correcting the data for potentiation, augmentation can be
estimated in a similar manner, and by repeating the process twice more estimates of the two components of facilitation can be obtained. For conditioning·
trains of only a few impulses, augmentation and potentiation are small and
approximate estimates of facilitation can be made directly.

THE PROCESSES THAT INCREASE TRANSMITTER RELEASE
ARE DISTINCT

A considerable amount of evidence suggests that the four processes listed in
Table I are separable and that all of these processes are present during repetitive
stimulation. For example, it has been established that facilitation retains its
kinetic properties during and following repetitive stimulation (Magleby, 1973b).
Evidence has also been presented for an expression factor which acts to
increase the magnitude of augmentation while having little or no effect on the

magnitude of potentiation, and a time constant factor which acts to increase the time constant of decay of potentiation while having little effect on the time constant of decay of augmentation (Magleby and Zengel, 1976c). It has also been shown that the addition of a small amount (0.1 mM) of Ba^{2+} to the bathing solution has a selective effect in increasing the magnitude of augmentation without changing its time constant of decay or changing transmitter release in the absence of repetitive stimulation. On the other hand, adding Sr^{2+} or replacing Ca^{2+} in the bathing solution with Sr^{2+} leads to a selective increase in the magnitude and time constant of decay of the second component of facilitation (Zengel and Magleby, 1977). These observations and the large differences in the time constants of decay of the processes all support the suggestion that some of the factors involved in facilitation, augmentation, and potentiation of transmitter release are different.

KINETIC PROPERTIES OF THE PROCESSES THAT AFFECT TRANSMITTER RELEASE

Our goal in studying the kinetic properties of transmitter release is 2-fold. First, to determine whether the four processes listed in Table I are sufficient to describe the effect of repetitive stimulation on transmitter release. If not, it will be necessary to look for further processes or to re-examine the properties of the four processes. Second, any model for the mechanism of transmitter release must be able to account for the properties of the four processes. At present, there are no models formulated in terms of mechanism that can account for the dynamic properties of transmitter release. Knowing the properties of the processes should help in formulating experiments and models to define the mechanism of transmitter release.

In studies of the kinetic properties of the processes that increase transmitter release, the assumption is usually made that each process arises from a change in some underlying factor in the nerve terminal. It is usually further assumed that each nerve impulse leads to an incremental change in this factor and that during repetitive stimulation the magnitude of each process increases as the increments sum up. On the basis of these assumptions, the observed magnitude of the processes depends on the magnitude of the increments added by each impulse, the rates of decay of the changes in the underlying factors between impulses, and the relationships between the underlying factors and the experimentally observed processes.

The question of whether the magnitude of an observed process is related to the first or fourth power of the underlying factor that changes in the nerve terminal is important since it may give some information about whether, in a simplified model, the processes are related to the first or fourth power of the concentration of Ca^{2+} (a possible factor) at some site involved in transmitter release. Facilitation has been described with a first power (Mallart and Martin, 1967; Magleby, 1973a) and fourth power model (Barrett and Stevens, 1972; Younkin, 1974). Potentiation is best described with a first power model (Magleby and Zengel, 1975b). Depending on the experimental conditions, augmentation has been described with either a first power or fourth power model (Magleby and Zengel, 1977). In kinetic studies of this type it should be kept in mind that the conclusions reached depend on the initial assumptions, which may be incorrect or oversimplified. Only after a large number of studies over a wide range of experimental conditions have been done will it be possible to come to a firm conclusion about the underlying kinetic properties of the processes.

POSSIBLE MECHANISMS OF THE PROCESSES THAT INCREASE TRANSMITTER RELEASE

Ca^{2+} is required for evoked transmitter release (Katz and Miledi, 1969) and some experimental evidence is compatible with the suggestion that an accumulation of Ca^{2+} in the nerve terminal may lead to facilitation (Katz and Miledi, 1968; Younkin, 1974) and potentiation (Rosenthal, 1969; Weinreich, 1971) of transmitter release. Increases in ionized Ca^{2+} in nerve cell bodies with repetitive stimulation have been observed (Stinnakre and Tauc, 1973). Accumulation of Na^+ in the nerve terminal may also potentiate transmitter release (Birks and Cohen, 1968; Atwood, Swenarchuk and Gruenwald, 1975). It appears, then, that these changes in Na^+ and Ca^{2+} distributions in the nerve terminal may give rise to some of the factors which affect transmitter release. However, practically nothing is known about how these ions might act. For example, it is not known whether Ca^{2+} may lead to facilitation because a residue of Ca^{2+} accumulates at specific sites, or whether Ca^{2+} activates a change in some factor in the nerve terminal and it is the change in the factor (release sites?, position of synaptic vesicles?) that accumulates and gives rise to facilitation. In the first case, the time course of facilitation might reflect the time course of the concentration of Ca^{2+} at its site of action, while in the second, the time course of facilitation might reflect the relaxation of the activated factor back to its control level. Detailed knowledge about the time course of Ca^{2+} entry, sequestering and extrusion in nerve terminals would be useful in understanding release mechanisms, as would information about the effects of Sr^{2+} and Ba^{2+} on the processes that affect Ca^{2+} regulation. Some of the effects of repetitive stimulation on the level of free Ca^{2+} in squid axons (Baker, Hodgkin and Ridgway, 1971) are similar to the effects of repetitive stimulation on augmentation (Magleby and Zengel, 1976b). Detailed studies relating structural changes to changes in the levels of facilitation, augmentation, and potentiation would also be useful in determining the mechanisms of these processes. For example, Quilliam and Tamarind (1973) have found a possible relationship between post-tetanic potentiation (potentiation) and closer clustering of synaptic vesicles about the release sites in sympathetic ganglia. Detailed studies of acetylcholine metabolism may also be useful in determining the mechanism of the various processes. For example, Birks (1977) has described a longlasting potentiation of transmitter release which appears to be related to an increase in transmitter stores in sympathetic ganglia.

THE QUANTAL HYPOTHESIS AND THE PROCESSES THAT INCREASE TRANSMITTER RELEASE

The quantal hypothesis of del Castillo and Katz (1954a) states that transmitter release (m) can be described as the product of two factors, usually described as the number of quanta available for release (n) and the average probability that a given quantum will be released (p), i.e. $m = np$. The multiplicative relationship between facilitation and potentiation (Magleby, 1973b) is similar to the multiplicative relationship between n and p. This observation is consistent with the suggestion that facilitation and potentiation may result from changes in the two quantal parameters. Rosenthal (1969) and Wilson and Skirboll (1974) have found that potentiation is associated with an apparent increase in the parameter p. Bennett and Fisher (1977) have suggested that the increased release of transmitter during short trains of impulses is primarily due to an increase in n. As facilitation would be the major factor increasing release during the first few impulses of a train, facilitation may be associated with an increase in n. Wernig (1975) has found that estimates of n increase with the length of the nerve terminal, suggesting that n may have a structural correlate such as the number of active zones in the terminal. Although progess is being made, more studies will be needed to determine the relationship between the processes that increase transmitter release and that statistical factors n and p. In these studies it will be necessary to correlate changes in n and p with actual estimates of facilitation, augmentation, and potentiation.

CONCLUSION

Fig. 1 presents a schematic diagram of some of the current questions that will have to be answered in order to understand the dynamic properties of transmitter release. It is known that Ca^{2+} leads to evoked transmitter release (m). However, the relationship between the structural and chemicals factors in the nerve terminal, the statistical parameters n and p from the quantal hypothesis, and the processes that increase transmitter release are not yet clear. Determining these relationships will be necessary to understand the mechanism of transmitter release.

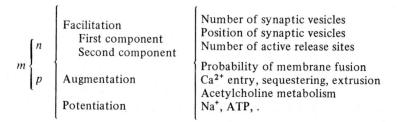

Fig. 1. The relationships between the structural and chemical factors in the nerve terminal that affect transmitter release, the processes that increase transmitter release, and the statistical parameters n and p from the quantal hypothesis will have to be determined to understand the effect of repetitive stimulation on transmitter release m.

SUMMARY

During repetitive stimulation of a neuromuscular junction under conditions of low quantal content, end-plate potentials progressively increase in amplitude. This increase is due to an increase in the number of quanta of transmitter released by each nerve impulse. A kinetic analysis of the changes in transmitter release during and following repetitive stimulation suggests that there are four processes that act to increase transmitter release: first and second components of facilitation that decay with time constants of about 50 and 30 msec, augmentation which decays with a time constant of about 7 s, and potentiation which decays with a time constant which ranges from about 30 s to min. These processes are separable on the basis of their kinetic and pharmacological properties. The mechanisms of these processes are not yet known, but some possibilities are briefly discussed in terms of structural, chemical, and statistical factors.

ACKNOWLEDGEMENT

I thank Dr. J. Zengel for helpful comments on the manuscript. Supported in part by NIH grant NS 10277 from the U.S. Public Health Service.

REFERENCES

Atwood, H.L., Swenarchuk, L.E. and Gruenwald, C.R. (1975) Long-term synaptic facilitation during sodium accumulation in nerve terminals. *Brain Res.*, 100, 198—204.

Baker, P.F., Hodgkin, A.L. and Ridgway, E.B. (1971) Depolarizaton and calcium entry in squid giant axons. *J. Physiol. (Lond.)*, 218, 709—755.

Barrett, E.F. and Magleby, K.L. (1976) Physiology of cholinergic transmission . In *Biology of Cholinergic Functions,* A.M. Goldberg and I. Hanin (Eds.), Raven Press, New York, pp. 29—100.

Barrett, E.F. and Stevens, C.F. (1972) The kinetics of transmitter release at the frog neuromuscular junction. *J. Physiol. (Lond.)*, 227, 691—708.

Bennett, M.R. and Fisher, C. (1977) The effect of calcium ions on the binomial parameters that control acetylcholine release during trains of nerve impulses at amphibian neuromuscular synapses. *J. Physiol. (Lond.)*, 271, 673—698.

Birks, R.I. (1977) A long-lasting potentiation of transmitter release related to an increase in transmitter stores in a sympathetic ganglion. *J. Physiol. (Lond.),* 271, 847—862.

Birks, R.I. and Cohen, M.W. (1968) The influence of internal sodium on the behavior of motor nerve endings. *Proc. roy. Soc.* B, 170, 401—421.

del Castillo, J. and Katz, B. (1954a) Quantal components of the end-plate potential. *J. Physiol. (Lond.)*, 124, 560—573.

del Castillo, J. and Katz, B. (1954b) Statistical factors involved in neuromuscular facilitation and depression. *J. Physiol. (Lond.)*, 124, 574—585.

Dodge, F.A. Jr. and Rahamimoff, R. (1967) Co-operative action of calcium ions in transmitter release at the neuromuscular junction. *J. Physiol. (Lond.)*, 193, 419—432.

Feng, T.P. (1941) Studies on the neuromuscular junction. XXVI. The changes of the end-plate potential during and after prolonged stimulation. *Chin. J. Physiol.*, 16, 341—372.

Hubbard, J.I. (1963) Repetitive stimulation at the mammalian neuromuscular junction, and the mobilization of transmitter. *J. Physiol. (Lond.)*, 169, 641—662.

Hubbard, J.I. (1973) Microphysiology of vertebrate neuromuscular transmission. *Physiol. Rev.*, 53, 674—723.

Katz, B. and Miledi, R. (1968) The role of calcium in neuromuscular facilitation. *J. Physiol. (Lond.)*, 195, 481—492.

182

Katz, B. and Miledi, R. (1969) Spontaneous and evoked activity of motor nerve endings in calcium ringer. *J. Physiol. (Lond.)*, 203, 689—706.

Liley, A.W. (1956) The quantal -components of the mammalian end-plate potential. *J. Physiol. (Lond.)*, 133, 571—587.

Magleby, K.L. (1973a) The effect of repetitive stimulation on facilitation of transmitter release at the frog neuromuscular junction. *J. Physiol. (Lond.)*, 234, 327—352.

Magleby, K.L. (1973b) The effect of tetanic and post-tetanic potentiation of facilitation of transmitter release at the frog neuromuscular junction. *J. Physiol. (Lond.)*, 234, 353—371.

Magleby, K.L. and Zengel, J.E. (1975a) A dual effect of repetitive stimulation on post-tetanic potentiation of transmitter release at the frog neuromuscular junction. *J. Physiol. (Lond.)*, 245, 163—182.

Magleby, K.L. and Zengel, J.E. (1975b) A quantitative description of tetanic and post-tetanic potentiation of transmitter release at the frog neuromuscular junction. *J. Physiol. (Lond.)*, 245, 183—208.

Magleby, K.L. and Zengel, J.E. (1976a) Augmentation: a process that acts to increase transmitter release at the frog neuromuscular junction. *J. Physiol (Lond.)*, 257, 449—470.

Magleby, K.L. and Zengel, J.E. (1976b) Long term changes in augmentation, potentiation, and depression of transmitter release as a function of repeated synaptic activity at the frog neuromuscular junction. *J. Physiol. (Lond.)*, 257, 471—494.

Magleby, K.L. and Zengel, J.E. (1976c) Stimulation-induced factors which affect augmentation and potentiation of transmitter release at the neuromuscular junction. *J. Physiol. (Lond.)*, 260, 687—717.

Magleby, K.L. and Zengel, J.E. (1977) Effect of repetitive stimulation on augmentation of transmitter release at the frog neuromuscular junction. *Fed. Proc.*, 36, 486.

Mallart, A. and Martin, A.R. (1967) An analysis of facilitation of transmitter release at the neuromuscular junction of the frog. *J. Physiol. (Lond.)*, 193, 679—694.

Martin, A.R. (1955) A further study of the statistical composition of the end-plate potential. *J. Physiol. (Lond.)*, 130, 114—122.

Peper, K., Dreyer, F., Sandri, C., Akert, K. and Moor, H. (1974) Structure and ultrastructure of the frog motor endplate: a freeze-etching study. *Cell Tiss. Res.*, 149, 437—455.

Quilliam, J.P. and Tamarind, D.L. (1973) Some effects of preganglionic nerve stimulation on synaptic vesicle populations in rat superior cervical ganglion. *J. Physiol. (Lond.)*, 235, 317—331.

Rosenthal, J. (1969) Post-tetanic potentiation at the neuromuscular junction of the frog. *J. Physiol. (Lond.)*, 203, 121—133.

Stinnakre, J. and Tauc, L. (1973) Calcium influx in active Aplysia neurons detected by injected aequorin. *Nature New Biol.*, 242, 113—115.

Thies, R.E. (1965) Neuromuscular depression and the apparent depletion of transmitter in mammalian muscle. *J. Neurophysiol.*, 28, 427—442.

Wernig, A. (1975) Estimates of statistical release parameters from crayfish and frog neuromuscular junctions. *J. Physiol. (Lond.)*, 244, 207—221.

Weinreich, D. (1971) Ionic mechanism of post-tetanic potentiation at the neuromuscular junction of the frog. *J. Physiol. (Lond.)*, 212, 431—446.

Wilson, D.F. and Skirboll, L.R. (1974) Basis for posttetanic potentiation at the mammalian neuromuscular junction. *Amer. J. Physiol.*, 227, 92—95.

Younkin, S.G. (1974) An analysis of the role of calcium in facilitation at the frog neuromuscular junction. *J. Physiol. (Lond.)*, 237, 1—14.

Zengel, J.E. and Magleby, K.L. (1977) Transmitter release during repetitive stimulation: selective changes produced by Sr^{2+} and Ba^{2+}. *Science*, 197, 67—69.

The Regulatory Role of Membrane Na^+-K^+-ATPase in Non-Quantal Release of Transmitter at the Neuromuscular Junction

FRANTIŠEK VYSKOČIL

Institute of Physiology, Czechoslovak Academy of Sciences, 142 20 Prague (Czechoslovakia)

INTRODUCTION

It was shown by Brooks in 1954 that acetylcholine (ACh) is released from the isolated diaphragm of the guinea-pig even in the absence of nerve stimulation. This fact was subsequently confirmed by several authors and spontaneous miniature end-plate potentials (MEPPs) were considered to be an electrophysiological correlate of the resting ACh output from skeletal muscle (see e.g. Krnjević and Mitchell, 1961).

As early as in 1963, however, Mitchell and Silver realized the serious discrepancy between the changes of biologically assayed ACh in their experiments and electrophysiologically recorded MEPPs under different experimental conditions. Thus, for example, when the K^+ concentration in the bathing fluid is increased from 5 to 30 mM, the MEPP frequency (representing the number of quanta released per unit time) is raised 200–300 times in the rat diaphragm (Liley, 1956) whereas the output of ACh increases only about 3-fold (Mitchell and Silver, 1963). A similar discrepancy appeared when the effect of temperature was studied. Mitchell and Silver therefore concluded that only a small fraction of the resting release derives from MEPPs and that, surprisingly, 97–99% of released ACh has a different origin.

A similar conclusion was also drawn by Fletcher and Forrester (1975) who estimated that only 2% of overall ACh release from rat diaphragm at rest could represent quantal release and suggested that the remainder has a non-synaptic source. In our recent observation on the mouse diaphragm (Vizi and Vyskočil, 1979), where we compared the total and quantal release ACh before and during stimulation, we estimated that 1% of the total release represents quantal release. The non-quantal release of ACh was considered by some workers to be of "non-synaptic" origin and was attributed to the preterminal regions of motor axons (Brooks, 1954; Mitchell and Silver, 1963; Fletcher and Forrester, 1975; cf. Evans and Saunders, 1974), too far away from the postsynaptic membrane to evoke detectable depolarisation. Another possibility suggested by Katz and Miledi (1965) was that a steady and diffuse leakage may occur from the terminals all over the junction. This continuous leakage, because of hydrolysis by cholinesterase, can barely be recorded electrophysiologically. Nevertheless, Katz and Miledi (1977) successfully tested this assumption on anti-esterase (diisopropylfluorophosphate, DFP) treated frog sartorius muscles where individual end-plates of superficial muscle fibres were subjected to a massive ionophoretic dose of (+)-tubocurarine (TC). During this local curarisation, various levels of hyperpolarisation were observed in different muscle fibres and this effect was attributed to the leakage of cytoplasmic ACh from nerve terminals causing a minute steady postsynaptic depolarisation, which is prevented by TC. The average depolarisation was only about 40 μV (at 18–24°C) in this preparation and such a change is so small that it does not provide a good opportunity for studying the characteristics of the steady leakage. We therefore decided to use the mouse diaphragm and to perform similar experiments on this preparation (Vyskočil and Illés, 1977, 1978), where the electrical responses to ACh (MEPPs in particular) are more pronounced apparently because of the small diameter and high input impedance of the muscle fibres and compact junctional area (Salpeter and Eldefrawi, 1973). As will be demonstrated later, mouse end-plates were hyperpolarized by 1–2 mV during local curarization in the presence of anticholinesterases at 32°C. We checked the idea that the membrane Na^+,K^+-ATPase (Skou, 1965) of the nerve terminal membrane regulates non-quantal ACh liberation (Paton, Vizi and Zar, 1971; Vizi, 1973. 1975, 1977) in this preparation, and also in frog sartorius muscle.

MOUSE DIAPHRAGM AND NON-QUANTAL ACh RELEASE

The whole diaphragm of a white mouse was dissected and divided into several strips, each of them pinned to a plastic plate and immersed in oxygenated perfusion saline (Liley, 1956). Intracellular recordings were made with 20–25 MΩ glass microelectrodes inserted into the muscle fibres close to the termination of small branches of the main intramuscular nerve trunk. In some cases, the bathing medium contained 6×10^{-6} M neostigmine methylsulphate as an anticholinesterase (Vyskočil and Illés, 1977), but experiments were also performed on diaphragm strips preincubated for 30 min in a solution containing an irreversible cholinesterase inhibitor Soman (10^{-4} M; Vyskočil and Illés, 1978). Tetrodotoxin (10^{-6} M) was present in the bath during experiments with anticholinesterases to eliminate spontaneous twitching.

When superficial muscle fibres were successfully impaled with a microelectrode and large (1–2 mV) MEPPs were observed on the oscilloscope screen, another pipette (50–100 μm tip diameter) containing physiological saline (of the same composition as that in the bath) saturated with TC (about 10^{-2} M) was immersed in the bath and rapidly placed close to the fibre. The effect of TC which diffused from the tip of the pipette was usually very rapid and MEPPs disappeared within seconds (Fig. 1A). In the normal solution, hyperpolarisation of the muscle fibre membrane (indicated in Fig. 1 as downward deflection) was routinely observed. It ranged in different fibres from 0.2 to 3.0 mV, the mean value (± S.E.M.) being 1.1 ± 0.27 mV for a resting membrane potential (RMP) of −71 mV (20 fibres) in the presence of neostigmine. In experiments with Soman, similar results were obtained (mean hyperpolarization 0.9 ± 0.30 mV at RMP −68 mV; Vyskočil and Illés, 1978). The similar value of hyperpolarization obtained in the presence of neostigmine and after

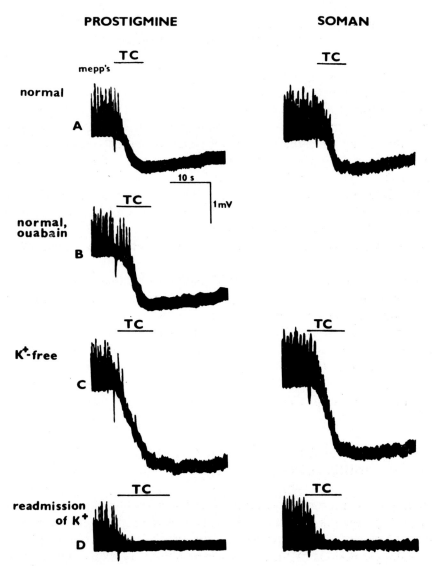

Fig. 1. Mouse diaphragm. The effect of local curarisation on MEPPs and membrane potential in the presence of 6×10^{-6} M neostigmine (left) and after pretreatment of the diaphragm with Soman (right). The muscle strips were incubated at $32°C$ in the normal solution (A), normal solution plus 2×10^{-5} M ouabain (B) or a K^+ free solution (C) respectively. The bottom records (D) were obtained 10 min after readmission of K^+ into the K^+ free solution. Horizontal bars indicate the time of tubocuranine (TC) diffusion from pipette located in the end-plate area (Vyskočil and Illés, 1978).

pretreatment with Soman makes it unlikely that the effect results from a direct postsynaptic depolarizing action of neostigmine, and is consistent with the interpretation (Katz and Miledi, 1977) that continuous ACh leakage is responsible.

It has been found (Vizi and Vyskočil, 1979) that total ACh output is affected by activation or inhibition of M-ATPase, and we have investigated whether this can be detected electrophysiologically.

INHIBITION AND ACTIVATION OF Na$^+$-K$^+$-ACTIVATED
MEMBRANE ATPase (M-ATPase)

When strips of neostigmine-treated diaphragm were immersed in a solution containing 2×10^{-5} M ouabain to block M-ATPase activity (Fig. 1B) the mean hyperpolarisation after TC increased to 1.5 ± 0.23 mV (mean RMP -65.5 mV). The suspension of muscles in a potassium-free solution (Fig. 1C) also led to substantially higher TC-hyperpolarisation than in control experiments (mean 1.88 ± 0.4 mV and 1.65 ± 0.4 mV in prostigmine and after Soman treatment respectively).

On the other hand, activation of M-ATPase by adding 5 mM K$^+$ after the muscles had been kept for $1-2$ h in K-free solution abolished the hyperpolarisation when TC was applied $10-20$ min after the readdition of K$^+$ ions (Fig. 1D) i.e. during the period of increased M-ATPase activity (Vizi and Vyskočil, 1979).

It has recently been found (Vizi and Vyskočil, 1979) that block of M-ATPase by ouabain leads to an approximately two-fold increase in total resting release of ACh from the mouse diaphragm. This finding correlates with the present results showing the potentiation of TC hyperpolarisation by a factor of $1.5-1.7$ during the blockade of membrane ATPase. Similarly, the activation of M-ATPase eliminated the electrophysiologically assayed non-quantal leakage of ACh (Vyskočil and Illés, 1977, 1978) and depressed the total release to one tenth of the control value (Vizi and Vyskočil, 1979).

FROG SARTORIUS AND NON-QUANTAL RELEASE DURING
INHIBITION AND ACTIVATION OF M-ATPase

It appeared of interest to ascertain whether a similar dependence of non-quantal leakage on the activity of M-ATPase also exists in frog muscle. This preparation allows continuous microelectrode recordings from single muscle fibres for longer periods of time. In several preliminary experiments, sartorius muscles were continuously perfused with a standard oxygenated Ringer solution with 1 mM glucose at a temperature of $32-35°$C, i.e. they were kept under conditions where one can expect a proper functioning of M-ATPase and higher output of ACh (Mitchell and Silver, 1963). To block cholinesterase the muscles were preincubated, as described above, in a solution containing either Soman or another irreversible inhibitor Armin (U.S.S.R.) (1×10^{-5} M). When the baseline had become sufficiently stable after impalement, a pipette (100 μm diameter) containing Ringer solution and 10^{-2} M ouabain was placed close to the junction for several minutes. As indicated in Fig. 2 a slight depolarisation of about 1 mV developed during this period. It was almost completely eliminated by local curarisation via a TC-containing pipette, which suggests that this depolarisation might be caused by ouabain-stimulated non-quantal leakage of ACh. After removal of both pipettes, the RMP and MEPPs usually recovered and in several cases the whole procedure was repeated (Fig. 2, bottom record).

If sartorius muscles were bathed in a K$^+$-free solution for $10-15$ min local

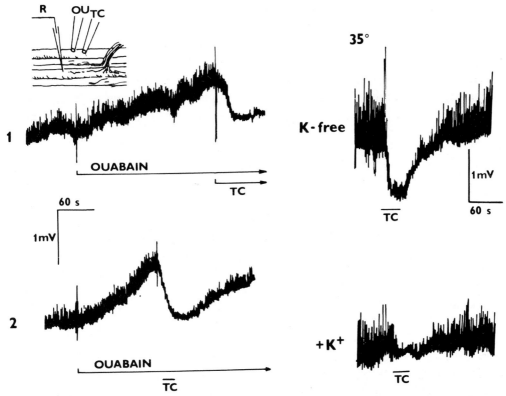

Fig. 2. The effect of ouabain on the neuromuscular junction of a frog sartorius muscle pretreated with Soman. Intracellular records from one muscle cell (starting RMP = −82 mV)
before and after local application of ouabain (OU) and tubocurarine (TC). Inset illustrates
the experimental situation and localisation of electrode and OU- and TC-filled pipettes. For
other details see text.

Fig. 3. TC-induced hyperpolarisation of postsynaptic membrane of one frog m. sartorius
fibre pretreated with Soman in a potassium free (K⁺-free) solution (RMP = 97 mV) and
7 min after readmission of 5 mM potassium (+K⁺) into the Ringer solution (RP = 100 mV)
at 35°C.

curarisation caused a hyperpolarisation similar to that seen in the mouse diaphragm ranging from 0.25−1 mV in four experiments (Fig. 3, top record). The
reactivation of M-ATPase after 1 h by addition of 5 mM K⁺ led to the disappearance of the hyperpolarising effect of TC (Fig. 3, bottom record). It thus
seems very likely that in the frog muscle ACh leakage is also dependent on the
activity of M-ATPase.

FINAL REMARKS

There is no longer any doubt that the largest part of the ACh released at the
resting neuromuscular junction, is in non-quantal form, probably originating
from cytoplasmic stores in nerve terminals (cf. Tauc et al., 1974). The question
arises about the physiological meaning of this spontaneous output of ACh.
Since it is released in relatively large amounts (Vizi and Vyskočil, 1979) the

possibility exists that it has a trophic influence on target muscle cells. Katz and Miledi (1977) suggested that it may induce or regulate junctional cholinesterase, but it may even be involved in more general "trophic" regulations (Thesleff, this volume). Whatever role other than impulse transmission is played by ACh it appears that the non-quantal release is controlled to a large extent by the activity of M-ATPase. The mechanism by which this membrane enzyme controls non-quantal liberation of transmitter is not clear yet and should be studied in more detail.

SUMMARY

The subsynaptic area of mouse diaphragm fibres was hyperpolarised by 1–2 mV during local curarisation of the junctional zone in the presence of neostigmine or after treatment of the muscle with organophosphate cholinesterase inhibitors. In a solution containing 5 mM K^+ the mean hyperpolarisation was 1.1 ± 0.27 mV. After adding 2×10^{-5} M ouabain the hyperpolarisation increased to 1.5 ± 0.25 mV. In a K^+-free bathing medium (i.e. blockade of membrane ATPase) the curare induced hyperpolarisation was also increased, to 1.8 ± 0.4 mV. Reactivation of membrane ATPase by addition of K^+ after a period in a K^+-free medium reduced the hyperpolarisation to zero. Similar results were also obtained in the frog sartorius muscle.

The spontaneous non-quantal leakage of acetylcholine at rest, manifested as hyperpolarisation during local curarisation, appears to be regulated by the activity of Na^+-K^+-ATPase of the nerve terminals directly, whereas the spontaneous quantal release (miniature end-plate potentials) is affected in the mouse preparation by the Na^+-K^+ pump only indirectly as a result of the slow changes in ionic distribution (Vizi and Vyskočil, 1979).

ACKNOWLEDGEMENT

We wish to thank Dr. Pavel Hník for his help in the preparation of manuscript.

REFERENCES

Brooks, V.B. (1954) The action of botulinum toxin on motor nerve filaments. *J. Physiol. (Lond.)*, 123, 501–515.

Evans, C.A.N. and Saunders, N.R. (1974) An outflow of acetylcholine from normal and regenerating ventral roots of the cat. *J. Physiol. (Lond.)*, 240, 15-32.

Fletcher, P. and Forrester, T. (1975) The effect of curare on the release of acetylcholine from mammalian nerve terminals and an estimate of quantum content. *J. Physiol. (Lond.)*, 251, 131–144.

Katz, B. and Miledi, R. (1965) The quantal release of transmitter substances. In *Studies in Physiology*, Springer, Berlin, pp. 118–125.

Katz, B. and Miledi, R. (1977) Transmitter leakage from motor nerve endings. *Proc. R. Soc. Lond. B*, 196, 59–72.

Krnjević, K. and Mitchell, J.F. (1961) The release of acetylcholine in the isolated rat diaphragm. *J. Physiol. (Lond.)*, 155, 246–262.

Liley, A.W. (1956) The effect of presynaptic polarisation on the spontaneous activity at the mammalian neuromuscular junction. *J. Physiol. (Lond.)*, 134, 427–443.

Mitchell, J.F. and Silver, A. (1963) The spontaneous release of acetylcholine from the denervated hemidiaphragm of the rat. *J. Physiol. (Lond.)*, 165, 117–129.

Paton, W.D.M., Vizi, E.S. and Zar, M.A. (1971) The mechanism of acetylcholine release from parasympathetic nerves. *J. Physiol. (Lond.)*, 215, 819–848.

Salpeter, M.M. and Eldefrawi, M.E. (1973) Sizes of end-plate compartments, densities of acetylcholine receptor and other quantitative aspects of neuromuscular transmission. *J. Histochem. Cytochem,*, 21, 769–778.

Skou, J.C. (1965) Enzymatic basis for active transport of Na^+ and K^+ across cell membrane. *Physiol. Rev.*, 45, 596–617.

Tauc, L., Hoffmann, A., Tsuji, S., Hinzen, D.H. and Faille, L. (1974) Transmission abolished on a cholinergic synapse after injection of acetylcholinesterase into the presynaptic neurone. *Nature*, 250, 496–498.

Vizi, E.S. (1973) Does stimulation of Na^+-K^+-Mg^{2+}-activated ATP-ase inhibit acetylcholine release from nerve terminals? *Brit. J. Pharmacol.*, 48, 346–347.

Vizi, E.S. (1975) Release mechanisms of acetylcholine and the role of Na^+-K^+-activated ATP-ase. In *Cholinergic Mechanisms*, P.G. Waser (Ed.), New York, Raven Press, pp. 199–211.

Vizi, E.S. (1977) Termination of transmitter release by stimulation of sodium-potassium activated ATPase. *J. Physiol. (Lond.)*, 267, 261–280.

Vizi, E.S. and Vyskočil, F. (1979) Changes in total and quantal release of acetylcholine in the mouse diaphragm during activation and inhibition of membrane ATPase. *J. Physiol. (Lond.)*, 286, 1–14.

Vyskočil, F. and Illés, P. (1977) Non-quantal release of transmitter at mouse neuromuscular junction and its dependence on the activity of Na^+-K^+-ATPase. *Pflügers Arch. ges. Physiol.*, 370, 295–297.

Vyskočil, F. and Illés, P. (1978) Electrophysiological examination of transmitter release in non-quantal form in the mouse diaphragm and the activity of membrane ATPase. *Physiol. bohemoslov*, 27, 449–455.

Stimulation of Cholinergic Synaptosomes Isolated from *Torpedo* Electric Organ

N. MOREL, M. ISRAËL, R. MANARANCHE and B. LESBATS

Département de Neurochimie, Laboratoire de Neurobiologie cellulaire du C.N.R.S.,
91190 Gif-sur-Yvette (France)

INTRODUCTION

Isolated nerve terminals (synaptosomes) (Gray and Whittaker, 1960; De Robertis et al., 1961) have proven to be useful tools to study the localization of substances and enzymes involved in neurotransmission, presynaptic metabolism and transmitter release processes (for review, see Jones, 1975). So far most synaptosomal preparations derive from the central nervous system which is heterogeneous regarding neurotransmitters. Recently it has been possible to isolate synaptosomes from the electric organ of *Torpedo* (Israël et al., 1976; Morel et al., 1977) which is purely cholinergic and richly innervated (Feldberg et al., 1940; Feldberg and Fessard, 1942). Nerve terminals from this peripheral tissue, analogous to neuromuscular junctions can be pinched off in a similar manner to brain nerve terminals despite their larger size and more complex shape. These separated nerve terminals (3.5 μm in diameter) retain all their cytoplasm. They are stable during long incubations without any leakage of contents as demonstrated by morphological, biochemical and osmotic controls. They synthesize (ATP) and acetylcholine (ACh) from extracellular precursors. We shall demonstrate here that a relatively small increase in the K^+ content of the medium (50 mM) triggers a Ca^{2+} dependent transmitter release. They are therefore markedly different from the empty particles (T sacs) isolated by Dowdall and Zimmermann (1977) or from the small nerve terminal fragments purified by Michaelson and Sokolovski (1978). The major advantage of Torpedo synaptosomes is that they are all cholinergic. This permitted the study of their metabolism and physiological properties without interferences of noncholinergic materials.

ISOLATION AND CHARACTERIZATION

The isolation procedure of *Torpedo* synaptosomes, published in full elsewhere (Israël et al., 1976; Morel et al., 1977), is schematically described in Fig. 1. The two main characteristics of the procedure are the gradual comminution of the tissue by successive filtrations and the use of centrifugation media having an ionic composition and osmolarity close to plasma. Figure 2 shows a

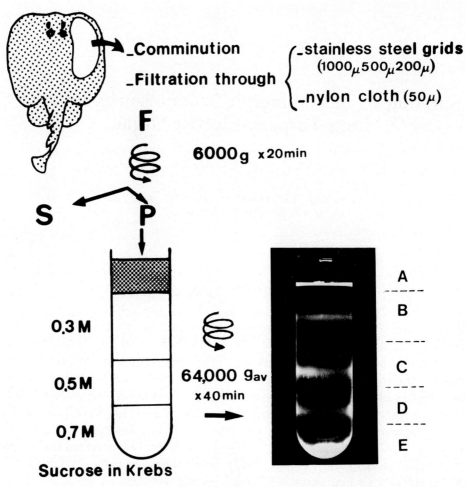

Fig. 1. Schematic description of synaptosome isolation from *Torpedo* electric organ. For more details, see Israël et al., 1976; Morel et al., 1977.

TABLE I

DISTRIBUTION OF BIOCHEMICAL MARKERS IN SUBCELLULAR FRACTIONS
OF THE ELECTRIC ORGAN

Results are means ± S.E.M. n, number of experiments. R_f is recovery in primary fractions, R_g is recovery on gradients. Primary fractions: F, total filtrate; S, supernate; P, pellet. Subfractions of P on density gradient (A, B, C, D, E), see Fig. 1. For details on the preparation of fractions and measurements, see Morel et al. (1977).

	n	F	S	P	R_f
Proteins (mg/g)	19	15.6 ± 0.6	11.3 ± 0.4	4.4 ± 0.2	100
ACh (nmol/g)	17	210 ± 18	56 ± 8	131 ± 10	89
ATP (nmol/g)	11	47.7 ± 7.3	8.6 ± 2.1	31.6 ± 4.4	84
Choline acetyltransferase (nmol/h per g)	14	1824 ± 170	1026 ± 108	489 ± 33	83
Lactate dehydrogenase ($\Delta E/min \cdot g \cdot ml$)	5	33.2 ± 13.8	21.3 ± 10.2	12.0 ± 0.9	100
Acetylcholinesterases (mmol/h · g)	5	12.5 ± 1.7	1.3 ± 0.2	6.5 ± 1.2	62
Succinate dehydrogenase ($\Delta E/h \cdot g \cdot ml$)	6	14.7 ± 2.6	0.44 ± 0.18	9.94 ± 2.04	71
5′ nucleotidase (nmol/h · g)	5	2384 ± 818	514 ± 136	1880 ± 670	100

very large field of fraction C showing the homogeneity and purity of the synaptosomal fraction. The distribution of biochemical markers (Table I) demonstrates that fraction C has a high specific activity for ACh (130 nmol/mg protein) and choline acetyltransferase (450 nmol/h · mg protein). The low acetylcholinesterase and succinate dehydrogenase (SDH) activities show that fraction C is hardly contaminated by postsynaptic membranes or mitochondria.

DESCRIPTION OF ELECTRIC ORGAN SYNAPTOSOMES

Figure 3a shows a synaptosome in conventionnal electron microscopy. It contains glycogen-like granules and synaptic vesicles. Mitochondria are not frequent in these synaptosomes, a fact in accordance with the low SDH content of fraction C and with the aspect of nerve terminals in situ. The synaptosome (Fig. 3a) is interesting because it shows a vesicular accumulation in contact with a thickening of the presynaptic membrane. In contrast to brain synaptosomes no postsynaptic membrane is seen in front of the so called "active zone". This is in good agreement with the poor acetylcholinesterase activity of fraction C (Table I). Figs. 3b and 3c show 2 synaptosomes visualized after freeze fracture. They were identified by their synaptic vesicles. An external view of the P face of the presynaptic membrane (Fig. 3c) shows many particles without special organization. *Torpedo* synaptosomes have a mean diameter of 3.5 μm and a mean volume of 22 μm^3 (Morel et al., 1978). They are therefore much bigger than brain synaptosomes (0.5 μm). A direct estimate of ACh and ATP synaptosomal concentrations in *Torpedo* gave respectively 20 and 3 mM (Morel et al., 1978). Such a high ACh concentration confirms previous indirect estimates which assumed that all the ACh is presynaptic (Dunant et al., 1974). It was possible to show that synaptosomal ACh was distributed among two different pools: vesicular ACh (about 70% of total ACh) and free ACh (30%) hydrolysed when the synaptosomal membrane is disrupted. With synaptosomes, the pools were characterized by successive freeze-thawing trials, a procedure very different from the homogenization which was employed on

TABLE I (continued)

A	B	C	D	E	R_g
0.18 ± 0.02	0.43 ± 0.03	0.35 ± 0.02	0.20 ± 0.01	3.00 ± 0.20	95
1 ± 0	11 ± 2	46 ± 5	14 ± 1	54 ± 6	96
0 ±	1.6 ± 0.3	8.6 ± 1.9	2.4 ± 0.3	10.2 ± 1.5	72
25 ± 5	69 ± 9	170 ± 21	43 ± 6	128 ± 20	89
0.19 ± 0.11	1.46 ± 0.18	2.68 ± 0.32	0.97 ± 0.19	6.23 ± 1.03	96
0.01 ± 0	0.14 ± 0.01	0.26 ± 0.05	0.09 ± 0.01	6.64 ± 1.61	110
0.04 ± 0.02	0.11 ± 0.05	0.13 ± 0.06	0.02 ± 0.01	8.85 ± 1.45	92
2 ± 2	120 ± 30	218 ± 86	22 ± 10	1486 ± 552	98

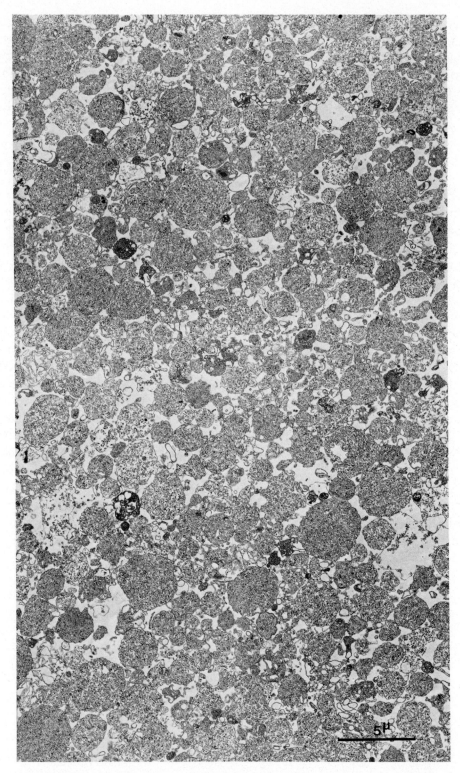

Fig. 2. *Torpedo* synaptosomes: fraction C. The very large field (1350 μm²) presented shows the homogeneity of fraction C. It should be emphasized that *Torpedo* electric organ synaptosomes are much bigger particles than brain synaptosomes (they occupy a 60 times larger surface).

Fig. 3. High magnification of *Torpedo* synaptosomes. (a) The cytoplasm contains synaptic vesicles and granules similarly to nerve endings in situ. An "active zone" shows an accumulation of vesicles at a presynaptic membrane thickening. No postsynaptic membrane remains attached to *Torpedo* electric organ synaptosomes. (b) Freeze fractured synaptosome. This technique also shows the presence of numerous synaptic vesicles within the particle. (c) Freeze fractured synaptosome showing numerous particles on the P face of the presynaptic membrane.

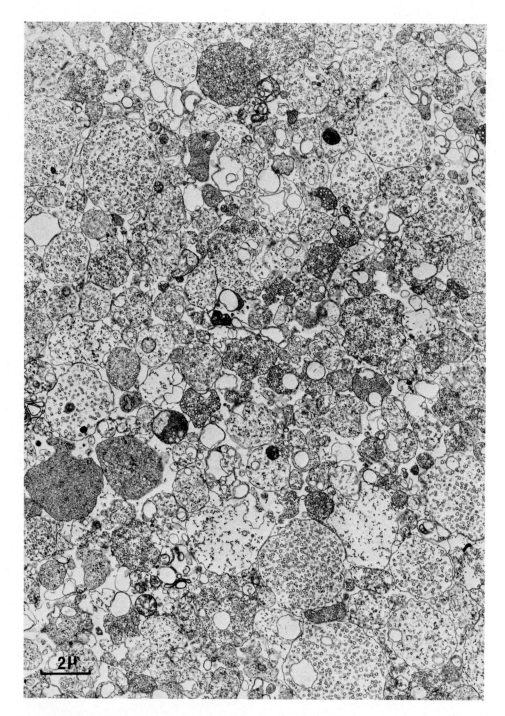

Fig. 4. Morphological control of incubated synaptosomes. No damage is seen after 2-h incubation at 20°C.

tissue samples but with similar results. The free pool in contrast to the vesicular pool has a rapid turnover during synaptic activity (see Israël and Dunant, this volume, p. 125).

INCUBATION OF SYNAPTOSOMES

Three different controls show that *Torpedo* synaptosomes can be incubated for several hours without damage. (1) Figure 4 demonstrates that their morphological appearance is unchanged after two hours incubation. (2) The ACh, ATP and choline acetyltransferase contents of synaptosomes are not

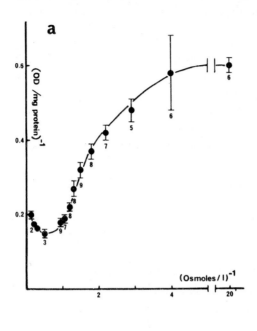

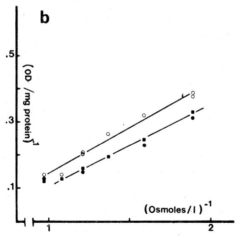

Fig. 5. Osmotic properties of synaptosomes. Aliquots (100 μl) of fraction C were injected into solutions of NaCl of various osmolarity. Light scattering was measured at 540 nm. The optical density was proportional to the number of synaptosomes. Blanks obtained without synaptosomes or after treating the particles with 0.2% Triton X 100 were equal. This technique is similar to the one described by Keen and White (1970) for brain synaptosomes. (a) Double reciprocal plot of optical density (OD) versus NaCl osmolarity (osmol/l). The curve is almost linear between 0.4 and 1 osmol/l (*Torpedo* plasma is close to 0.8 osmol/l). (b) Effect of incubation on the linear part of curve 5a. The linearity shows that synaptosomal osmotic properties are not altered by a 3 h incubation at 20°C. The small upward shift reflects a 10—20% loss of particles. ●, controls; ○, 3-h incubation at 20°C; ■, 3-h incubation at 2°C.

diminished during incubation. We have however observed a 10 to 20% loss of activities during the initial warming and dilution of the fraction (Morel et al., 1977). (3) Osmotic properties of synaptosomes were tested (Fig. 5). Figures 5a and 5b are double reciprocal plots of optical density against NaCl osmolarity. Between 0.4 and 1 osmol/l Fig. 5a shows that there is a linear relationship due to a diminution of the volume of synaptosomes. An osmotic shock occurs below 0.4 osmol/l (which is far below the plasma osmolarity of *Torpedo* of 0.8 osmol/l). After 3 h incubation at 20°C (Fig. 5b), the synaptosomes still show their osmotic properties – demonstration that the synaptosomal membrane is impermeable to Na^+, Cl^- or to the two ions together. The upward shift of the line is due to the 10 to 20% hydrolysis of synaptosomes occurring on warming and dilution.

METABOLISM OF ELECTRIC ORGAN SYNAPTOSOMES

Torpedo synaptosomes are able to synthesize ATP from extracellular adenosine. A facilitated diffusion mechanism for adenosine uptake has been described (K_m = 2.4 μM, V_{max} = 1036 pmol/h · mg protein). Most of the adenosine taken up is converted to ATP (Meunier and Morel, 1978). As expected from many previous studies on brain synaptosomes, we have also characterized a high affinity uptake of choline in *Torpedo* synaptosomes (Morel et al., 1977). Twenty percent of the incorporated choline is converted to ACh. Acetate, which is a good precursor of ACh in *Torpedo* electric organ (Israël and Tuček, 1974) and in neuromuscular junctions (Dreyfus, 1975), is efficiently incorporated into the ACh of synaptosomes. For extracellular acetate concentrations lower than 50 μM, as much as 95% of incorporated radioactivity was found as ACh (Morel et al., 1977). This was established by the fact that is was extractable by tetraphenylboron, migrated with the ACh spot on TLC chromatography and was completely hydrolysed by acetylcholinesterases. Since all the synaptosomal radioactivity is found as ACh, it should be possible to follow the release of ACh by measuring the efflux of radioactivity.

STIMULATION OF TORPEDO ELECTRIC ORGAN SYNAPTOSOMES

After their isolation, synaptosomes were incubated for two hours with (1-^{14}C)acetate (40 μM final concentration). They were then concentrated 20 times by centrifugation. An aliquot (0.1–0.2 ml) of the suspension was injected into the perfusion chamber shown in Fig. 6 (upper insert). The chamber is sealed by a Millipore filter (pore diameter 1.2 μm) protected by a small cotton plug. The perfusion rate was 1.6 ml/min. Considering that nearly all the synaptosomal radioactivity is ACh (see above), we can assume that the efflux of radioactivity will represent the released transmitter. The high perfusion rate, renewing the suspension volume 8 times in a minute, allows us to neglect acetate reuptake. Figure 6 shows an increased efflux of radioactivity in the perfusion fluid immediately (within 10 sec) after substituting K^+ (50 mM)

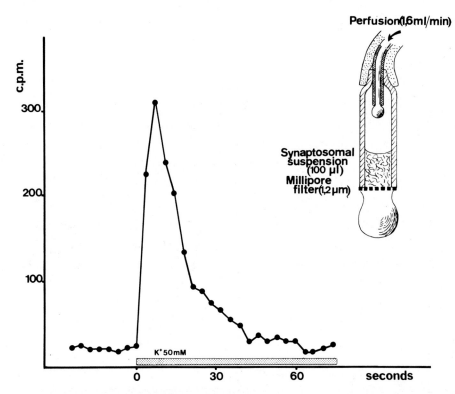

Fig. 6. Stimulation of *Torpedo* electric organ synaptosomes. Synaptosomal ACh was labelled using [1-^{14}C]acetate (40 μM final concentration) for 2 h at 20°C. All the incorporated radioactivity was found as ACh. Synaptosomes were concentrated 20 times by centrifugation (12,000 $\times$ g, 30 min) and 0.1−0.2 ml aliquots introduced into the perfusion chamber (upper insert). Radioactivity was determined in each drop of the effluent. After a washing period (15 min), a constant background radioactivity was obtained. The depolarization was started by perfusing a solution in which 50 mM KCl substitutes for an equivalent concentration of NaCl.

for Na$^+$. The peak of radioactivity reaches 10 to 15 times the background. The radioactivity decreases to base-line in about a minute even if K$^+$ depolarization is maintained. In Fig. 7, synaptosomes were washed with a Ca^{2+} free medium for 15 min. In that case K$^+$ (50 mM) was unable to trigger transmitter release unless Ca^{2+} was added. This demonstrates that transmitter release is fully Ca^{2+} dependent. The kinetic of the response shown in Fig. 7 is similar to that in Fig. 6. After washing the K$^+$ out for 10 min, synaptosomes can be restimulated by a new K$^+$ increase. The evoked release of radioactivity represents only 4.5% ± 1.0% (10 experiments) of the total radioactivity in the chamber. These facts demonstrate that the decay of evoked release during K$^+$ depolarization is not due to exhaustion of the total synaptosomal ACh but to that of a small releasable pool.

It is advantageous to study transmitter release on synaptosomal suspensions since such preparations allow control of the composition of the extracellular medium without diffusion delays, and avoid interactions with other cells as happens with tissue slices. Unfortunately, suspensions of such particles are difficult to handle. Several techniques have been described using either synaptosomal deposits (De Belleroche and Bradford, 1972; Levy et al., 1973) or

200

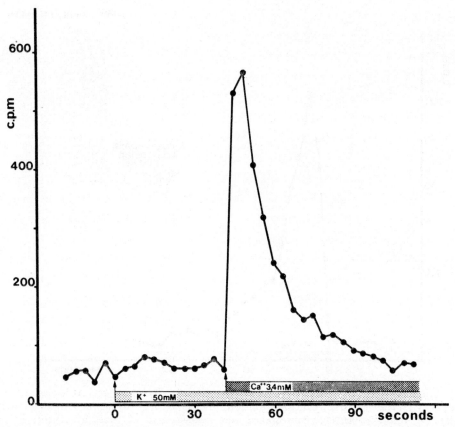

Fig. 7. Ca²⁺ dependency of the evoked ACh release. Similar experiment as in Fig. 6 but with a Ca²⁺ free medium. K⁺ depolarization becomes effective only after introducing Ca²⁺ (3.4 mM) in the perfusate.

synaptosomal suspensions (Mulder et al., 1975, Murrin et al., 1977). All these techniques have been, up to now, applied to brain synaptosomes containing various transmitters. In a recent report, Michaelson and Sokolovski (1978) have measured the effect of K⁺ (200 mM) depolarization on the ACh content of *Torpedo* synaptosomes. The curves obtained show a Ca²⁺ dependent loss of ACh (30%). Their observations agree with the present results in spite of the very different techniques used to follow ACh release. We must however point out that the synaptosomes we isolate derive from entire pinched off nerve endings, mean diameter: 3.5 μm (Morel et al., 1978). The smaller size of Michaelson's preparation, as estimated from their electron microscope picture would suggest a fragmentation of nerve endings during homogenization, a step which is avoided in our technique. The perfusion technique described enabled us to follow the ACh release with a time resolution of 4–5 sec. The time course suggests that the release of radioactivity reflects not the exhaustion of the synaptosomal ACh content but the depletion of a small releasable pool (4.5% of total radioactive ACh). It is therefore smaller than the free or vesicular ACh pools (Morel et al., 1977). The release of various transmitters from brain synaptosomes exhibits similar kinetics with a return to background levels in spite of the maintained K⁺ depolarization (Levy et al., 1973; Mulder et al.,

1975; Murrin et al., 1977). Richter (1976) and Richardson and Szerb (1974) have discussed this time course, also obtained with depolarized tissue slices, in terms of a small compartment of releasable transmitter. The Ca^{2+} entry after K^+ depolarization (Blaustein, 1975; Goddard and Robinson, 1976) is presumably important to mobilize this releasable pool.

SUMMARY

The isolation and characterization of pure cholinergic synaptosomes from the electric organ of *Torpedo* is described. A technique has been developed to study the release of ACh from these synaptosomes. K^+ depolarization triggers a Ca^{2+} dependent release. The good time resolution (4—5 sec) permits further investigations on the mechanisms of ACh release.

ACKNOWLEDGEMENTS

We are grateful to Dr. T. Gulik-Krzywicki for the freeze fracture of synaptosomes, and to Dr. Y. Morot-Gaudry for determing SDH activities in fractions. We thank P. Mastour-Frachon for her skilled technical assistance. This work was supported by a grant (No. 77.7.0251) from the D.G.R.S.T.

REFERENCES

Blaustein, M.P. (1975) Effect of potassium, veratridine and scorpion venom on calcium accumulation and transmitter release by nerve terminals in vitro. *J. Physiol. (Lond.)*, 247, 617—655.

De Belleroche, J.S. and Bradford, H.F. (1972) Metabolism of beds of mammalian cortical synaptosomes: response to depolarizing influences. *J. Neurochem.*, 19, 585—602.

De Robertis, E., Pellegrino de Iraldi, A., Rodrigues de Lores Arnaiz, G. and Salganicoff, L. (1961) Electron microscope observations of nerve endings isolated from rat brain. *Anat. Rec.*, 139, 220—221.

Dowdall, M.J. and Zimmermann, H. (1977) The isolation of pure cholinergic nerve terminal sacs (T-sacs) from the electric organ of juvenile *Torpedo*. *Neuroscience*, 2, 405—421.

Dreyfus, P. (1975) Identification de l'acétate comme précurseur du radical acétyl de l'acétylcholine des jonctions neuromusculaires du rat. *C.R. Acad. Sci. (Paris)*, 280, 1893—1894.

Dunant, Y., Gautron, J., Israël, M., Lesbats, B. and Manaranche, R. (1974) Evolution de la décharge de l'organe électrique de la Torpille et variations simultanées de l'acétylcholine au cours de la stimulation. *J. Neurochem.*, 23, 635—643.

Feldberg, W. and Fessard, A. (1942) The cholinergic nature of the nerves to the electric organ of the *Torpedo*. *J. Physiol. (Lond.)*, 101, 200—216.

Feldberg, W., Fessard, A. and Nachmansohn, D. (1940) The cholinergic nature of the nervous supply to the electrical organ of the *Torpedo*. *J. Physiol. (Lond.)*, 97, 3P.

Goddard, G.A. and Robinson, J.D. (1976) Uptake and release of calcium by rat brain synaptosomes. *Brain Res.*, 110, 331—350.

Gray, E.G. and Whittaker, V.P. (1960) The isolation of synaptic vesicles from the central nervous system. *J. Physiol. (Lond.)*, 153, 35P-37P.

Israël, M., Manaranche, R., Mastour-Frachon, P. and Morel, N. (1976) Isolation of pure cholinergic nerve endings from the electric organ of *Torpedo marmorata*. *Biochem. J.*, 160, 113—115.

Israël, M. and Tuček, S. (1974) Utilization of acetate and pyruvate for the synthesis of

202

"total", "bound" and "free" acetylcholine in the electric organ of *Torpedo. J. Neurochem., 22,* 487–491.

Jones, D.G. (1975) Synapses and synaptosomes. ed. by Chapman and Hall Ltd., London.

Keen, P. and White, T.D. (1970) A light-scattering technique for the study of the permeability of rat brain synaptosomes in vitro. *J. Neurochem., 17,* 565–571.

Levy, W.B., Redburn, D.A. and Cotman, C.W. (1973) Stimulus-coupled secretion of γ-aminobutyric acid from rat brain synaptosomes. *Science, 181.* 676–678.

Meunier, F.M. and Morel, N. (1978) Adenosine uptake by cholinergic synaptosomes from *Torpedo* electric organ. *J. Neurochem., 31,* 845–851.

Michaelson, D.M. and Sokolovski, M. (1978) Induced acetylcholine release from active purely cholinergic *Torpedo* synaptosomes. *J. Neurochem., 30,* 217–230.

Morel, N., Israël, M. and Manaranche, R. (1978) Determination of ACh concentration in *Torpedo* synaptosomes. *J. Neurochem., 30,* 1553–1557.

Morel, N., Israël, M., Manaranche, R. and Mastour-Frachon, P. (1977) Isolation of pure cholinergic nerve endings from *Torpedo* electric organ. Evaluation of their metabolic properties. *J. Cell Biol., 75,* 43–55.

Mulder, A.H., van den Berg, W.B. and Stoof, J.C. (1975) Calcium-dependent release of radiolabelled catecholamines and serotonin from rat brain synaptosomes in a superfusion system. *Brain Res., 99,* 419–424.

Murrin, L.C., De Haven, R.N. and Kuhar, M.J. (1977) On the relationship between [3H]-choline uptake activation and [3H]acetylcholine release. *J. Neurochem., 29,* 681–687.

Richardon, I.W. and Szerb, J.C. (1974) The release of labelled acetylcholine and choline from cerebral cortical slices stimulated electrically. *Brit. J. Pharmacol., 52,* 499–507.

Richter, J.A. (1976) Characteristics of acetylcholine release by superfused slices of rat brain. *J. Neurochem., 26,* 791–797.

Dual Effect of Noradrenaline on Cholinergic Transmission in Guinea-Pig Ileum

ONDŘEJ KADLEC and KAREL MAŠEK

Institute of Pharmacology, Czechoslovak Academy of Sciences, 12800 Prague
(Czechoslovakia)

INTRODUCTION

For many years it has been generally accepted that the activity of the gastro-intestinal tract is under reciprocal control of parasympathetic excitatory and sympathetic inhibitory fibers. However, some evidence suggests a more complicated relationship between these sections of the autonomic system. Under certain experimental conditions, the sympathetic neurotransmitter, noradrenaline (NA), inhibits the release of acetylcholine (ACh) from parasympathetic nerve terminals in guinea-pig ileum (Paton and Vizi, 1969). Our own work has suggested that, in addition to this inhibitory effect of NA, this amine can also facilitate ACh release in the intestine; the present experiments test this, and attempt to investigate our previous suggestion (Kadlec et al., 1977) that such an effect might be mediated through stimulation of endogenous synthesis of prostaglandins (PGs).

EFFECT OF NA ON PHOSPHOLIPID TURNOVER AND ON THE OUTPUT OF PGs

Isolated guinea-pig ileum with the attached sympathetic nerves was incubated (2 min) with ^{32}P (NaH_2PO_4). After washout of the label during the next 10 min period, the sympathetic nerves in the experimental group were stimulated (10 Hz, 1 ms, 10 V, 10 min). The labelling of phosphatidyl choline was determined after tissue homogenization, extraction, and thin-layer chromatography (TLC). In 6 pairs of preparations the amount of the labelled phosphatidylcholine found in the tissue of the stimulated group was $63.5 \pm 8.5\%$ ($P < 0.01$) of the unstimulated control group. Although there are many other possible interpretations for this result, it might indicate that NA released from sympathetic nerve endings increases the turnover of membrane phospholipids which in turn might cause an increase of PG synthesis through the release of their precursors, unsaturated fatty acids.

Longitudinal muscle-myenteric plexus preparations from the guinea-pig ileum were placed in an isolated organ bath. The amount of PGE-like substances released from the tissue into the bathing fluid was determined, after its

204

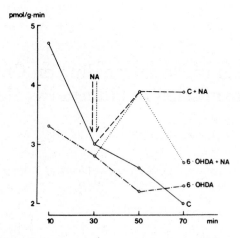

Fig. 1. Output of PG-like substances in terms of PGE₂ equivalents from isolated longitudinal strip of guinea-pig ileum. PG-like substances were collected during 10 min stimulation periods (5 Hz); 10 min resting periods followed where the amount of PG-like substances was lower or even undetectable and is not indicated. NA (5 μM) was added after the second collection period to one half of the preparations taken from either control (C) or sympathectomized (6-OHDA) animals. Each point represents 6–20 determinations.

extraction and TLC separation, by bioassay on the rat stomach strip (Kadlec et al., 1978). In the preparations taken from both control and sympathectomized animals (6-OHDA, 0.1 g/kg; given i.v. in two doses 40 and 24 h before killing), the addition of NA (5 μM) caused an increase of the output of PG-like substances (Fig. 1).

EFFECTS OF PGs ON CONTRACTIONS AND ACh OUTPUT

Contractions of the longitudinal muscle of isolated guinea-pig ileum were evoked either directly by adding ACh or indirectly by electric stimulation of the postganglionic cholinergic fibers (Kadlec et al., 1974). PGE₂ (6 nM) increased both directly and indirectly evoked contractions. Bennet et al. (1968) ascribed the potentiating effect of PGE mainly to its direct action upon smooth muscle. However, in our experiments, indomethacin (IND; 1 μM), an inhibitor of PG synthesis (Vane, 1971) decreased the contractions evoked indirectly and this decrease was reversed by PGE₂ (Fig. 2). IND induced some depression of the response to ACh, but this was not reversed by PGE₂. From these results it was tentatively concluded that PGE₂ may modulate cholinergic transmission in the myenteric plexus of guinea-pig ileum in addition to its direct action on smooth muscle.

Hadházy et al. (1973) and Botting and Salzmann (1974) reported no distinct effect of either PGE₁ (57 nM) or IND (28 μM) on the output of ACh from guinea-pig ileum. In our experiments the effect of PGE₂ (6 nM) on the output of ACh from this preparation was correlated with the initial level of ACh output (Kadlec et al., 1978). The output of ACh during nerve stimulation (0.1 Hz) was collected in the presence of the cholinesterase inhibitor physostigmine (salicylate or sulphate; base at 4 μM) concentration to prevent its enzymatic

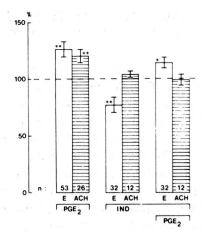

Fig. 2. Relative changes in contractions of guinea-pig isolated ileum treated with PGE_2 (6 nM) and/or IND (1 μM). The contractions were evoked by electrical (E) supramaximal coaxial stimulation (0.4 ms pulse width, 0.1 Hz) or by ACh (0.6−60 nM) and were compared with the responses before the drug treatment (100%). Values significantly different from controls at $P < 0.05$ and $P < 0.005$ are denoted by one or two asterisks respectively; n, number of individual measurements in each group of at least 5 experiments.

degradation; ACh was determined by a bioassay (Paton and Vizi, 1969). It was observed that in the preparations with high initial output of ACh the addition of PGE_2 did not evoke any further increase; however, in the preparations with low initial output of ACh, especially after pretreatment with IND, a measurable increase of this output was observed following the addition of PGE_2 (Fig. 3).

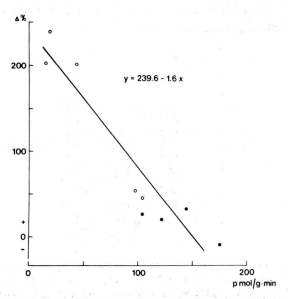

Fig. 3. Correlation between the initial output of ACh (pmol/g · min) during stimulation at a frequency of 0.1 Hz and its relative changes ($\Delta\%$) induced by PGE_2 (6 nM). Guinea-pig ileum; full circles, control preparations; open circles, preparations where the initial output of ACh was reduced by pretreatment with IND (1 μM). The regression coefficients were calculated by the method of least squares. The correlation ($r = -0.858$) was significant ($P < 0.01$).

EFFECT OF NA ON THE OUTPUT, CONTENT AND RATE OF ACh SYNTHESIS AND ON CONTRACTIONS

In control preparations, NA (5 μM) inhibited the output of ACh during low frequency stimulation of cholinergic neurons (0.1 Hz; control output: 163 pmol/g · min); however, during 5 Hz stimulation (control output: 1593 pmol/g · min), no effect of NA on the output of ACh was observed (Fig. 4a). Destruction of the sympathetic system by 6-hydroxydopamine (6-OHDA) pre-treatment decreased the ACh output during 5 Hz stimulation (Fig. 4b), but did not change output during 0.1 Hz stimulation. The addition of NA to the sympathectomized preparations (Fig. 4c) inhibited the output of ACh during 0.1 Hz stimulation, but the output during 5 Hz stimulation was increased. This increase was absent in preparations that were incubated in the presence of IND (1 μM). From these experiments it appeared that in sympathectomized preparations the addition of NA resulted in an increase of the output of ACh during 5 Hz stimulation, i.e. under the condition where the direct inhibitory effect of NA on ACh output was weak.

Tissue content of ACh was determined by bioassay after extraction in tri-chloracetic acid. Table 1 shows the accumulation of ACh in the tissue incubated in the bath for 4 h in the presence of physostigmine (4 μM); this increase might represent the accumulation of "surplus" ACh (MacIntosh and Collier, 1976). ACh accumulation was significantly slower ($P < 0.005$) in sympathectomized preparations (6-OHDA) compared to controls. The presence of NA (5 μM) during the incubation caused a significant increase of ACh content only in sympathectomized preparations ($P < 0.05$). As the accumulation of ACh may have a bearing on the rate of ACh synthesis (Szerb, 1975), these results suggested that it could be decreased by sympathectomy and restored by subsequent addition of NA.

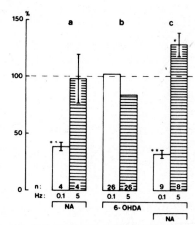

Fig. 4. Relative changes in the output of ACh from isolated longitudinal strip of guinea-pig ileum. Preparations were taken from either control or sympathectomized (6–OHDA) animals and treated with NA (5 μM) when indicated. The stimulated output was measured either at a frequency of 0.1 Hz (7 min periods) or at a frequency of 5 Hz (1 min periods). The output of ACh in the presence of NA (a, c) was compared with output before the drug treatment (100%) in control and sympathectomized groups respectively. The output of ACh in the absence of NA in sympathectomized group (b) was compared with the output of untreated control group (100%). Values significantly different from controls at $P < 0.05$ and $P < 0.005$ are denoted by one or two asterisks respectively; n, number of experiments.

TABLE 1

CONTENT OF ACh IN LONGITUDINAL STRIP OF GUINEA-PIG ILEUM

	Control	*6-OHDA*
After dissection	70.4 ± 4.4 (28)	74.8 ± 4.4 (21)
After 4 h + physostigmine	149.6 ± 9.4 (20)	110.0 ± 7.7 (16)
After 4 h + + physostigmine + NA	129.3 ± 27.0 (6)	150.2 ± 12.8 (4)

Means with their S.E. are expressed as nmol/g. Numbers of experiments are shown in parentheses.

TABLE 2

EFFECT OF 6-OHDA, NA AND PGE_2 ON ACh OUTPUT FROM AND CONTENT OF LONGITUDINAL STRIP FROM GUINEA-PIG ILEUM AND CALCULATED RATES OF ACh SYNTHESIS

	n	*ACh output* $(nmol/g \cdot h)$	*Change of ACh content* $(nmol/g \cdot h)$	*Rate of ACh synthesis* $(nmol/g \cdot h)$
Control	9	11.0 ± 1.8	+30.3 ± 8.1	31.3 ± 8.8
6-OHDA	4	15.3	−9.5	5.9
NA/6-OHDA	3	6.7	+11.6	18.2
PGE_2/6-OHDA	3	28.1	−4.6	23.5

Thus, the rate of ACh synthesis was calculated in experiments where double preparations were used. The content of ACh was determined in the first half of such preparation, then the output of ACh from the second half was measured during 0.1 Hz stimulation and finally its ACh content was determined. The rate

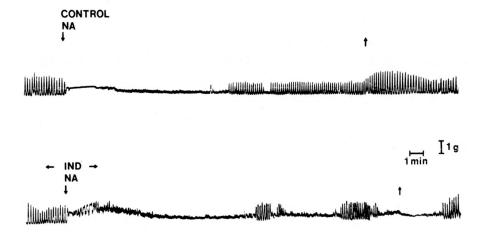

Fig. 5. Contractions of longitudinal muscle during peristaltic reflex activity of isolated guinea-pig ileum. The tension of the muscle was elevated (3 g) and intraluminal pressure increased (40 mm water column) throughout the experiment. NA (0.1 μM) was present in the bath as indicated by the arrows. Control preparation, the upper record; IND (2 μM) pre-treated preparation, the lower record.

of ACh synthesis was calculated as a sum of total ACh released plus the net change of its content during that interval (Paton et al., 1971). From our preliminary results (Table 2) it appears that sympathectomy decreased the rate of ACh synthesis and that it was restored either by NA or by PGE_2.

Peristaltic reflex of guinea-pig ileum was also registered in our experiments (Van Nueten et al., 1973). The cholinergic contractions of longitudinal muscle (Fig. 5) were inhibited by the addition of NA (0.1 μM). Upon the washout of NA, when its direct inhibitory effect faded, the amplitude of contractions was increased above the level before the addition of NA. This rebound phenomenon was not observed in the preparations pretreated with IND (2 μM). It is therefore suggested that NA contributed to the observed increase of cholinergic contractions via the release of tissue PG's.

SUMMARY

A hypothesis is proposed that NA, via a release of tissue PGs, may positively influence cholinergic transmission in myenteric plexus-longitudinal muscle preparation of guinea-pig ileum. Evidence is presented that NA increased the release of tissue PG's and that PGE_2 increased the output of ACh from parasympathetic nerve terminals. An increase of ACh output and of cholinergic contractions following NA addition could be demonstrated only under special conditions.

ACKNOWLEDGEMENT

PGE_2 was kindly supplied by Dr. J. Pike of the Upjohn Co.

REFERENCES

Bennett, A.J., Eley, K.G. and Scholes, G.B. (1968) Effect of prostaglandins E_1 and E_2 on human, guinea-pig and rat isolated small intestine. Brit. J. Pharmacol. Chemother. 34, 630–638.

Botting, J.H. and Salzmann, R. (1974) The effect of indomethacin on the release of prostaglandin E_2 and acetylcholine from guinea-pig isolated ileum at rest and during field stimulation. Brit. J. Pharmacol. 50, 119–124.

Hadházy, P., Illés, P. and Knoll, J. (1973) The effects of PGE_1 on responses to cardiac vagus nerve stimulation and acetylcholine release. Europ. J. Pharmacol. 23, 251–255.

Kadlec, O., Mašek, K. and Šeferna, I. (1974) A modulating role of prostaglandins in contractions of the guinea-pig ileum. Brit. J. Pharmacol. 51, 565–570.

Kadlec, O., Mašek, K. and Šeferna, I. (1977) A new possible relationship between the sympathetic and parasympathetic system. Physiol. bohemoslov. 26, 451.

Kadlec, O., Mašek, K. and Šeferna, I. (1978) Modulation by prostaglandins of the release of acetylcholine and noradrenaline in guinea-pig isolated ileum. J. Pharmacol. exp. Ther. 205, 635–644.

MacIntosh, F.C. and Collier, B. (1976) Neurochemistry of cholinergic terminals. In Neuromuscular Junction, E. Zaimis (Ed.), pp. 99–228, Springer Verlag, Berlin.

Paton, W.D.M. and Vizi, E.S. (1969) The inhibitory action of noradrenaline and adrenaline on acetylcholine output by guinea-pig ileum longitudinal muscle strip. Brit. J. Pharmacol. 35, 10–28.

Paton, W.D.M., Vizi, E.S. and Zar, M.A. (1971) The mechanism of acetylcholine release from parasympathetic nerves. *J. Physiol. (Lond.)* 215, 819–848.

Szerb, J.C. (1975) Endogenous acetylcholine release and labelled acetylcholine formation from [^{3}H]choline in the myenteric plexus of the guinea-pig ileum. *Can. J. Physiol. Pharmac.* 53, 566–574.

Vane, J.R. (1971) Inhibition of prostaglandin synthesis as a mechanism of action for aspirin-like drugs. *Nature New Biol.,* 231, 232–235.

Van Nueten, J.M., Geivers, H., Fontaine, J. and Janssen, P.A.J. (1973) An improved method for studying peristalsis in the isolated guinea-pig ileum. *Arch. int. Pharmacodyn.* 203, 411–414.

PART III

POSTSYNAPTIC ACTION OF ACETYLCHOLINE ELECTROPHYSIOLOGY

Drug-Receptor Interaction at the Frog Neuromuscular Junction

F. DREYER, K. PEPER, R. STERZ, R.J. BRADLEY and K.-D. MÜLLER

II. Physiologisches Institut Universität des Saarlandes, D.6650 Homburg/Saar (F.R.G.)

A. MORPHOLOGY

Quantitative application of neurotransmitters and their derivatives by iono-phoresis is only possible if the topology of the end-plate region is well known. For this purpose, it is necessary to use Nomarski-interference optics during the experiment, since with this system it is possible to visualize the living end-plate. At the frog neuromuscular junction, a single myelinated nerve approaches the muscle fiber, (see Fig. 1), loses its myelin, and contacts the muscle for several 100 μm in the form of a thin (1–3 μm) terminal nerve branch. Often, side-branches are present, and a complicated network of terminals forms the end-plate. The postsynaptic membrane underneath the terminal branches is charac-terized by the junctional folds and the presence of acetylcholine (ACh) recep-tors. The pattern of postjunctional folds can be directly observed in the scanning electron microscope, after the nerve branch has been removed by a treatment with collagenase (McMahan et al., 1972; Peper and McMahan, 1972). Figure 2 gives an example. Here the regular pattern of folds can be seen

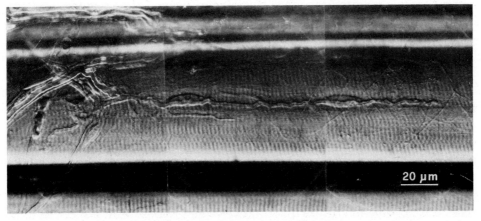

Fig. 1. Living frog neuromuscular junction viewed in the Nomarski-interference microscope. M. cutaneous pectoris preparation.

* This work has been supported by the DFG, Sonderforschungsbereich 38, project N.

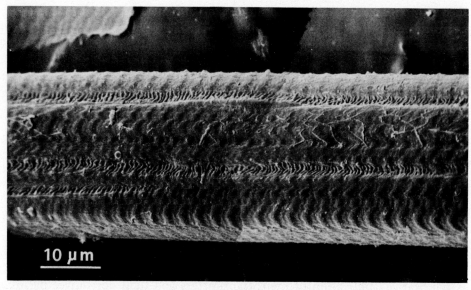

Fig. 2. Postsynaptic membrane of a frog muscle fiber with junctional folds as viewed in the scanning electron microscope. The presynaptic nerve terminals were removed after colla- genase treatment. M. cutaneus pectoris preparation.

perpendicular to the longitudinal axis of a branch. This regular pattern corre- lates with that of the presynaptic active zones apparent in freeze fracture tech- niques (Dreyer et al., 1973; Peper et al., 1974; Heuser et al.., 1974). For an experiment, it is absolutely necessary to select only those end-plates which exhibit a single branch, separated from other branches by at least 50–80 μm. It is then possible, under magnification of 400–500, to accurately position the tip of an ionophoresis pipette 18 μm above the end of the branch. This arrange- ment is necessary for a quantitative ionophoretic application of a drug as will be shown in the next section. Also, the direct visualization of the end-plate under high magnification is helpful in order to insert a voltage sensing and a current feeding pipette for voltage clamping of the end-plate region.

B. QUANTITATIVE IONOPHORESIS

Once the topology of the end-plate is known and the pipettes are put in an adequate position in relation to the end-plate branch, it is possible to apply a drug to the end-plate branch (in a quantitative manner). The drug, being released by a short pulse through the ionophoretic pipette, diffuses to the end- plate and causes a conductance change of the clamped postsynaptic membrane, which can be recorded (Fig. 3). Although the concentration of the drug is not uniformly distributed along the terminal branch and changes continuously with time, it is possible to analyze quantitatively the response with respect to several parameters. It is, for example, known that the amplitude of a response does not rise linearly with increasing agonist concentration but instead the amplitude rises with a power of 2.7 (for ACh). This number is called the Hill coefficient n_H. It can be shown that even a single response like that in Fig. 3 can be

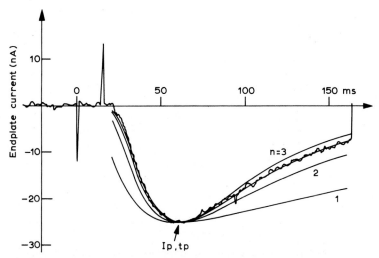

Fig. 3. ACh-response of a voltage-clamped end-plate (thick line). Theoretical curves (thin lines) calculated for different Hill coefficients according to eqn. 1 in the text. Best fit for $n_H = 2.6$. Time to peak 62 ms, peak end-plate current 26 nA. Distance between ACh-pipette and end-plate 18 μm. ACh-pulse 1850 pC. Holding potential -80 mV, temperature 23°C.

analyzed for n_H. The theoretical time course of the response $I_{sum}(t)$ is given (Peper et al., 1975) by the equation

$$I_{sum}(t) = I_p \left\{ \frac{t_p}{t} \cdot \exp\left(1 - \frac{t_p}{t}\right) \right\}^{1.5 n_H - 0.5} \tag{1}$$

for a linear terminal where I_p is the peak amplitude and t_p the time to peak of the response. Note that the Hill coefficient is the only unknown parameter which can be determined independently from distance of release, pipette transport number etc..

When the distance r of the pipette is increased in the experiment, then the time to peak of the response becomes larger. The equation describing this dependence is (Peper et al., 1975)

$$r^2 = \left(6 - \frac{2}{n_H}\right) \cdot D \cdot t_p \tag{2}$$

After the determination of n_H, the diffusion coefficient D can be determined from a plot of t_p vs. distance r of release. There is a deviation from the theoretical values when the distance is made smaller and the time to peak reaches values compatible with the difussion delay in the synaptic cleft. Therefore, the same plot also gives an indication where the condition of steady state is violated. We found that a distance of at least 15 μm must be observed for the application of most drugs in order to ensure the steady state condition.

With the pipette in a correct position, the amount of applied drug can be changed (by variation of the drug pulse amplitude and duration), and the resulting changes in time to peak and peak amplitude of the response observed. Dose response curves are then computed on a log–log plot (Fig. 4). At low doses, the peak end-plate current displays a strong dependence on the dose, the slope being equal to the Hill coefficient. With higher doses, the peak current

216

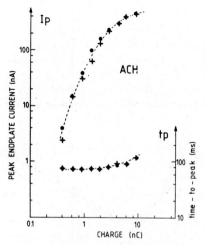

Fig. 4. Dose—response curve for ACh at a frog end-plate. Peak end-plate current I_p and time to peak t_p as a function of charge through the drug pipette. The parameters determined from standard dose—response curves are $n_H = 2.7$, $K = 30\ \mu M$, and $I_{max} = 22\ nA/\mu m$ for the cooperative model (for the exact procedure of evaluating the parameters see Dreyer et al., 1978). Distance between pipette and end-plate = $20\ \mu m$. Holding potential -80 mV, temperature $22°C$.

increases less, leading to a bend in the curve. Simultaneously the time to peak increases. These are signs that ACh-receptors located directly underneath the pipette begin to saturate. Complete saturation (i.e. horizontal dose—response curve) cannot be obtained under the chosen experimental conditions, because the terminal is long in comparison to the activated receptor area and higher drug doses always reach areas of the branch which are not yet saturated. By using the experimentally determined values of I_p and t_p it is possible to deduce the maximum current I_{max} per $1\ \mu m$ length of the terminal, and the apparent dissociation constant K (for the exact procedure see Dreyer et al., 1978). The maximum number of ion channels per μm which can be opened by high agonist concentrations, can be calculated from the value of I_{max}. These values are given in Table I. It should be noted here that the dependence of t_p on the dose is an important criterion of whether or not the response is in steady state and whether the analysis of the dose-response curve is valid. Deviations from the normal dependence my occur by desensitization or by channel blocking (see section D).

Several models have been proposed for the explanation of Hill coefficient

TABLE I

Drug	ACh	CCh	SubCh
Hill coefficient	2.7	2.2	2.4
Diffusion coefficient 10^{-6} cm^2 s^{-1}	10.7	7.7	–
Max. conductance per 1 μm, g_{max}	170	150	80
Max. number of open ionic channels per 1 μm *	9100	8000	4300
Apparent dissociation constant (μM)	28	336	18

* For a single channel conductance of 18.6 pS. ACh, acetylcholine; CCh, carbachol; SubCh, suberyldicholine.

>1. It is generally assumed that more than one agonist molecule has to act on a receptor in order to open it. There might also be a functional reason for the existence of a sigmoid dose—response curve. It is known and recently discussed in the literature that a considerable amount of ACh is released in a nonquantal manner by the nerve terminal independent of excitation. If we assume that it is the quantal release which triggers the muscle contraction, then receptors have to discriminate between quantal and nonquantal release. This may be done by the introduction of a threshold, since it can be expected that quantal release will almost saturate nearby receptors, whereas nonquantal release is speculated to be more evenly distributed. A high Hill coefficient would create such a threshold, since the shape of the response curve is sigmoid, i.e. at low doses the response is almost undetectable, but then increases very steeply with higher doses before saturation is reached. In the following section some results of quantitative ionophoresis will be discussed.

C. pH-EFFECTS

In order to obtain a better insight into the molecular events governing the interaction between agonist and receptor, we have studied some of the parameters on which the interaction may depend. The pH of the external fluid is one such condition (Scuka, 1975, 1977; Mallart and Molgo, 1978).

We found that, for example, the parameters g_{max}, K, and the Hill coefficient were increased with higher pH, whereas the mean conductance of a single channel was apparently unchanged. A strong dependence on pH was observed for the mean open time of the channel, τ. This is especially interesting, since it is well known that τ also depends on the membrane potential, and it was of considerable interest to find out if the effects of pH can be obviated by a shift in membrane potential. We have studied, therefore, the dependence of τ on the membrane potential at different pH values. A a convenient measure of τ we have taken the decay time constant of nerve evoked end-plate currents. The dependence on membrane potential and pH is shown in Fig. 5. It is obvious

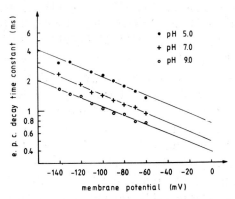

Fig. 5. Dependence of the end-plate current decay time constant on membrane potential for different pH. Lines are least-square fits to the experimental points assuming an exponential relationship. Ethylene glycol treated frog cutaneus pectoris muscle preparation. Temperature 22°C.

218

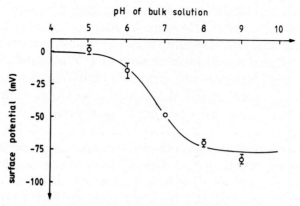

Fig. 6. Dependence of surface potential on pH. The points were obtained from experiments like that shown in Fig. 5 assuming a zero surface potential for low pH. The line is a least-square fit of a titration curve to the data.

that the relationship is linear, i.e., τ depends exponentially on the membrane potential. The slope indicates an e-fold increase in τ with an hyperpolarization of about 100 mV. When the bulk fluid was made more acid, the τ-values were increased. However, the dependence on membrane potential was the same, as can be seen from the parallel shift with respect to the control curve. In the case of pH 9.0 the same parallel shift was observed toward smaller decay time constants.

This result indicates that in the range of pH and membrane potential used here, both effects obviously have a common mechanism. The simplest model to explain the observed facts is the assumption that surface charges exist which can be titrated by protons. When either the surface potential is changed by titration or the membrane potential is altered, the electric field of the membrane changes. Our results indicate that the mechanism which determines τ is dependent on this electric field. With this assumption, an estimate of how the surface potential depends on different pH can be made (see Fig. 6). Assuming that the surface potential is almost zero at pH 5.0, a surface potential of −62 mV at pH 7.4 and a maximum of −76 mV at pH 9.0 were determined. The values agree quite well with values given for other biological membranes (see for example Drouin and Neumcke, 1974; McLaughlin, 1977). The points can be fitted by a curve which is used to describe the titration of a weak acid.

The measured values of the Hill coefficient in relation to pH could also be fitted by a titration curve similar to that shown in Fig. 6. The minimum Hill coefficient was reached at pH 5.0, the maximum at pH 9.0. Surprisingly, the effects of pH on the Hill coefficient could not be counteracted by changes in membrane potential. In fact, there was no change in Hill coefficient with membrane potential. The study of pH dependence led us to the conclusion that the mechanism determining τ is located somewhere inside the membrane, whereas that of the Hill coefficient is possibly located on the membrane surface.

D. THE EFFECT OF PREDNISOLONE

Prednisolone and other related steroids have been used in the successful treatment of myasthenia gravis. Although the immunosuppressive properties of these drugs may account for their potency, some investigators have reported a direct action of prednisolone on the neuromuscular junction. Some experiments suggest a facilitation of transmission, whereas others indicate an inhibitory effect. We studied the effect of 40 μM prednisolone on the voltage-clamped frog end-plate by measuring the dose–response curve in the absence and presence of the drug (Fig. 7). As can be seen, there is a gradual reduction of the end-plate current, and, surprisingly, a decrease in the time to peak t_p for high agonist concentrations. This effect can also be shown by comparing the time courses of single responses before and after drug application (Fig. 8): the amplitude is depressed and the time to peak occurs earlier compared to the control. This indicates that in the presence of prednisolone the drug reactions are not in steady state and the evaluation of dose–response curve parameters by the method described above is no longer valid. Therefore, a more appropriate method of investigation must be applied.

Similar changes in the time course of such responses have been observed when barbiturates (Adams, 1976) or curare (Manalis, Dreyer, Peper, and Sterz, unpublished observation) were present in the bath solution. An explanation of this fact was to assume that such drug molecules cause a reversible blocking of open ion channels as has been proposed by Adams (1976) for the action of certain barbiturates.

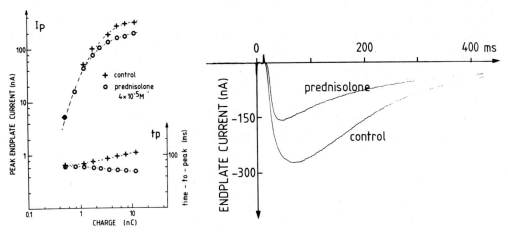

Fig. 7. Dose–response curve (log–log plot) for ACh in the absence and presence of prednisolone. The parameters for the control curve are n_H = 2.7, K = 28 μM, and I_{max} = 14 nA/μm for the cooperative model. Note especially the differences in the dose-dependence of the time to peak values. Distance between pipette and end-plate 18 μm, membrane holding potential −80mV, temperature 22°C. Prednisolone was dissolved in a small amount of ethanol prior to dilution in Ringer solution. This concentration of ethanol has no effects on the parameters investigated.

Fig. 8. Comparison of the ACh-induced end-plate current with and without 40 μM prednisolone in the bath solution. The responses are taken from the data in Fig. 7 (charge Q = 7.5 nC).

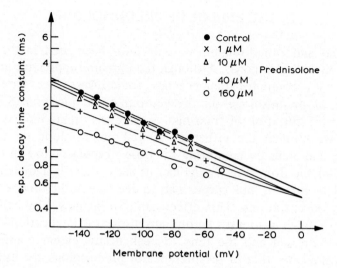

Fig. 9. The effect of several prednisolone concentrations on the decay time constant of nerve evoked end-plate currents at different membrane potentials. The solid lines are least-square fits to the experimental points assuming an exponential dependence on the membrane potential. To avoid muscle contraction the preparation was pretreated with 1 M ethylene glycol for 60 min. Temperature 22°C.

The following reaction scheme for channel blocking is in good agreement with experimental results (Adams, 1976, 1977; Ruff, 1977):

$$AT \underset{\alpha}{\overset{\beta'}{\rightleftharpoons}} AR + B \underset{b}{\overset{f}{\rightleftharpoons}} ARB$$

After binding of an agonist molecule A to a receptor T whose conformation is associated with a closed ion channel, the receptor will convert into an active state AR associated with an open channel with the conductance γ. Assuming that a prednisolone molecule B binds to AR only, the receptor changes to a state ARB whose conductance is assumed as zero. f and b are the forward and backward rate constants of the blocking and unblocking reaction.

Such reaction schemes for channel blocking can be easily studied by mea-

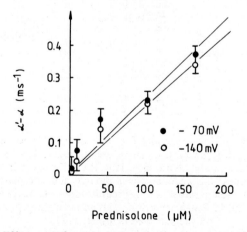

Fig. 10. Plot of the difference of rate constants in the presence and absence of prednisolone against the drug concentration for two different membrane potentials. The values are taken from the data in Fig. 9. The solid curves are least-square regression lines to the values.

10 nA

50 ms

Fig. 11. Recovery of the end-plate current after a conditioning pulse in the presence of 100 μM prednisolone. Equal amounts of ACh were applied ionophoretically directly at the end-plate in the conditioning and test pulse. Pulse 100 nA/1 ms, holding potential −80 mV, temperature 22°C.

suring the time course of the falling phase of nerve evoked end-plate currents for different drug concentrations (Adams, 1976; Beam, 1976a,b). Figure 9 shows the result of an experiment where different prednisolone concentrations were applied. In a drug-free Ringer solution the end-plate currents follow a simple exponential time course over most of their falling phase. The decay time constant became larger with hyperpolarization and was changed approximately e-fold for each 100 mV change in membrane potential. In the presence of prednisolone both the decay time constant and its membrane potential dependence were affected, but still the falling phase of the end-plate currents could be described by a single exponential curve, in contradiction to the observed double exponential time courses in the presence of local anaesthetics (see Beam, 1976a). This can be explained by assuming a slow dissociation of prednisolone molecules from the blocked channel. The new decay rate constant of the end-plate current will be approximately $\alpha' = \alpha + c \cdot f$, where c is the prednisolone concentration and α the rate constant of the control; comment on f is given later in this paragraph. In Fig. 10 the difference $\alpha' - \alpha$ is plotted against the prednisolone concentration for two different membrane potentials. In a first approximation the relationship is linear and identical for both membrane potentials. f can be obtained from the slope of the line, and seems to be potential independent as one would expect from the fact that prednisolone molecules are uncharged. f has a value of about $2.3 \cdot 10^6 \ M^{-1} \ s^{-1}$.

The unblocking of the state ARB can be demonstrated with an experiment as shown in Fig. 11. The first response was elicited with a short ionophoretic pulse of ACh. Some of the opened ion channels will be blocked by prednisolone molecules, so that a second ionophoretic ACh-pulse will cause a reduced response if it follows a short time after the first one. If this time is prolonged, more ion channels become unblocked and, therefore, the second response becomes larger, finally reaching the amplitude of the first response. From such experiments the unblocking rate b could be estimated to be 14 s^{-1}, corresponding to a time constant of 70 ms. The value did not depend on the membrane potential. From f and b it is possible to estimate the equilibrium binding constant to the open ion channel. The value $b/f = 6 \ \mu M$ is in the same range as reported for other drugs like barbiturates and procaine (Adams, 1976, 1977), local anaesthetics (Neher and Steinbach, 1978), and curare (Colquhoun, Dreyer and Sheridan, 1978). The fact that such different drugs (charged and uncharged molecules) have similar blocking rate constants suggests a common action at the receptors with their ion channels.

Prednisolone, however, has presynaptic effects as well. Already in a concentration as low as 1 μM, which has practically no effect on the postsynaptic site, the nerve evoked end-plate currents are increased by a factor of 2, thus, facilitating neuromuscular transmission. Since this is not observed with ionophoretic application of the drug, it is clear that prednisolone affects the presynaptic release of transmitter. Therefore, the action of the drug in situ may be a very complicated combination of effects on both the presynaptic and postsynaptic membrane depending on the concentration of the drug and transmitter as well as on the duration of agonist action.

SUMMARY

When the topology of an end-plate is known, it is possible to apply agonists to the postsynaptic membrane in a quantitative manner. As a consequence, complete dose—response curves can be obtained and the parameters determining the equilibrium reaction between agonist and receptor can be evaluated. These are the Hill coefficient, the apparent dissociation constant, and the maximum number of open channels. Studies of membrane potential and surface charge titration in relation to pH have led to the conclusion that the Hill coefficient may be determined by surface charges whereas the end-plate current decay time is controlled by electrical forces inside the membrane.

The decay time for nerve-induced end-plate currents is a convenient experimental parameter for testing models which explain the mechanism of action of drugs on the end-plate receptors. It appears that many molecules both charged and uncharged even including steroids cause a blockade of the open ion channel at the nicotinic acetylcholine receptor inducing a premature closure of this channel and a consequent alteration of its future behaviour. This phenomenon depends on the rates of channel blocking and unblocking for the molecule of drug in question.

REFERENCES

Adams, P.R. (1976) Drug blockade of open end-plate channels. *J. Physiol. (Lond.)*, 260, 531–552.

Adams, P.R. (1977) Voltage jump analysis of procaine action at frog end-plate. *J. Physiol. (Lond.)*, 268, 291–318.

Beam, K.G. (1976) A voltage-clamp study of the effect of two lidocaine derivatives on the time course of end-plate currents. *J. Physiol. (Lond.)*, 258, 279–300.

Beam, K.G. (1976) A quantitative description of end-plate currents in the presence of two lidocaine derivatives. *J. Physiol. (Lond.)*, 258, 301–322.

Colquhoun, D., Dreyer, F., and Sheridan, R.E. (1978) The action of tubocurarine at the neuromuscular junction. *J. Physiol. (Lond.)*, 284, 198P.

Dreyer, F., Peper, K., Akert, K., Sandri, C., and Moor, H. (1973) Ultrastructure of the "active zone" in the frog neuromuscular junction. *Brain Res., 62, 373–380.*

Dreyer, F., Peper, K., and Sterz, R. (1978) Determination of dose–response curves by quantitative ionophoresis at the frog neuromuscular junction. *J. Physiol. (Lond.), 281,* 395–419.

Drouin, H. and Neumcke, B. (1974) Specific and unspecific charges at the sodium channels of the nerve membrane. *Pflügers Arch., 351, 207–229.*

Heuser, J.E., Reese, T.S., and Landis, D.M.D. (1974) Functional changes in frog neuromuscular junctions studied with freeze-fracture. *J. Neurocytol., 3, 109–131.*

Mallart, A. and Molgo, J. (1978) The effects of pH and curare on the time course of endplate currents at the neuromuscular junction of the frog. *J. Physiol. (Lond.), 276,* 343–352.

McLaughlin, S. (1977) Electrostatic potentials at membrane-solution interfaces. In *Current Topics in Membrane and Transport, Vol. 9,* F. Bronner and A. Kleinzeller (Eds.), Academic Press, New York, pp. 71–144.

McMahan, U.J., Spitzer, N.C., and Peper, K. (1972) Visual identification of nerve terminals in living isolated skeletal muscle. *Proc. roy. Soc. B., 181, 421–430.*

Neher, E. and Steinbach, J.H. (1978) Local anaesthetics transiently block currents through single acetylcholine-receptor channels. *J. Physiol. (Lond.), 277, 153–176.*

Peper, K. and McMahan, U.J. (1972) Distribution of acetylcholine receptors in the vicinity of nerve terminals on skeletal muscle of the frog. *Proc. roy. Soc. B., 181, 431–440.*

Peper, K., Dreyer, F., Sandri, C., Akert, K., and Moor, H. (1974) Structure and ultrastructure of the frog motor end-plate. *Cell Tiss. Res., 149, 437–455.*

Peper, K., Dreyer, F., and Müller, K.-D. (1975) Analysis of cooperativity of drug-receptor interaction by quantitative iontophoresis at frog motor end-plates. *Cold Spring Harb. Symp. quant. Biol., 40, 187–192.*

Ruff, R.L. (1977) A quantitative analysis of local anaesthetic alteration of miniature endplate currents and end-plate current fluctuations. *J. Physiol. (Lond.), 264, 89–124.*

Scuka, M. (1975) The amplitude and the time course of the end-plate current at various pH levels in the frog sartorius muscle. *J. Physiol. (Lond.), 249, 183–195.*

Scuka, M. (1977) The effect of pH on the conductance change evoked by iontophoresis in the frog neuromuscular junction. *Pflügers Arch., 369, 239–244.*

The Time Course of Postsynaptic Currents in Fast and Slow Junctions and its Alteration by Cholinesterase Inhibition

L.G. MAGAZANIK, V.V. FEDOROV and V.A. SNETKOV

Sechenov Institute of Evolutionary Physiology and Biochemistry, Academy of Sciences of the U.S.S.R., Leningrad (U.S.S.R.)

INTRODUCTION

Under normal conditions both end-plate currents (EPCs) evoked by motor nerve stimulation and spontaneous miniature end-plate currents (MEPCs) are composed of a short pulse of inward current that has a rapid growth phase and relatively slower decay phase. It is known that these currents reflect the time course of membrane permeability changes induced by the interaction of acetylcholine (ACh) with receptors on the postjunctional membrane.

A quantitative model for the kinetics of ionic channel opening and closing has been proposed by Magleby and Stevens (1972a,b), Anderson and Stevens (1973) and Dionne and Stevens (1975). This model gives a quantitative description of the time course of EPCs, and there is strong evidence that the rate of decay of EPCs and MEPCs reflects the first order relaxation process of the ionic channels. However, it is clear that single channel kinetics can not be the only factor limiting the rate of decay of MEPCs and EPCs in the presence of cholinesterase inhibitors (anti-AChE) (see reviews: Colquhoun, 1975; Rang, 1975, Barrett and Magleby, 1976; Gage, 1976; Magazanik, 1976; Steinbach and Stevens, 1976; Neher and Stevens, 1977).

There are many examples of synapses in which the postsynaptic responses are very prolonged. Thus, in frog fast and slow muscle fibres the end-plate potentials (EPPs) differ significantly in duration (Kuffler and Vaughan Williams, 1953; Oomura and Tomita, 1960; Nasledov and Fedorov, 1966).

The main aim of the experiments to be described was to provide a quantitative analysis of the time course of currents in these two kinds of junction in frog and rat muscle in order to assist the elucidation of rate-limiting processes responsible for the observed difference in duration.

Another aim of this study was to investigate the effect of anti-AChEs on the time course of EPCs and MEPCs in fast and slow junctions and to measure the dependence of this effect on the ionic content of the medium.

It was hoped that the investigation of anti-AChE action and comparison of postsynaptic responses in fast and slow junctions might give more information about the nature of the process which determines the temporal characteristics of the postsynaptic conductance change caused by ACh.

METHODS

M. sartorius and m. ileofibularis from the frog *Rana temporaria* and m. extensor digitorum longus and m. soleus from the rat were used. Experiments were done throughout the year at room tempeature (18–22°C) on frog muscles and at 37°C on rat muscles. The bathing solutions had the composition (mM): I (for frog), NaCl, 115; KCl, 2.5; $CaCl_2$, 1.8; $NaHCO_3$, 2.4; II (for rat), NaCl, 135; KCl, 5; $CaCl_2$, 2; $MgCl_2$, 1; $NaHCO_3$, 12; NaH_2PO_4, 0.5; glucose, 11; bubbled with 95% O_2 +5% CO_2.

The nerves were stimulated by square pulses of 10–100 μsec duration. To prevent muscle contraction during depolarization, sartorius preparations were bathed for about 60 min in standard solution to which 0.75 M ethylene glycol had been added. The muscles were then transferred to solution containing 5 mM Ca^{2+} and 5 mM Mg^{2+} (Gage and Eisenberg, 1967) and allowed to recover for 60 min before being transferred to standard solution. The resting potential of surface fibres typically ranged from 50 to 70–80 mV. Some of them were capable of generating action potentials, so the quantum content was slightly depressed by adding $MgCl_2$ to standard Ringer solution.

M. ileofibularis and rat muscles were not detubulated, but transmission was blocked by decreasing the concentration of Ca^{2+} to 0.5 mM and by adding $MgCl_2$.

EPCs and spontaneous miniature end-plate currents (MEPCs) were recorded either with a voltage-clamp technique (Takeuchi and Takeuchi, 1959; Magleby and Stevens, 1972a,b), or with extracellular microelectrodes filled with 1 M NaCl in agar, placed close to the end-plate region (Katz and Miledi, 1973a).

Fast and slow muscle fibres of the frog were distinguished by their different types of response to nerve stimulation and by measurement of their input impedance (Burke and Ginsborg, 1956).

RESULTS

Effects of acetylcholinesterase inhibition on end-plate currents at the frog neuromuscular junction

We used anticholinesterase drugs of different chemical structure and mode of interaction with the active site of the enzyme. Galanthamine (alkaloid from *Galanthus Woronowi*, nonalkylating competitive inhibitor), prostigmine (noncompetitive carbamylating reversible inhibitor), and armin (diethoxy-*p*-nitrophenyl phosphate, phosphorylating irreversible inhibitor) were used in concentrations which have been shown to effectively eliminate AChE activity. The treatment by collagenase (Sigma type I, 72 unit/ml for 30 min) was also used. All experiments were performed on voltage clamped, detubulated sartorius muscle fibres.

The relationship between the time constant of EPC decay (τ) and membrane potential (V) was found to be exponential, and can be described by the equation (Magleby and Stevens, 1972a; Gage and McBurney, (1975):

$$\tau(V) = \tau(0) \exp(V/H)$$

TABLE I

EFFECTS OF ANTI-AChE AND COLLAGENASE ON THE EPC DECAY TIME CONSTANT AND ITS VOLTAGE-DEPENDENCE IN FROG SARTORIUS NEUROMUSCULAR JUNCTIONS

Inhibitor	Concentration (M)	Decay time constant, $\tau(0)$ (ms)	Voltage dependence, H (mV)
Control	–	0.87 ± 0.17 (11)	125 ± 22 (13)
Prostigmine	1×10^{-6}	1.96 ± 0.34 (4)	106 ± 21 (4)
Armin	5×10^{-7}	2.33 ± 0.38 (4)	143 ± 8 (4)
Galanthamine	2×10^{-5}	2.33 ± 0.28 (4)	122 ± 15 (4)
Collagenase (Sigma I)	72 u/ml, 30 min	2.08 ± 0.22 (4)	123 ± 18 (4)

Means ± S.E.M., number of experiments in parentheses. For definition of $\tau(0)$ and H see text. Temperature, 18–20°C.

where $\tau(0)$ is the time constant at holding potential of zero and H is a constant which indicates the potential dependence of τ.

Table I shows the results obtained. All procedures for abolishing AChE activity produced approximately the same effects: the amplitude of EPCs increased, and the decay remained exponential for most of its length. $\tau(0)$ was increased, but H was unchanged, confirming the results of Magleby and Stevens (1972a) and Gage and McBurney (1975) with neostigmine. It seems quite unlikely that the effect of different inhibitors and collagenase treatment is caused by the same direct action on gating molecules in the postjunctional membrane, as proposed by Magleby and Stevens (1972a) for neostigmine.

Thus the prolongation of life-time of ACh molecules in the synaptic cleft changes neither the kinetics of individual ionic channels (Katz and Miledi, 1973a) nor the potential dependence of τ.

Influence of changes of ionic content on the effect of anti-AChE

The end-plate responses are prolonged in prostigmine treated muscles after replacement of a proportion of Na^+ by sucrose (Fatt and Katz, 1951; Kordaš, 1968b; Katz and Miledi, 1973a). This was commonly interpreted as a result of a decrease in the ionic strength of the medium.

In addition to sucrose, we used also Tris, lithium and hydrazinium ions as substituents for Na^+ (Table II). When AChE is active, the lowering of $[Na]_0$ does not affect τ significantly. After treatment with armin the lowering of $[Na]_0$ induces and increase of $\tau(0)$ if Na^+ is replaced by sucrose or $Tris^+$. Replacement of Na^+ by Li^+ or hydrazinium ions does not prolong the current decay. The similarity between the effects of Na^+ replacement by sucrose and by $Tris^+$ suggests that it is not only the change of ionic strength which is responsible for the change of $\tau(0)$. The difference between the effects of sucrose and $Tris^+$ as compared to Li^+ and hydrazinium$^+$ seems to be in accordance with their permeability through ACh-activated channels. A shift of reversal potential (E_r, control value = -1 ± 3 mV, n = 10) was observed when Na^+ was partly replaced by $Tris^+$ ($E_r = -20 \pm 8$ mV, n = 10) or by sucrose ($E_r = -16 \pm 5$ mV, n = 9). After

TABLE II

EFFECTS OF REPLACING SODIUM BY SOME SUBSTANCES ON THE TIME COURSE
AND VOLTAGE DEPENDENCE OF EPC DECAY IN FROG SARTORIUS
NEUROMUSCULAR JUNCTIONS

Solution	Control		After armin (5×10^{-7} M) treatment	
	$\tau(0)$ (ms)	H (mV)	$\tau(0)$ (ms)	H (mV)
Standard Ringer	0.87 ± 0.17 (11)	125 ± 22 (13)	2.33 ± 0.38 (4)	143 ± 8 (4)
0 Na (Li)	1.00 ± 0.27 (4)	141 ± 35 (8)	2.13 ± 0.27 (4)	122 ± 9 (4)
0 Na (Hydrazinium)	0.73 ± 0.18 (5)	149 ± 38 (5)	2.08 ± 0.35 (5)	119 ± 13 (5)
40 Na (Sucrose)	0.74 ± 0.13 (3)	115 ± 17 (3)	4.76 ± 1.29 (8) *	102 ± 16 (7)
40 Na (Tris)	0.99 ± 0.21 (5)	96 ± 13 (7) *	5.26 ± 1.28 (5) *	99 ± 11 (5) *

Solutions: Concentrations of the residual sodium (mM) and the substituents (used in concentrations required for isoosmolarity) are given in the first column.
Mean ± S.E.M., number of experiments in brackets.
* This value differs significantly ($P < 0.05$) from corresponding control.

complete replacement of Na^+ by Li^+ or hydrazinium$^+$, E_r did not differ significantly from the control value.

Comparison of the EPC and MEPC time course in fast and slow muscle fibres of the frog

It is known that postsynaptic potentials have a slower time course in slow than in fast frog muscle fibres (Burke and Ginsborg, 1956; Oomura and Tomita, 1960; Nasledov and Fedorov, 1966). This is partly due to the high input impedance, and hence long membrane time constant, of slow muscle fibres (Burke and Ginsbrog, 1956; Magazanik and Nasledov, 1971), but voltage-clamp records also show a marked difference in the time course of EPCs and MEPCs in slow and fast muscles (Fig. 1). Multiple synaptic contacts occur in slow muscle and this can produce artefacts due to spatial non-uniformity of the membrane potential. In some records, steps occurred in the growth phase of the EPCs in slow muscle, presumably representing activation of adjacent junctions; such records were not used for analysis. Focal extracellular recording avoided this problem, and showed the same marked difference in MEPC time course in fast and slow muscles (Fig. 1).

A comparison of the characteristics of EPCs and MEPCs in fast and slow

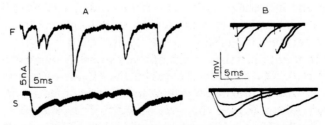

Fig. 1. Miniature end-plate currents recorded in voltage-clamped (A) fibres or extracellularly (B) in fast (F) and slow (S) junctions of frog ileofibularis muscle.

TABLE III

MEPCs AND EPCs RECORDED IN FAST (F) AND SLOW (S) JUNCTIONS IN FROG ILEOFIBULARIS MUSCLE

	Control				Anti-AChE present			
	n	Rise-time (ms)	Half-decay time (ms)	Amplitude (nA)	n	Rise-time (ms)	Half-decay time (ms)	Amplitude (nA)
Voltage-clamp recording								
F MEPCs	10	0.37 ± 0.03	1.00 ± 0.05	2.3 ± 0.1	10	0.64 ± 0.08	2.70 ± 0.18	3.7 ± 0.2
EPCs	17	0.55 ± 0.03	1.45 ± 0.05		13	1.00 ± 0.12	3.51 ± 0.20	
S MEPCs	9	1.45 ± 0.15	4.41 ± 0.23	1.4 ± 0.1	8	2.22 ± 0.21	9.6 ± 1.1	3.0 ± 0.3
EPCs	11	2.13 ± 0.21	6.36 ± 0.43		19	2.72 ± 0.23	10.4 ± 0.6	
Extracellular recording								
F MEPCs	14	0.47 ± 0.02	1.43 ± 0.06		50	0.68 ± 0.02	3.75 ± 0.11	
EPCs	30	0.65 ± 0.03	1.54 ± 0.04		50	0.86 ± 0.03	4.20 ± 0.12	
S MEPCs	7	1.05 ± 0.11	5.35 ± 0.34		14	1.56 ± 0.07	8.39 ± 0.33	
EPCs	7	1.50 ± 0.11	6.20 ± 0.80		40	1.80 ± 0.09	9.61 ± 0.26	

Mean ± S.E.M., n, number of fibres. Anti-AChE treatment by armin (5–10 × 10^{-7} M), 30 min. Temperature, 19–21°C.
Holding potential: F, –90 mV; S, –70 mV.

230

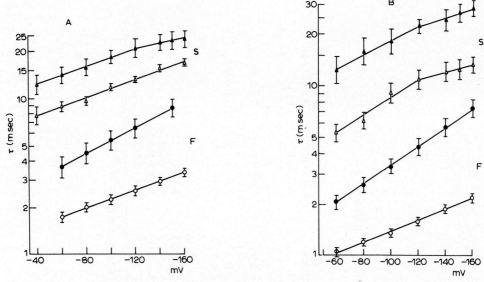

Fig. 2. The decay time constant of EPCs (A) and MEPCs (B) recorded in fast (F) and slow (S) junctions as a function of membrane potential before (open symbols) and after (filled symbols) treatment of muscles with armin MEPCs (1×10^{-6} M). Means ± S.E.M.

muscles recorded by both techniques is shown in Table III. The mean amplitude of "slow" MEPCs was smaller, on average by 40%, than that of fast MEPCs, but the rise-time and half-decay time were longer (3.6-fold and 4.6-fold respectively, with voltage-clamp recording). The decay phase of both slow and fast MEPCs was usually exponential, though a minority of slow MEPCs showed a markedly non-exponential decay.

The effects of membrane potential on EPCs and MEPCs were examined in voltage-clamped fibres only (Fig. 2). The range of membrane potential was from −60 to −160 mV and the fibres were not detubulated. The exponential relationship between decay time constant and membrane potential (Magleby and Stevens, 1972a; Anderson and Stevens, 1973; Gage and McBurney, 1975; Dionne and Stevens, 1975) was confirmed in our experiments on fast junctions. In slow junctions this relationship was exponential when EPCs but not MEPCs were estimated, the deviation from exponentiality becoming more noticeable under strong hyperpolarization. The growth phase was less sensitive to changes of holding potential in both kinds of junction.

Effects of AChE inhibition on EPCs and MEPCs in fast and slow junctions

Presumably the difference in the postjunctional currents time course is due partly to a difference in the rate of enzymatic hydrolysis of ACh in the two kinds of junction. There is some evidence for a lower activity of AChE in frog slow muscle fibres (Csillik et al., 1961; Heaton et al., 1972). The properties of EPCs and MEPCs in fast and slow muscle fibres of m. ileofibularis were examined after AChE inhibition (Table III). It can be seen that inhibition of AChE by armin (and in few experiments by prostigmine) caused an enhancement and prolongation of postsynaptic currents in both kinds of junctions.

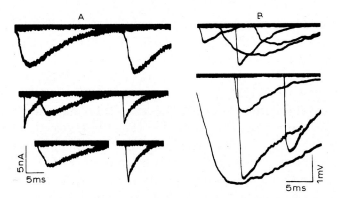

Fig. 3. A selection of MEPCs recorded in voltage-clamped slow muscle fibres (A) or by extracellular recording (B) after armin (1×10^{-6} M) treatment, showed marked variability in the decay rates.

The prolongation of half-decay time was more pronounced in fast fibres. The effect of AChE inhibition on the rise time of EPCs and MEPCs was small compared to the effect on the duration of the decay phase. These results show that the difference in the rate of ACh hydrolysis in fast and slow junctions cannot be the only factor responsible for the difference in the time course of post-

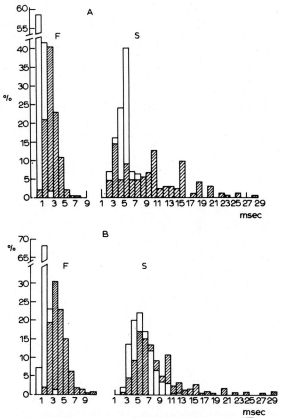

Fig. 4. Histograms showing half-decay time of MEPCs recorded under voltage clamp (A) or extracellularly (B) in fast (F) and slow (S) muscle fibres of frog ileofibularis muscle. Blank columns — before, and hatched columns — after armin (1×10^{-6} M) treatment. 700–1800 individual MEPCs were estimated in each series of experiments.

232

junctional currents, because after AChE inhibition the half-time of MEPC decay in slow junctions is 2.2—3.6 times larger than that in fast junctions.

As well as becoming longer after AChE inhibition, the MEPCs recorded became highly variable in duration (Fig. 3). This variability was particularly pronounced in slow fibres (Fig. 4).

We found no correlation between the amplitude and the time course of MEPCs in slow junctions. The duration of MEPCs growth phase did not correlate with the duration of decay phase: MEPCs with extremely slow decay may have either normal (~1.5 ms) or quite slow (up to 5.0—8.0 ms) growth phase. AChE inhibition did not affect significantly the exponentiality of MEPC decay and the voltage sensitivity of τ in both kinds of junctions.

Lowering of Na^+ concentration did not prolong MEPC decay at fast or slow junctions if AChE was active. After AChE inhibition the substitution of sucrose for a proportion of Na^+ resulted in prolongation of MEPCs and EPCs at both types of junction, the effect being more pronounced at fast junctions. Very varied results were obtained on slow fibres, some showing the same prolongation as fast junctions (1.5—2.3 times) and others showing no effect. The effect appeared to be more pronounced when the muscle was soaked in low-sodium solution before being treated with anti-AChE than when the order of the treatments was reversed.

It is known that receptor blockade by tubocurarine shortens as well as reduces MEPCs in fast junctions if ACh hydrolysis is inhibited (Katz and Miledi, 1973a). This effect of tubocurarine also occurred in slow fibres (Fig. 5). The distribution of decay half-time shifted towards smaller values both in fast and slow junctions, but all of the extremely slow MEPCs recorded in slow junctions disappeared completely.

Katz and Miledi (1973a) calculated the fraction (p) of the released ACh

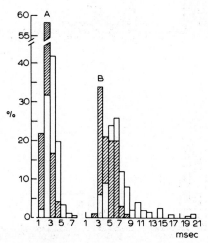

Fig. 5. Histograms showing the effect of tubocurarine (iontophoretically applied) on the half-decay time of MEPCs recorded extracellularly in fast (A) and slow (B) muscle fibres of m. ileofibularis treated by armin (1 × 10⁻⁶ M). Blank columns, control; hatched columns, during the tubocurarine application. Note the disappearance of extremely slow MEPCs in B during tubocurarine action.

molecules bound to receptors, from

$$p = \frac{t_n - t_c}{t_n - \alpha t_c}$$

where t_n and t_c are the time constants of MEPC decay before and during tubo-curarine action respectively, α is the ratio of reduced to normal amplitudes of MEPCs.

Calculated values of p in our experiments were 0.48 ± 0.03 (n = 8) in fast junctions and 0.51 ± 0.05 (n = 5) in slow ones. These values of p did not differ significantly and are close to those found by Katz and Miledi (1973a) (0.66); Magleby and Terrar (1975) (0.58) and Colquhoun et al. (1977) (0.52). Thus about the same fraction of released ACh is bound to receptors both in fast and slow junctions. The value of p obtained, however, reflects the mean fraction of bound ACh molecules. In slow muscle the great variability suggests that different quanta have quite different values of p, possibly reflecting nonuniformity of the mutual disposition of receptors and sites of transmitter release in these junctions.

Effect of atropine on the time course of EPCs in fast and slow junctions

It is known that atropine reduces the amplitude, shortens the time course and drastically reduces the potential dependency of EPC and MEPC decay recorded at frog sartorius muscle (Beránek and Vyskočil, 1968; Kordaš, 1968a; Magazanik and Vyskočil, 1969; Adler and Albuquerque, 1976; Feltz and Large, 1976).

Atropine reduced the amplitude and decreased the duration of the growth phase and the potential dependence of decay in a similar way in fast and slow junctions. The shorterning of EPC decay half-time in slow junctions was

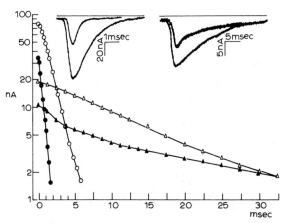

Fig. 6. Effect of atropine on EPCs recorded in fast (circles) and slow (triangles) junctions (voltage-clamped fibres of m. ileofibularis). Insets: left, "fast" and right, "slow" responses before and in the presence of 5×10^{-5} M atropine. The decay of EPCs was plotted semilogarithmically as a function of time from the experiments illustrated in insets. An example of non-exponential decay of "slow" EPCs was selected. Open symbols, control; filled symbols, in the presence of atropine.

234

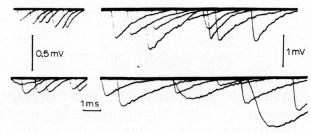

Fig. 7. MEPCs recorded extracellularly in end-plates of rat EDL (upper row of records) and SOL (lower row) before (left) and after armin (1 × 10⁻⁶ M) treatment (right). Temperature, 37°C.

approximately twice less, however, both in normal and anti-AChE-treated muscles. The non-exponentiality of EPC decay was noticed in 6 out of 14 experiments on slow junctions after atropine treatment (Fig. 6).

Comparison of the time course of synaptic currents in fast and slow muscle fibres of the rat

Rat muscles are known to contain several types of muscle fibres (Hess, 1970; Close, 1972; Ellisman et al., 1976), but the correlation between sub-types and the electrophysiological properties of different fibres is still not clear.

We have compared the characteristics of EPCs and MEPCs recorded extracellularly or by voltage clamp in superficial fibres of the extensor digitorum longus muscle (EDL), which contains mainly fast fibres, and of the soleus muscle (SOL), which contains mainly slow ones. Typical examples of MEPCs recorded in fast and slow muscles are shown in Fig. 7. There was a small difference in the time course of MEPCs in EDL and SOL (Table IV) but the mixed character of both muscles may result in an underestimation of this difference.

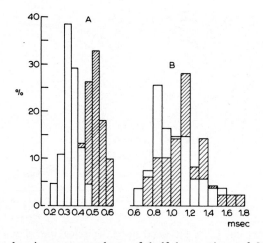

Fig. 8. Histograms showing mean values of half-decay time of MEPCs recorded extracellularly in end-plates of EDL (blank columns) and SOL (hatched columns) before (A) and after (B) armin (1 × 10⁻⁶ M) treatment. Temperature, 37°C. The mean of half-decay time of 50–70 MEPCs was estimated in each end-plate studied. Ordinates: percentages of all end-plates. From 60 to 70 end-plates was studied in each series of experiments.

TABLE IV

MEPCs RECORDED IN EXTENSOR DIGITORUM LONGUS (EDL) AND SOLEUS (SOL) MUSCLES OF THE RAT

	Control				Anti-AChE present			
	n	Rise-time (ms)	Half-decay time (ms)	Amplitude (nA)	n	Rise-time (ms)	Half-decay time (ms)	Amplitude (nA)
Voltage-clamp recording								
EDL	50	0.20 ± 0.02	0.33 ± 0.04	3.2 ± 0.3	130	0.26 ± 0.05	0.85 ± 0.16	5.4 ± 0.3
SOL	54	0.24 ± 0.03	0.42 ± 0.06	2.7 ± 0.2	102	0.31 ± 0.06	1.00 ± 0.22	4.2 ± 0.3
Extracellular recording								
EDL	132	0.20 ± 0.02	0.40 ± 0.05		110	0.26 ± 0.03	1.04 ± 0.22	
SOL	122	0.25 ± 0.02	0.55 ± 0.06		96	0.30 ± 0.04	1.15 ± 0.22	

Mean ± S.E.M.; n, number of fibres. Anti-AChE treatment by armin ($5-10 \times 10^{-7}$ M), 30 min. Temperature, 37°C. Holding potential: −80 mV.

Distribution of mean values of half-decay time examined in 60—70 individual junctions in EDL and SOL is presented in Fig. 8. There is an obvious difference between extreme values which is still observed after anti-AChE treatment: the fastest MEPCs are not seen in SOL and the slowest MEPCs are not seen in EDL.

DISCUSSION

Some of our results concerning the influence of anti-AChE on the time course of EPCs and MEPCs in frog fast junction are in general agreement with previous observations. Prolongation was never as great as reported by Katz and Miledi (1973a) but was of the same order as the effects reported by Kordaš (1972), Magleby and Stevens (1972a), Gage and McBurney (1975) and Magleby and Terrar (1975). The simple exponential decay and the voltage sensitivity of its time constant also appears unaffected by AChE inhibition. The same effects were seen after removal of AChE activity by collagenase and some anti-AChE drugs. The anti-AChE drugs used in our experiments are known to attack the active site of AChE different mechanisms, whereas collagenase is thought to detach AChE molecules from the basal membrane into the junction. Our results are in agreement with Magleby and Terrar (1975) who found that various types of anti-AChE treatment all affected EPCs in the same manner, prolonging their time course but not affecting the voltage sensitivity of the rate of decay.

The uniform action of such different agents argues strongly against the involvement of any side effect (e.g. a direct action on the gating mechanism or conductance of individual ionic channels) and in favour of the basic effect being a prolongation of ACh action in the synaptic cleft.

The slow time course of EPCs after AChE inhibition implies that the rate of diffusion of ACh within the synaptic cleft must be much lower than in free solution. Analysis of the growth phase of MEPCs led Gage (1976) to suggest that the diffusion of ACh is retarded by some diffusion barrier so that the effective diffusion coefficient is 2—3 orders of magnitude less than in free solution. The binding of ACh-molecules to receptors also effectively retards their removal (Katz and Miledi, 1973a) and can contribute to the prolongation of EPC decay which occurs when AChE is blocked. Consequently drugs that reduce the number of free receptors, (e.g. tubocurarine) should shorten the MEPC and EPC decay after AChE inhibition, an effect which has been confirmed by many workers, and was also shown in our experiments.

The effects of other procedures which can prolong EPCs in the presence of anti-AChE, however, such as addition of ACh to a solution flowing over the end-plate (Magleby and Terrar, 1975) or lowering $[Na]_0$ (Kordaš, 1968b; Katz and Miledi, 1973a) are not easily explained by the buffered ACh removal hypothesis.

Thus is was suggested that increasing of ACh concentration in the synaptic cleft leads to (1) an increase in the number of successful ACh-receptor collisions as ACh diffuses from the synaptic cleft, or (2) an increase in t, the time that an ACh molecule is retained on the receptor, or both. In terms of cooperative hypothesis it means that more than one ACh molecule is required to open the channel or to increase its mean life-time (Magleby and Terrar, 1975). This idea is rather plausible, but it seems puzzling that the cooperative effect can be demonstrated only in the presence of anti-AChE, especially if the relatively small enhancement of postsynaptic responses induced by anti-AChE is taken into account.

Prolongation of the EPC time course by the lowering of Na^+ in the presence of anti-AChE was explained by an increse of repetitive binding due to the reduction of the ionic strength of the medium (Katz and Miledi, 1973a). We have found, however, that this phenomenon does not depend on the ionic strength, because the replacement of Na^+ by $Tris^+$ ions is as effective as the replacement by sucrose. The prolongation seems to correlate with some other properties of the substituents for Na^+.

Full replacement of sodium by Li^+ and hydrazinium$^+$ did not induce the prolongation of EPC decay. Evidently, both ions can pass through open synaptic channels since the replacement does not induce any shift of reversal potential. The mechanism of the observed correlation between the ability of ions to permeate through the activated postjunctional membrane and the time course of the EPCs is quite unclear. How can passing ions regulate the life-time of channels' opening or the cooperative action of ACh molecules, or repetitive binding of ACh molecules to the receptor? It is clear now that some properties of the receptor depend upon the ionic composition of medium, and more work is needed to elucidate the mechanism of this dependence.

The main result of the present study is the finding that the end-plates of slow frog and mammalian muscles generate rather slow EPCs and MEPCs in the absence of AChE inhibition. Elucidation of the factors controlling the time course of these responses may by fruitful. MEPCs recorded extracellularly in frog slow junctions had the growth and decay phases 2.8 and 3.6 times slower than those in fast junctions. The growth phase of MEPCs is mainly determined by two processes: (1) release of ACh from the nerve terminal and its diffusion across the synaptic cleft; (2) binding of ACh molecules to their receptors and conformational change of receptors from the inactive to the active state. The low voltage and temperature sensitivity of the growth phase is the basis of the view that diffusion of ACh mainly determines the growth phase of the end-plate conductance change (Gage and McBurney, 1975). In these respects there was no difference between fast and slow junctions, and it is reasonable to assume that ACh diffusion governs the growth phase in both types of junction.

According to the hypothesis discussed above the rate of decay of synaptic currents normally reflects the rate at which receptors undergo a conformational change from the open to the closed configuration (Magleby and Stevens, 1972a,b). Do the ionic channels in slow junctions have a longer life-time than in the fast ones? The inability of tubocurarine to shorten MEPC decay in slow junctions when AChE is intact, or of lowering the sodium concentration to prolong it, is consistent with this suggestion, and we have some preliminary results obtained by spectral analysis of ACh-induced current noise to suggest that the mean life-time of ionic channels activated by AC h is 2—3 folds longer in the slow junctions than in the fast ones. Neher and Sakmann (1975) and Dreyer et al. (1976) have shown a similar difference in the life-time of channels induced by ACh application to junctional and extrajunctional receptors in frog muscle fibres.

It is known that atropine shortens the life-time of ionic channels activated by ACh (Katz and Miledi, 1973b; Feltz and Large, 1976). A smaller shorterning effect of atropine on EPC decay in slow junction was observed in our experiments. In some slow fibres the EPC decay consisted of two components as after local anesthetics treatment (see reviews: Magazanik, 1976; Neher and Stevens, 1977). Evidently, ionic channels in fast and slow junctions differ not only in the kinetics but also in some pharmacological properties.

The inhibition of AChE prolonged responses both in fast and slow junctions. The duration of growth phase increased slightly (1.1—1.4 times), but the main effect was on the decay phase which was prolonged 2.5—2.8 times in fast junctions and 1.6—2.2 times in slow junctions. On the other hand enhancement of MEPC amplitude was greater in slow junctions (2.4 times) than in fast ones (1.6 times). It seems that the functional role of AChE is similar in both kinds of junctions. On the other hand, inhibition of AChE caused a much greater dispersal of MEPC decay rates in slow than in fast fibres, with decay half-times ranging from 2 to 30 ms. The appearance of such extremely slow and scattered MEPC after AChE inhibition and their high sensitivity to the shortening action of tubocurarine may reflect different conditions at different parts of slow junctions; the diffusion of ACh is perhaps not uniform in all parts of the junction.

It is known that in the fast junction the sites of quantal release and patches of high receptor density are located opposite each other (Peper et al., 1974), so that the diffusional pathway of ACh molecules is the same throughout the junction. The fine structure of slow junctions has not been investigated so thoroughly, but it is known that postsynaptic folds are absent. If structural correlation between the sites of ACh release and the receptors is poor in slow junction, this could be associated with different diffusional pathways for ACh molecules in different parts of the junction.

We suggest that two factors control the time course of MEPCs in slow junctions. Under normal conditions when AChE is active the decay of MEPCs reflects the rate of closing of ionic channels which is 2—3 times slower than in fast junctions. After AChE inhibition the second factor becomes apparent: long and non-uniform diffusional pathways which may augment the functional role of buffered removal.

It is unclear whether the considerations discussed above for frog slow and fast junctions may be used to explain also the differences in the MEPC time course observed in rat muscles. The problem is complicated by the heterogeneity of fibre composition of rat muscles. The analysis of the time course of currents recorded in two rat muscles, fast EDL and slow SOL, allows to assume the existence of different types of junctions predominating either in fast or in slow muscles.

SUMMARY

End-plate currents (EPCs) and miniature end-plate currents (MEPCs) were recorded in fast (m. sartorius, m. ileofibularis) and slow (m. ileofibularis) muscle fibres of the frog or in m. extensor digitorium longus and m. soleus of the rat by the voltage-clamp or focal extracellular recording techniques.

The duration of EPC rise time and decay in frog fast junctions was quite different from that in the slow ones, and the difference remained after AChE inhibition. Atropine shortened the time course of EPCs more redily in fast junctions than in the slow ones. If AChE was inhibited, the iontophoretic application of tubocurarine also shortened the decay of MEPCs in fast junctions more than in the slow ones.

EPC and MEPC decay remained to be exponential, with time constant dependent upon the membrane potential after AChE inhibition by reversible or irreversible inhibitors or by collagenase treatment. If AChE was inhibited, the substitution of sucrose or unpermeant $Tris^+$ ions for Na^+ resulted in prolongation of EPCs and MEPCs in both types of junctions, the effect being more pronounced at fast junctions. When Na^+ was replaced by permeant ions (Li^+ or hydrazinium$^+$) no prolongation of currents was observed.

The factors determining the time course of EPCs (activity of AChE, life-time of ionic channels, length of diffusional pathways, repetitive binding of acetylcholine molecules) and their importance in various neuromuscular junctions are discussed.

REFERENCES

Adler, M. and Albuquerque, E.X. (1976) An analysis of the action of atropine and scopolamine on the end-plate current of frog sartorius muscle. *J. Pharmacol. exp. Ther.*, 196, 360–372.

Anderson, C.R. and Stevens, C.F. (1973) Voltage-clamp analysis of acetylcholine produced end-plate current fluctuations at frog neuromuscular junction. *J. Physiol. (Lond.)*, 235, 655–691.

Barrett, E.F. and Magleby, K.L. (1976) Physiology of cholinergic transmission. In *Biology of Cholinergic Function*, A.M. Goldberg and I. Hanin (Eds.), Raven Press, New York, pp. 29–100.

Beránek, R. and Vyskočil, F. (1968) The effect of atropine on the frog sartorius neuromuscular junction. *J. Physiol. (Lond.)*, 195, 493–503.

Burke, W. and Ginsborg, B.L. (1956) The electrical properties of the slow muscle fibre membrane. *J. Physiol. (Lond.)*, 132, 586–598.

Close, R. (1972) Dynamic properties of mammalian sceletal muscles. *Physiol. Rev.*, 52, 129–197.

Colquhoun, D. (1975) Mechanisms of drug action at the voluntary muscle end-plate. *Ann. Rev. Pharmacol.*, 15, 307–325.

Colquhoun, D., Large, W.A. and Rang, H.P. (1977) An analysis of the action of a false transmitter at the neuromuscular junction. *J. Physiol. (Lond.)*, 266, 361–395.

Csillik, B., Schneider, J. and Kalman, G. (1961) Uber die histochemische Structur in tetanischer und tonischer myoneuralen Synapsen. *Acta Neuroveg. (Wien)*, 22, 212–224.

Dionne, V.E. and Stevens, C.F. (1975) Voltage dependence of agonist effectiveness at the frog neuromuscular junction: Resolution of a paradox. *J. Physiol. (Lond.)*, 251, 245–270

Dreyer, F., Walter, C. and Peper, K. (1976) Junctional and extrajunctional acetylcholine receptors in normal and denervated frog muscle fibres. Noise analysis experiments with different agonists. *Pflügers Arch.* 366, 1–9.

Ellisman, M.H., Rash, J.E., Staehelin, A. and Porter, K.R. (1976) Studies of excitable membranes. II. A comparison of specializations at neuromuscular junctions and nonjunctional sarcolemmas of mammalian fast and slow twitch muscle fibres. *J. Cell Biol.*, 68, 752–774.

Fatt, P. and Katz, B. (1951) An analysis of the end-plate potential recorded with an intracellular electrode. *J. Physiol. (Lond.)*, 115, 320–370.

Feltz, A. and Large, W.A. (1976) Effects of atropine on the decay of miniature end-plate currents at the frog neuromuscular junction. *Brit. J. Pharmacol.,* 56, 111–113.

Gage, P.W. (1976) Generation of end-plate potentials. *Physiol. Rev.,* 56, 177–247.

Gage, P.W. and Eisenberg, R.S. (1967) Action potentials, after potentials, and excitation-contraction coupling in frog sartorius muscle without transverse tubules. *J. gen. Physiol.,* 53, 298–310.

Gage, P.W. and McBurney, R.N. (1975) Effects of membrane potential, temperature and neostigmine on the conductance change caused by a quantum of acetylcholine at the toad neuromuscular junction. *J. Physiol. (Lond.),* 244, 385–407.

Heaton, J., Buckley, J.A. and Evans, R.H. (1972) The cholinesterase activity of myoneural junctions from frog twitch and tonic muscles. *Experientia,* 28, 503–504.

Hess, A. (1970) Vertebrate slow muscles fibres. *Physiol. Rev.,* 50, 40–62.

Katz, B. and Miledi, R. (1973a) The binding of acetylcholine to receptors and its removal from the synaptic cleft. *J. Physiol. (Lond.),* 231, 549–574.

Katz, B. and Miledi, R. (1973b) The effect of atropine on acetylcholine action at the neuro-muscular junction. *Proc. roy. Soc. B,* 184, 221–226.

Kordaš, M. (1968a) The effect of atropine and curarine on the time course of the end-plate current in frog sartorius muscle. *Int. J. Neuropharmacol.,* 7, 523–530.

Kordaš, M. (1968b) A study of the end-plate potential in sodium deficient solutions. *J. Physiol. (Lond.),* 198, 81–90.

Kordaš, M. (1972) An attempt at an analysis of the factors determining the time course of the end-plate current. I. The effects of prostigmine and of the ratio of Mg^{2+} to Ca^{2+}. *J. Physiol. (Lond.),* 224, 317–332.

Kuffler, S.W. and Vaughan Williams, E.M. (1953) Small-nerve junctional potentials. *J. Physiol. (Lond.),* 121, 318–340.

Magazanik, L.G. (1976) Functional properties of postjunctional membrane. *Ann. Rev. Pharmacol.,* 16, 161–175.

Magazanik, L.G. and Nasledov, G.A. (1971) Equilibrium potentials of postsynaptic membrane of frog tonic muscle fibre during the change of extracellular ionic medium. *Biofizika,* 16, 450–456 (in Russian).

Magazanik, L.G. and Vyskočil, F. (1969) Different action of atropine and some analogues on the end-plate potentials and induced acetylcholine potentials. *Experientia,* 25, 618–619.

Magleby, K.L. and Stevens, C.F. (1972a) The effect of voltage on the time course of end-plate currents. *J. Physiol. (Lond.),* 223, 151–171.

Magleby, K.L. and Stevens, C.F. (1972b) A quantitative description of end-plate currents. *J. Physiol. (Lond.),* 223, 173–197.

Magleby, K.L. and Terrar, D.A. (1975) Factors affecting the time course of decay of end-plate currents: a possible cooperative action of acetylcholine on receptors at the frog neuromuscular junction. *J. Physiol. (Lond.),* 244, 467–495.

Nasledov, G.A. and Fedorov, V.V. (1966) Summation of synaptic responses to repetitive stimulation of twitch and tonic muscle fibres. *Fiziol. Zh. USSR,* 52, 757–764 (in Russian).

Neher, E. and Sakmann, B. (1975) Noise analysis of voltage clamp currents induced by different cholinergic agonists in normal and denervated frog muscle fibres. *Pflügers Arch.,* 355, R63.

Neher, E. and Stevens, C.F. (1977) Conductance, fluctuations and ionic pores in membranes. *Ann. Rev. Biophys. Bioeng.,* 6, 345–381.

Oomura, Y. and Tomita, T. (1960) Study on properties of neuromuscular junction. In *Electrical Activity of Single Cell,* Y,. Katsuki (Ed.), Igaku Shoin, Tokyo, pp. 181–205.

Peper, K., Dreyer, F., Sandri, C., Akert, K. and Moor, H. (1974) Structure and ultrastructure of the frog motor end-plate. A freeze-etching study. *Cell Tiss. Res.,* 149, 437–455.

Rang, H.P. (1975) Acetylcholine receptors. *Quart. Rev. Biophys.,* 7, 283–399.

Steinbach, H.N. and Stevens, C.F. (1976) Neuromuscular transmission. In *Neurobiology of the Frog,* R. Llinas and W. Precht (Eds.), Springer, Berlin.

Takeuchi, A. and Takeuchi, N. (1959) Active phase of frog's end-plate potential. *J. Neurophysiol.,* 22, 395–412.

Activation of Acetylcholine Receptors in Mammalian Sympathetic Ganglion Neurones

A.A. SELYANKO and V.I. SKOK

Bogomolets Institute of Physiology, Kiev (U.S.S.R.)

INTRODUCTION

The ionic mechanisms underlying the excitatory action of acetylcholine (ACh) on the subsynaptic membrane of autonomic ganglion have been studied extensively in amphibian sympathetic (Nishi and Koketsu, 1960; Blackman et al., 1963; Koketsu, 1969) and parasymapathetic (Dennis et al., 1971) ganglia using intracellular electrodes. Only occasional attempts have been made to study such action using intracellular electrodes in mammalian autonomic ganglion (see Skok, 1973; Dun et al., 1976). The aim of this work was to investigate the action of the synaptically released transmitter and exogenous ACh on the superior cervical ganglion of the rabbit and to compare this action of ACh with that produced by some cholinomimetic drugs. For preliminary report see Skok and Selyanko (1977).

METHODS

The experiments were performed on superior cervical ganglia rapidly isolated from rabbits killed by air embolism. The ganglion was pinned out upon Sylgard transparent plastic in the chamber shown in Fig. 1A (see McMahan and Kuffler, 1971). The chamber was filled with solution warmed to $37°C$ and equilibrated with O_2 (95%) and CO_2 (5%), at pH 7.4. A non-wettable Teflon ring prevented the overflow of solution. Neurones were stimulated orthodromically by single stimuli $0.1-0.5$ ms in duration applied through a suction electrode containing the cervical sympathetic nerve trunk. Drugs were applied to the cell surface iontophoretically through double or single barrelled micropipettes. Leakage was prevented with breaking currents $1-2$ nA. One barrel was filled with ACh and another with a cholinomimetic (2.5 M). The intensity and duration of iontophoretic currents ranged from 1×10^{-8} A to 3×10^{-7} A and from 2 ms to 20 ms. In a few experiments ACh was added to the perfusion fluid at concentrations of 1 to 300 mM. In these experiments all solutions contained neostigmine (2.9×10^{-6} M) and atropine (1.4×10^{-6} M).

The intracellular electrodes were conventional glass micropipettes filled with 2.5 M KCl or 2 M potassium citrate. Resistances were between 90 and 200 MΩ. The impalement of a cell with a microelectrode and the placement of the microiontophoretic pipette were observed under Nomarski differential inter-

242

ference optics at a total magnification of 400X (a 40X water immersion objective with a 1.6-mm working distance). Fig. 1B shows the neurones visible through the differential interference microscope. Current was passed through the neuronal membrane through the recording electrode using a conventional bridge circuit or through a second intracellular electrode inserted into the same cell.

The perfusion fluid was of the following composition (mmol): NaCl, 137.9; KCl, 4.0; $CaCl_2$, 2.0; $MgCl_2$, 0.5; KH_2PO_4, 1.0; $NaHCO_3$, 12.0; glucose, 11.0 (Eccles, 1955). In sodium-deficient solution 76 mM of NaCl were substituted by 152 mM sucrose. In potassium-rich solution 8 mM of NaCl were substituted by KCl. In potassium-deficient solution 4 mM of KCl were substituted by

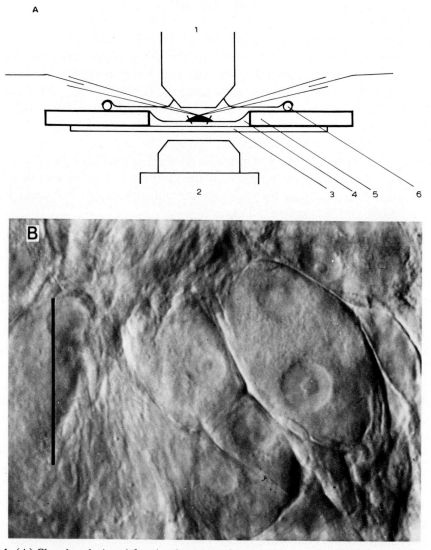

Fig. 1. (A) Chamber designed for simultaneous observation of single cells by Nomarski optics and electrophysiological recording from the neurones of isolated mammalian sympathetic ganglion and (B) photomicrographs of neurones of the rabbit superior cervical ganglion viewed through the Nomarski optics. (1), objective lens; (2), condenser lens; (3), glass slide; (4), Sylgard; (5), plastic material; (6), Teflon ring. Calibration bar (B) is 50 μm.

8 mM sucrose. In chloride-deficient solution 137.9 mM of NaCl were substituted by sodium glutamate. The drugs used were acetylcholine chloride (ACh), tetramethylammonium bromide (TMA), carbamylcholine chloride (CCh), atropine sulphate and neostigmine.

RESULTS

Effects of ACh on membrane potential and membrane resistance

All 14 neurones studied by the addition of 0.5 to 3×10^{-4} M ACh to the perfusion fluid responded with depolarization followed by hyperpolarization. The overall duration of such two-phase responses caused by 0.5−1.5 min perfusion with ACh was about 3 min. The depolarizations ranged from 11 mV to 25 mV and were accompanied by a decrease in membrane resistance which ranged from 23% to 93% of the resting value. The hyperpolarization was not

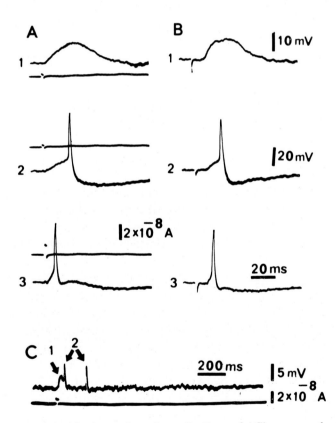

Fig. 2. Responses evoked by iontophoretic applications of ACh compared with the responses evoked in the same cell by single orthodromic stimuli. (A), the response evoked by iontophoretic applications of ACh. The current is monitored in the lower trace in (1) and in the upper trace in (2) and (3), (B), the responses evoked by single orthodromic stimuli. The iontophoretic current (A) and the orthodromic stimulus (B) are subthreshold in (1) and increase in their intensities from (2) to (3). Note the same time scale in (A) and (B) shown in (B3). Voltage scale for (1) is shown in (B1), and for the rest of the records in (B2). (C), a depolarizing response evoked by iontophoretic application of ACh (1) and spontaneous EPSPs (2). The iontophoretic current is monitored in the lower trace.

accompanied by a decrease in membrane resistance. Since in 61 neurones in which ACh was locally applied by iontophoresis the response was a pure depolarization, the hyperpolarization was attributed to stimulation of the electrogenic sodium pump by the entrance of large numbers of sodium ions (cf. Brown and Scholfield, 1974).

Only when ACh was applied iontophoretically to the most sensitive areas of the soma did the time course of the potential evoked by ACh approach that of the excitatory postsynaptic potential (EPSP) (cf. Figs. 2A,B). Since increase in the dose of ACh above threshold did not increase the rise time of the ACh evoked potential it is suggested that there is a high density of cholinoceptive sites in the activated area of the membrane and that they cannot be saturated by the doses of ACh used in these experiments (cf. Katz and Miledi, 1964; Feltz and Mallart, 1971; Dreyer and Peper, 1974).

The potential evoked by iontophoretic application of ACh always appeared within a few milliseconds and reached peak 10 to 50 ms later. We never observed long latency responses with peak time of hundreds of milliseconds observed in amphibian autonomic ganglion neurones (Koketsu, 1969; Hartzell et al., 1977). Thus the soma of rabbit sympathetic ganglion neurones probably does not have muscarinic receptors.

Sometimes the iontophoretic application of ACh was followed by spontaneous EPSPs (Fig. 2C). This may indicate that ACh depolarizes presynaptic terminals. Similar phenomenon observed in the eserinized amphibian sympathetic ganglion led to the suggestion that presynaptic terminals possessed ACh receptors (Koketsu and Nishi, 1968; Nishi, 1970).

Dose—response relationships

Although it is difficult to measure the concentration of ACh near the cell membrane, relative changes in this concentration can be estimated from the changes in electrical charge passed through the microelectrode (Carslaw and Jaeger, 1947; del Castillo and Katz, 1955; Hartzell et al., 1976). The number of receptors activated by ACh is thought to be proportional to the increase in membrane conductance produced by ACh (see Werman, 1969), and in turn the conductance increase is thought to be proportional to the membrane depolarization if the latter does not exceed approximately 10 mV (Takeuchi and Takeuchi, 1959). So the relationship between the concentration of ACh and the number of receptors activated by ACh may be similar to the relationship between the electric charge passed through the ACh micropipette and the amplitude of potential evoked by ACh if this does not exceed 10 mV. The slope of such a relationship plotted on a double logarithmic scale is the Hill number (n_H) and indicates the lower limit of the number of ACh molecules combining with one ACh receptor (Werman, 1969; Colquhoun, 1975).

In four neurones studied the dose—response relationship gave a mean value of 2.4 ± 0.6 for the n_H. The results obtained from one neurone are shown in Fig. 3. Although the relationship between the dose of ACh and the amplitude of the potential evoked by ACh was not linear (A), it could be linearized by using a double logarithmic scale (B). The non-linearity of high amplitude potentials evoked by ACh is probably due to the generation of local responses.

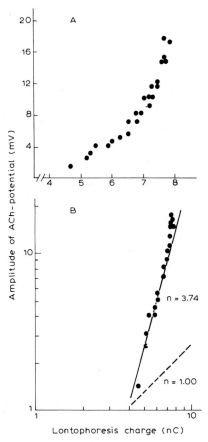

Fig. 3. Dose–response curves which show the relationship in a single cell between the iontophoretic charge used to release ACh and the amplitude of potential evoked by ACh. The same data is plotted on a linear scale in (A) and in a double logarithmic scale in (B). Hill number (n_H) calculated as a slope in (B) is in this case 3.74. A slope of 1.0 is indicated by the interrupted line.

From these results one can suggest that at least two molecules of ACh combine with one ACh receptor in rabbit superior cervical ganglion neurones. Similar relationships have been found at other sites containing nicotinic receptors (Ziskind and Werman, 1975; Bregestovski et al., 1976; Dreyer et al., 1976; Peper et al., 1976).

Effects of cholinomimetic drugs

It has been suggested that the efficacy of agonists on the ACh receptor depends on how their molecular structure compares with that of ACh receptor (Michelson and Zeimal, 1973). It thus seemed interesting to compare the action of CCh and TMA with that of ACh. CCh or TMA were applied through one barrel of the same double barrelled micropipette used to apply ACh. The duration of the iontophoretic current was kept constant and its intensity was adjusted so that the depolarization evoked by CCh or TMA was approximately equal in amplitude to that evoked by ACh. As shown in Fig. 4, both the peak and the decay of the potentials evoked by CCh and TMA were much delayed

246

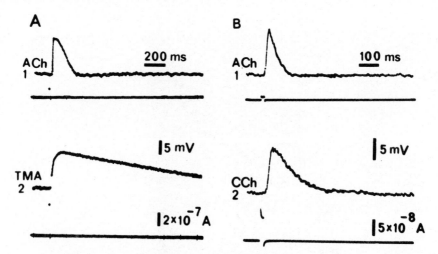

Fig. 4. In two cells A and B responses to TMA (A2) and CCh (B2) are compared with the responses obtained in the same cells with ACh (A1 and B1). 5 ms in (A) and 20 ms in (B) pulses of iontophoretic current are monitored in the lower trace. In the same cell the current intensities were adjusted until the amplitude of the potential evoked by TMA or CCh is approximately equal to that evoked by ACh. Note different time scales in (A) and (B).

when compared with those evoked by ACh. The difference was attributed to the fact that acetylcholinesterase does not hydrolyse CCh and TMA and thus these drugs were able to activate more remote ACh receptors.

Since the transfer numbers for ACh and the cholinomimetic drugs used in this study do not differ significantly (Dionne, 1976), their concentrations near the membrane should be proportional to the electric charges passed through the drug barrels and thus their efficacies can be compared by comparing the charges which produced depolarizations of similar amplitudes. Using this method, it was found that in evoking a depolarization of about 5 mV CCh was 7.8 times less effective than ACh (mean obtained from three cells) and TMA was 4.6 times less effective than ACh (mean obtained from six cells). Lower efficacies for CCh and TMA have also been found at other nicotinic sites (Magazanik, 1968; Bregestovski et al., 1975; del Castillo and Webb, 1977). However in the rabbit superior cervical ganglion, TMA was more effective than CCh.

Reversal potential for the potential evoked by ACh and for the EPSP

In most neurones the reversal potential for the EPSP (E_{EPSP}) and for the potential evoked by ACh (E_{ACh}) were measured by the extrapolation method. Only in a few neurones which were impaled with two intracellular electrodes was it possible to observe actual reversion of EPSP or of the potential evoked by ACh. Fig. 5 illustrates an experiment in which the neurone was depolarized to a level where both EPSP and the potential evoked by ACh disappeared at −15 mV. Clearly in this cell E_{EPSP} was identical to E_{ACh}. The reversal potentials estimated in this particular cell by extrapolation and the direct method were similar.

The neurones studied solely by the extrapolation method could be divided into two groups. In the first group E_{EPSP} was -14.4 ± 1.6 mV (mean $\pm$ S.E.; n = 23), and E_{ACh} was -16.5 ± 1.2 mV (n = 21), and in the second -41.6 ± 0.8 mV (n = 13) and -43.0 ± 2.3 mV (n = 6) respectively. The difference between E_{EPSP} and E_{ACh} within groups was statistically insignificant. Thus the ionic mechanisms underlying the effect of the transmitter and the effect of the exogenously applied ACh seem to be similar which confirms the conclusion made by many other investigators that ACh is an excitatory transmitter in mammalian sympathetic ganglion.

However, the difference in E_{EPSP} and in E_{ACh} between the two groups of neurones requires further analysis. One possible explanation is that the input resistance inthe cell does not remain constant at different membrane potential levels, and changes when current is passed into the cell. The amplitudes of the EPSP and the potentials evoked by ACh depend on the cell input resistance according to the equation:

$$\Delta V = \frac{\Delta g}{\dfrac{1}{R_{inp}} + \Delta g} (E_{rev} - E_m) \,,$$

where ΔV is the amplitude of EPSP or of the potential evoked by ACh, R_{inp} is cell input resistance, Δg is the change in the membrane conductance produced by a transmitter or applied ACh, E_{rev} is a reversal potential for EPSP or for the potential evoked by ACh, and E_m is the membrane potential level.

When the input resistance was measured at different membrane potential levels in four neurones of the first group it was found that the input resistance did not change when the membrane was hyperpolarized to -100 mV. However, when in two neurones of the second group the membrane was hyperpolarized to this level, the input resistance was 3.8 and two times greater than that measured at their resting potential level. One example of such increase in input resistance is shown in Fig. 6C; measurements of E_{EPSP} and E_{ACh} showed that this particular neurone belonged to second group (Figs. 6A,B).

Thus it is assumed that the more negative values of E_{EPSP} and E_{ACh} in the second group of neurones are probably due to an inherent error of the extrapolation method which occurs when the increase in input resistance accompanies a hyperpolarization of the cell membrane by current passed through the membrane. Similar increase in input resistance has been observed by others (Roper, 1976; Hartzell et al., 1977).

This assumption is also supported by the fact that we have never observed an actual reversal of the EPSP or the potential evoked by ACh in the neurones of the second group at the E_{EPSP} or E_{ACh} level determined by the extrapolation method.

An alternative explanation of the differences in E_{EPSP} and E_{ACh} between the first and the second group of neurones is that in the second group the number of channels opened by ACh is voltage dependent (see Stevens, 1976).

Our observations thus indicate that the true values of E_{EPSP} and E_{ACh} are those observed in the first group of neurones. These values are in close agreement with those observed in amphibian sympathetic (Nishi and Koketsu, 1960) and parasympathetic (Dennis et al., 1971) ganglion neurones and in amphibian neuromuscular junction (Takeuchi and Takeuchi, 1960).

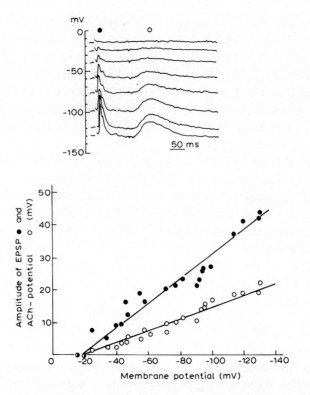

Fig. 5. Reversal potentials for the EPSP and the potential evoked by ACh recorded in the same sweep. In this case, separate intracellular electrodes were used for current passing and recording the membrane potential. Both the EPSP and the potential evoked by ACh, disappeared at −15 mV. The graph below shows a large number of data obtained from the same cell. Closed and open circles represent single EPSPs and potential evoked by ACh respectively. Resting membrane potential was −72 mV.

Difficulties in the estimation of E_{EPSP} in the neurones of the second group using the extrapolation method explain unusually low average value of E_{EPSP} obtained with this method in mammalian sympathetic ganglion neurones in earlier experiments (Skok, 1973).

Effect of a change in external ionic concentration on E_{EPSP} and E_{ACh}

The nature of the ionic channels opened by the transmitter and by exogenous ACh was investigated further by studying the effect of changes in the external ionic concentration on E_{EPSP} and E_{ACh}. Results obtained from six cells in which the E_{EPSP} and E_{ACh} were estimated by the extrapolation method before and 5 min after the perfusing solution was changed are illustrated by Fig. 7.

A two-fold decrease in sodium concentration or a five-fold decrease in potassium concentration resulted in 21 mV (mean; n = 8) and 13 mV (n = 4) shifts in E_{EPSP} towards more negative levels. A three-fold increase in potassium concentration studied in two neurones shifted E_{EPSP} to level 14 mV and 24 mV more positive than in the normal solution. Almost complete removal of chloride ions did not change the E_{EPSP} (n = 3). E_{ACh} shifted in similar way as

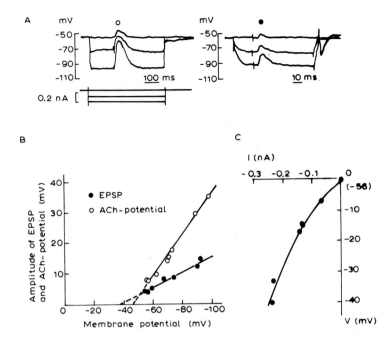

Fig. 6. (A) Reversal potentials for the EPSP and the potential evoked by ACh measured in the same cell using the extrapolation method; in this case the same intracellular electrode was used to pass the current through the membrane and to record the membrane potential. E_{EPSP} was −37 mV and E_{ACh} was −46 mV. In (B) some of the data illustrated in (A) is presented graphically. The potentials evoked by ACh are indicated by open circles and EPSPs by closed circles. In (C) the current-voltage relationship is shown for the same cell. The current pulses were 0.6 s in duration. Some of them are shown in (A), bottom trace, and the corresponding changes in membrane potential in the top trace. Resting membrane potential was −56 mV.

E_{EPSP} when sodium or potassium concentrations were altered; like E_{EPSP}, it was not affected by a change in the chloride concentration ($n = 7$).

These results indicate that ACh increases the membrane permeability to sodium and potassium ions and not to chloride. ACh has the same effect at the amphibian neuromuscular junction (Takeuchi and Takeuchi, 1960) and in amphibian sympathetic ganglion neurones (Nishi and Koketsu, 1960).

SUMMARY

(1) The action of the excitatory transmitter, exogenous acetylcholine (ACh) and cholinomimetic drugs carbachol (CCh) and tetramethylammonium (TMA) on the neuronal membrane was studied in the isolated superior cervical ganglion of the rabbit with the intracellular electrodes.

(2) Judging by the latency and time course of potentials evoked by the iontophoretical application of ACh to the soma membrane, nicotinic receptors only were activated. It appears that the soma membrane of the neurones in rabbit superior cervical ganglion does not have muscarinic receptors. Sometimes spontaneous excitatory postsynaptic potentials (EPSPs) appeared as a result of ACh application indicating excitation of presynaptic terminals.

250

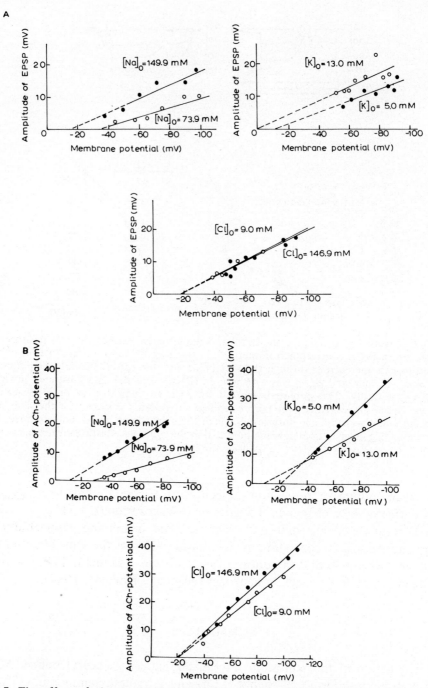

Fig. 7. The effect of changes in the external ionic concentration on E_{EPSP} (A) and E_{ACh} (B). Each of six graphs corresponds to a different cell tested with the ionic changes indicated.

(3) The Hill number found from dose—response relationship was 2.4 ± 0.6.

(4) The receptors of soma membrane were 7.8 times less sensitive to CCh and 4.6 times less sensitive to TMA than to ACh. Depolarizations evoked by CCh and TMA had much slower time courses than similar depolarizations evoked by ACh.

(5) Reversal potentials for the EPSP (E_{EPSP}) and for potentials evoked by exogenous ACh (E_{ACh}) were -14.4 ± 1.6 mV and -16.5 ± 1.2 mV respectively with no statistically significant difference between them. Much more negative values were obtained in one group of neurones probably due to an inaccuracy in the extrapolation method.

(6) A decrease in the external sodium or potassium concentrations made E_{EPSP} and E_{ACh} more negative, and an increase in external potassium concentration made them more positive than in normal solution. Al alteration of the external chloride concentration did not change E_{EPSP} and E_{ACh}.

(7) The results indicate that at least two ACh molecules combine with one nicotinic receptor and that the excitatory transmitter and exogenous ACh trigger identical ionic mechanisms in the membrane resulting in an increase in sodium and potassium permeabilities while chloride permeability remains unchanged.

REFERENCES

Blackman, J.G., Ginsborg, B.L. and Ray, C. (1963) On the quantal release of the transmitter at a sympathetic synapse. *J. Physiol. (Lond.)*, 167, 355–373.

Bregestovski, P.D., Vulfius, E.A. and Veprintsev, B.N. (1975) Possible mechanisms of desensitization. In *Nature of Cholinoreceptor and Structure of its Active Center*, B.N. Veprintsev and E.A. Vulfius (Eds.), Pushchino, pp. 113–139 (in Russian).

Bregestovski, P.D., Il'in, V.I., Veprintsev, B.N., Vulfius, E.A. and Yurchenko, O.P. (1976). Mechanisms of the activation of cholinoreceptive membrane. In *Proc. 7th neurochem. Conf.*, Leningrad, p. 67 (in Russian).

Brown, D.A. and Scholfield, C.N. (1974). Movements of labelled sodium ions in isolated rat superior cervical ganglia. *J. Physiol. (Lond.)*, 242, 321–351.

Carslaw, H.S. and Jaeger, J.C. (1947) *Conduction of Heat in Solids*. Clarendon Press, Oxford.

Colquhoun, D. (1975) Mechanisms of drug action at the voluntary muscle endplate. *Ann. Rev. Pharmacol.*, 15, 307–325.

Del Castillo, J. and Katz, B. (1955). On the localization of acetylcholine receptors. *J. Physiol. (Lond.)*, 128, 157–181.

Del Castillo, J. and Webb, G.D. (1977). Rapid desensitization of acetylcholine receptors of eel electroplaques following iontophoretic application of agonist compounds. *J. Physiol. (Lond.)*, 270, 271–282.

Dennis, M.J., Harris, A.J. and Kuffler, S.W. (1971). Synaptic transmission and its duplication by focally applied acetylcholine in parasympathetic neurons in the heart of the frog. *Proc. roy. Soc. Lond. B.*, 177, 509–539.

Dionne, V.E. (1976). Characterization of drug iontophoresis with a fast microassay technique. *Biophys. J.*, 16, 705–717.

Dreyer, F. and Peper, K. (1974). The acetylcholine sensitivity in the vicinity of the neuromuscular junction of the frog. *Pflügers Arch.*, 348, 273–286.

Dreyer, F., Müller, K.D., Peper, K. and Sterz, R. (1976). M. omohyoideus of the mouse as a convenient mammalian muscle preparation. *Pflügers Arch.*, 367, 115–122.

Dun, N., Nishi, S. and Karczmar, A.G. (1976). Alteration in nicotinic and muscarinic responses of rabbit superior cervical ganglion cells after chronic preganglionic denervation. *Neuropharmacology*, 15, 211–218.

Eccles, R.M. (1955) Intracellular potentials recorded from a mammalian sympathetic ganglion. *J. Physiol. (Lond.)*, 130, 572–584.

Feltz, A. and Mallart, A. (1971) An analysis of acetylcholine responses of junctional and extralunctional receptors of frog muscle fibres. *J. Physiol. (Lond.)*, 218, 85–100.

Hartzell, H.C., Kuffler, S.W. and Yoshikami, D. (1976). The number of acetylcholine molecules in a quantum and the interaction between quanta at the subsynaptic membrane of the skeletal neuromuscular synapse. In *Cold Spring Harbor Symp. quant. Biol.* Vol. 40, pp. 175–186.

Hartzell, H.C., Kuffler, S.W. Stickgold, R. and Yoshikami, D. (1977) Synaptic excitation and inhibition resulting from direct action of acetylcholine on two types of chemoreceptors on individual amphibian parasympathetic neurons. *J. Physiol. (Lond.)*, 271, 817–846.

Katz, B. and Miledi, R. (1964). Further observations on the distribution of acetylcholine — reactive sites in skeletal muscle. *J. Physiol. (Lond.)*, 170, 379–388.

Koketsu, K. (1969) Cholinergic synaptic potentials and the underlying ionic mechanisms. *Fed. Proc.* 28, 101–112.

Koketsu, K., Nishi, S. (1968) Cholinergic receptors at sympathetic preganglionic nerve terminals. *J. Physiol. (Lond.)*, 196, 293–310.

Magazanik. L.G. (1968) On the mechanism of the desensitization of the postsynaptic membrane in muscle fibre. *Biofizika*, 13, 199–203 (in Russian).

McMahan, U.J. and Kuffler, S.W.(1971) Visual identification of synaptic boutons on living ganglion cells and of varicosities in postganglionic axons in the heart of the frog. *Proc. roy. Soc. Lond. B.*, 177, 485–508.

Michelson, M.Ya. and Zeimal, E.V. (1973) *Acetylcholine. An Approach to the Molecular Mechanism of Action.* Pergamon Press, Oxford.

Nishi, S. (1970) Cholinergic and adrenergic receptors of sympathetic preganglionic nerve terminals. *Fed. Proc.*, 29, 1957–1965.

Nishi, S. and Koketsu, K. (1960). Electrical properties and activities of single sympathetic neurons in frogs. *J. cell. comp. Physiol.*, 55, 15–30.

Peper, K., Dreyer, F. and Müller, K.-D. (1976) Analysis of cooperativity of drug-receptor interaction. In *Cold Spring Harbor Symp. quant. Biol.* Vol. 40, pp. 187–192.

Roper, S. (1976). An electrophysiological study of chemical and electrical synapses on neurons in the parasympathetic cardiac ganglion of the mudpuppy Necturus maculosus: evidence for intrinsic ganglionic innervation. *J. Physiol. (Lond.)*, 254, 427–454.

Skok, V.I. (1973) *Physiology of Autonomic Ganglia.* Igaku Schoin, Tokio.

Skok, V.I. and Selyanko, A.A. (1977) Membrane permeability changes caused in mammalian sympathetic ganglion neurones by transmitter and iontophoretic applications of acetylcholine and TMA. In *Proc. 27th Congr. physiol. Sci.*, Vol. 12, Paris, p. 559.

Stevens, Ch.F. (1976) Molecular basis for postjunctional conductance increases induced by acetylcholine. In *Cold Spring Harbor Symp. quant. Biol.*, Vol. 40, pp. 169–174.

Takeuchi, A. and Takeuchi, N. (1959). Active phase of frog's endplate potential. *J. Neurophysiol.*, 22, 395–411.

Takeuchi, A. and Takeuchi, N. (1960) On the permeability of endplate membrane during the action of transmitter. *J. Physiol. (Lond.)*, 154, 52–87.

Werman, R. (1969) An electrophysiological approach to drug-receptor mechanisms. *Comp. Biochem. Physiol.*, 30, 997–1017.

Ziskind, L. and Werman, R. (1975). At least three molecules of carbamylcholine are needed to activate a cholinergic receptor. *Brain Res.*, 88, 177–180.

Acetylcholine as an Excitatory and Inhibitory Transmitter in the Mammalian Central Nervous System

J.S. KELLY, JANE DODD and R. DINGLEDINE *

M.R.C. Neurochemical Pharmacology Research Unit, Department of Pharmacology, Medical School, Cambridge (United Kingdom)

INTRODUCTION

When acetylcholine (ACh) and various other cholinomimetic agents were applied to single neurones by micro-iontophoresis the resultant changes in neuronal activity rarely fit the clear-cut pharmacological groupings demonstrated so readily in peripheral tissues. For instance, the larger neurones of the feline neocortex (Krnjević and Phillis, 1963; Crawford and Curtis, 1966) and most neurones in the caudate nucleus (McLennan and York, 1966) appear to be particularly sensitive to muscarinic agents whereas the Renshaw cells of the spinal cord (Curtis and Ryall, 1966a,b) and the larger neurones of the ventrobasal complex of the thalamus (Andersen and Curtis, 1964a,b; McCance et al., 1968), the pyriform cortex (Legge et al., 1966) the cerebellum (McCance and Phillis, 1964; Crawford et al., 1966) and the brainstem (Bradley et al., 1964; Bradley and Dray, 1976) are equally responsive to both nicotinic and muscarinic agents. Although all nicotinic responses described to date have been excitatory in nature (see Dingledine and Kelly, 1978) muscarinic receptors can be either excitatory or inhibitory depending on the location of the cell under study.

In this chapter therefore, we would like to draw attention to our studies of the muscarinic actions of ACh in the hippocampus and thalamus. In the hippocampus they have led us to conclude that the excitatory action of ACh on hippocampal pyramidal cells is mediated by a decrease in potassium permeability which at first sight appears to develop too slowly to be involved in the generation of the excitatory postsynaptic potentials seen during stimulation of the septo-hippocampal pathway. In the thalamus, extracellular studies show the excitatory and inhibitory actions of ACh to be located on two separate populations of neurones in a way which is consistent with the participation of cholinergic pathways from the midbrain reticular formation, in arousal.

* Present address: Epilepsy Unit and Department of Physiology, Duke University, Durham, N.C. 27710 (U.S.A.).

ACh-EVOKED EXCITATION IN THE HIPPOCAMPAL
IN VITRO SLICE PREPARATION

In each experiment the tip of a multibarrelled microiontophoretic pipette was placed just deep to the surface of the in vitro hippocampal slice in the region of the pyramidal cell body layer, and a second intracellular micropipette used to penetrate the soma of a near by neurone (see Dingledine et al., 1977a,b). The proximity of the iontophoretic pipette to the cell under study was then tested by showing the cell to be depolarized and excited by relatively

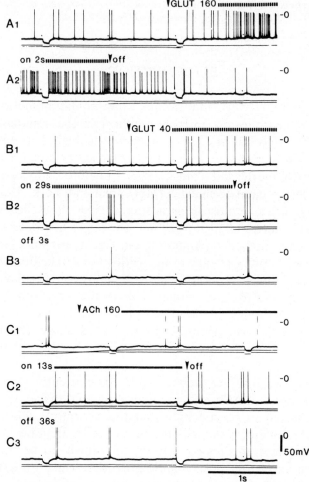

Fig. 1. Intracellular records from a CA1 pyramidal cell to show the onset and offset of the depolarization evoked by glutamate and ACh. In (A) a relatively large iontophoretic application of glutamate (GLUT) with an iontophoretic current of 160 nA, caused a rapid and marked depolarization and excitation which was followed by an equally rapid recovery. In (B) recovery was equally rapid after a lesser degree of depolarization and excitation had been maintained for almost 30 s by a smaller application of glutamate with a current of 40 nA. The peak effect also occurred within a few seconds of the onset of the glutamate application. In contrast, in (C), a smaller level of depolarization and excitation occurred when 4 times more current was applied to an adjacent barrel of the electrode containing ACh and the peak effect occurred some 15 s after the onset of the current application. Even after a fairly modest depolarization and excitation recovery was still not complete 36 s after the ACh application was terminated.

small and brief applications of L-glutamate (Figs. 1A and 1B). Although in the majority of cells examined in this way, a significant degree of excitation and depolarization could be attained with glutamate-injecting currents in the region of 50 nA applied for between 20 and 30 s the current required to eject sufficient ACh to evoke a similar level of depolarization or excitation on the same neurone was usually 5 or 6 times greater. Often the peak ACh evoked effect did not occur unless the application was maintained for approximately 4 times as long as the equipotent application of glutamate.

On average, the peak level of depolarization evoked by ACh occurred 89 s after the first detectable depolarization and the peak level of excitability was reached some 62 s after the first sign of an increase in excitability. In contrast, the peak depolarization and excitations evoked by glutamate occurred approximately 70 s earlier at 19 and 16 s respectively. Furthermore, the slowness with which the depolarization and excitation evoked by ACh developed was exaggerated by a delay of approximately 16 s between the onset of the ACh ejecting current and the appearance of the depolarising response.

Not only was there a striking difference between the time of onset and the rate with which the two excitations developed but quite modest depolarizations

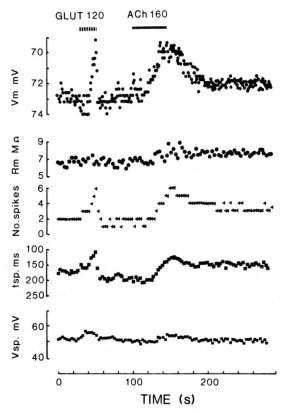

Fig. 2. Comparison of the responses evoked by ACh and glutamate. Even in a neurone in which responses of equal amplitude could be evoked by the application of similar iontophoretic currents to the adjacent barrels of a microelectrode containing glutamate and ACh, the peak depolarization and excitation evoked by ACh, and the subsequent recovery, were very much delayed when compared with the responses evoked by glutamate. Vm, is the membrane potential; Rm, the membrane resistance; tsp, the time to the first spike; Vsp, the firing threshold of the first spike.

256

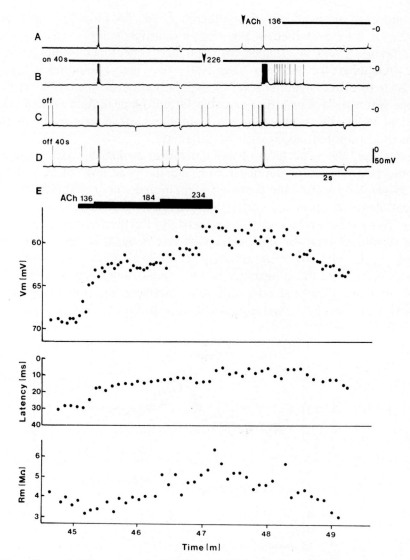

Fig. 3. The effect of ACh on the excitability of CA1 cells and the increase in membrane resistance which accompanies the depolarization and increase in excitability evoked by ACh. In (A–D), intracellular records show the additional spikes evoked by ACh to occur predominantly in association with the depolarizing pulses used to test the excitability of the membrane. In (B) the depolarization evoked by the pulses is prolonged by the intensity of the evoked firing. In (E) the graphs from a subsequent application of ACh to the same cell show how the depolarization and increase in excitability is accompanied by an increase in membrane resistance. Note the temporal dispersion between these parameters which is discussed in the text. See Fig. 2 for abbreviations.

and excitations evoked by ACh were followed by an extremely slow recovery of the control levels of potential and excitability. In Fig. 1C for instance the relatively small excitation and depolarization evoked by a modest application of ACh with a current of 160 nA applied for 15 s was still present 40 s after the ACh application ended whereas a similar excitation reached within a few seconds of the start of a glutamate application with a current of 40 nA (Fig. 1B) vanished within a second or so of terminating the ejecting current even

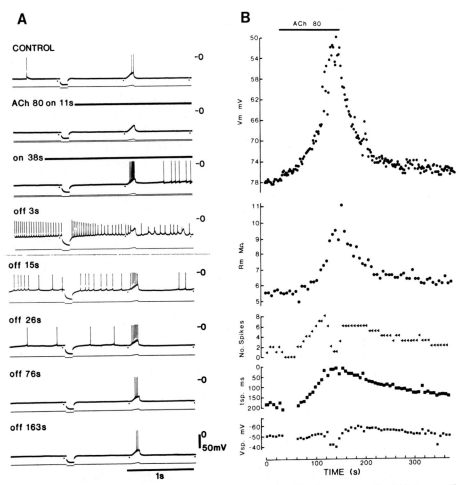

Fig. 4. More detailed study of the action of ACh on CA1 neurone. In (A) intracellular records show how the membrane resistance and excitability were tested alternately by passing hyperpolarizing and depolarizing pulses of current through the intracellular micro-electrodes. The increase in excitability is shown most clearly by the increase in the number of spikes evoked by the depolarizing ramps of current from 2 to 8, and the increase in resistance by the increase in amplitude of the hyperpolarizing pulse which reached a peak 3 s after the ACh application was terminated. In (B) the graphs show the relationship between the depolarization evoked by ACh and the increase in membrane excitability and resistance. See Fig. 2 for abbreviations.

though this level of excitability was maintained at the same level for 30 s. Often the end of the glutamate-evoked depolarizations and excitations were not only abrupt but often accentuated by a short period of reduced excitability. Occasionally, after particularly intense excitations, or following the use of larger doses of glutamate, this interval of decrease excitability was prolonged and accompanied by a slight hyperpolarization and a reduction in membrane resistance. On the other hand, prolonged periods of increased excitability with high levels of spontaneous firing were a consistent feature of ACh evoked depolarizations. Even on the few occasions when peak responses of equal magnitude were obtained on applying similar currents to the glutamate and ACh barrels, the slow onset and prolonged recovery of the ACh evoked response was still a striking feature (Fig. 2).

EFFECT OF ACh ON MEMBRANE RESISTANCE

The relationship between the onset of the ACh-evoked depolarization and the increase in membrane excitability and resistance was investigated in more detail by passing pulses of current through the intracellular microelectrode (Figs. 3 and 4). The excitability was tested by means of depolarizing ramps of just sufficient magnitude to evoke one or two action potentials near the summit of the ramp during the control period and the resistance with rectangular hyperpolarizing pulses of similar amplitude. As shown in Fig. 4 the initial depolarization evoked by ACh was often accompanied by a decrease in excitability which in this instance lasted for approximately 25 s. Only after a further delay of approximately 7 s did the excitability increase above the control level. Often the associated increase in the membrane resistance (Figs. 3 and 4) could be detected only after a further delay of approximately 22 s. During the peak of the depolarizations evoked by ACh the amplitude of the action potentials often declined and the firing threshold rose, presumably as a result of depolarization block. Slight broadening of the action potentials was a consistent feature of the peak depolarization evoked by ACh. As in many other experiments the restoration of the potential, excitability and membrane resistance following the termination of the ACh application was greatly delayed. Consistently, the recovery of all three parameters follow a similar time course (Figs. 2, 3 and 4).

EQUILIBRIUM POTENTIAL OF THE ACh-EVOKED DEPOLARIZATION

Even though during the initial stages of the ACh-evoked response the depolarization and the increase in membrane resistance were out of synchrony (Figs. 2, 3 and 4), the peak depolarizations (ΔV) evoked by ACh were a hyperbolic function of the ratio of the membrane resistance of the cell measured before (R), to the increase in resistance which occurred during, the peak of the depolarization evoked by ACh (ΔR). Indeed, even though as shown in Fig. 5A the individual values of the product $\Delta V(R/\Delta R)$ fluctuated from cell to cell, the mean value of 29 mV $\pm$ 2.9 does not differ significantly from the product 34.6 calculated from the equation log $\Delta V = [(R/\Delta R) - 7.00 \pm 1.29]/3.87 \pm 0.89$ (cf. Fig. 5B). The fact that the product calculated for each individual cell approximates to the mean value for all the cells is consistent with the idea that the excitation is mediated by the closure of ionic channels normally open in the absence of ACh, since the sum of the product $\Delta V(R/\Delta R)$ and the resting potential V, is the reversal level of the ionic species whose movement is altered by the action of ACh (Ginsborg, 1967; 1973; Krnjević et al., 1971; Ginsborg et al., 1974; Adams and Brown, 1975). Our results, therefore, not only confirm the observation of Krnjević et al. (1971) that the depolarization of feline cortical neurones by ACh in vivo is accompanied by an increase in membrane resistance, but the value of (34.6 + 74 mV) obtained in this study for the reversal level of the ionic species involved is not significantly different from that obtained by Krnjević et al. (1971). Since it is difficult to believe that another process such as anomalous rectification could be responsible for the

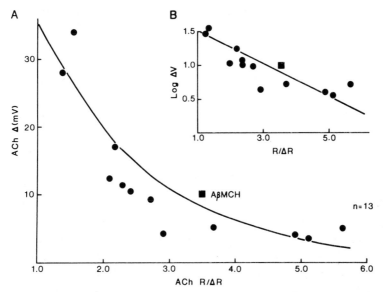

Fig. 5. The peak depolarization evoked by ACh as a function of the increase in membrane resistance. The peak depolarization evoked by ACh and acetyl-β-methylcholine (Aβ MCh) (ΔV) in 13 neurones (mean resting potential 73.6 mV), as a function of the ratio of the membrane resistance (R) measured before the application of ACh to the increase in membrane resistance (ΔR) evoked by ACh and Aβ MCh. In (B) the hyperbolic function has been linearized by comparing log ΔV with R/ΔR. The mean product of the value ΔV (R/ΔR) was calculated to be 34.5 mV. The mean equilibrium potential for the response evoked by ACh or acetyl-β-methylcholine was therefore 108.2 mV.

constancy of the product ΔV(R/ΔR) the unexplained lag between the onset of the increase in resistance and the depolarization must in part be attributed to the distribution of the ACh receptors on peripheral dendrites and the location of the intracellular electrode in the cell body. Presumably if the action of ACh is confined to the peripheral dendrites, somatic changes will occur only when a considerable proportion of the dendritic field has become involved in the interaction with ACh. However, only a small fraction of the delay in the onset of the various responses can be explained by the time required for the ejection and diffusion of sufficient ACh to dentritically located receptors. Of course, as discussed in some detail by Krnjević et al. (1971) the rate at which a depolarization involving the closure of ionic channels will develop is dependent on both the magnitude of the change in membrane resistance and the resting resistance of the cell in the absence of ACh. However, in the present experiments the rate of depolarization was 0.2 mV/s, several orders of magnitude slower than predicted. A discrepancy of this magnitude cannot reflect differences between the onset of the somatic response and the response on the peripheral dendrites and must indicate the slowness with which ACh interacts with receptors or, alternatively, the slowness with which the receptors bring about the closure of the ionic channels involved. However, it should be pointed out that recent work in other situations (Kirkwood and Sears, 1978) suggests that a depolarization of the cell body by as little as 3 μV may accompany the synchrony of the motoneurones that drive the intracostal muscles. Clearly not only would a depolarization of this magnitude develop much more rapidly but it would go undetected with our present techniques. The slowness of the response evoked by iontophoretically applied ACh should not therefore be used as an argument to exclude ACh as the likely mediator of the EPSP's evoked by stimulation of the septo-hippocampal pathway. The possible role of intermediates such as cyclic GMP is discussed elsewhere (Dingledine and Kelly, 1978).

LOCALIZATION OF ACh-INHIBITED CELLS IN THE THALAMUS

In most regions of the mammalian brain only a small percentage of the neurones detected with multi barrelled micropipettes are inhibited by iontophoretic ACh. Indeed, the much more frequent occurrence of cells excited by rather small doses of ACh (see Krnjević, 1974), together with the finding that both the inhibitory and excitatory actions of ACh are often antagonized by the same pharmacological agents has led to the suggestion that the inhibitory action of ACh is mediated indirectly through the excitation of neighbouring cholinoceptive interneurones whose inhibitory transmitter is a substance other than ACh (Duggan and Hall, 1975; Randić et al., 1964). A number of authors (Randić et al., 1964; McLennan, 1970; Phillis and York, 1967a,b; 1968; Jordan and Phillis, 1972) have tried to overcome this difficulty by working in the most superficial layers of the cerebral cortex where a strong inhibitory action of ACh predominates and ACh excitation has seldom been reported.

However, in our own studies the feline thalamus proved much more suitable and short iontophoretic pulses of ACh were found to inhibit the spontaneous discharge of virtually every cell found to lie within the region bounded by the stereotaxic coordinates AP 9.5–11.0, L 7.0–8.5 and H 4.0–7.0 (Ben-Ari et al., 1976a,b; Dingledine and Kelly, 1977). Subsequent histological analysis showed that microelectrode tracks in which only ACh-inhibited cells were encountered lay solely within the dorsolateral nucleus reticularis of the thalamus (nucR). On the other hand, in microelectride tracks in which ACh-excited cells were found to lie deep to the ACh-inhibited cells, histology showed that the microelectrode had passed through the shoulder of the nucR and penetrated the ventrobasal complex of the thalamus, a region long known to contain ACh-excited cells (Andersen and Curtis, 1964a,b) (Fig. 6A,B).

Our conclusion that the ACh-inhibited cells lay within the nucR was also stengthened by their pattern of discharge. The average firing rate of the ACh-inhibited cells in the dorsolateral nucR was 22 Hz (S.E. = 1.4; n = 114) and ongoing activity was interrupted by characteristic long duration, high frequency bursts in which the maximum frequency often exceeded 400 Hz. The firing pattern of the ACh-inhibited cells was therefore typical of that attributed to cells of the nucR by many other groups of workers (see Negishi et al., 1962; Mukhametov et al., 1970; Schlag and Waszak, 1971; Lamarre et al., 1971; Steriade and Wyzinski, 1972; Waszak, 1974). On the other hand the deeper located cells found to be excited by ACh had a significantly lower average firing frequency of 6.0 Hz (S.E. = 0.8; n = 46) and the highest frequency obtained by their rather less prominent periods of burst activity rarely exceeded 100 Hz. Very much earlier Andersen and Curtis (1964a) drew atten-

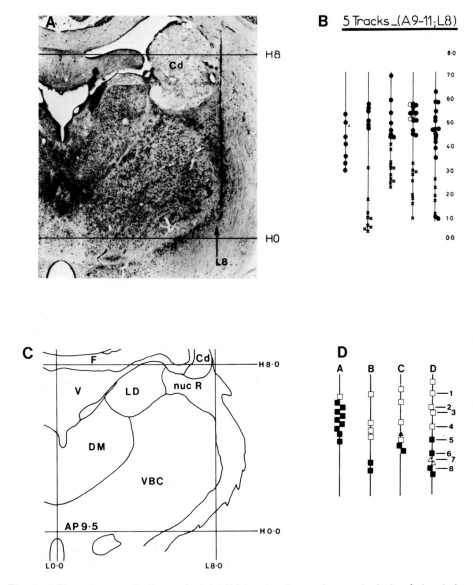

Fig. 6. ACh and synaptically evoked inhibition in the nucleus reticularis of the thalamus. The photomicrograph in (A) is from an experiment in which only ACh-inhibited cell were encountered. Subsequent histological analysis showed the microelectrode tract to lie solely within the nucleus reticularis. Cd, head of caudate nucleus. The five vertical lines in (B) schematically show the depth distribution of cells which were either inhibited (● or ○) or excited (x) as microelectrodes penetrated (●) or were withdrawn (○) from the brain of 4 cats along the coordinates AP 9.0–11.0 and L 7.0 or 8.0 (Snider and Niemer, 1961). (C) is a schematic line drawing of the thalamus at stereotaxic plane AP 9.5. (D) shows the responses of neurones encountered in four different cats during microelectrode tracks, along the coordinate AP 9.5 L 8.0, penetrating the dorsolateral nucleus reticularis and underlying ventrobasal complex. Each symbol presents the results obtained from a single neurone tested first for its response to iontophoretic ACh and then for its initial response to high frequency stimulation (usually at 200 Hz, repeated at 0.5 Hz) of the ipsilateral midbrain reticular formation (MRF). □, neurones inhibited by both ACh and the MRF; ■, neurones excited by both ACh and the MRF; ▲, neurones inhibited by ACh but excited by MRF; △, neurones excited by ACh but inhibited by the MRF. The most superficially-located neurones in each track are inhibited by ACh and the MRF, and conversely the deeper neurones are excited by ACh and the MRF. NucR, nucleus reticularis; VBC, ventrobasal complex; DM, dorsal medial nucleus; LD, dorsal lateral nucleus, F, fornix; Cd, head of caudate; V, ventricle.

tion to the slow firing and the short duration of the burst activity of the ACh-excited cellls in the ventrobasal thalamus.

NATURE OF THE ACh-EVOKED INHIBITION

In contrast to the rather long latency of onset (4–8 s) for the ACh-evoked excitation of cells in the ventrobasal thalamus, the onset of the ACh-evoked inhibition of the spontaneously active cells in nucR was rapid. Indeed on many occasions the latency of onset of the ACh-evoked inhibition was as short as that for the onset, on the same cell of a glutamate evoked excitation or of an inhibition evoked by equally potent doses of GABA.

EFFECT OF ATROPINE

The ACh-evoked inhibitions proved extremely sensitive to iontophoretic applications of atropine revealing their rather strong muscarinic character. On 12 of 13 cells atropine completely and reversibly blocked the ACh-evoked inhibition. On three of four cells tested the atropine-sensitive ACh-evoked inhibition was also reversibly blocked by an iontophoretic application of dihydro-β-erythroidine (DHβE) suggesting that the ACh receptor mediating inhibition in the nucR may be of mixed muscarinic nicotinic character. However, the ratio of the current applied to the DHβE barrel to that applied to the atropine barrel for effective blockade of ACh-evoked inhibitions in the nucR was consistently larger than that for the same microelectrode when antagonizing ACh-evoked excitations in the ventrobasal complex (VBC). In the cortex Jordan and Phillis (1972) also found ACh-evoked inhibitions to be more readily blocked by atropine than by DHβE.

CORRELATION BETWEEN EFFECTS OF ACh AND STIMULATION OF THE MIDBRAIN RETICULAR FORMATION (MRF)

High frequency stimulation of the MRF is accompanied by facilitated transmission through the major thalamic relay nuclei, including the lateral geniculate (Singer, 1973), medial geniculate (Symmes and Anderson, 1967), ventro-posteriolateral nucleus (Steriade, 1970) and the ventrolateral nucleus (Purpura et al., 1966). In the case of the ventrolateral and lateral geniculate nuclei intracellular recording showed such facilitation to result from an attenuation of recurrent inhibitory potentials (Purpura et al., 1966; Singer, 1973), and so the MRF evoked facilitation was attributed to disinhibition.

The observation that the MRF evokes disinhibitory potentials in the thalamo-cotrical relay cells raises the possibility that the MRF stimulation will inhibit the neurones of the nucR, and indeed there is some evidence that stimulating the MRF can lead to a decrease in unit activity in the rostral pole of nucR (Schlag and Waszak, 1971; Waszak, 1974; Yingling and Skinner, 1975). Anatomically a direct pathway from the MRF to nucR has been identified (Scheibel and Scheibel, 1958; Edwards and de Olmos, 1976), which may well be cholinergic (Shute and Lewis, 1967).

Post-stimulus interval histograms (PSIH) showed that most neurones in the dorsolateral nucR and adjacent VBC responded to repetitive stimulation of the MRF (usually 3 pulses at 200 Hz, repeated every 1 or 2 s) with alternating periods of inhibition and excitation. In our study (Dingledine and Kelly, 1977) there was a highly significant correlation between the direction of the *initial* response to MRF stimulation and the response of the cell to ACh. Eighty-six percent of the ACh-inhibited cells in the nucR were initially inhibited by MRF stimulation, and conversely 78% of the ACh-excited cells of the VBC were initially excited by the stimulus (Fig. 6C,D).

ATROPINE AND THE RETICULAR INPUT TO THE nucR

If the MRF sends a cholinergic projection to the nucR or the VBC one might expect the MRF-evoked responses to be blocked by atropine, since the ACh-evoked responses on the same cellls are very sensitive to atropine. Unfortunately, we could not consistently antagonize the MRF evoked responses with massive iontophoretic (24 cells) or intravenous (4 cells) doses of atropine. The iontophoretic dose of atropine used, was far larger than that needed to block completely the ACh-evoked inhibition of the cells in the nucR. On the other hand, in 10 cells atropine reduced slightly the magnitude of a late inhibition occurring in a polyphasic response to stimulation of the MRF, and in four of these cases there was some degree of recovery. Since the dendrites of cells in the dorsolateral nucR are extremely long (up to 1–2 mm) and run in densely packed bundles (Scheibel and Scheibel, 1972), it is possible that inhibitory cholinergic synapses present on distant dendrites of nucR cells are relatively inaccessible to atropine. With the notable exception of the nicotinic synapse on the Renshaw cell (Curtis and Ryall, 1966a,b), it has proved very difficult to block other putative cholinergic synapses in the CNS with cholinergic antagonists, even in the habenulo-interpeduncular (Lake, 1973) and septo-hippocampal (Salmoiraghi and Stefanis, 1967) pathways, which are now known to have all the biochemical properties expected of cholinergic tracts (Lewis et al., 1967; Katasaka et al., 1973; Kuhar et al., 1975; Leranth et al., 1975; Storm-Mathisen, 1975).

IONIC BASE OF THE ACh-EVOKED INHIBITION

The ionic mechanisms responsible for the ACh-evoked inhibition in the nucR will be revealed only by the use of intracellular electrodes. However, the exaggerated burst activity that occurs during the ACh-evoked inhibition (Ben-Ari et al., 1976b) but not during inhibition of the same cells to a similar degree by GABA and glycine could be regarded as tentative evidence that changes in ionic conductance evoked by ACh must differ from the large increases in Cl^- conductance that are associated with the actions of GABA (Dreifuss et al., 1969) and glycine (Ten Bruggencate and Engberg, 1971).

Although the rapid onset of the ACh-evoked inhibition of cells of the nucR is quite unlike the slow onset of the muscarinic inhibition of frog ganglion cells described by Weight and Padjen (1973), the high firing rate of the neurones in the nucR could be the result of an abnormally high Na^+ permeability. In this situation even small reductions in Na^+ permeability could cause the immediate onset of inhibition. Indeed Krnjević (1974) has already suggested that the often seen initial depressant action of ACh on ACh-excited cells is a special feature of cells with a high spontaneous firing rate.

SUMMARY

Using intracellular techniques and a in vitro slice preparation of the hippocampus we have confirmed earlier work which shows the excitatory action of acetylcholine on cortical neurones to be slow in onset. The depolarization, which occurs during the increase in excitability, is a hyperbolic function of the substantial increase in membrane resistance which always accompanies these changes, as could be predicted if the change were mediated by a decrease in potassium conductance.

In the feline thalamus we have been able to show, by the use of extracellular recording techniques and micro-iontophoresis, that both acetylcholine and electrical stimulation of a putative cholinergic pathway from the midbrain reticular formation have an inhibitory action on almost every cell encountered in the nucleus reticularis. A few millimetres away, in the ventrobasal complex, the opposite effect occurs and every cell is excited. These two opposing actions of acetylcholine seem to be equally important as mediators of the increase in synaptic efficiency which accompanies arousal.

ACKNOWLEDGEMENTS

We are grateful for the excellent technical assistance of George R. Marshall. Jane Dodd is a MRC scholar.

REFERENCES

Adams, P.R. and Brown, D.A. (1975) Actions of γ-aminobutyric acid on sympathetic ganglion cells. *J. Physiol. (Lond.)*, 250, 85–120.

Andersen, P. and Curtis, D.R. (1964a) The excitation of thalamic neurones by acetylcholine. *Acta physiol. scand.*, 61, 85–99.

Andersen, P. and Curtis, D.R. (1964b) The pharmacology of the synaptic and acetylcholine-induced excitation of ventrobasal thalamic neurones. *Acta physiol. scand.*, 61, 100–120.

Ben-Ari, Y., Dingledine, R., Kanazawa, I. and Kelly, J.S. (1976a) Inhibitory effects of acetylcholine on neurones in the feline nucleus reticularis thalami. *J. Physiol. (Lond.)*, 261, 647–671.

Ben-Ari, Y., Kanazawa, I. and Kelly, J.S. (1976b) Exclusively inhibitory action of iontophoretic acetylcholine on single neurones of feline thalamus. *Nature (Lond.)*, 259, 327–330.

Bradley, P.B and Dray, C. (1976) Observations on the pharmacology of cholinoceptive neurones in the rat brain stem. *Brit. J. Pharmacol.*, 57, 599—602.

Bradley, P.B., Dhawan, B.N. and Wolstencroft, J.H. (1964) Some pharmacological properties of cholinoceptive neurones in the medulla and pons of the cat. *J. Physiol. (Lond.)*, 170, 59—60P.

Crawford, J.M. and Curtis, D.R. (1966) Pharmacological studies on feline Betz cells. *J. Physiol. (Lond.)*, 186, 121—138.

Crawford, J.M., Curtis, D.R., Voorhoeve, P.E. and Wilson, V.J. (1966) Acetylcholine sensitivity of cerebellar neurones in the cat. *J. Physiol. (Lond.)*, 186, 139—165.

Curtis, D.R. and Ryall, R.W. (1966a) The excitation of Renshaw cells by cholinomimetics. *Exp. Brain Res.*, 2, 49—65.

Curtis, D.R. and Ryall, R.W. (1966b) The acetylcholine receptors of Renshaw cells. *Exp. Brain Res.*, 2, 66—80.

Dingledine, R., Dodd, J. and Kelly, J.S. (1977a) Intracellular recording from pyramidal neurones on the in vitro transverse hippocampal slice. *J. Physiol. (Lond.)*, 169, 13—15P.

Dingledine, R., Dodd, J. and Kelly, J.S. (1977b) ACh-evoked excitation of cotrical neurones. *J. Physiol. (Lond.)*, 273, 79—80P.

Dingledine, R. and Kelly, J.S. (1977) Brain stem stimulation and the acetylcholine-evoked inhibition of neurones in the feline nucleus reticularis thalami. *J. Physiol (Lond.)*, 271, 135—154.

Dingledine, R. and Kelly, J.S. (1978) Cholinergic processes at synaptic junctions. In *Intracellular Junctions and Synapses: Receptors and Recognition, Series B, Vol. 2*, J. Feldman, N.B. Gilula and J.D. Pitts (Eds.), Chapman and Hall, London, pp. 141—180.

Dreifuss, J.-J., Kelly, J.S. and Krnjevic, K. (1969) Cortical inhibition and γ-aminobutyric acid. *Exp. Brain Res.*, 9, 137—154.

Duggan, A.W. and Hall, J.G. (1975) Inhibition of thalamic neurones by acetylcholine. *Brain Res.*, 100, 445—449.

Edwards, S.B. and De Olmos, J.S. (1976) Autoradiography studies of the projections of the midbrain reticular formation: ascending projections of nucleus cuneiformis. *J. comp. Neurol.*, 165, 417—432.

Ginsborg, B.L. (1967) Ion movements in junctional transmission. *Pharmac. Rev.*, 19, 289—316.

Ginsborg, B.L. (1973) Electrical changes in the membrane in junctional transmission. *Biochim. biophys. Acta*, 300, 289—317.

Ginsborg, B.L., House, C.R. and Silinsky, E.M. (1974) Conductance changes associated with the secretory potential in the cockroach salivary gland. *J. Physiol. (Lond.)*, 236, 723—737.

Jordan, L.M. and Phillis, J.W. (1972) Acetylcholine inhibition in the intact and chronically isolated cerebral cortex. *Brit. J. Pharmacol. Chemother.*, 45, 584—595.

Katasaka, K., Nakamura, Y. and Hassler, R. (1973) Habenulointerpeduncular tract; A possible cholinergic neuron in rat brain. *Brain Res.*, 62, 264—267.

Kirkwood, P.A. and Sears, T.A. (1978) The synaptic connexions to intercostal motoneurones as revealed by the average common excitation potential. *J. Physiol. (Lond.)*, 275, 103—134.

Krnjevic, K. (1974) Chemical nature of synaptic transmission in vertebrates. *Physiol. Rev.*, 54, 418—540.

Krnjevic, K. and Phillis, J.W. (1963) Pharmacological properties of acetylcholine sensitive cells in the cerebral cortex. *J. Physiol. (Lond.)*, 166, 328—350.

Krnjevic, K., Pumain, R. and Renaud, L. (1971) The mechanism of excitation by acetylcholine in the cerebral cortex. *J. Physiol. (Lond.)*, 215, 247—268.

Kuhar, M.J., DeHaven, R.N., Yamamura, H.I., Rommelspacher, H. and Simon, J.R. (1975) Further evidence for cholinergic-habenulointerpeduncular neurons: pharmacologic and functional characteristics. *Brain Res.*, 97, 265—275.

Lake, N. (1973) Studies of the habenulo-interpeduncular pathway in cats. *Exp. Neurol.*, 41, 113—132.

Lamarre, Y., Filion, M. and Coreau, J.P. (1971) Neuronal discharges of the ventrolateral nucleus of the thalamus during sleep and wakefulness in the cat. I. Spontaneous activity. *Exp. Brain Res.*, 12, 480—498.

Legge, K.F., Randic, M. and Straughan, D.W. (1966) The pharmacology of neurones in the pyriform cortex. *Brit. J. Pharmacol.*, 26, 87—107.

Leranth, C.S., Brownstein, M., Zabrosaky, L., Jaranyi, Z.S. and Palkovits, M. (1975)

Morphological and biochemical changes in the rat interpeduncular nucleus following the transection of the habenulo-interpeduncular tract. *Brain Res., 99*, 124–128.

Lewis, P.R., Shute, C.C.D. and Silver, A. (1967) Confirmation from choline acetylase of a massive cholinergic innervation to the rat hippocampus. *J. Physiol. (Lond.), 191*, 215–224.

McCance, I. and Phillis, J.W. (1964) The action of acetylcholine on cells in cat cerebellar cortex. *Experientia, 20*, 217–218.

McCance, I., Phillis, J.W., Tebecis, A.K. and Westerman, R.A. (1968) The pharmacology of acetylcholine-excitation of thalamic neurons. *Brit. J. Pharmacol., 32*, 652–662.

McLennan, H. (1970) Inhibition of long duration in the cerebral cortex. A quantitative difference between excitatory amino acids. *Exp. Brain Res., 10*, 417–426.

McLennan, H. and York, D.H. (1966) Cholinergic mechanisms in the caudate nucleus. *J. Physiol. (Lond.), 187*, 163–175.

Mukhametov, L.M., Rizzolatti, G. and Tradardi, V. (1970) Spontaneous activity of neurones of nucleus reticularis thalami in freely moving cats. *J. Physiol. (Lond.), 210*, 651–667.

Negishi, K., Lu, E.S. and Verzeano, M. (1962) Neuronal activity in the lateral geniculate body and the nucleus reticularis of the thalamus. *Vision Res., 1*, 343–353.

Phillis, J.W. and York, D.H. (1967) Cholinergic inhibition in the cerebral cortex. *Brain Res., 5*, 517–520.

Phillis, J.W. and York, D.H. (1967) Strychnine block of neuronal and drug-induced inhibition in the cerebral cortex. *Nature (Lond.), 216*, 922–923.

Phillis, J.W. and York, D.H. (1968) Pharmacological studies on a cholinergic inhibition in the cerebral cortex. *Brain Res., 10*, 297–306.

Purpura, D.P., McMurtrey, J.C. and Maekawa, K. (1966) Synaptic events in ventrolateral thalamic neurons during suppression of recruiting responses by brain stem reticular stimulation. *Brain Res., 1*, 63–76.

Randic, M., Siminoff, R. and Straughan, D.W. (1964) Acetylcholine depression of cortical neurones, *Exp. Neurol., 9*, 236–242.

Salmoiraghi, G.C. and Stefanis, C.N. (1967) A critique of iontophoretic studies of central nervous system neurons. *Int. Rev. Neurobiol., 10*, 1–30.

Scheibel, M.E. and Scheibel, A.B. (1958) Structural substrates for integrative patterns in brain stem reticular core. In *Reticular Formation of the Brain*, H.H. Jasper, L.D. Proctor, R.S. Knighton, W.C. Noshay and R.T. Costello (Eds.), Little Brown, Boston, pp. 31–55.

Scheibel, M.E. and Scheibel, A.B. (1972) Specialized organization patterns within the nucleus reticularis thalami of the cat. *Exp. Neurol., 34*, 316–322.

Schlag, J. and Waszak, M. (1971) Electrophysiological properties of units of the thalamic reticular complex. *Exp. Neurol., 32*, 79–97.

Shute, C.C.D. and Lewis, P.R. (1967) The ascending cholinergic reticular system: neocortical. olfactory, and subcortical projections. *Brain, 90*, 497–520.

Singer, W. (1973) The effect of mesencephalic reticular stimulation on intracellular potentials of cat geniculate neurons. *Brain Res., 61*, 35–54.

Snider, R.S. and Niemer, W.T. (1961) A Stereotaxic Atlas of the Cat Brain, University of Chicago Press, Chicago.

Steriade, M. (1970) Ascending control of thalamic and cortical responsiveness. *Int. J. Neurobiol., 12*, 87–144.

Steriade, M. and Wyzinski, P. (1972) Cortically elicited activities in thalamic reticularis neurones. *Brain Res., 42*, 514–520.

Storm-Mathisen, J. (1975) Choline acetyltransferase and acetylcholine in fascia dentata following lesion of the entorhinal afferents. *Brain Res., 80*, 181–197.

Symmes, D. and Anderson, K.V. (1967) Reticular modulation of higher auditory centers in monkey. *Exp. Neurol., 18*, 161–176.

Ten Bruggencate, G. and Engberg, I. (1971) Iontophoretic studies in Deiters' nucleus of the inhibitory actions of GABA and related amino acids and the interaction of strychnine and pictrotoxin. *Brain. Res., 25*, 431–448.

Waszak, M. (1974) Firing pattern of neurones in the rostral and ventral part of nucleus reticularis thalami during EEG-spindles. *Exp. Neurol., 43*, 38–59.

Weight, F.F. and Padjen, A. (1973) Acetylcholine and slow synaptic inhibition in frog sympathetic ganglion cells. *Brain Res., 55*, 225–228.

Yingling, C.F. and Skinner, J.F. (1975) Regulation of unit activity in nucleus reticularis thalami by the mesencephalic reticular formation and the frontal granular cortex. *Electroenceph. clin. Neurophysiol., 39*, 635–642.

Studies on a False Transmitter at the Neuromuscular Junction

W.A. LARGE and H.P. RANG

*Department of Pharmacology, St. George's Hospital Medical School,
London SW17 (United Kingdom)*

INTRODUCTION

Much work has been done on false transmitters synthesized by adrenergic neurones when they are supplied with analogues of the appropriate precursors, but until recently this approach had not been used in the study of cholinergic neurones. In the last few years, however, abundant evidence has been obtained of the ability of cholinergic nerve terminals to take up, acetylate and release as false transmitters, a variety of analogues of choline. The first suggestion of this possibility came from early work by Reitzel and Long (1953) who found that not only choline, but also some analogues, such monoethylcholine, were capable of partly restoring neuromuscular transmission after it had been blocked by hemicholinium. Subsequently Bowman et al. (1962) suggested that triethylcholine blocked neuromuscular transmission by substituting acetylcholine (ACh) in the nerve terminals with an inactive false transmitter, acetyltriethylcholine. Biochemical studies on transmitter release at the neuromuscular junction are difficult because of the small amount of ACh involved and the large preponderance of non-synaptic tissue in muscle preparations. Biochemical evidence for the formation and release of false transmitters at cholinergic synapses has therefore come from work on other systems, such as brain slices, sympathetic ganglion, electric organ and the myenteric plexus. Choline analogues clearly shown to give rise to false transmitters include the sulfonium analogue (Frankenberg et al., 1973), the pyrrolidinium analogue, pyrrolcholine (Collier et al., 1976; Kilbinger, 1977; Zimmermann and Dowdall, 1977) and the N-ethylated derivaties, mono- and triethylcholine (Collier et al., 1976). Although these analogues appear generally to be slightly less good substrates than choline for both the high affinity choline uptake mechanism and for acetylation by choline acetyltransferase (Barker and Mittag, 1975), the amount of transmitter stored and released does not seem to be greatly affected (Collier et al., 1976; Kilbinger, 1977).

The neuromuscular junction is so far the only cholinergic synapse at which an electrophysiological study of the release and postsynaptic action of a false transmitter has been carried out, and this paper summarizes some experimental findings that have already been published (Large and Rang, 1976; 1978; Colquhoun et al., 1977) together with some new observations. The experiments were carried out on rat phrenic nerve-diaphragm preparations bathed in Krebs' solution at $20° - 21°C$. Fibres were impaled at the end-plate region with a single electrode (for voltage recording) or with two electrodes (for measurement of end-plate currents under voltage clamp).

RESULTS AND DISCUSSION

Preliminary observations

Addition of 0.1 mM monoethylcholine (MECh) to the solution bathing a rat phrenic nerve-diaphragm preparation produces no obvious effect so long as the rate of transmitter release is low. If, however, the nerve is stimulated at 3 Hz or more the amplitude of the end-plate potential (EPP) decreases progressively and at the same time its time course becomes shorter, with both the time to peak and the half-decay time decreasing by about 25%. A decrease in the EPP amplitude could result from transmitter depletion without the release of any false transmitter, but a shortening of the EPP time course is not compatible with transmitter depletion alone. Thus inhibitors of ACh synthesis, such as hemicholinium or triethylcholine, which produce either no false transmitters or transmitters that have no postsynaptic actions, produce under the same circumstances a reduction in EPP amplitude with no change in the time course. Furthermore, prolonged stimulation in the presence of MECh causes no further change in EPP amplitude or timecourse (presumably because ACh has been fully replaced by the false transmitter).

The postsynaptic action of acetylmonoethylcholine

To find out whether the shortening of the EPP time course was consistent with the replacement of ACh by acetylmonoethylcholine (AMECh) as the transmitter we made measurements of the time course of the conductance change produced by an individual quantum of transmitter, under voltage clamp conditions. Under normal conditions the postsynaptic current (MEPC) generated by a single quantum rises rapidly to a peak and then decays exponentially, (Fig. 1) with a time constant (τ_{MEPC}) that varies with the membrane potential. In rat muscle at 21°C and −80 mV membrane potential, the mean of τ_{MEPC} was 1.84 ms (see Table I). After nerve stimulation in the presence of MECh τ_{MEPC} decreased, the mean value being 1.06 ms (Table I), which is 58% of the normal value. The greater degree of shortening of the synaptic current compared with the end-plate potential comes about because the potential change is governed by the passive electrical properties of the membrane as well as by the time course of the underlying synaptic current, and is thus a less sensitive index of changes in transmitter action. For this reason voltage clamp recording was used wherever possible in this study.

The shortening of the MEPC shown in Fig. 1 and Table I was the maximum that could be achieved by nerve stimulation or addition of KCl (20–40 mM) to the medium in order to evoke ACh release, in the presence of MECh. To test whether this change corresponded to complete substitution of ACh by AMECh

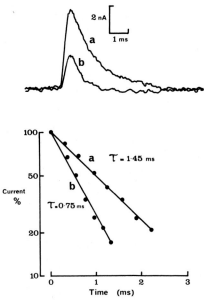

Fig. 1. Normal and false MEPCs in rat muscle. Upper panel: (a) normal MEPC, (b) "false" MEPC recorded after nerve stimulation at 3 Hz for 60 min in the presence of 0.1 mM MECh. Holding potential −80 mV. Lower panel: Semilogarithmic plots of records (a) and (b) showing the change in the time constant of the exponential decay. Current is expressed as a percentage of the peak current. (Reproduced with permission from Colquhoun et al., 1977.)

we compared the reduction in τ_{MEPC} to the channel lifetime determined for ACh and AMECh by noise analysis (Katz and Miledi, 1972; Anderson and Stevens, 1973). Spectral analysis of end-plate current fluctuations produced by adding AMECh or ACh to the bathing solution (Colquhoun et al., 1977) showed that the mean ratio of channel lifetimes, τ_{AMECh}/τ_{ACh}, in 15 cells was 0.56 ± 0.05, which is the same as the ratio $\tau_{false}/\tau_{normal}$ for MEPCs. The calculated channel conductance, γ, from the noise measurements was the same (25–27 pS) for ACh and AMECh. It is known that both τ_{MEPC} and τ_{ACh} increase by the same amount when the cell is hyperpolarized and we found that τ_{MEPC} for false MEPCs and τ_{AMECh} behaved in the same way.

TABLE I

CHARACTERISTICS OF NORMAL AND FALSE MEPCs IN RAT MUSCLE

		Amplitude (nA)		τ_{MEPC} (msec)		$\tau_{false}/\tau_{normal}$
No anticholinesterase	Normal	3.30	0.10 (11)	1.84	0.07 (11)	0.58
	False	2.18	0.13 (n)	1.06	0.04 (11)	
With anticholinesterase	Normal	3.80	0.09 (6)	3.60	0.23 (6)	0.61
	False	2.43	0.32 (5)	2.18	0.13 (5)	

Figures show mean ± S.E.M. The number of muscles tested is shown in brackets; measurements were made on 4–8 fibres in each muscle. "False" MEPCs were recorded after a 30–60 min nerve stimulation at 3Hz in the presence of 0.1 mM MECh.

270

It is known that AMECh is susceptible to hydrolysis by acetylcholinesterase (Collier et al., 1976) and false MEPCs were found to be lengthened about two-fold in the presence of anticholinesterase drugs (Table I). Even in the presence of anticholinesterases, however, when diffusion of transmitter rather than channel lifetime is thought to determine τ_{MEPC} (see Gage, 1976) false MEPCs were still shorter than normal MEPCs, which was at first sight surprising, because the diffusion coefficient for AMECh is presumably very similar to that of ACh. The explanation for this comes from an analysis of the interaction between binding to receptor sites and the rate of diffusion of the transmitter out of the synaptic cleft (Katz and Miledi, 1973). If a large fraction of the released transmitter molecules are bound to receptors, diffusional loss of transmitter from the synaptic cleft will be retarded. Reduction of the binding (e.g. by adding an inhibitor such as tubocurarine) reduces this retardation and correspondingly reduces τ_{MEPC}.

With certain assumptions (see Colquhoun et al., 1977), the fraction b of released transmitter molecules that are bound to receptors is given by

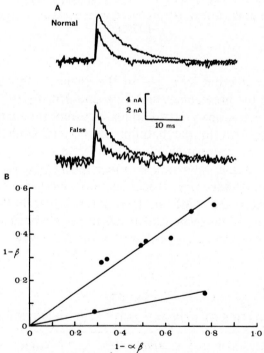

Fig. 2. Effect of (+)-tubocurarine on MEPCs in rat muscle in the presence of 3 μM neostigmine. (A) Records of normal and false MEPCs. The larger record of each pair was recorded before application of (+)-tubocurarine. The smaller record was recorded after coarse pipette containing 20 mM (+)-tubocurarine had been brought close to the endplate for few minutes and then removed. The gain of the recording system was doubled for the false MEPCs. The reduction in amplitude caused by (+)-tubocurarine was accompanied by a greater shortening of normal than of false MEPCs. Holding potential −80 mV. (B) Plot of $(1 − \beta)$ against $(1 − \alpha\beta)$ for normal MEPCs (upper line) and false MEPCs (lower line). The slope of these lines provides a measure of b, the fraction of transmitter molecules bound (see text).
(Reproduced with permission from Colquhoun et al., 1977.)

b = $(1 - \beta)/(1 - \alpha\beta)$ where β is the ratio of τ_{MEPC} in the presence of tubo-curarine to τ_{MEPC} in its absence (anticholinesterase being present thoughout) and α is the corresponding ratio for MEPC amplitudes. Results from fibres showing normal and false MEPCs are shown in Fig. 2. From the plot it can be seen that b is smaller for false MEPCs (0.21 ± 0.04; 7 experiments) than for normal MEPCs (0.52 ± 0.04; 7 experiments). This difference, reflecting a lower affinity of AMECh than ACh for the receptors, exactly accounts for the difference between τ_{false} and τ_{normal} in the presence of anticholinesterase. If we assume that equilibrium is reached between the receptor occupancy and the concentration of free transmitter in the cleft during the end-plate response, the values of b for ACh and AMECh can be used to calculate their relative affinities for the end-plate receptors. This calculation shows that the affinity of AMECh for the receptors is 0.25 times that of ACh, which is similar to the ratio of the potencies of these two agents on the frog rectus abdominis muscle (Holton and Ing, 1949).

The kinetics of transmitter exchange

The fact that ACh in motor nerve terminals can be replaced by AMECh, whose presence in the released transmitter can be sensitively detected by the reduction of τ_{MEPC}, provides a means of following electrophysiologically the turnover of transmitter in cholinergic nerve terminals under various conditions. In resting muscles incubated in 0.1 mM MECh for up to 22 h at 20°C no shortening of τ_{MEPC} occurred, though the usual shortening, to 58% of the control value, occurred when these muscles were subsequently soaked in 40 mM K$^+$ plus MECh for 1 h. Thus no incorporation of AMECh into the transmitter stores occurred at rest. On the other hand, if muscles were preloaded with AMECh by soaking them in 40 mM K$^+$ plus 0.1 mM MECh for 1—2 h and then transferred to normal Krebs' solution containing 0.1 mM choline, the MEPCs gradually lengthened as ACh replaced AMECh; after 4 h the conversion back to ACh was about 60% complete and after 20 h it was complete. If the medium contained 0.1 mM MECh instead of 0.1 mM choline, no recovery occurred (Fig. 3).

Thus is appears that the replacement of AMECh by ACh occurs at about the rate that would be expected from Potter's (1970) measurements of ACh turnover in rat diaphragm. He found that [^{14}C]choline was incorporated into the ACh store at a rate of about 0.5% per minute in resting muscle, corresponding to a half-time for exchange of 140 min. The fact that the reverse process (replacement of ACh by AMECh) does not occur at rest implies that one or more steps in the mechanism of transmitter synthesis must discriminate quite strongly in favour of choline.

The main steps in transmitter synthesis are (1) uptake of choline, (2) acetylation by choline acetyltransferase and (3) packaging of transmitter into vesicles. The study of Barker and Mittag (1975) shows that, in brain synaptosomes, the uptake of MECh has the same K_m and V_{max} as choline, and the fraction of each precursor that is acetylated within the synaptosomes is also the same, so there is no evidence for a large difference in choline and MECh handling by processes (1) and (2). The third possibility is that packaging of AMECh into releasable

272

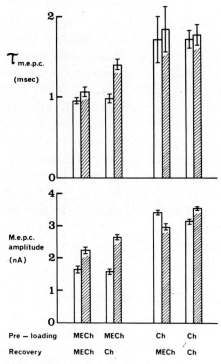

Fig. 3. Effect of incubation of resting muscles in the presence of choline or MECh (0.1 mM) on the time constant (above) and amplitude (below) of MEPCs. In the "preloading" period the muscles were exposed to 40 mM K^+ for 60–90 min in the presence of 0.1 mM MECh or 0.1 mM choline. MEPCs were then recorded (open bars) a few minutes after returning to normal solution containing choline or MECh. The muscles were then kept for 4 h at rest in the "recovery" solution and a further series of MEPCs recorded (hatched bars). Note that muscles loaded with MECh and then exposed to choline show an increase in τ_{MEPC} whereas muscles loaded with choline and then exposed to MECh show no change in τ_{MEPC}.

quanta occurs less readily than with ACh. Some evidence in support of this comes from the finding (Fig. 3) that if a muscle is soaked in 46 mM K^+ plus MECh for 90 min, then transferred to normal Krebs' solution plus MECh, the MEPC amplitude is initially small, but increases by 38% after 4 h, even though, from the constancy of τ_{MEPC}, the transmitter composition is not changing. If the same experiment is done with choline present throughout, the change in MEPC amplitude is much smaller, thus the transmitter content of the quanta appears to be reduced when rapid release is evoked in the presence of MECh, and increases slowly as the muscle is rested, whereas this happens to a much smaller extent with choline. This would be compatible with a relatively low rate of packaging of AMECh into vesicles compared with ACh, but to test this directly requires a technique for separating cytoplasmic from vesicular transmitter, which is not yet possible in muscle.

VARIABILITY OF QUANTA RELEASED DURING INCORPORATION OF FALSE TRANSMITTER

The possibility of distinguishing electrophysiologically the transmitter composition of individual quanta enables an interesting question to be approached.

During the process of exchange of ACh for AMECh do individual quanta contain either pure ACh or pure AMECh, or do they contain a mixture of both transmitters? According to the vesicle hypothesis one might expect at the half-way stage of exchange to see two populations of quanta, giving a bimodal distribution of τ_{MEPC}, corresponding to pre-existing ACh-containing vesicles and newly formed AMECh-containing vesicles. On the other hand, if, as suggested by some (see Marchbanks, 1977, Tauc, 1977) quanta come from a cytoplasmic pool of transmitter, no heterogeneity would be expected.

In these experiments partial exchange of transmitter was usually produced by nerve stimulation at 3 Hz for 15 min in the presence of 0.1 mM MECh, which caused τ_{MEPC} to decline on average to 81 ± 2% of the control value, corresponding to 45% replacement of ACh by AMECh. Stimulation for a further 60 min caused complete replacement. The coefficient of variation of τ_{MEPC}, CV_τ, was inititally 0.09 ± 0.01 (mean and standard error of 54 cells in 12 muscles). At the intermediate stage of exchange CV_τ increased significantly to 0.15 ± 0.01 (32 fibres; 7 muscles) but the distribution of τ_{MEPC} never became bimodal. After complete exchange CV_τ decreased again to 0.13 ± 0.01 (29 fibres; 5 muscles). In the absence of MECh a similar pattern of stimulation had no effect on CV_τ. A typical result with MECh is shown in Fig. 4; it can be seen that partial exchange increased the variability of τ_{MEPC} without causing a bimodal distribution. The average increase in CV_τ was only about 30% of the increase that would have been expected if the quanta had formed two distinct populations with the characteristics of normal and false MEPCs respectively.

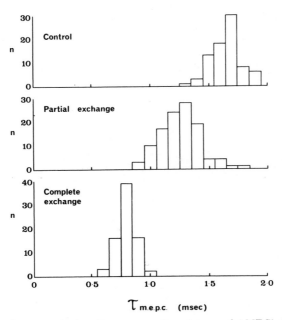

Fig. 4. Variability of τ_{MEPC} during the course of exchange of AMECh for ACh. Histograms of spontaneous MEPCs from three different fibres in the same muscle in the presence of 0.1 mM MECh. Upper: control; middle: after nerve stimulation, 3 Hz for 15 min; lower: after further period of stimulation, 3 Hz for 90 min, to bring about full exchange. At the halfway stage some increase in variability occurs, but no clearly bimodal distribution.

The observed increase in CV_τ at the intermediate stage of exchange would be difficult to account for on the basis of release from a cytoplasmic pool of transmitter, but is clearly smaller than would occur if vesicles formed two distinct populations. One possibility is that a fairly rapid exchange of transmitter occurs between individual vesicles. In two experiments we therefore monitored CV_τ as a function of time after the end of the period of stimulation. Following the initial increase it was found that CV_τ decreased with time, returning to the control value in about $2\frac{1}{2}$ h. This strongly suggests that transmitter exchanges spontaneously between synaptic vesicles causing the initially heterogeneous population to become homogeneous over the course of 1–2 h. Even allowing for this process, the initial increase in CV_τ appears to be smaller than is consistent with the existence of two separate populations of quanta

One hypothesis that would accommodate these results is that a substantial fraction of the transmitter is stored as a cytoplasmic pool, whose composition changes progressively from ACh to AMECh, and from which vesicles are filled as they are formed. This would cause the vesicle population to become somewhat heterogeneous because "old" vesicles would contain less AMECh than "new" vesicles, but would not cause a sharp division into two separate populations. This would be consistent with measurements of transmitter compartmentation in electric tissue (Israël et al., 1970), but more experiments will be needed to test whether it is correct for motor nerve terminals.

SUMMARY

Monoethylcholine (MECh) has been shown to be taken up, acetylated and released as a false transmitter, acetylmonoethylcholine (AMECh) by mammalian motor nerve terminals. AMECh can be distinguished electrophysiologically from ACh by its briefer postsynaptic action, which results in a shortening of the decay time constant (τ) of miniature end-plate currents recorded with a voltage clamp. This shortening corresponds with the mean channel lifetime (determined by noise analysis) for AMECh, which is 44% shorter than that of ACh.

We have investigated whether, at intermediate stages of exchange, two populations of quanta (corresponding to preexisting ACh quanta and newly synthesized AMECh quanta) can be detected. We found no evidence of two distinct populations, but a clear increase in variability of quanta, indicative of some heterogeneity of transmitter composition.

A possible interpretation of these results is that the major part of the presynaptic transmitter store is in the form of a free pool and only a small part is in the immediately releaseable vesicular store.

ACKNOWLEDGEMENT

This work was supported by the Medical Research Council.

REFERENCES

Anderson, C.R. and Stevens, C.F. (1973) Voltage clamp analysis of acetylcholine produced end-plate current fluctuations at frog neuromuscular junction. *J. Physiol. (Lond.),* 235, 655–691.

Barker, L.A. and Mittag, T.W. (1975) Comparative studies of substrates and inhibitors of choline transport and choline acetyltransferase. *J. Pharmac. exp. Ther.,* 192, 86–95.

Bowman, W.C., Hemsworth, B.A. and Rand, M.J. (1962) Triethylcholine compared with other substances affecting neuromuscular transmission. *Brit. J. Pharmacol.,* 19, 198–218.

Collier, B., Barker, L.A. and Mittag, T.W. (1976) The release of acetylated choline analogues by a sympathetic ganglion. *Molec. Pharmacol.,* 12, 340–344.

Colquhoun, D., Large, W.A. and Rang, H.P. (1977) An analysis of the action of a false transmitter at the neuromuscular junction. *J. Physiol. (Lond.),* 266, 361–395.

Frankenberg, L., Heimbürger, G., Nilsson, C. and Sörbo, B., (1973) Biochemical and pharmacological studies of the sulfonium analogues of choline and acetylcholine. *Europ. J. Pharmacol.,* 23, 37–46.

Gage, P.W. (1976) Generation of end-plate potentials. *Physiol. Rev.,* 56, 177–247.

Holton, P. and Ing, H.R. (1949) The specificity of the trimethylammonium group in acetylcholine. *Brit. J. Pharmacol.,* 4, 190–196.

Israël, M., Gautron, J. and Lesbats, B. (1970) Fractionnement de l'organe électrique de la torpille: localization subcellulaire de l'acétylcholine. *J. Neurochem.,* 17, 1441–1450.

Katz, B. and Miledi, R. (1972) The statistical nature of the acetylcholine potential and its molecular components. *J. Physiol. (Lond.),* 224, 665–699.

Katz, B. and Miledi, R. (1973) The binding of acetylcholine to receptors and its removal from the synaptic cleft. *J. Physiol. (Lond.),* 231, 549–574.

Kilbinger, H. (1977) Formation and release of acetylpyrrolidinecholine (*N*-methyl, *N*-acetoxyethylpyrrolidinium) as a false cholinergic transmitter in the myenteric plexus of the guinea-pig small intestine. *Nauyn-Schmiedeberg's Arch. Pharmacol.,* 296, 153–158.

Large, W.A. and Rang, H.P. (1976) A false transmitter at the neuromuscular junction. *J. Physiol. (Lond.),* 258, 105–106P.

Large, W.A. and Rang, H.P. (1978) Incorporation of acetylmonoethylcholine into the transmitter pool at the mammalian neuromuscular junction. *J. Physiol. (Lond.),* 275, 61–62P.

Marchbanks, R.M. (1977) Turnover and release of acetylcholine. In *Synapses.* G.A. Cottrell and P.N.R. Usherwood (Eds.), Blackie, Glasgow-London, pp. 81–101.

Potter, L.T. (1970) Synthesis, storage and release of [^{14}C]acetylcholine in isolated rat diaphragm muscles. *J. Physiol. (Lond.),* 206, 145–166.

Reitzel, N.L. and Long, J.P. (1953) Hemicholinium antagonism by choline analogues. *J. Pharmac. exp. Ther.,* 127, 15–21.

Tauc, L. (1977) Transmitter release at cholinergic synapses. In *Synapses,* G.A. Cottrell and P.N.R. Usherwood (Eds.), Blackie, Glasgow-London, pp. 64–78.

Zimmermann, H. and Dowdall, M.J. (1977) Vesicular storage and release of a false cholinergic transmitter (acetylpyrrolcholine) in the *Torpedo* electric organ. *Neuroscience,* 2, 731–739.

PART IV

POSTSYNAPTIC ACTION OF ACETYLCHOLINE BIOCHEMISTRY

The Nicotinic Acetylcholine Receptor from *Torpedo* Electroplax

R.D. O'BRIEN, R.E. GIBSON * and KATUMI SUMIKAWA

Section of Neurobiology and Behavior, Cornell University, Ithaca, N.Y. 14853, and
* Section of Radiochemistry and Radiopharmacology, George Washington
University Medical Center, Washington, DC 20037 (U.S.A.)

INTRODUCTION

Because of the small space available to review this large subject, we will only discuss selected features of this receptor, which is the only receptor extensively studied in membrane fragments and also in a highly purified form. On the basis of evidence given below, we assume that the physiological receptor has two parts: a recognition component, which recognizes and binds to acetylcholine (ACh) and related drugs, and an ionophore, which is separable from but attached to the recognition component. When ACh binds to the recognition component, it modifies the ionophore, which opens to permit the passage of Na^+ and K^+. The combination of these two parts we call the receptor complex.

TIME DEPENDENT CHANGES

Physiological work implies that there are at least two time-dependent changes which occur when a ligand such as ACh attaches to its receptor. One occurs when a brief pulse of ACh is applied to the receptor in the membrane. What one sees is a correspondingly brief permeability to sodium and potassium ions, with a time course in the millisecond range. Clearly there must be configurational changes in the receptor complex which lead to this transient ion permeability. Secondly, when the receptor is exposed to a lengthy pulse of ACh, it goes into a relatively passive state in which it no longer responds effectively to brief pulses of ACh. This phenomenon of desensitization was elegantly demonstrated by Katz and Thesleff (1957) and is shown in Fig. 1. In muscle its onset has a half-time of about 1 sec. The effect is reversible with a half-life of about 5 sec. Thus we should expect to find time-dependent changes with a comparable reversal time when we look at suitably isolated ACh receptor. However, in electric eel (Lester et al., 1975) and *Torpedo marmorata* (Sugiyama et al., 1976) the desensitization is much slower, taking several minutes for completion.

Katz and Thesleff (1957) showed that their findings were compatible with the desensitization being related to an increase of affinity of the receptor for ACh, the increase being about 20-fold.

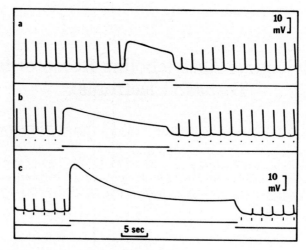

Fig. 1. Desensitization: the normal muscle response to a fixed application of acetylcholine is sharply reduced if the muscle is briefly exposed to a high concentration of acetylcholine. (Drawn from data of Katz and Thesleff, 1957)

Biochemists have therefore looked for at least two time-dependent changes corresponding to excitation and desensitization, and have hoped to find them reflected in agonist binding. There is no reason to rule out the possibility of several other time-dependent changes, which may prepare the molecule to go through the configurational change leading to the two physiologically evident events.

Time-dependent changes in binding were first observed by Weber et al. (1975), who incubated membrane fragments from *Torpedo* electroplax with certain agonists, such as ACh and carbamylcholine. They found that the longer the incubation with agonist, the more effectively did agonist protect against subsequent binding to the snake venom component, α-Naja toxin. This important pioneering study nevertheless has two problems. One is that the use of interference with toxin binding is a rather indirect approach to measuring agonist binding. Secondly, the toxin technique cannot be carried out very quickly. We remedied this situation by developing a relatively fast filtration technique in which we could study the binding of ACh to *Torpedo* electroplax membranes which occurs within seconds of its application. As Fig. 2 shows, one can see that the amount of ACh which binds to the receptor increases sharply in the early seconds, and levels out with a slower increase thereafter. Studies with 1 μM ACh showed that affinity increased at a rate constant of 0.03 sec^{-1}, corresponding to a half-time of 23 sec. The change in affinity for ACh is not precisely calculable, but the computed line for 10^{-6} M acetylcholine which is shown in Fig. 2 uses a factor of 250-fold, that is an increase from 3 μM to 12 nM, both of which values are compatible with direct affinity measurements. Comparing this with the toxin studies, their half-time of desensitization is about 1 min, in rough agreement with us. But Table I shows that the change in affinity for carbamylcholine (for instance) is only 3.5-fold, a very small increase.

Other techniques have also been used to study time-dependent changes, including the effects of agonists on the fluorescent agent quinacrine (Grun-

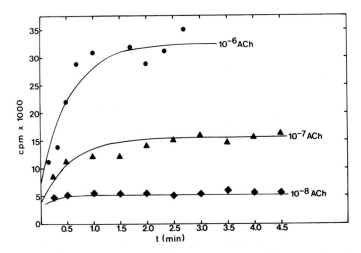

Fig. 2. Time-dependent changes in binding of various concentrations of [³H]acetylcholine to membranes from *Torpedo californica* electroplax. The method involves fast filtration on glass fiber paper. (From O'Brien, Test and Gibson, unpublished).

hagen and Changeux, 1976) and fairly similar half-times were found. Bonner et al. (1976) have studied changes in the native fluorescence of *Torpedo* membrane fragments, and also describe agonist-induced changes for example with a half-life of 9.7 sec with 4 μM ACh. Probably all these techniques correspond to the desensitization phenomenon. One conclusion we can draw is that a fairly close mimic of the physiological event can be seen biochemically. Another conclusion is that the extensive binding studies done by equilibrium dialysis, which is a slow technique, and designed to measure the binding constants of various agonists and antagonists, all describe properties of the desensitized receptor.

We have only described results relating to the relatively slow changes involved in desensitization. In order to unravel the fast changes involved in normal excitation, special fast methods are required. Stopped-flow kinetic studies have been described using the fluorescent agent quinacrine, which is believed to bind to the ionophore (Grunhagen et al., 1977). The constants describing the binding of ligand to receptor, followed by possible isomerization step, have been measured, but the constants do not bear an obvious relationship to physiological events.

TABLE I

ANALYSIS OF DATA OF WEBER, DAVID-PFEUTY AND CHANGEUX (1975)

Found: Velocity of toxin reaction decreases 3.43× between t^0 and t^∞ for carbamylcholine at $c = 0.5$ μM.

Assume: (Popot, Sugiyama and Changeux, 1976) K_d for carbamylcholine at $t^0 = 46$ μM. Fractional occupancy by carbamylcholine = $1/(1 + K_d/c)$.

If decrease in toxin reactivity is due entirely to increase in carbamylcholine occupancy caused by increase in K_d for carbamylcholine, ∴ K_d for carbamylcholine at $t^\infty = 13.1$ μM. Increase in affinity = $46/13.1 = 3.5×$.

THE RECOGNITION COMPONENT

We now discuss the detailed composition of what we call the recognition component, i.e. the component which binds α-bungarotoxin, ACh and related ligands, and is probably distinguishable from the ionophore.

Studies on sucrose-gradient separation of either crude solubilized or highly-purified *Torpedo californica* receptor show (Fig. 3) that one can obtain two proteins (Gibson et al., 1976; Chang and Bock, 1977). There is excellent evidence that the heavy component (H) is a dimer of the light component (L). For instance, Chang and Bock (1977) showed that treatment with the reducing agent DTT converted almost all of the H to the L form. In addition, we have found (Fig. 4) that the peptide fingerprints of tryptic digests of separated H and L forms (after performic acid oxidation) are identical (Sumikawa and O'Brien, 1978). Our improved calculations of the molecular weight of the dimer, H, is 414,000 daltons. We think it probably that the dimer is the form that is present in the membrane-bound receptor, because much of the protein solubilized from the membrane-bound receptor is alreay in the H form, and it seems extraordinarily unlikely that two precisely matching halves, with complementary SH groups available for bonding, exist in the isolation from one another.

The purified recognition component from *Torpedo californica* can be separated into its subunits by electrophoresis in sodium dodecyl sulfate (SDS). We find four principal subunits (Fig. 5). Other laboratories have suggested that some of these subunits may be artefacts caused by proteolytic degradation during purification. However, we have now harvested all four subunits and

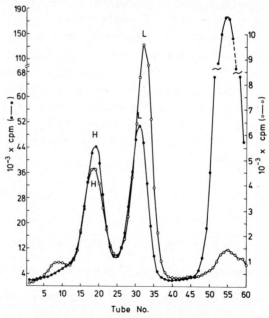

Fig. 3. Demonstration of heavy (H) and light (L) oligomers in *Torpedo californica* receptor solubilized by Triton X-100, labeled with α-[^{3}H]bungarotoxin and centrifuged in a sucrose gradient. ○——○, purified receptor; ●——●, total electroplax membrane extract. The peak on the right is unreacted free toxin.

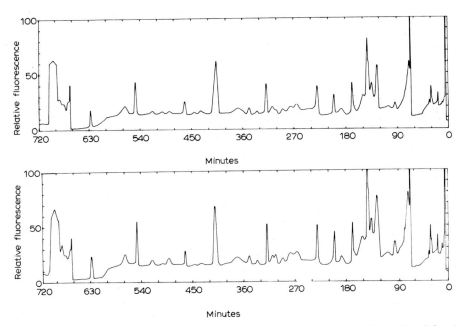

Fig. 4. Peptide fingerprints of H and L oligomers from purified receptor from *T. californica*. (From Sumikawa and O'Brien, 1978).

studied the peptide fingerprint of tryptic digests of each after performic acid oxidation, and found (Fig. 6) that each one is unique. This rules out the possibility that any one is derived from degradation of the other.

There is general agreement that the smallest subunit α contains the recognition site for ACh. The evidence is that the reagent MBTA, which probably binds like ACh but does so covalently so that the binding survives in SDS gels, can be shown to bind exclusively to the smallest subunit (Karlin et al., 1976). There is also agreement that the largest, or δ, subunit is involved in the S–S bridge which binds together the two halves of the dimer to make up the total recognition component (Hamilton et al., 1977; Hucho et al., 1978).

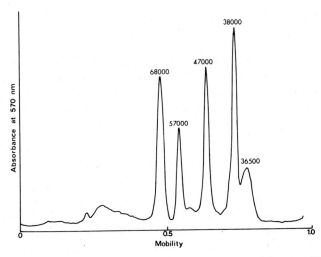

Fig. 5. Electrophoresis in SDS gels of purified receptor from *T. californica*. (From Sumikawa and O'Brien, 1978).

284

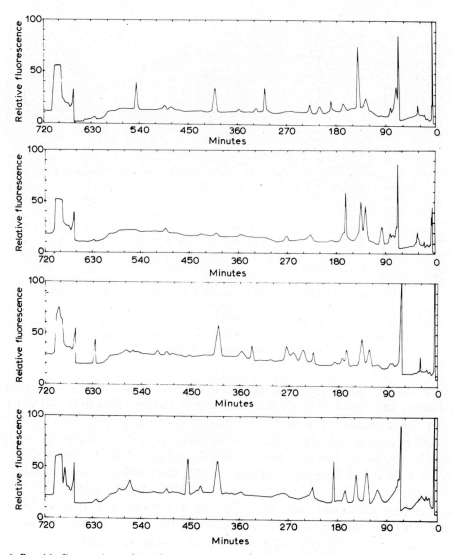

Fig. 6. Peptide fingerprints of the four subunits (α, β, γ and δ) separated from purified receptor from *T. californica*. (From Sumikawa and O'Brien, 1978).

The weights of the four subunits α, β, γ and δ shown in Fig. 5 are (in thousands) 38, 47, 57 and 68. A theoretical protein with one of each subunit would have an molecular weight of 210,000 and thus the light monomer (L), whose molecular weight we estimate roughtly at 258,000 (Sumikawa and O'Brien, 1978) probably has only one of the subunits represented in duplicate; there is not room for more. Data from several laboratories (e.g. Lindstrom et al., 1978) suggest that the α-subunit is the most plentiful. If the duplicate subunit was the α-subunit, so that the total formula was α_2, β, γ, δ, the L monomer would have an molecular weight of 248,000, in good agreement with our calculation of 258,000. Each L monomer would than have two ACh binding sites, and so the complete dimer, H, which we postulate to be in the physiological receptor complex, would have four ACh binding sites. This is in good agreement with the early report of one ACh binding site per 100,000 mol. wt. (Eldefrawi and Eldefrawi, 1973).

THE IONOPHORE

The structure described so far is that which is revealed when the purified proteins with recognition sites are studied. But to the physiologist, the binding of ACh is only the first of a number of important steps, and the ACh receptor is the whole complex of molecules which has the ability to couple the binding of ACh into some form of consequent change in ionic permeability. Once the recognition component had been isolated, the question arose as to whether this protein was the recognition component alone, or included the ionophore, or conceivably both along with other entities.

Toxins and local anesthetics have now been explored which block the ACh response in physiological experiments, buth without blocking ACh binding measured directly in biochemical experiments. Such toxins might bind to the ionophore itself. One example is a poison derived from that found in the skins of certain frogs. It is called perhydrohistrionicotoxin. Its binding to *Torpedo* membranes is not influenced by nicotinic drugs such as curare or nicotine (Eldefrawi et al., 1977). Another example is ceruleotoxin (Bon and Changeux, 1977) which has very little action on binding of ACh to microsacs under conditions in which it blocks Na^+ efflux from the microsacs. The macromolecules from *Torpedo* electroplax which bind perhydrohistrionicotoxin and ACh, and perhaps correspond to the ionophore and recognition complex respectively, have been partially separated, and also it has been shown that the binding of perhydrohistrionicotoxin to electroplax membranes is unaffected by agents for the ACh recognition site, such as α-bungarotoxin and d-tubocurarine (Eldefrawi et al., 1977). Because the separation of histrionicotoxin binding and α-bungarotoxin binding components was achieved easily, i.e. by Sephadex chromatography, it is likely that the isolated recognition complex described above does not contain the ionophore. Unfortunately there is evidence of extensive non-specific binding of perhydrohistrionicotoxin, which can bind even to detergents.

This year, Sobel et al. (1978) found that after dissolving *Torpedo* membrane fragments in nonionic detergents, they could centrifuge in a sucrose gradient and precipitate an aggregated protein which, on electrophoresis in SDS, proved to contain a single 43,000 mol. wt. peptide. The protein was therefore called the "43,000-protein". Because it bound quinacrine, local anesthetics and histrionicotoxin, it may well be the ionophore. It was particularly interesting to note that in the whole receptor complex carbamylcholine, which presumably binds to the recognition component, profoundly influences the fluorescence of quinacrine, which probably binds to the ionophore. Buth when the "43,000-protein" was separated from the recognition complex, the fluorescence seen when quinacrine bound to the "43,000-protein" was insensitive to carbamylcholine.

SPECIES DIFFERENCES

Important differences exist between receptors from different organisms, even when we consider only receptor in electroplax. Thus the purified recogni-

286

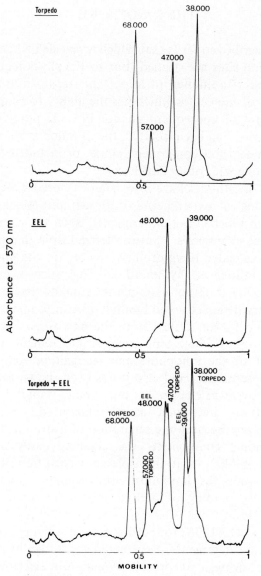

Fig. 7. Mixture of two subunits from electric eel and four subunits from *Torpedo california*, separated on gels containing SDS (From Sumikawa and O'Brien, unpublished).

tion component of electric eel has only two subunits (39,000 and 48,000 mol. wt.) instead of the four in *Torpedo*. [But Karlin et al. (1976) report three subunits for eel.] Although these two subunits have molecular weights very close to two of the subunits present in *Torpedo californica*, we found on mixing purified eel with purified *T. californica* recognition components that all 6 subunits migrated differently (Fig. 7), so the two eel subunits are not related to any of the *Torpedo* subunits. In addition, eel only has the L oligomer, probably because it lacks the heavy subunit with its S–S bonding possibility.

Even more surprising are differences found between different species of *Torpedo*. Thus whole membrane fragments of *T. californica* electroplax showed, on SDS electrophoresis, a 60,000 peptide lacking in *T. marmorata*

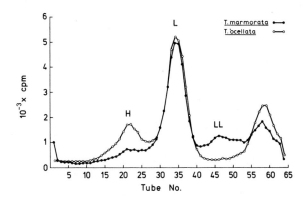

Fig. 8. Differences in ratios of H and L oligometer in unpurified *Torpedo ocellata* and *marmorata* receptor, studied as in Fig. 3. (From Gibson et al., 1976).

(Flanagan et al., 1976). And the 43,000 peptide derived from the 43,000-protein described above for *T. marmorata* and believed to be associated with the ionophore, is not found in whole membrane fragments from *T. californica* (Flanagan et al., 1976). Furthermore, *T. mamorata* extracts display very little of the H oligomer; in *T. ocellata* the H oligomer is clearly seen but is far less abundant than the L form; whereas in *T. marmorata* the H and L are present in comparable amounts (Figs. 8 and 3) (Gibson et al., 1976). Electric eel has only one oligomer, probably corresponding to the L form (Meunier et al., 1972).

RECONSTITUTION

Several investigators have been successful in mimicking some important properties of the receptor in the so-called microsacs, which are vesicles formed directly from electroplax membranes. The microsacs are loaded with $^{22}Na^+$, whose efflux is measured (Kasai and Changeux, 1971). Unfortunately the microsacs have a good deal of spontaneous leakiness to Na^+, but nevertheless can be stimulated by agonists to leak the Na^+ much faster, and this promoted rate responds to appropriate agonists and antagonists at concentrations similar to those effective physiologically. Unfortunately the ion channel conductances calculated from these findings were 10^5 times smaller than those found from physiological studies (Katz and Miledi, 1972).

Attempts have been made to reconstitute purified receptor by insertion into synthetic vesicles but until recently no reliable and repeatable results have been obtained. Recent work (Briley and Changeux, 1978) has suggested that in solubilization of the membranes prior to attempted reconstitution, the ionophore is lost, although some key properties of the recognition component can be successfully retained, especially the time-dependent change in binding affinity indicating an ability to be densensitized. Schiebler and Hucho (1978) have been able to achieve fairly good reconstitution using cholate solubilized receptor, which is then roughly purified by sucrose gradient centrifugation and contains little protein but the α, β, γ and δ subunits already discussed. This preparation was incorporated into synthetic vesicles formulated from lipids

288

whose composition was similar to that found in electroplax membranes. The resultant vesicles behaved substantially like the original microsacs. If these experiments are confirmed, they may imply that the ionophore is a part of the α-β-γ-δ system, and is not a distinct and different component.

THE WHOLE RECEPTOR COMPLEX

Recently Ross et al. (1977) have used x-ray diffraction and image-enhanced electron microscopy with *Torpedo californica*. They show (Figs. 9, 10) that the receptor complex looks like a horseshoe volcano sticking out some 55 Å into the synaptic cleft, with a crater of about 8Å in diameter, and with 15 Å jutting out into the cytoplasmic side. Our rough reconstruction of their data (Fig. 11) leads us to assume an approximate molecular weight of 641,000 daltons for the whole complex. If this estimate is true, and if the ionophore and recognition complex are different entities, then there is only room for one H protein of 414,000 daltons (the dimer of the recognition complex) in the receptor complex, and the remainder, of 227,000 daltons, must be made up of the ionophore. If indeed the ionophore is made up exclusively of units of 43,000 daltons, about 5 such units could be present per receptor complex. By contrast,

Fig. 9. Electron micrographs, enhanced by phase averaging, of receptor in *Torpedo californica*. (From Ross et al., 1977).

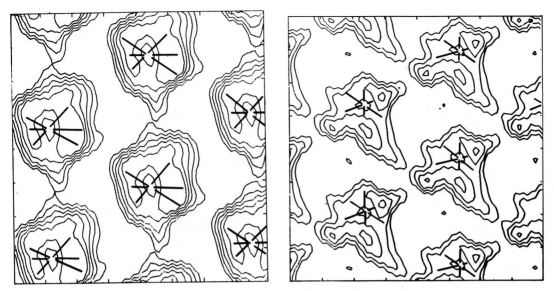

Fig. 10. Contour plots of phase-averaged images of receptor in *Torpedo californica*. Top: synaptic cleft side. Bottom: cytoplasmic side. (From Ross et al., 1977).

if Schiebler and Hucho (1978) are correct, and the ionphore is not a separate entity, but is a part of the α_2-β-γ-δ complex, then one H protein plus one L protein would comfortably account for the whole system as observed by Ross et al. (1977).

In conclusion, we may expect in the next few years to have the complete molecular description of the nicotinic receptor complex from electroplax. The major new step must be the purification and characterization of the ionphore. We shall probably also find time-dependent changes in its character which

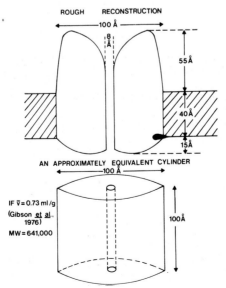

Fig. 11. Reconstruction from data of Ross et al. (1977) of approximate structure of acetyl-choline receptor.

parallel the physiological events. It is likely that we shall be able to insert purified recognition components plus ionophore into a synthetic membrane, and bestow on the membrane the essential porperties (including appropriate time-dependent changes) of the native receptor. But a solution of the most difficult task may by many years in the future: to account for the molecular mechanism by which binding of ACh opens the ionophore.

SUMMARY

Time-dependent changes in acetylcholine (ACh) binding to electroplax membranes have been observed which correspond to the physiological event of desensitization. The recognition component of the cholinergic receptor is probably made up of a protein with the subunit composition α_2-β-γ-δ, with α as the ACh binding site; these subunits constitute the light protein, L. Reconstruction of the receptor complex from electron micrographs suggests that its molecular weight is about 640,000. One possibility is that the complex is made up of one molecule of protein H, which is a dimer of L, and of 5 subunits of a molecular weight of 43,000, each representing a separate ionophore. Another possibility is that one molecule of protein H is combined with one molecule of protein L.

ACKNOWLEDGEMENTS

The portions of this work dealing with reseach from this laboratory were funded in part by National Institutes of Heath grant No. NS-09144. We are indebted to Stephen Jones for his advice in preparing the manuscript.

REFERENCES

Bon, C. and Changeux, J.-P. (1977) Ceruleotoxin: a possible marker of the cholinergic ionophore. Europ. J. Biochem., 74, 43—51.

Bonner, R., Barrantes, F.J. and Jovin, T.M. (1976) Kinetics of agonist-induced intrinsic fluorescence changes in membrane-bound acetylcholine receptor. Nature (Lond.), 263, 429—431.

Briley, M.S. and Changeux, J.-P. (1978) Recovery of some functional properties of the detergent-extracted cholinergic receptor protein from Torpedo marmorata after reintegration into a membrane environment. Europ. J. Biochem., 84, 429—439.

Chang, H.W. and Bock, E. (1977) Molecular forms of acetylcholine receptor. Effects of calcium ions and sulfhydryl reagent on the occurrence of oligomers. Biochem., 16, 4513—4520.

Eldefrawi, M.E. and Eldefrawi, A.T. (1973) Purification and molecular properties of the acetylcholine receptor from Torpedo electroplax. Arch. Biochem., 159, 362—373.

Eldefrawi, A.T., Eldefrawi, M.E., Albuquerque, E.X., Oliveira, A.C., Mansour, N., Adler, M., Daly, J.W., Brown, G.B., Burgermeister, W. and Witkop, B. (1977) Perhydrohistrionicotoxin: A potential ligand for the ion conductance modulator of the acetylcholine receptor. Proc. nat. Acad. Sci. (Wash.), 74, 2172—2176.

Flanagan, S.D., Barondes, S.H. and Taylor, P. (1976) Affinity partitioning of membranes. Cholinergic receptor-containing membranes from Torpedo californica. J. Biol. Chem., 251, 858—865.

Gibson, R.E., O'Brien, R.D., Edelstein, S.J. and Thompson, W.R. (1976) Acetylcholine receptor oligomers from electroplax of *Torpedo* species. *Biochem.*, 15, 2377—2383.

Grunhagen, H.-H. and Changeux, J.-P. (1976) Studies on the electrogenic action of acetylcholine with *Torpedo marmorata* electric organ, V. *J. Mol. Biol.*, 106, 517—535.

Grunhagen, H.-H. Iwatsubo, M. and Changeux, J.-P. (1977) Fast kinetic studies on the interaction of cholinergic agonists with the membrane-bound acetylcholine receptor from *Torpedo marmorata* as revealed by quinacrine fluorescence. *Europ. J. Biochem.*, 80, 255—242.

Hamilton, S.L., McLaughlin, M. and Karlin, A. (1977) Disulfide bond cross-linked dimer in acetylcholine receptor from *Torpedo californica*. *Biochem. Biophys. Res. Commun.*, 79, 692—699.

Hucho, F., Bandini, G. and Suarez-Isla, B.A. (1978) The acetylcholine receptor as part of a protein complex in receptor-enriched membrane fragments from *Torpedo californica* electric tissue. *Europ. J. Biochem.*, 83, 335—340.

Karlin, A., Weill, C.L., McNamee, M.G. and Valderrama, R. (1976) Facets of the structures of acetylcholine receptors from *Electrophorus* and *Torpedo*. *Cold Spring Harbor Symp. quant. Biol.*, 40, 203—210.

Kasai, M. and Changeux, J.-P. (1971) In vitro excitation of purified membrane fragments by cholinergic agonists. II. The permeability change caused by cholinergic agonists. *J. Membrane Biol.*, 6, 24—57.

Katz, B. and Miledi, R. (1972) The statistical nature of the acetylcholine potential and its molecular components. *J. Physiol. (Lond.)*, 224, 665—699.

Katz, B. and Thesleff, S. (1957) A study of the 'desensitization' produced by acetylcholine at the motor end-plate. *J. Physiol. (Lond.)*, 138, 63—80.

Lester, H.A., Changeux, J.-P. and Sheridan, R.E. (1975) Conductance increases produced by bath application of cholinergic agonists to *Electrophorus* electroplaques. *J. Gen Physiol.*, 65, 797—816.

Lindstrom, J., Einarson, B. and Merlie, J. (1978) Immunization of rats with polypeptide chains from *Torpedo* acetylcholine receptor causes an autoimmune response to receptors in rat muscle. *Proc. nat. Acad. Sci. (Wash.)*, 75, 769—773.

Meunier, J.C., Olsen, R.W. and Changeux, J.-P. (1972) Studies on the cholinergic receptor protein from *Electrophorus electricus*. Effect of detergents on some hydrodynamic properties of the receptor protein in solution. *FEBS Lett.*, 24, 63—68.

Ross, M.J., Klymkowsky, M.W., Agand, D.A. and Stroud, R.M. (1977) Structural studies of a membrane-bound acetylcholine receptor from *Torpedo californica*. *J. Mol. Biol.*, 116, 635—659.

Schiebler, W. and Hucho, F. (1978) Membranes rich in acetylcholine receptor: characterization and reconstitution to excitable membranes from exogenous lipids. *Europ. J. Biochem.*, 85, 55—63.

Sobel, A., Heidmann, T., Hofler, J. and Changeux, J.-P. (1978) Distinct protein components from *Torpedo marmorata* membranes carry the acetylcholine receptor site and the binding site for local anesthetics and histrionicotoxin. *Proc. nat. Acad. Sci. (Wash.)*, 75, 510—514.

Sugiyama, H., Popot, J.-L. and Changeux, J.-P. (1976) Studies on the electrogenic action of acetylcholine with *Torpedo marmorata* electric organ. *J. Mol. Biol.*, 106, 485—496.

Sumikawa, K. and O'Brien, R.D. (1978) Composition of the subunits of acetylcholine receptor from *Torpedo californica*. *Biochem.*, submitted.

Weber, M., David-Pfeuty, T. and Changeux, J.-P. (1975) Regulation of binding properties of the nicotinic receptor protein by cholinergic ligands in membrane fragments from *Torpedo marmorata*. *Proc. nat. Acad. Sci. (Wash.)*, 72, 3443—3447.

Acetylcholine Receptor of *Limnaea Stagnalis* Neurones. Study of the Active Site Functional Groups

CATHERINE A. VULFIUS, OLGA P. YURCHENKO, V.I. IL' IN, P.D. BREGESTOVSKI and B.N. VEPRINTSEV

Laboratory of Nerve Cell Biophysics, Institute of Biological Physics, USSR Academy of Sciences, 142292, Pushchino (USSR)

INTRODUCTION

It is now well established that acetylcholine receptors (AChRs) are specific membrane proteins. This fact permits one to apply the method of chemical modification of protein (Baker, 1967; Singer, 1967; Shaw, 1970) to studies of AChR structure and function (Chavin, 1971; Hall, 1972; Vulfius, 1975). The method consists of treatment of tissue or protein with a group-specific reagent possessing affinity for the active site. The reagent interacts with certain protein side-chains under mild conditions. The reaction leads to formation of a covalent bond. Therefore the reagent labels the protein in its active site or near it. According to the changes in protein properties a conclusion may be drawn as to the role of the modified amino acid residues in the function of the protein. The possibility of protecting the protein against modification with some substrate or inhibitor gives evidence for specificity of the reaction with the protein studied and may indicate the localization of modified groups.

We have applied this method to identify essential groups in the AChR of *Limnaea stagnalis* neurones and to reveal conformational transition at different functional states of the receptor.

MATERIALS AND METHODS

Experiments were carried out on identified neurones from right and left parietal ganglia of *Limnaea stagnalis* isolated by the method of Kostenko et al. (1972, 1974). A neurone was placed into the chamber with a continuous flow of Ringer solution. Two microelectrodes filled with 2.5 M KCl were inserted into the neurone. All the reagents and drugs were dissolved in Ringer solution and added to the perfusion fluid. The effect of any reagent was estimated from the changes in the neuronal response to the same acetylcholine (ACh) concentration before and after treatment with the reagent. ACh action was measured as an increase in membrane conductance under voltage-clamp conditions.

Modifying reagents tested were: 1-cyclohexyl-3[N-(methylmorpholino)-ethyl] carbodiimide p-toluene sulphonate (CMEC), methyl glycine ester hydrochloride (MGE), dithiothreitol (DTT), dithiobischoline perchlorate, bromo-acetylcholine perchlorate, N-ethylmaleimide (NEM), some N-alkyl derivatives of NEM, several β-halogenoethylamines with the general formula

$$R_3\overset{+}{N}(CH_2)_n\underset{\underset{Bz}{|}}{N}CH_2CH_2Cl \cdot HBr \text{ , where } R = -CH_3 \text{ , } -C_2H_5 \text{ ; } n = 3,5$$
$$Br^-$$

and di-2-[N-(benzyl-2-bromoethyl)]aminoethyl ester of suberic acid hydrobromide. A stock solution of β-halogenoethylamine was left for 30 min at room temperature to form reactive ethylenimmonium ion. Then it was diluted with Ringer solution to the desired concentration and applied to the neurone. Reversible cholinergic agonists and antagonists were also used.

RESULTS AND DISCUSSION

None of the reagents tested affected the steady-state current-voltage relation of the neuronal excitable membrane or the reversal potential of responses to ACh.

The action of carbodiimide: identification of carboxyl group

Water-soluble carbodiimides are known to interact with various protein groups (carboxyl, imidazole, thiol, amino groups) in neutral and slightly alkaline medium; the ease of reaction with carboxyl groups far exceeds that with other protein groups at acidic pH (Riehm and Scheraga, 1966; Carraway and Koshland, 1972). The reaction product with carboxyl groups is usually unstable and hydrolyzes quickly with regeneration of the original groups. Addition of a nucleophilic agent such as MGE to the reaction mixture stabilizes irreversible modification of carboxyl groups but does not affect the reaction of carbodiimide with other protein groups (Hoare and Koshland, 1967).

CMEC alone decreased neuronal response to ACh reversibly, the sensitivity to ACh being restored some minutes after removal of CMEC from the bathing solution (Fig. 1). When the neurone was treated with the mixture of CMEC and MGE the response to a standard ACh concentration recovered only partially and was maintained at a low constant level for some hours, in spite of extensive washing with fresh Ringer solution (Fig. 1). The effect of this mixture was stronger at pH 6.0 than in neutral medium. MGE without CMEC was ineffective (Il'in et al., 1976).

The reversible block of ACh action induced by CMEC and its conversion into an irreversible one when the *Limnaea* neurone is treated with the mixture of CMEC and MGE, as well as the pH-dependence of this inactivation, suggest a reaction of CMEC with a carboxyl group essential for ACh action.

In some experiments the neurones were treated with CMEC and MGE in the

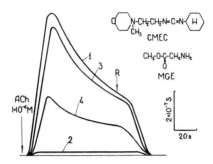

Fig. 1. The decrease in neuronal sensitivity to ACh due to treatment with CMEC. (1-4) Responses of a neurone to $1 \cdot 10^{-6}$ M ACh: (1), before treatment; (2), in the presence of $7 \cdot 10^{-3}$ M CMEC; 3–10 min after removal of CMEC from the bathing solution; (4), after 6-min treatment with the mixture of $7 \cdot 10^{-3}$ M CMEC and $4 \cdot 10^{-3}$ M MGE and washing the reagents out. The times of ACh and Ringer solution (R) addition are marked by arrows, in this and subsequent figures.

presence of a cholinergic ligand. The antagonist (+)-tubocurarine and TEA) and agonists, including TMA, prevented the modification. The data suggest the carboxyl group to be located at or near the anionic subsite of the AChR active site.

Our results are consistent with the suggestion of Edwards et al. (1970) that a carboxyl group plays an important role in the function of AChR at the frog neuromuscular junction.

The action of dithiothreitol:identification of a disulphide bond and evidence for conformational changes of AChR

A short treatment with dithiothreitol (DTT), a specific disulphide reagent (Cleland, 1964), markedly depressed neuronal responses to ACh (Fig. 2). The sensitivity to ACh was restored completely after some minutes exposure of the neurone to dithiobischoline, which is known to be an SH-oxidizing agent (Bartels et al., 1970). If any SH-specific alkylating agent (N-ethylmaleimide (NEM) or some of its derivaties) or bromoacetylcholine was applied after DTT, the inhibition of the cholinoreceptive membrane could not be reversed by dithiobischoline (Bregestovski et al., 1977).

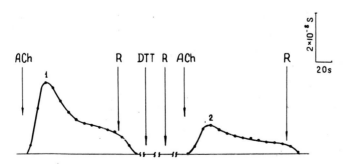

Fig. 2. The decrease in neuronal sensitivity to ACh due to DTT treatment. (1 and 2) Responses of a neurone to $2 \cdot 10^{-6}$ M ACh before and after 2 min exposure of the neurone to $1 \cdot 10^{-3}$ M DTT.

296

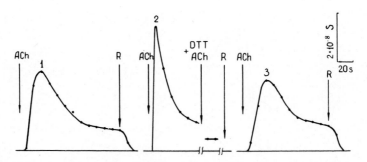

Fig. 3. Protection of AChR against DTT by a high ACh concentration. (1 and 3) Responses to $2 \cdot 10^{-6}$ M ACh before and after exposure of the neurone to $1 \cdot 10^{-3}$ M DTT for 2 min in the presence of $5 \cdot 10^{-5}$ M ACh; (2), response to conditioning ACh concentration; the duration of pretreatment with ACh was about 2 min.

These results indicate a disulphide bond to be essential in ACh-induced conductance changes of the neuronal membrane and concur with the data of Karlin and other authors who have shown the existence of an important disulphide bond in nicotinic receptors of electroplaques (Karlin and Bartels, 1966; Karlin, 1969; Weill et al., 1974), vertebrate skeletal muscles (Rang and Ritter, 1971; Ben-Haim et al., 1973; Lindstrom et al., 1973) and Aplysia neurones (Sato et al., 1976).

We tested cholinergic agonists and antagonists for the ability to protect the neurones against DTT, so as to obtain information on the location of the disulphide bond (Bregestovski et al., 1977). ACh and other agonists, e.g. tetramethylammonium (TMA) and decamethonium, at concentrations inducing maximal effects, prevented DTT action (Fig. 3). The results give evidence that the disulphide bond is located in the AChR molecule. In contrast none of the antagonists tested, e.g. (+)-tubocurararine, phenyltrimethylammonium (PhTMA) * and diphenyldecamethonium, when used at concentrations which strongly inhibited responses to ACh, were effective in protecting the receptors against reduction of disulphide bonds by DTT.

One explanation of the discrepancy between the high protective capacity of the agonists and the inefficacy of antagonists may be that these agents interact with different sites in AChR so that agonists produce a steric hindrance to the reaction of DTT with the disulphide bond while antagonists do not. This suggestion seems unlikely because of the protective action of TMA and decamethonium, taken together with the absolute inability of antagonists with very similar structure, i.e. PhTMA and diphenyldecamethonium, to prevent DTT action. It is difficult to imagine that such drugs bind to different sites of the receptor surface.

An alternative explanation for this discrepancy is that binding of agonist molecule to the AChR active site might evoke a conformational transition of the receptor molecule which would mask the disulphide bond; binding of antagonisis would not induce such a conformational change. The experiments

---

* In contrast to its effect on nicotinic AChRs of electroplaques and vertebrate skeletal muscles, PhTMA does not excite *Limnaea* neurone AChRs at concentrations up to $1 \cdot 10^{-3}$ M, but blocks them effectively.

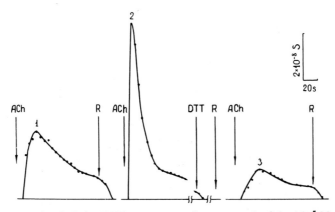

Fig. 4. Inactivation of AChR by DTT treatment after removal of $1 \cdot 10^{-5}$ M ACh from the bathing solution. (1 and 3) Responses to $2 \cdot 10^{-6}$ M ACh before and after treatment; (2), response to the conditioning ACh ($1 \cdot 10^{-5}$ M) dose. When the DTT was added no ACh was in the perfusion fluid, this being 2 min after the prior ACh addition. Control experiments showed that responses to a standard ACh concentration were very much reduced 1.5–2 min after the washout of $1 \cdot 10^{-5}$ M ACh, due to desensitisation.

revealed a direct correlation between the excitatory activity of the agonists and their ability to prevent the modifying action of DTT. Therefore agonists seem to protect AChR by inducing a transition to another conformation in which the disulphide bond becomes inaccessible to DTT.

It is seen from Fig. 3 that in the experiments in which agonists were used as protectors, DTT was added to the perfusion fluid when almost all the AChRs were desensitised. Possibly the conformational state in which the disulphide bond is protected against reduction corresponds to the desensitised state. To check this possibility we tested DTT in the absence of ACh, after 2-min exposure of the neurones to a high ACh concentration causing desensitisation of the AChR population (Fig. 4). Under these conditions, DTT inactivated the AChR as strongly as in the complete absence of ACh. Thus, desensitised AChR without an agonist can be easily reduced by DTT. Evidently, the AChR-agonist complex must exist for successful protection (Bregestovski et al., 1977).

The action of β-halogenoethylamines: indication of AChR oligomeric structure

All the β-halogenoethylamines caused irreversible decreases in responses to ACh (Andrianova et al., 1977). Antagonists such as (+)-tubocurarine and pentamethonium prevented this effect. Probably it is just the AChR molecule that these reagents modify. Agonists ($5 \cdot 10^{-5}$ M ACh, $5 \cdot 10^{-4}$ M TMA) were unable to protect the receptors. This lack of effect may be due to the high reactivity of β-halogenoethylamines and to the low stability of the AChR-agonist complex, so that agonists, at the concentrations used, cannot compete with the covalent reagents.

β-halogenoethylamines react with various protein groups, which is due to their high reactivity (Streitwieser, 1956). We did not manage so far to identify the group modified by these reagents. The possibility of protecting the AChR with antagonists, and the structural similarity of the ethylenimmonium ion (the reactive form of β-halogenoethylamines) to the trimethylammonium

298

grouping of ACh, suggest that the group is situated in the anionic subsite of the AChR site or in some peripheral anionic site essential for AChR function.

Detailed ACh-concentration-action curves were determined with β-halogenoethylamines (Fig. 5). The extent of the decrease in neuronal sensitivity to ACh after β-halogenoethylamine treatment was dependent on the reagent concentration and on the time of exposure. The most prominent features were a lowering of the maximal response to ACh and a flattening of the curve; a parallel shift of the curve to the right was never observed (Andrianova et al., 1977). The pattern of the changes in the curve indicates that there are no spare AChRs in the *Limnaea* neuronal membrane.

The neuronal conductance dependence on ACh concentration, as expressed in a Hill plot, is a straight line with a slope of 1.7 (mean of 33 experiments). After treatment with 5-[N-benzyl-2-chloroethyl)]aminopentyltriethylammonium (see general formula in Methods, R = $-C_2H_5$, n = 5), which contains one reactive group per molecule, the slope of the Hill plot was reduced to 1.0 (mean of 20 experiments). In three experiments with a bifunctional derivative of suberyldicholine, the di-2-[N-(benzyl-2-bromoethyl)]aminoethyl ester of suberic acid, the sensitivity to ACh was strongly decreased but the Hill coefficient did not change. These findings, along with a low Hill coefficient for the response of the neurone to suberyldicholine (0.84, mean of 10 experiments) support the suggestion of a dimeric or oligomeric structure for the AChR in the *Limnaea* neurone membrane (Ger et al., 1971; Yurchenko et al., 1973). If this is really so, suberyldicholine and its alkylating derivative interact with two receptor subunits simultaneously. Whereas treatment with the alkylating derivative of suberyldicholine does not reduce cooperative interaction between the AChR monomers, it seems that monofunctional reagents alkylate only one subunit in the dimer and reduce the cooperative interaction between two monomers (for more details see Andrianova et al., 1977).

Our data are in agreement with the hypothesis on oligomeric structure of

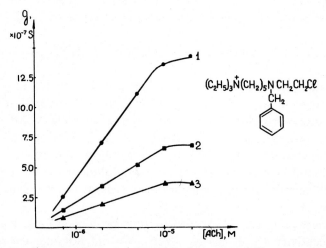

Fig. 5. Dependence of the membrane conductance changes on ACh concentration before and after treatment of a neurone with $1 \cdot 10^{-4}$ M 5-[N-(benzyl-2-chloroethyl)]aminopentyltriethylammonium bromide. (1), Control; (2), after 5-min treatment, and wash-out of the reagent; (3), after a further 5-min treatment.

nicotinic AChRs advanced by Khromov-Borisov and Michelson (1966), Karlin (1967) and Changeux and Podleski (1968).

General scheme of AChR conversion from one functional state into another

Our interpretations of the results are summarized in Fig. 6.

(I) Resting AChR

When there is no agonist, the ionic channel of AChR is closed. The carboxyl group in the anionic subsite can be modified by the mixture of CMEC and MGE, the disulphide bond is accessible to reduction by DTT, and some group Y^- is accessible to alkylation by β-halogenoethylamine. AChR which has been modified by any of these reagents can no longer be activated by ACh.

(II) Activated AChR

If an agonist is added to the bathing solution before the reagent, AChR is activated, the ionic channel which is chloride-seletive, (see Kislov and Kazachenko, 1974) opens, and we see an increase in membrane conductance. The agonist molecule produces steric hindrance of the reaction of CMEC with the carboxyl group, but cannot compete with β-halogenoethylamine for the group Y^-. Binding of the agonist to the active site evokes a conformational transition of AChR and the disulphide bond becomes inaccessible to DTT. Thus in the presence of an agonist AChR can be modified only by β-halogeno-ethylamines.

(III) Desensitised AChR in the presence of agonist

The receptor occupied by an agonist becomes desensitised. The channel is closed and the membrane conductance is low, but the carboxyl group is still

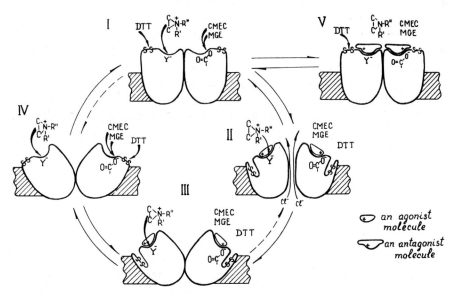

Fig. 6. Scheme of proposed AChR conversions from one functional state into another. The accessibility of functional groups to the modifying reagents is marked by arrows. See the text for details.

protected against CMEC and the conformation masking the disulphide bond from DTT is still maintained. Thus, only β-halogenoethylamines can alkylate AChR in its desensitised state if the agonist is present.

(IV) Desensitised AChR in the absence of agonist

After removal of the agonist from the bathing solution the AChR-agonist complex dissociates, but the AChR is maintained in its desensitised state for some time, and the responses to ACh are less than in the control. The reactions with CMEC or β-halogenoethylamines can proceed. The conformation masking the disulphide bond is not maintained any more, and the disulphide bond can be reduced by DTT. Thus, in the desensitised state without agonist, AChR can be modified by all three classes of reagents.

(V) AChR in the presence of antagonist

If an antagonist is added to the bathing solution before a reagent, a complex with the active site forms. The antagonist molecule hinders both CMEC reaction with the carboxyl group and alkylation by β-halogenoethylamines of the group Y^-. This molecule is too small, however, to produce steric hindrance for the reaction of DTT with the disulphide bond. Thus antagonists protect AChR only against CMEC and β-halogenoethylamines, but not against DTT.

SUMMARY

Essential groups of the acetylcholine receptors (AChRs) of *Limnaea stagnalis* neurones were identified and conformational transitions of different receptor states were revealed, by the method of chemical modification of proteins with the help of the voltage-clamp technique. All the reagents used modified AChRs specifically. Experiments with carbodiimide, dithiothreitol (DTT) and SH-alkylating agents showed: (1) there is a carboxyl group in the anionic subsite or near it, and an essential disulphide bond in the AChR molecule; (2) agonists induce a conformational transition of AChR which masks the disulphide bond against DTT attack; (3) two different conformations of the desensitised AChR can exist depending on whether the active site is occupied by an agonist or free. In the former state the disulphide bond is inaccessible to DTT. The high value of the Hill coefficient for responses to ACh and its lowering after treatment with β-halogenoethylamines suggest an oligomeric structure of *Limnaea* neurone AChRs and positive cooperativity between subunits.

ACKNOWLEDGEMENTS

The authors are indebted to Drs. Irina P. Andrianova, Emilia A. Gracheva and Galina V. Gridasova for the synthesis of β-halogenoethylamines and bromo-acetylcholine; Dr. V.Z. Kampel for the synthesis of N-alkyl derivaties of maleimide, and Mrs. Olga Shvirst for the help in preparing the manuscript.

REFERENCES

Andrianova, I.P., Bregestovski, P.D., Veprintsev, B.N., Vulfius, C.A., Gracheva, E.A., Gridasova, G.V., Il'in, V.I. and Yurchenko, O.P. (1977). Investigation of acetylcholine receptors of *Limnaea stagnalis* neurones with the help of chemical modification by β-chloroethylamines. In Comparative Pharmacology of Synaptic Receptors, *Nauka, Leningrad*, pp. 166–170 (in Russian).

Baker, B.R. (1967) *Design of Active-Site Directed Irreversible Enzyme Inhibitors,* Wiley, New York.

Bartels, E., Deal, W., Karlin, A. and Mautner, H.G. (1970) Affinity oxidation of the reduced acetylcholine receptor. *Biochim. biophys. Acta,* 203, 568–571.

Ben-Haim, D., Landau, E.M. and Silman, I. (1973) The role of a reactive disulphide bond in the function of the acetylcholine receptor at the frog neuromuscular junction. *J. Physiol. (Lond.),* 234, 305–325.

Bregestovski, P.D., Il'in, V.I., Yurchenko, O.P., Veprintsev, B.N. and Vulfius, C.A. (1977) Acetylcholine receptor conformational transition on excitation masks disulphide bonds against reduction. *Nature (Lond.),* 270, 71–73.

Carraway, K.L. and Koshland, D.E. (1972) Carbodiimide modification of proteins. In *Methods in Enzymology, Vol. 25,* Academic Press, New York, pp. 616–623.

Changeux, J.-P. and Podleski, T.R. (1968) On the excitability and cooperativity of the electroplax membrane. *Proc. nat. Acad. Sci. (Wash.),* 59, 944–950.

Chavin, S.I. (1971) Isolation and study of functional membranes proteins. *FEBS Lett.,* 14, 269–282.

Cleland, W.W. (1964) Dithiothreitol. A new protective reagent for SH groups. *Biochemistry,* 3, 480–482.

Edwards, C., Bunch, W., Marfey, P., Marois, R. and Van-Meter, D. (1970) Studies on the chemical properties of the acetylcholine receptor site of the frog neuromuscular junction. *J. Membrane Biol.,* 2, 119–126.

Ger, B.A., Zeimal, E.V. and Kvitko, I.J. (1971) The investigation of the cholinoreceptors of *Limnaea stagnalis* neurones with the use of bisquaternary compounds. *Comp. gen. Pharmacol.,* 2, 225–246.

Hall, Z.W. (1972) Release of neurotransmitters and their interaction with receptors. *Ann. Rev. Biochem.,* 41, 925–952.

Hoare, D.J. and Koshland, D.E. (1967) A method for the quantitative modification and estimation of carboxylic acid groups in proteins. *J. biol. Chem.,* 242, 2447–2453.

Il'in, V.I., Bregestovski, P.D., Vulfius, C.A. and Girshovich, A.S. (1976) Identification of carboxyl groups in the acetylcholine receptor of *Limnaea stagnalis* neurones. *Dokl. Akad. Nauk. SSSR,* 227, 1469–1471 (in Russian).

Karlin, A. (1967) On the application of "a plausible model" of allosteric proteins to the receptor for acetylcholine. *J. theor. Biol.,* 16, 306–320.

Karlin, A. (1969) Chemical modification of the active site of the acetylcholine receptor. *J. gen. Physiol.,* 54, 245–264.

Karlin, A. and Bartels, E. (1966) Effects of blocking sulfhydryl groups and of reducing disulfide bonds on the acetylcholine-activated permeability system of the electroplax. *Biochim. Biophys. Acta,* 126, 525–535.

Khromov-Borisov, N.V. and Michelson, M.J. (1966) The mutual disposition of cholinoreceptors of locomotor muscles and the changes in their disposition in the course of evolution. *Pharmacol. Rev.,* 18, 1051–1090.

Kislov, A.N. and Kazachenko, V.N. (1974) Ion currents of excited cholinergic membrane of the giant isolated snail neurones. In *Biophysics of a Living Cell, Vol. 4,* Pushchino, pp. 39–44 (in Russian).

Kostenko, M.A. (1972) The isolation of single nerve cells of the brain of the mollusc *Lymnaea stagnalis* for their further cultivation in vitro. *Cytology,* 14, 1274–1279 (in Russian).

Kostenko, M.A., Geletyuk, V.I. and Veprintsev, B.N. (1974) Completely isolated neurons of the mollusc, *Lymnaea stagnalis.* A new object for nerve cell biology investigation. *Comp. Biochem. Physiol.,* 49A, 89–100.

Lindstrom, J.M., Singer, S.J. and Lennox, E.S. (1973) The effects of reducing and alkylating agents on the acetylcholine receptor activity of frog sartorius muscle. *J. Membrane Biol.,* 11, 217–226.

Rang, H.P. and Ritter, J.M. (1971) The effect of disulfide bond reduction on the properties

302

of cholinergic receptors in chick muscle. *Mol. Pharmacol., 7*, 620–631.

Riehm, J.P. and Scheraga H.A. (1966) Structural studies of ribonuclease. XXI. The reaction between ribonuclease and a water-soluble carbodiimide. *Biochemistry, 5*, 99–115.

Sato, T., Sato, M. and Sawada, M. (1976) Effects of disulfide bond reduction on the excitatory and inhibitory postsynaptic response of Aplysia ganglion cells. *Jap. J. Physiol., 26*, 471–485.

Shaw, E. (1970) Selective chemical modification of proteins. *Physiol. Rev., 50*, 244–296.

Singer, S.J. (1967) Covalent labeling of active sites. In *Advances in Protein Chemistry, Vol. 22* Academic Press, New York, pp. 1–54.

Streitwieser, A. (1956) Solvolytic displacement reactions at saturated carbon atoms. *Chem. Rev., 56*, 571–752.

Vulfius, C.A. (1975) Study of acetylcholine receptors by the method of chemical modification. In *Acetylcholine Receptor Nature and Structure of its Active Site*, Pushchino, pp. 22–44 (in Russian).

Weill, C.L., McNamee, M.G. and Karlin, A. (1974) Affinity labeling of purified acetylcholine receptor from *Torpedo californica. Biochem. biophys. Res. Commun., 61*, 997–1003.

Yurchenko, O.P., Vulfius, C.A. and Zeimal, E.V. (1973) Cholinesterase activity in ganglia of Gastropoda, *Lymnaea stagnalis* and *Planobarius corneus.* I. Effect of anticholinesterase agents on giant neurone depolarization by acetylcholine and its analogues. *Comp. Biochem. Physiol., 45A*, 45–60.

Biochemical Characterization of Muscarinic Receptors: Multiplicity of Binding Components

JEAN MASSOULIÉ, SUSAN CARSON and GABOR KATO *

Laboratoire de Neurobiologie, Ecole Normale Supérieure, 75230 Paris (France)

MUSCARINIC RESPONSES-INTERVENTION OF METABOLIC INTERMEDIARY EVENTS

The responses of vertebrate cells sensitive to acetylcholine (ACh) have long been classified as nicotinic or muscarinic. These two classes of responses differ in their pharmacological sensitivity to agonists and antagonists, and in their physiological characteristics: the onset of the muscarinic response shows a longer delay (approximately 100 ms) and a longer duration (more than 0.5 s) than the nicotinic response with a delay of less than 1 ms and a duration of less than 100 ms (Purves, 1976). Muscarinic effects include a large variety of responses: depolarisation of cortical neurones, which appears to result from a lowering of the permeability to K^+ ions (Krnjević, 1975); inhibitory action on the cardiac atrium, resulting apparently from an increase in the permeability to K^+ ions; contraction of smooth muscle; stimulation of exocrine activity in the parotid glands, of protein secretion in the pancreas, and of thyroid activity. An efflux of K^+ ions has been measured in smooth muscle and in parotid glands, but, unlike the nicotinic receptors which seem to act through changes in membrane polarity by direct coupling with a specific ionophore (Changeux, 1975), muscarinic receptors appear to affect ionic permeability only indirectly. In a number of cases, it has been shown that ACh increases the cellular concentration of cyclic GMP, by activating a guanyl cyclase, which in turn induces the cellular response, possibly through changes in the phosphorylation of key proteins (Greengard, 1976). A metabolic effect on inositide phospholipid turnover has also been known for a long time to accompany muscarinic responses. Michell (1975) has argued that since this effect does not require Ca^{2+}, in contrast with the increase in cyclic GMP, it must represent a more direct consequence of muscarinic receptor activation. The primary effect would be the splitting of a phosphoinositol moiety from phosphatidylinositol, by a membrane-bound enzyme. It appears possible that this would in turn increase the Ca^{2+} permeability of the membrane, thereby activating guanylate cyclase, and thus inducing indirectly the specific physiological response. This view is quite

---

* Present address: Institut Battelle, 7, Route de Drize, Geneva (Switzerland).

consistent with the fact that some cells respond to several mediators: for instance, the parotid gland is stimulated by ACh, α-adrenergic agonists, and substance P, and all these responses seem to be mediated by Ca^{2+} ion efflux (Putney, 1977). The relative slowness of muscarinic responses is thus probably due to the fact that they are mediated by a series of metabolic reactions, and it is quite likely that muscarinic receptors of different cells might be coupled in different ways to various effectors.

MUSCARINIC BINDING SITES IN SMOOTH MUSCLE

The binding properties of muscarinic receptors have been studied by taking advantage of their saturability and of their pharmacological specificity. The methods used by various authors involve binding of labelled reversible antagonists and of agonists, as well as alkylation by irreversible agonists, and displacement by unlabelled agents, as summarized by Massoulié et al. (1977).

Antagonists such as atropine possess a high affinity (in the nanomolar range) towards muscarinic receptors, and this has been used to study their binding properties in smooth muscle strips by Paton and Rang (1965), who observed a high affinity binding component, which seemed to be related to the contraction response. This corresponded to the binding of 180 pmol of atropine per gram of tissue, but only 90 pmol of methylatropine. Occupancy of a fraction of the receptor sites by an antagonist results in a parallel shift of the dose—response curve (contraction as a function of the concentration of ACh, or other agonist), without depression of the maximal response until a high proportion of the receptors have been occupied. This is interpreted in terms of a large "receptor reserve" whereby the binding of agonists to a small fraction of receptors appears sufficient to elicit a maximal effect.

Gill and Rang (1966) have introduced an alkylating muscarinic antagonist BCM (benzilylcholine mustard) which allows an irreversible or slowly reversible blockade of the receptors. Young et al. (1972) have examined various analogues of this compound and have shown that the N-propyl homologue, PrBCM (propylbenzilylcholine mustard) yields a more stable derivative. These compounds exhibit an atropine-sensitive binding component which corresponds well to the atropine high affinity sites (220 pmol/g tissue; Fewtrell and Rang, 1973).

MULTIPLE BINDING SITES FOR ANTAGONISTS: ALLOSTERIC EFFECTS

Gupta, Moran and Triggle (1976) observed that low concentrations of BCM provoked a shift in the dose—response curve for the contraction of ileum strips, elicited by ACh, without depression of the maximal response. This shift reached dose-ratios as high as 10^3, but was partly reversible upon washing, although previous studies (Gill and Rang, 1966) had led to the conclusion that the rate of alkylation was much faster than the rate of dissociation of the intermediate reversible complex. The dose-ratio was stabilized at a level of 30 after washing and no further decrease was observed. Concentrations of BCM higher than 30 nM produced a depression of the maximal response, leading eventually to complete blockade, but the contractions reappeared with a half-time of 45 min. After recovery of the response, the dose-ratio was maintained at a value of 390. The authors conclude that two types of sites can be alkylated by BCM: the first class of sites (rapidly reacting) produces a shift in the dose—response curve, while the second class produces a depression in the response, and is de-alkylated relatively rapidly. They suggest that the first class corresponds to allosteric sites while the second represents the actual active receptor sites. Alkylation of these sites would block the effect of agonists, while alkylation of the first class would only decrease their affinity. In this view, the antagonist binding sites are heterogeneous, and the receptors can exist in two different states. This provides an alternative to the hypothesis of a very large receptor reserve in smooth muscle. It is interesting that Cuthbert and Young (1973), studying the chick amnion muscle, have concluded that the receptor reserve is small in this non-innervated organ. If the hypothesis of an allosteric effect of the bound antagonist is correct one would expect to obtain the same dose ratios if the receptors were identical, and therefore it seems possible that the difference observed reflects at least in part a molecular difference between mammalian ileum and chick amniotic receptors.

Clark and Mitchelson (1976) have reported that gallamine induces a shift in the dose—response of guinea pig heart (inotropic effect of ACh or carbamyl-choline) but does not depress the maximal response, and does not appear to occupy the same sites as the agonist or as atropine. These results are thus in agreement with the hypothesis of allosteric sites modulating the affinity of muscarinic receptors for their agonists.

ANTAGONIST BINDING TO SUBCELLULAR PREPARATIONS

The binding of atropine and of the agonist muscarone to brain subcellular fractions was studied by Farrow and O'Brien (1973) who characterized a high affinity binding site (with a dissociation constant of 0.6 nM for atropine, 90 nmol bound/g of tissue). Yamamura and Snyder (1974a,b) have used the reversible antagonist quinuclidinyl benzilate (QNB) for characterising the muscarinic receptor sites. They have shown that it binds to rat brain (1974a) or guinea-pig ileum (1974b) homogenates, exhibiting a saturable component, sensitive to muscarinic agents, and with a dissociation constant of less than 0.1 nM. The density of these sites was 65 nmol/g of brain and 190 pmol/g of smooth muscle. Using this compound, they have studied the distribution of receptors in the brain of rat (Yamamura et al., 1974a; Kuhar and Yamamura, 1975, 1976) and of monkey (Yamamura et al., 1974b). The binding sites were present exclusively in those regions and organs which were expected to contain muscarinic receptors. Similar observations were reported by Burgen et al. (1974a,b) who used the alkylating mustard PrBCM on small intestine and on brain homogenates.

The binding of other antagonists to subcellular preparations of various tissues has been studied. Beld and Ariëns (1974) have made use of the stereo-specificity of the muscarinic effect to evaluate the "non-specific" component of binding: of the two enantiomers of benzetimide, dexetimide is active, while levetimide is inactive. They observed that bovine trachea homogenates bind approximately twice as much atropine (21 pmol/g of tissue) as dexetimide (11 pmol/g of tissue). This observation is similar to that reported by Paton and Rang (1965) for guinea-pig ileum strips, and may be related to an heterogeneity of binding sites: it is conceivable that atropine binds to a class of sites which do not accept methylatropinium or dexetimide and therefore appears less specific than these ligands of higher affinity.

Working with a bovine brain particulate preparation, we have also observed that the number of high affinity QNB binding sites (0.6 nmol/g of protein) was lower than that of atropine binding sites (2 nmol/g of protein) (Carson et al., 1977). The binding of both antagonists followed a simple saturation curve, in agreement with other reports for QNB binding to heart cell membranes (Cavey et al., 1977), and for benzilylcholine binding to rat brain synaptosomal preparations. Birdsall et al. (1976) found that the level of binding at saturation was the same for all antagonists and that mutual displacements indicated the existence of a single uniform set of sites.

The pharmacological properties of muscarinic receptors in ileum, bronchial muscle, and iris appear indistinguishable (Barlow et al., 1972) and, moreover, the receptor sites of smooth muscle, brain, and parotid gland have been found to possess identical binding properties (Beld et al., 1975; Birdsall and Hulme, 1976). However, it has been reported that, although the L isomer of QNB is the active enantiomer which is specifically bound to rat brain membrames (Baumgold et al., 1977), there is a non-stereospecific binding of QNB to erythrocyte membranes, which is, however, specifically displaced by muscarinic agents. In addition the mydriatic effect on rat iris also does not appear stereospecific (Aronstam et al., 1977a). It was also reported that antagonists carrying a quaternary ammonium group have more affinity than the tertiary compounds for the ileum receptors but not for the cardiac atrium receptors (Barlow et al., 1976).

HETEROGENEITY OF AGONIST BINDING TO SUBCELLULAR PREPARATIONS

Apart from the possible heterogeneity among antagonist binding sites, which has already been discussed, there seems to exist a distinct heterogeneity which is revealed only in agonist binding. Whereas the most potent antagonists contain aromatic rings and probably bind at least partly through hydrophobic interactions it is not so for agonists which exhibit less affinity for the binding sites. The dissociation constants for ACh, for example for rat brain homogenates, are 3.3×10^{-7} M and 2.5×10^{-5} M (Birdsall et al., 1976) while its half-maximal effect on ileum contraction is obtained at 2.7×10^{-7} M (Burgen and Spero, 1968). Similar observations have been reported by other authors (Birdsall and Hulme, 1976; Carson, et al., 1977; Cavey et al., 1977).

The demonstration of a binding heterogeneity for agonists but not for antagonists parallels in a striking manner an earlier observation of Burgen and Spero (1968). These authors measured two different physiological effects of muscarinic agonists on ileum strips, namely contraction and K^+ or Rb^+ efflux. They found that the efflux response occurred at higher agonist concentrations and that the ratio of the corresponding effective concentrations was different for different agonists. This implied that the receptor sites involved were not identical and excluded an explanation based on a difference in receptor reserve for the two effects. However, except for benzhexol, antagonists shifted both of the dose–response curves by the same factor; reversible antagonists as well as BCM, did not differentiate the two receptors. These experiments suggest that an agonist-binding heterogeneity might be related to the coupling of receptor sites with different effector systems, responsible respectively for contraction and cation efflux.

It is therefore possible that muscarinic receptors exist in distinct states, either because of intrinsic molecular differences, or because of their association with different effector systems. In this connection it is interesting that Baumgold et al. (1977) have observed an activating action of acidic phospholipids on QNB binding by a rat brain synaptosomal preparation: phosphatidyl serine increased binding up to 40% and phospholipases A and C decreased it by 100% and 60% respectively (Aronstam et al., 1978a). It has also been recently reported that reaction of sulfhydryl groups of the receptors could alter their affinity for agonists and antagonists (Aronstam et al., 1977b; Aronstam et al., 1978b). The results indicated that at least two types of SH groups were involved. Reaction of the first set of groups with p-chloromercuribenzoate (pCMB), which could be prevented by agonists, abolished both antagonist and agonist binding. Reaction of a second class of groups with pCMB was not inhibitory; their reductive alkylation by N-ethylmaleimide (NEM) on the contrary converted the receptor into a high affinity state for agonists, and this effect was increased if NEM was allowed to react in the presence of agonists. There was no concomitant change in the affinity towards antagonists. The heterogeneity for agonist binding, which has been previously described seems, therefore, to reflect the existence of two allosteric-type states of the receptor which are controlled by SH alkylation.

The same authors have observed that the balance between the high and low agonist affinity states was not identical in different regions of the brain (Aronstam et al., 1978d). In addition to the occurrence of multiple affinity states for agonists, it has recently been claimed that binding of an antagonist can shift the receptor into a high affinity state for antagonists (Kloog and Sokolovsky, 1977).

SOLUBILISATION OF MUSCARINIC RECEPTORS

The studies on the binding of muscarinic agents by smooth muscle strips or tissue homogenates and subcellular particulate preparations, which we have discussed so far, reveal an heterogeneity of binding sites. It appears possible that such an heterogeneity is at least in part due to environmental effects on the receptor protein, and in order to understand the precise functioning of muscarinic transmission it would be important to study the receptor in an isolated state. Unfortunately, unlike nicotinic receptors which do not lose their binding properties upon solubilisation with non-ionic detergents (Changeux, 1975), muscarinic receptors seem to be totally inhibited by these compounds (Beld and Ariëns, 1974) and the solubilised proteins failed to show any muscarinic binding (Fewtrell and Rang, 1973).

Solubilisation of proteins with muscarinic binding characteristics has however been described by Bartfai et al. (1974), who submitted a membrane preparation to high salt concentration, and the binding proteins were partially purified (Alberts and Bartfai, 1976). In contradiction with the observations of Aronstam, Abood and Baumgold (1978a) who observed an inhibition of binding to membranes by phospholipases, these enzymes were used by Bartfai et al. (1976) during the purification of the membrane preparation and in this case did not appear to interfere with binding to the soluble protein. Recently, Aronstam et al. (1978c) have reported the solubilisation of muscarinic receptors from bovine brain by a combination of non-ionic detergent, salt and a metal chelator. In these two cases, the solubilised proteins were reported to possess the same affinity towards antagonists such as atropine.

By extracting a crude brain synaptosomal preparation with solutions of high salt concentration, we have obtained a protein solution which specifically bound muscarinic agents, but its affinity towards antagonists appeared markedly lower than that of the particulate preparation. In addition, the binding of agonists appeared to correspond exclusively to the high affinity site (Carson et al., 1977). In this case, it would seem that the receptor undergoes a structural change upon solubilisation as evidenced by antagonist binding. The binding of agonists implies either that only the high affinity sites have been solubilised, or that an agonist high affinity state is stabilized under these conditions.

A different approach to the biochemical characterisation of the receptor protein consist in labelling the muscarinic sites in the membrane-bound state with an alkylating antagonist, before solubilisation. Fewtrell and Rang (1973) using [^{3}H]BCM have characterised two polypeptide bands in sodium dodecyl sulphate polyacrylamide gel electrophoresis, with molecular weights of 23,000 and 50,000. Birdsall and Hulme (1976) using the same procedure with [^{3}H]PrBCM have obtained a band of higher molecular weight, 87,000. Possibly some proteolysis took place in the small intestine preparation of Fewtrell and Rang which could account for the difference in molecular weights.

We have solubilized a [^{3}H]PrBCM-labelled brain membrane preparation with Triton X-100 and analysed it by equilibrium isopycnic centrifugation in a cesium chloride gradient. The label was distributed into two peaks, one of which was at the top of the gradient, and represented non-specific binding since it was not depressed in the presence of atropine in excess. The second peak seemed to correspond nearly exclusively to specific binding, and equilibrated around a density of 1.25 g/cm^3. This peak was not homogeneous however, since its width was markedly greater than that of standard proteins. In gel filtration chromatography the labelled proteins were resolved into two components, with Stokes radii of approximately 4 and 5 nm. Centrifugation in sucrose gradients also indicated an heterogeneous population of molecules, sedimenting between 5 S and 15 S with a broad peak between 7–8 S. These results indicate that the "specific" labelling, as defined under experimental conditions, corresponds to several protein components. It remains to be examined whether these components are all related to the muscarinic response of the cells, and if so, what is the significance of their multiplicity.

SUMMARY

The slow time course of the muscarinic responses appears to indicate that the physiological effects are indirectly produced through a series of reactions which possibly involve the synthesis of cyclic GMP, or the metabolism of inositol phospholipids. This suggests that muscarinic receptors corresponding to different responses might be coupled to various effectors which would then introduce an heterogeneity among the receptor sites. The binding of antagonists and agonists to subcellular preparations of smooth muscle and brain and the effect of antagonist binding on the contraction of smooth muscle and on ionic responses of this tissue are discussed in this light. Attempts to solubilize the receptors are also discussed.

REFERENCES

Alberts, P. and Bartfai, T. (1976) Muscarinic acetylcholine receptor from rat brain: partial purification and characterization. *J. Biol. Chem.*, 251, 1543–1547.

Aronstam, R.S., Abood, L.G. and MacNeil, M.K. (1977a) Muscarinic cholinergic binding in human erythrocyte membranes. *Life Sci.*, 20, 1175–1180.

Aronstam, R.S., Hoss, W. and Abood, L.G. (1977b) Conversion between configurational states of the muscarinic receptor in rat brain. *Europ. J. Pharmacol.*, 45, 279–282.

Aronstam, R.S., Abood, L.G. and Baumgold, J. (1978a) The role of phospholipids in muscarinic binding by neural membranes. *Biochem. Pharmacol.*, 26, 1689–1695.

Aronstam, R.S., Abood, L.G., Hoss, W. and Kellogg, C. (1978b) SH reagents and heavy metals on brain muscarinic acetylcholine receptor. *Trans. Amer. Sco. Neurochem.*, 9, 122.

Aronstam, R.S., Eldefrawi, M.E. and Schuessler, D. (1978c) Preliminary characterization of the solubilized muscarinic receptor protein. *Fed. Proc.*, 37.

Aronstam, R.S., Kellogg, C. and Abood, L.G. (1978d) Development of muscarinic cholinergic receptors in inbred strains of mice: identification of receptor heterogeneity and relation to audiogenic seizure susceptibility. *Brain Res.*, in press.

Barlow, R.B., Franks, F.M. and Pearson, J.D.M. (1972) A comparison of the affinities of antagonists for acetylcholine receptors in the ileum. bronchial muscle and iris of the guinea pig. *Brit. J. Pharmacol.*, 46, 300–314.

Barlow, R.B., Berry, K.J., Glenton, P.A.M., Nikolaou, N.M. and Soh, K.S. (1976) A comparison of affinity constants for muscarinic-sensitive acetylcholine receptors in guinea-pig atrial pacemaker cells at 29°C and in ileum at 29°C and 37°C. *Brit. J. Pharmacol.*, 58, 613–620.

Bartfai., T., Anner, J., Schultzberg, M. and Montelius, J. (1974) Partial purification and characterization of a muscarinic acetylcholine receptor from rat cerebral cortex. *Biochem. Biophys. Res. Commun.*, 59, 725–733.

Bartfai, T., Berg, P., Schultzberg, M. and Heilbronn, E. (1976) Isolation of a synaptic membrane fraction enriched in cholinergic receptors by controlled phospholipase hydrolysis of synaptic membranes. *Biochim. Biophys. Acta*, 426, 186–197.

Baumgold, J., Abood, L.G. and Aronstam, R. (1977) Studies on the relationship of binding affinity to psychoactive and anticholinergic potency of a group of psychotomimetic glycolates. *Brain Res.*, 124, 331–340.

Beld, A.J., and Ariens, E.J. (1974) Stereospecific binding as a tool in attempts to localize and isolate muscarinic receptors. Part II. Binding of (+)-benzetimide, (−)-benzetimide

310

and atropine to a fraction from bovine tracheal smooth muscle and to bovine caudate nucleus. *Europ. J. Pharmacol.,* 25, 203–209.

Beld, A.J., Van den Hoven, S., Wouterse, A.C. and Zegers, M.A.P. (1975) Are muscarinic receptors in the central and peripheral nervous systems different? *Europ. J. Pharmacol.,* 30, 360–363.

Birdsall, N.J.M., Burgen, A.S.V., Hiley, C.R. and Hulme, E.C. (1976) The binding of agonists and antagonists to muscarinic receptors. *J. Supramolec. Struct.,* 4, 367–376.

Birdsall, N.J.M. and Hulme, E.C. (1976) Biochemical studies on muscarinic receptors. *J. Neurochem.,* 27, 7–16.

Burgen, A.S.V., Hiley, C.R. and Young, J.M. (1974a) The binding of [^{3}H]propylbenzilylcholine mustard by longitudinal muscle strips from guinea pig small intestine. *Brit. J. Pharmacol.,* 50, 145–151.

Burgen, A.S.V., Hiley, C.R., and Young, J.M. (1974b) The properties of muscarinic receptors in mammalian cerebral cortex. *Brit. J. Pharmacol.,* 51, 279–285.

Burgen, A.S.V. and Spero, L. (1968) The action of acetylcholine and other drugs on the efflux of potassium and rubidium from smooth muscle of the guinea-pig intestine. *Brit. J. Pharmacol.,* 34, 99–115.

Carson, S., Godwin, S., Massoulié, J. and Kato, G. (1977) Solubilisation of atropine-binding material from brain. Nature (Lond.), 266, 176–178.

Cavey, D., Vincent, J.P., and Lazdunski, M. (1977) The muscarinic receptor in heart cell membranes. Association with agonists, antagonists and anti-arrhythmic agents. *FEBS Lett.,* 84, 110–114.

Changeux, J.P. (1975) The cholinergic receptor protein from fish electric organ. In *Handbook of Psychopharmacology, Vol. 6,* Iversen, Iversen and Snyder (Eds.), Plenum, New York, pp. 235–301.

Clark, A.L. and Mitchelson, F. (1976) The inhibitory effect of gallamine on muscarinic receptors. *Brit. J. Pharmacol.,* 58, 323–331.

Cuthbert, A.W. and Young, J.M. (1973) The number of muscarinic receptors in chick amnion muscle. *Brit. J. Pharmacol.,* 49, 498–505.

Farrow, J.T. and O'Brien, R.D. (1973) Binding of atropine and muscarone to rat brain fractions and its relation to the acetylcholine receptor. *Molec. Pharmacol.,* 9, 33–40.

Fewtrall, C.M.S. and Rang, H.P. (1973) The labelling of cholinergic receptors in smooth muscle. In *Drug Receptors,* H.P. Rang (Ed.), MacMillan, London, pp. 211–224.

Gill, E.W. and Rang, H.P. (1966) An alkylating derivative of benzilylcholine with specific and long lasting parasympatholytic activity. *Molec. Pharmacol.,* 2, 284–297.

Greengard, P. (1976) Possible role for cyclic nucleotides and phosphorylated membrane proteins in postsynaptic actions of neurotransmitters. *Nature (Lond.),* 260, 101–108.

Gupta, S.K., Moran, J.F. and Triggle, D.J. (1976) Mechanism of action of benzilylcholine mustard at the muscarinic receptor. *Molec. Pharmacol.,* 12, 1019–1026.

Kloog, Y. and Sokolovsky, M. (1977) Muscarinic acetylcholine receptor interactions: Competition binding studies with agonists and antagonists. *Brain Res.,* 134, 167–172.

Krnjević, K. (1975) Acetylcholine receptor in vertebrate CNS. In *Handbook of Psychopharmacology, Vol. 6,* Iversen, Iversen and Snyder (Eds.), Plenum, New York, pp. 97–126.

Kuhar, M. and Yamamura, H.I. (1975) Light autoradiographic localisation of cholinergic muscarinic receptors in rat brain by specific binding of a potent antagonist. *Nature (Lond.),* 253, 560–561.

Kuhar, M.J. and Yamamura, H.I. (1976) Localisation of cholinergic receptors in rat brain by light microscopic autoradiography. *Brain. Res.,* 110, 229–243.

Massoulié, J., Godwin, S., Carson, S. and Kato, G. (1977) Biochemical studies on muscarinic receptors. *Biomedicine,* 27, 250–259.

Michell, R.H. (1975) Inositol phospholipids and cell surface receptor function. *Biochem. Biophys. Acta,* 415, 81–147.

Paton, W.D.M. and Rang, H.P. (1965) The uptake of atropine and related drugs by intestinal smooth muscle of the guinea pig in relation to acetylcholine receptors. *Proc, roy. Soc. B.,* 163, 1–44.

Purves, R.D. (1976) Function of muscarinic and nicotinic receptors. *Nature (Lond.),* 261, 149–151.

Putney, J.W. (1977) Muscarinic, α-adrenergic and peptide receptors regulate the same calcium influx sites in the parotid gland. *J. Physiol. (Lond.),* 268, 139–149.

Yamamura, H.I. and Snyder, S.H. (1974a) Muscarinic cholinergic binding in rat brain. *Proc. nat. Acad. Sci. (Wash.),* 71, 1725–1729.

Yamamura, H.I. and Snyder, S.H. (1974b) Muscarinic cholinergic receptor binding in the longitudinal muscle of the guinea pig ileum with [^{3}H]quinuclidinyl benzilate. *Molec. Pharmacol.*, 10, 861—867.

Yamamura, H.I., Kuhar, M.J. and Snyder, S.H. (1974a) In vivo identification of muscarinic cholinergic receptor binding in rat brain. *Brain Res.*, 180, 170—196.

Yamamura, H.I., Kuhar, M.J., Greenberg, D. and Snyder, S.H. (1974b) Muscarinic cholinergic receptor binding: regional distribution in monkey brain. *Brain Res.*, 66, 541—546.

Young, J.M., Hiley, R. and Burgen, A.S.V. (1972) Homologues of benzilylcholine mustard. *J. Pharmacol.*, 24, 950.

Cholinesterases of the Motor End-Plate of the Rat Diaphragm

MIRO BRZIN and TOMAŽ KIAUTA

Institute of Pathophysiology, University of Ljubljana, 61105 Ljubljana (Yugoslavia)

INTRODUCTION

A number of cytochemical procedures are used for the localization of the cholinesterases (ChE) in the motor end-plate (MEP) of skeletal muscle, but these do not give uniform results (see Friedenberg and Seligman, 1972).

Procedures for the accurate ultrastructural localization of enzyme activity involve a physico-chemical system which is extremely demanding so far as the rates of capturing reactions, leading to a rapid and quantitative precipitation of an insoluble and electron-dense product in the tissue, are concerned. In addition, subcellular localization of ChE in the elements of the MEP is even more complicated owing to the extremely high focal activity of the enzyme (Brzin and Zajiček, 1958; Brzin and Zeuthen, 1961; Salpeter, 1967, 1969). This high activity ($2-8 \times 10^{-4}$ μmol acetylcholine/MEP $\cdot$ h) would supply enough of the hydrolytic product to indicate the site of enzyme activity after only a very short incubation time if the capturing reaction were fully efficient.

The capturing reaction in the Cu-thiocholine procedure of Koelle and Friedenwald (1949), in which acetylthiocholine iodide is the substrate, was believed to be the formation of a cupric copper-thiocholine complex. More recently the detailed chemistry of Cu-thiocholine was elucidated by Souchay and Tsuji (Souchay and Tsuji, 1970; Tsuji, 1974) and, using their reaction scheme, Grubić and Brzin (1978) have found the capturing reaction to be quantitative. It has also been reported that only CNS^- or CN^- ions can be substituted for I^- in the capturing reaction (Brzin and Pucihar, 1976).

While the above considerations apply only to the formation of the primary precipitate, the original method (Koelle and Friedenwald, 1949) as well as most of its modifications (see Pearse, 1972; Silver, 1974) use a second step, i.e., the conversion of the primary precipitate into CuS. Since this conversion involves the dissolution of the primary precipitate and reprecipitation (Malmgren and Sylvén, 1955; Holmstedt, 1957), the dangers of a possible translocation of the primary precipitate and of possible diffusion artefacts are obvious, particularly in the case of the ultrastructural localization of enzyme activity (Klinar and Brzin, 1977). The method used in the present work is, however, a one-step procedure and the primary precipitate, deposited close to the sites of enzyme activity, remains there throughout the subsequent steps of the procedure.

Furthermore, the fact that all the components of the incubation medium, with the exception of the substrate, are present in the preincubation medium, is also of considerable importance.

While the precipitate obtained in the presence of I^- tends to aggregate into large crystals, particularly at the sites of high enzyme activity, the precipitates formed in the presence of CNS^- or CN^- are more finely granulated, thus making a higher resolution possible (Brzin and Pucihar, 1976).

The aim of the present study was to demonstrate that the one-step Cu-thiocholine procedure with CNS^- or CN^- as capturing anions is a useful source of information when used to complement quantitative measurements of enzyme activity in interpreting the involvement of ChE in synaptic transmission. The subcellular localization of acetylcholinesterase (AChE) and butyrylcholinesterase (BuChE) was examined in the MEP of the normal diaphragm of adult and developing rats. In addition, the AChE localization in the MEP of the denervated rat diaphragm was studied. Concomitant quantitative measurements of AChE and BuChE activities in similar preparations were made by using a radiometric method.

MATERIALS AND METHODS

Male Wistar strain albino rats weighing between 100 and 160 g were used.

Denervation procedure. Animals were anesthetized with ether and the left hemidiaphragm denervated by transection of the phrenic nerve about 0.5 cm proximal to the muscle.

Cytochemistry. Adult (normal and denervated) diaphragms were perfused in situ through the inferior vena cava with a fixative (5% formaldehyde in veronal-acetate buffer, pH 7.4). In developing diaphragms, isolated either before or after birth, this stage was omitted. After isolation the muscles were immersed into the same fixative for 2 h at 4°C and after fixation rinsed in Ringer's solution (containing NaCl 150 mM; KCl 3.7 mM; $CaCl_2$ 1.2 mM; and $MgCl_2$ 2.7 mM) for 1 to 2 h. Then the MEP-containing regions were dissected under a dissecting microscope. The muscle bands containing MEPs were sliced with a razor blade parallel to the muscle fibers into sections about 0.2 mm thick.

The sections were first preincubated for 20 min at 0–4°C in the following preincubation medium: glycine 20 mM; $CuSO_4$ 10 mM; KCNS or KCN 5 mM; NaCl 100 mM; acetate buffer 50 mM, pH 5.5, and either tetramonoisopropyl-pyrophosphortetramide (iso-OMPA) (20 μM) or 1:5-bis(4-allyldimethyl-ammoniumphenyl)-pentan-3-one dibromide (BW284C51) (10 μM) for the localization of AChE and BuChE, respectively.

The samples were then transferred into the incubation medium and incubated for 1.5 to 4 min at 0–4°C. The composition of the incubation medium was identical to that of the preincubation medium, except for the addition of substrate (3 mM acetylthiocholine chloride, AThChCl, or 3 mM butyrylthiocholine chloride, BuThChCl, for the localization of AChE and BuChE, respectively).

After incubation, the samples were rinsed quickly with distilled water, postfixed in 1% $KMnO_4$ for 1 h, dehydrated in ethanol, embedded in Epon (Serva), thin sectioned, contrasted with lead citrate and viewed in a Siemens Elmiskop I electron microscope.

In some experiments, I⁻ instead of CN⁻ or CNS⁻ was used for comparison.

Quantitative measurements. The adult diaphragms, both intact and denervated, were perfused in situ with Ringer's solution, the muscles were removed and MEP-containing regions were separated under a dissecting microscope from those without MEPs. Some intact adult diaphragms, used for comparison with developing diaphragms, were not dissected. The developing diaphragms were not perfused nor were the MEP-containing regions dissected.

The muscle samples were homogenized in Ringer's solution with small hand-operated glass homogenizers and AChE and BuChE activities in the muscle homogenates determined radiometrically by using specific substrates and inhibitors. For the separation of the labelled reaction product from the labelled substrate, thin-layer chromatography (Lewis and Eldefrawi, 1974) was used. Details of the procedure have been published elsewhere (Kiauta et al., 1977).

RESULTS

Fig. 1 shows an adult MEP with the reaction product obtained in the presence of I⁻ as the capturing anion in the preincubation and incubation media for the localization of AChE. The resolution of the possible sites of enzyme activity is very poor because the large crystalline aggregates bridge the space between the presynaptic and the postsynaptic membrane, thus giving the impression that the enzyme activity might be associated with the whole synaptic region as well as with the Schwann cell-nerve ending interspace. A more satisfactory localization is obtained when CNS⁻ or CN⁻ instead of I⁻ is the capturing agent, as demonstrated in the following figures. The localization of AChE is demonstrated in Fig. 2. The reaction product can be seen adhering to both synaptic membranes. It is also located in the primary and secondary synaptic clefts as well as in the Schwann cell-nerve ending interspace. When the incubation time is prolonged from 1.5 min (Fig. 2) to 3 min (Fig. 3), the density of the end-product deposits in the clefts increases considerably, suggesting either that the cleft substance exhibits AChE activity or that the end-product observed in the clefts originates at the highly active synaptic membranes and just fills the space available.

When the localization of BuChE (Fig. 4) is compared with that of AChE, the deposits in the synaptic region seem much sparser, in spite of the longer incubation time (4 min). This is in agreement with the quantitatively determined BuChE and AChE activity in the adult MEP-containing region of the diaphragm (Kiauta et al., 1977).

In the newborn rats, however, the ratio between AChE and BuChE activity of the diaphragm is quite different from that in the adult animals (Table I). During postnatal development AChE activity increases somewhat and is 30% higher in adult than in newborn animals. During the same period BuChE activity decreases rather drastically so that in the adult, the level is only about 20% of that at birth. Cytochemistry of the junctional regions of newborn rats revealed a distinct activity of both enzymes located at the same sites as in the

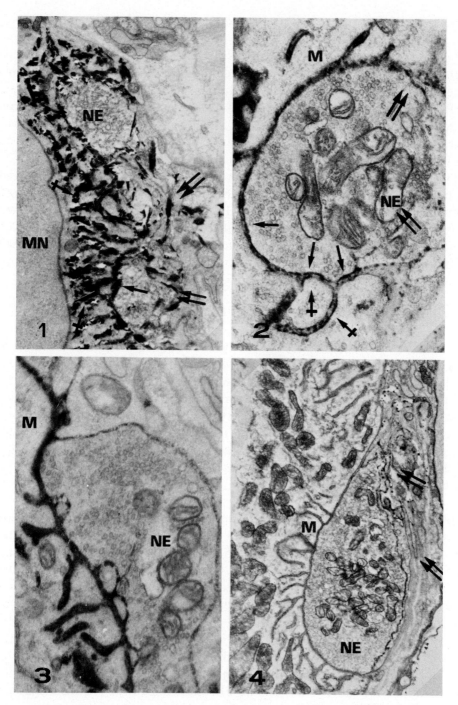

Fig. 1. Motor end-plate of rat diaphragm. Localization of AChE activity. Incubation: 2 min in the presence of I⁻. A large mass of the crystalline reaction product fills the primary (arrow) and secondary (crossed arrows) synaptic clefts and the Schwann cell-axolemma interspace (double arrows). The membranes are distorted and no fine resolution is possible. NE, nerve ending; MN, muscle cell nucleus. ×19,800.

Fig. 2. Motor end-plate of rat diaphragm. Localization of AChE activity. Incubation: 1.5 min in the presence of CNS⁻. The reaction product indicating sites of AChE activity is localized at the presynaptic (arrows) and postsynaptic (crossed arrows) membranes as well as in the Schwann cell-axolemma interspace (double arrows). Some of the grains of the reaction

TABLE I

AChE AND BuChE ACTIVITY IN THE DIAPHRAGM OF NEWBORN AND ADULT RATS

Stage of development	AChE activity	BuChE activity	AChE/BuChE
Newborn (n = 8)	23.2 ± 4.6 *	34.7 ± 4.9	0.67
Adult (n = 8)	30.2 ± 3.9	7.4 ± 1.2	4.08

* Values are 10^{-5} mol substrate hydrolyzed/g protein · h (mean ±S.D.).

MEPs of adult animals. In addition, the reaction product was found at the perinuclear membrane of Schwann cells (Figs. 5 and 6), a location never observed in adult MEPs.

Fig. 7 shows an adult MEP 3 days after nerve section. The activity of AChE is predominantly confined to the synaptic clefts and only a few particles of the reaction product can be seen at the degenerating nerve ending. The deposits of the reaction product in the clefts seem sparser than in the innervated MEPs incubated for the same period of time (2 min). This difference was observed in more than 30 denervated MEPs. Quantitative measurements show that after denervation the AChE activity decreases to about 30% of the control value within 3 days after denervation and remains relatively stable until the 14th day after denervation (Kiauta et al., 1977).

Twelve days after nerve section, when the nerve ending is fully degenerated, AChE activity can still be demonstrated in the surviving muscular component of the MEP (Fig. 8). Even 5 weeks after denervation, when the degeneration of muscle cells is well advanced, AChE activity persists and the reaction product can be seen not only in the secondary synaptic clefts, but also at the perinuclear membrane and in the cisternae of the sarcoplasmic reticulum (Fig. 9).

DISCUSSION

The cytochemical procedure used in this work was designed to take into account all the known facts about the chemistry of the capturing reaction. The extremely high focal activity of the enzyme was considered as well. Since it has not been recognized until recently that I⁻ (or any other substituting capturing

---

product in the clefts do not seem to be attached to membranes. NE, nerve ending, containing mitochondria and synaptic vesicles; M, muscle cell. ×51,600.

Fig. 3. Motor end-plate of rat diaphragm. Localization of AChE activity. Incubation: 3 min in the presence of CN⁻. Localization essentially as in Fig. 2, but the reaction product fills the entire synaptic space owing to the longer incubation. NE, nerve ending; M, muscle cell. ×36,200.

Fig. 4. Motor end-plate of rat diaphragm. Localization of BuChE activity. Incubation: 3 min in the presence of CNS⁻. The reaction product can be seen between Schwann cell infoldings and in the Schwann cell-nerve ending interspace (double arrows). A thin but distinct deposit of the reaction product can be observed at the presynaptic and postsynaptic membranes as well as in the synaptic cleft. NE, nerve ending; M, muscle cell. ×14,000.

318

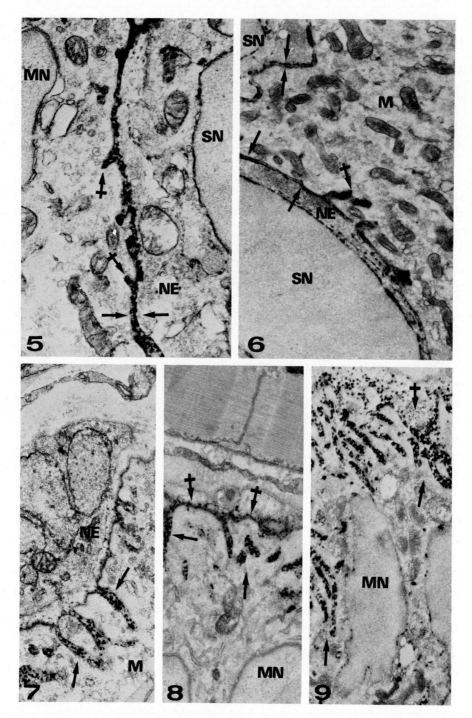

Fig. 5. Developing neuromuscular junction of a 2-day-old rat. Localization of AChE activity. Incubation: 2 min in the presence of CNS⁻. The reaction product is localized at both synaptic membranes, in the primary synaptic cleft (arrows) and immature secondary synaptic clefts (crossed arrows). The reaction product can be seen along the membrane of the Schwann cell nucleus (SN) whereas at the perinuclear membrane of the muscle cell (MN) no reaction product is found. NE, nerve ending. ×48,000.

Fig. 6. Developing motor end-plate of a 2-day-old rat. Localization of BuChE activity. Incubation: 2 min in the presence of CNS⁻. Localization identical to that in Fig. 5. Arrows:

anion) is absolutely necessary for the formation of the precipitate, most Cu-thiocholine procedures do not include any of them in the preincubation medium. It is often prescribed that I⁻, present accidentally in the incubation medium as a part of the substrate, should be eliminated by prior precipitation (see Pearse, 1972; Silver, 1974). Fortunately (or unfortunately!) it is rather difficult to remove I⁻ so that the concentration remaining is usually sufficient for a positive histochemical reaction, even though the concentration might be too low for a fully efficient capturing reaction.

The present one-step cytochemical procedure based on the Cu-thiocholine reaction can be used for the study of a wide variety of cholinergic structures. The information obtained in this way could be pertinent for the understanding of the functional interplay between the individual components of the ACh system during synaptic transmission.

Cytochemical evidence, however, often raises more questions than it answers, e.g., what is the function of AChE localized at the presynaptic membranes or what is the function of BuChE in the elements of the MEP? When the ultrastructural localization of AChE in the MEP is considered in the light of direct microgasometric (Brzin and Zajiček, 1958) and indirect autoradiographic (Salpeter, 1967, 1969; Barnard, 1974) measurements of the AChE activity in the MEP, the synaptic space seems to have an AChE activity comparable to that of a purified commercial AChE preparation. If ACh is supposed to diffuse through this synaptic space for a distance of at least 500 Å and reach the post-synaptic membrane unhydrolysed, a mechanism preventing its hydrolysis must be postulated.

The presence of BuChE in the MEP-containing region of the adult rat muscle was established a long time ago. The results of our experiments concerning its ultrastructural localization agree with those published previously (Teräväinen, 1967). The localization of BuChE in the adult MEP is thus similar to that of AChE, but its activity, as judged by the amount of the reaction product deposited, seems to be much lower than that of AChE. This cyto-chemical observation has been confirmed by quantitative measurements (Kiauta et al., 1977).

---

reaction product at both synaptic membranes; crossed arrows: developing secondary synaptic clefts. M, muscle cell; SN, Schwann cell nucleus; NE, nerve ending. ×32,700.

Fig. 7. Motor end-plate of rat diaphragm 3 days after nerve section. Localization of AChE. Incubation: 2 min in the presence of CNS⁻. The reaction product can be seen in the secondary synaptic clefts (arrows) whereas in the region of the degenerating nerve ending and in the enlarged primary synaptic cleft few particles can be observed. NE, nerve ending; M, muscle cell. ×21,500.

Fig. 8. Motor end-plate of rat diaphragm 12 days after nerve section. Localization of AChE. Incubation: 2 min in the presence of CNS⁻. The nerve ending has completely degenerated. The reaction product can be seen in synaptic folds (arrows) and at the muscle surface of the former junction (crossed arrows). MN, muscle cell nucleus. ×21,700.

Fig. 9. Motor end-plate of rat diaphragm 5 weeks after nerve section. Localization of AChE. Incubation: 2 min in the presence of CNS⁻. The reaction product is in secondary synaptic clefts (arrows) and diffusely distributed at the sarcolemma (crossed arrow) of the former junction. Deposits of the reaction product can be seen in the sarcoplasmic reticulum and at the perinuclear membrane of the degenerating muscle cell (MN). ×24,700.

In the diaphragm of newborn rats, the activity of BuChE exceeds that of AChE. The functional significance of both the low BuChE activity in the adult and the 4–5-fold greater BuChE activity at the time of birth, remains obscure. BuChE does not seem to be involved in neuromuscular transmission since inhibition of this enzyme does not obviously impair neuromuscular transmission (Heffron, 1972). According to a recent attractive hypothesis (Koelle et al., 1977a, b), BuChE and AChE may be metabolically linked in such a way that BuChE actually serves as a precursor for AChE at cholinergic synapses. This hypothesis must, however, await further experimental confirmation.

Denervation of a mammalian skeletal muscle may result in either an increase or a decrease in the activities of the muscle ChE, depending on the animal species (Brzin and Majcen-Tkačev, 1963; Tennyson et al., 1977). After phrenic nerve section in the rat, the AChE activity of the diaphragm decreases, that in the MEP region falling to about 30% of the control value 3 days after denervation (Kiauta et al., 1977). Cytochemically, the presynaptic contribution to the total MEP activity practically disappears. The question, however, arises whether all the missing enzyme activity (70%) should be attributed to the degenerating presynaptic elements or whether an accompanying decrease in AChE activity of postsynpatic origin should be assumed. The decreased deposits of the reaction product in the primary and secondary synaptic clefts suggest the latter possibility. Cytochemistry, however, cannot decide this question without quantitative confirmation.

As far as the more advanced stages of denervation are concerned, it can be seen that the postsynaptic part of the neuromuscular junction persists relatively unchanged both morphologically, and with regard to the remaining AChE demonstrable by cytochemistry, for a very long time after nerve section.

The cytochemical method can also be combined with another type of method to provide information which neither method would yield when used alone. For instance, the combination of the cytochemical technique described above with the solubilization of AChE by specific solubilizing agents (Betz and Sakmann, 1971; Hall and Kelly, 1971) can provide additional information on the mode of attachment of AChE to the structures of the MEP (Sketelj and Brzin, 1978). These experiments suggest that AChE molecules are not attached to all the reacting structures of the MEP in the same way. Specifically, a collagenous component may be important for the attachment of AChE molecules in the synaptic cleft, but not in the Schwann cell-nerve ending interspace.

SUMMARY

The ultrastructural localization of AChE and BuChE was studied in normal adult, developing, and denervated adult motor end-plates of the rat diaphragm, using the CNS^- or CN^- modification of the one-step Cu-thiocholine procedure. In normal adult end-plates, AChE activity was observed on pre- and post-synaptic membranes, and was probably associated with the cleft substance as well. The reaction product was also present in the Schwann cell-nerve ending interspace. BuChE exhibited a similar distribution but in the motor end-plate proper the deposits of end-product were much less pronounced. The end-plate at early developmental stages, however, showed a BuChE activity which considerably exceeded the activity of AChE as revealed by quantitative measurements. After denervation AChE activity persisted on the postsynaptic side of the junction whereas the activity on the membranes of degenerating nerve endings was almost completely lost.

The possible functional significance of the localization and activity of cholinesterases at the motor end-plate is discussed.

ACKNOWLEDGEMENTS

This investigation was supported in part by the Slovene Research Community and in part by National Institutes of Health Grant No. 02-008-1, Z-ZF-6. The authors wish to thank Mrs. Simona Pucihar and Mr. Vasilij Loboda for excellent technical assistance.

REFERENCES

Barnard, E.A. (1974) Neuromuscular transmission — enzymatic destruction at acetylcholine. In *The Peripheral Nervous System*, J.I. Hubbard (Ed.), Plenum, New York, pp. 201–224.

Betz, W. and Sakmann, B. (1971) "Disjunction" of frog neuromuscular synapses with proteolytic enzymes. *Nature New Biol.*, 232, 94–95.

Brzin, M. and Majcen-Tkačev, Ž. (1963) Cholinesterase in denervated end plates and muscle fibres. *J. Cell Biol.*, 19, 349–358.

Brzin, M. and Pucihar, S. (1976) Iodide, thiocyanate and cyanide ions as capturing reagents in one-step copper-thiocholine method for cytochemical localization of cholinesterase activity. *Histochemistry*, 48, 283–292.

Brzin, M. and Zajiček, J. (1958) Quantitative determination of cholinesterase activity in individual end-plates of normal and denervated gastrocnemius muscle. *Nature (Lond.)*, 181, 626.

Brzin, M. and Zeuthen, E. (1961) Quantitative evaluation of the thiocholine method for cholinesterase as applied to single end-plates from mouse gastrocnemius muscle. *C. R. Trav. Lab. Carlsberg*, 32, 139–153.

Friedenberg, R.M. and Seligman, A.M. (1972) Acetylcholinesterase at the myoneural junction: Cytochemical ultrastructure and some biochemical considerations. *J. Histochem. Cytochem.*, 20, 771–792.

Grubić, Z. and Brzin, M. (1978) Quantitative evaluation of the trapping reaction of copper-thiocholine histochemical procedures for localization of cholinesterases. *Histochemistry*, 56, 213–220.

Hall, Z.W. and Kelly, R.B. (1971) Enzymatic detachment of end-plate acetylcholinesterase in muscle. *Nature New Biol.*, 232, 62–63.

Heffron, P.F. (1972) Actions of a selective inhibitor of cholinesterase tetramonoisopropyl pyrophosphortetramide on the rat phrenic nerve-diaphragm preparation. *Brit. J. Pharmacol.*, 46, 714–724.

Holmstedt, B. (1957) A modification of the thiocholine method for the determination of cholinesterase. II. Histochemical application. *Acta physiol. scand.*, 40, 331–337.

Kiauta, T., Brzin, M. and Dettbarn, W.-D. (1977) Synthesis of neuromuscular cholinesterases in innervated and denervated rat diaphragm. *Exp. Neurol.*, 56, 281–288.

Klinar, B. and Brzin, M. (1977) A comparison between the one-step and the two-step copper thiocholine for the cytochemical localization of cholinesterases. *Histochemistry*, 50, 313–318.

Koelle, G.B. and Friedenwald, J.S. (1949) A histochemical method for localizing cholinesterase activity. *Proc. Soc. exp. Biol. Med.*, 70, 617–622.

Koelle, W.A., Smyrl, E.G., Ruch, G.A., Siddons, V.E. and Koelle, G.B. (1977a) Effects of

322

protection of butyrylcholinesterase on regeneration of ganglionic acetylcholinesterase. *J. Neurochem.*, 28, 307–311.

Koelle, G.B., Koelle, W.A. and Smyrl, E.G. (1977b) Effects of inactivation of butyrylcholinesterase on steady state and regenerating levels of ganglionic acetylcholinesterase. *J. Neurochem.*, 28, 313–319.

Lewis, M.K. and Eldefrawi, M.E. (1974) A simple, rapid, and quantitative radiometric assay of acetylcholinesterase. *Anal. Biochem.*, 57, 588–592.

Malmgren, H. and Sylvén, B. (1955) On the chemistry of the thiocholine method of Koelle, *J. Histochem. Cytochem.*, 3, 441–445.

Pearse, A.G.E. (1972) *Histochemistry: Theoretical and Applied*, 3rd Edn., Churchill, London.

Salpeter, M.M. (1967) Electron microscope radioautography as a quantitative tool in enzyme cytochemistry. I. The distribution of acetylcholinesterase at motor endplates of a vertebrate twitch muscle. *J. Cell Biol.*, 32, 379–389.

Salpeter, M.M. (1969) Electron microscope radioautography as a quantitative tool in enzyme cytochemistry. II. The distribution of DFP-reactive sites at motor endplates of a vertebrate twitch muscle. *J. Cell Biol.*, 42, 122–134.

Silver, A. (1974) *The Biology of Cholinesterases*, North-Holland, Amsterdam.

Sketelj. J. and Brzin, M. (1978) Attachment of acetylcholinesterase to structures of motor endplate. In preparation.

Souchay, P. and Tsuji, S. (1970) Contribution à l'étude de la réaction sels de cuivre-thiocholine utilisée dans la détection histochimique des cholinésterases. *Ann. Histochim.*, 15, 263–271.

Tennyson, V.M., Kremzner, L.T. and Brzin, M. (1977) Electron microscopic-cytochemical and biochemical studies of acetylcholinesterase activity in denervated muscle of rabbits. *J. Neuropathol. exp. Neurol.*, 36, 245–275.

Teräväinen, H. (1967) Electron microscopic localization of cholinesterases in the rat myoneural junction. *Histochemie*, 10, 266–271.

Tsuji, S. (1974) On the chemical basis of thiocholine methods for demonstration of acetylcholinesterase activities. *Histochemistry*, 42, 99–110.

PART V

DEVELOPMENT OF CHOLINERGIC SYNAPSES
NEUROTROPHIC REGULATION
INTERCELLULAR RELATIONS

The Life History of Acetylcholine Receptors

DOUGLAS M. FAMBROUGH, PETER N. DEVREOTES *, JOHN M. GARDNER and
DIANA J. CARD **

*Department of Embryology, Carnegie Institution of Washington, Baltimore,
MD 21210 (U.S.A.)*

INTRODUCTION

The acetylcholine (ACh) receptor is the major integral membrane glycoprotein known to occur in the postsynaptic membrane at the neuromuscular
junction. ACh receptors also occur in extrajunctional regions of embryonic and
denervated muscle fibers (Fig. 1). Thus, the study of ACh receptor metabolism
yields information pertinent to various scientific problems, including the
mechanisms of membrane biogenesis, myogenesis, formation of synapses, and
regulation of spatial distribution of cell surface functions. These matters have
been reviewed in detail recently (Fambrough, 1979). Here we will describe
briefly the life-history, as best we know it, of extrajunctional ACh receptor
molecules. Presumably, the events in the life of junctional ACh receptors
include most of the same events (plus some others), although the timing and
duration of events may be different.

SITE OF BIOSYNTHESIS AND PROCESSING

Newly synthesized ACh receptors are first detected in the Golgi apparatus
(Fig. 2) (Fambrough and Devreotes, 1978). At this stage the receptor units are
already large glycoprotein molecules intimately associated with lipid bilayer
(Devreotes and Fambrough, 1975; Patrick et al., 1977). Detergent solubilized
receptor molecules at this stage resemble cell surface ACh receptors in size,
affinity for ACh and (+)-tubocurarine and precipitability with concanavalin A
and with antireceptor antisera.

Receptor molecules probably appear in the Golgi apparatus within about 15
min of the time of biosynthesis of their constituent polypeptide chains
(Fambrough and Devreotes, 1978). Biosynthesis and assembly of receptor subunits to form complete receptors involves both protein synthesis and glycosylation. When the lipid intermediate pathway of protein glycosylation is blocked

---

* Present address: Department of Biochemistry, University of Chicago, Chicago, IL (U.S.A.).
**Present address: Department of Pharmacological and Physiological Sciences, University of
Chicago, Chicago, IL (U.S.A.).

326

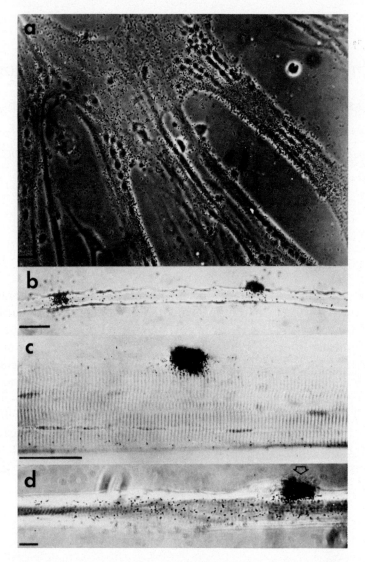

Fig. 1. Light-microscope autoradiograms showing the approximate location of cholinergic receptors in embryonic, adult and denervated adult muscles, as revealed by ^{125}I-labeled α-bungarotoxin binding. (a) Embryonic chick skeletal muscle in tissue culture. (b) Portion of single muscle fibre from 19-day embryonic chick anterior latissimus dorsi muscle, illustrating extrajunctional receptors and clusters of receptors at the periodic sites of multiple innervation. (C) End-plate region of single muscle fibre from normal adult human deltoid muscle. (d) Portion of muscle fibre from 10-day denervated rat diaphragm muscle, illustrating the continued presence of dense receptor clustering at former postsynaptic site (arrow) and also nearby extrajunctional receptor sites. Magnification bars represent 20 μm in (a), (b), and (d), 50 μm in (c). (From Fambrough et al., 1977).

by tunicamycin, receptor formation ceases, but receptors already in the Golgi apparatus continue their transport to the cell surface (Fambrough, 1977). As yet it is unknown whether biosynthesis of receptor subunits involves rough endoplasmic reticulum, which is a very minor organelle in muscle.

The intracellular pool of newly synthesized receptors contains those receptors which will appear in the plasma membrane during the next three hours

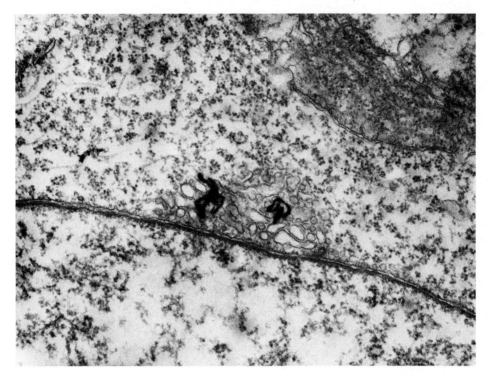

Fig. 2. Electron microscope autoradiograph illustrating the association of ACh receptors with the Golgi apparatus, as revealed by the binding of iodinated α-bungarotoxin to fixed, saponin-treated chick skeletal muscle cells in tissue culture.

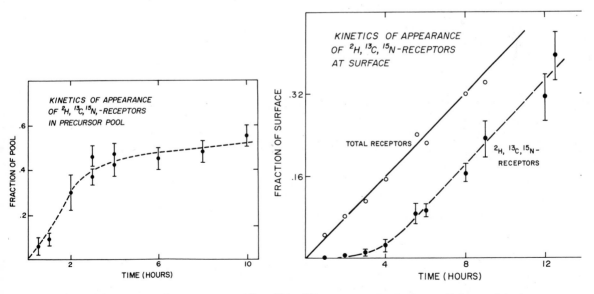

Fig. 3. (a) Kinetics of appearance of ^{2}H-, ^{13}C-, ^{15}N-receptors in the intracellular pool during incubation in medium containing ^{2}H-, ^{13}C-, ^{15}N-amino acids. (b) Kinetics of appearance of total (^{1}H-, ^{12}C-, ^{14}N-, plus ^{2}H-, ^{13}C-, ^{15}N-) and density-shifted (^{2}H-, ^{13}C-, ^{15}N-) receptors on myotube surfaces during incubation in medium containing ^{2}H-, ^{13}C-, ^{15}N-amino acids. (From Devreotes et al, 1977).

328

(Fig. 3) (Devreotes et al., 1977). In young myotubes in vitro, this pool contains 10 to 15% as many receptors as are present on the plasma membrane. Transport of ACh receptors to the plasma membrane possibly makes some non-essential use of microtubular elements: 3 μM colchicine slows the rate of receptor appearance in the plasma membrane about 50% while lumicolchicine has no effect and higher concentrations of colchicine are no more inhibitory (Rotundo and Fambrough, unpublished observations).

INCORPORATION INTO PLASMA MEMBRANE

Receptor molecules appear in the plasma membrane about 3 h after biosynthesis (Devreotes et al., 1977). This appearance probably involves fusion of membrane vesicles containing ACh receptor molecules with the plasma membrane. Before this event, receptor molecules are located in membrane-bound structures with the receptor sites protected from interaction with ligands (presumably because the receptor sites occur on the inner surface of closed vesicles) (Devreotes and Gardner, unpublished observations).

Incorporation of new receptors into the plasma membrane is an energy-requiring process as judged by rapid inhibition by compounds which interfere with ATP production. Low temperature also prevents incorporation (Fig. 4). In young myotubes the sites of incorporation are well distributed over the cell surface (Hartzell and Fambrough, 1973). In older myotubes in vitro there may be preferential incorporation at sites of ACh receptor clustering (Fischbach et al., 1976).

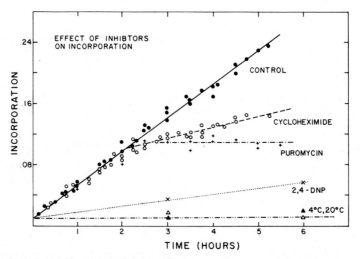

Fig. 4. Incorporation of new ACh receptors into plasma membranes of cultured chick skeletal muscle after blockage of old receptors with unlabelled α-bungarotoxin. Unlabelled α-bungarotoxin was removed and inhibitors added at t = 0. Number of newly incorporated receptors has been normalized to total number of original surface receptors at t = 0. Data are from 5 experiments for cycloheximide, 3 for puromycin, 2 for 2,4-dinitrophenol, 1 for each temperature. Data for each individual experiment have been normalized so that the control rate is 4.5%/h. Actual control rates ranged from 3.8%/h to 5.1%/h. (From Fambrough et al. 1977).

Most extrajunctional ACh receptor molecules are mobile in the plasma membrane (Axelrod et al., 1976). However, clustered extrajunctional ACh receptors and also synaptic receptors are stationary on a time-scale of minutes to hours (Axelrod et al., 1976; Fambrough and Pagano, 1977). It is unknown whether the mechanisms limiting receptor mobility in these two cases are the same.

DEGRADATION

Average extrajunctional receptor lifetimes have been estimated as 8–30 h for ACh receptors in various muscles and muscle cultures from homeothermic species. For cultured chick (Gardner and Fambrough, 1978) and calf (Merlie et al., 1976) skeletal muscle in vitro, receptor lifetimes (approximately 18 h) have been determined by pulse-chase experiments involving direct labeling of receptor molecules with isotopically labelled amino acids. In other cases, estimates are based upon the loss of radioactivity from the muscle after iodinated α-bungarotoxin was bound to ACh receptor sites. The bound α-bungarotoxin may slightly decrease the turnover rate (Fig. 5).

Studies on the mechanism of degradation have, for the most part, involved labelling of receptor sites with iodinated α-bungarotoxin and then following the

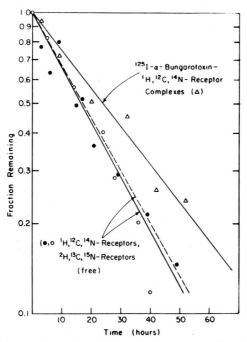

Fig. 5. Degradation of normal and density shifted receptors without bound α-bungarotoxin, employing a pulse chase experimental regime. The open and closed circles represent two experiments combining the degradation data for [1]H-, [12]C-, ^{14}N-receptors and [2]H-, [13]C-, ^{15}N-receptors from the pulse and chase phase of the experiments respectively. The results of one parallel experiment measuring the decay of ^{125}I-α-bungarotoxin receptor complexes from an identically treated set of cultures is also shown (△———△). (From Gardner and Fambrough, 1978).

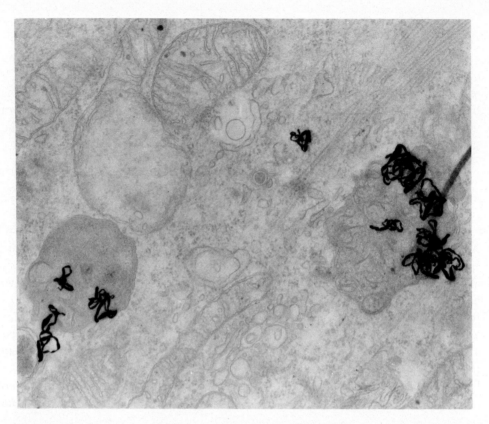

Fig. 6. Electron microscope autoradiogram showing silver grains over secondary lysosomes of cultured chick muscle. Cells were first labelled with ^{125}I-α-bungarotoxin, fixed in glutaraldehyde, and prepared for EM autoradiography. The appearance of grains over secondary lysosomes is interpreted as evidence for the uptake of bungarotoxin receptor complexes into the cytoplasm, and their intracellular degradation. (From Devreotes and Fambrough, 1976a).

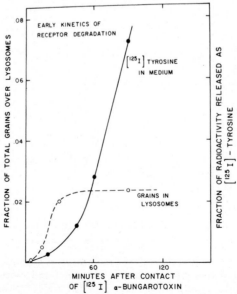

Fig. 7. Kinetics of transport of radioactivity to secondary lysosomes (open circles) and kinetics of liberation of iodo-tyrosine from chick muscle cultures (solid circles) after brief exposure of muscle to ^{125}I-α-bungarotoxin. Data obtained from electron microscopic autoradiograms and from chromatographic analysis of culture medium. (From Fambrough et al., 1977).

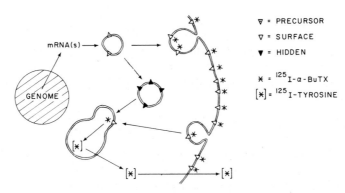

HYPOTHETICAL "LIFE-CYCLE" of ACETYLCHOLINE
RECEPTORS

Fig. 8. Schematic life history of ACh receptors in cultured skeletal muscle. The figure depicts a cross section through a myotube with ACh receptors symbolized as triangles in membrane profiles. The newly synthesized receptors ($\triangledown$) occur in Golgi apparatus and post-Golgi vesicles and constitute about 10–15% of total ACh receptors. ACh receptors in the plasma membrane (∇) are depicted as associated with iodinated α-bungarotoxin (*) as would occur in an experiment to estimate turnover rate. Iodotyrosine, the radioactive product of proteolysis of the iodinated α-bungarotoxin, is symbolized by [*]. About 70% of the ACh receptors are located in the plasma membrane as individual molecules and aggregates. Only about 2% of receptors are in the degradation pathway at any instant. The remaining ACh receptors ($\blacktriangledown$) occur on internal membranes and have been referred to as "hidden" because they are not labelled by extracellular α-bungarotoxin. The average lifetime of the hidden sites is about the same as for receptors in the plasma membrane.

fate of the radioactivity (Berg and Hall, 1974; Chang and Huang, 1975; Devreotes and Fambrough, 1975, 1976a). The degradation process is energy-dependent and not tightly coupled to receptor biosynthesis. α-Bungarotoxin-receptor complexes are internalized and transported to secondary lysosomes (Fig. 6) where proteolysis occurs. Proteolytic breakdown can be blocked by long-term treatment of muscle with Trypan blue. Treated cells can still internalize receptors and transport them to lysosomes, but breakdown does not occur and radioactivity accumulates to high levels in the lysosomal compartment (over 20% in 8 h). The kinetics of transport to lysosomes have been estimated roughly by electron microscope autoradiography (Fambrough et al., 1977), and they seem to be consistent with other kinetic data on receptor turnover (Fig. 7).

Figure 8 is a cartoon illustrating the main features of ACh receptor metabolism. The biosynthetic route is essentially that used in other cells for secretion of secretory proteins. This is probably true for muscle as well. Rotundo (Rotundo and Fambrough, 1977, and unpublished observations) has studied the kinetics of secretion of ACh esterase in detail and proposed that secretion of this glycoprotein and the biogenesis of plasma membrane involve the same mechanism in skeletal muscle. Thus any perturbation of ACh receptor biosynthesis, transport and incorporation into plasma membrane is accompanied by an equal perturbation of ACh esterase biosynthesis and secretion.

CONTROL OF EXTRAJUNCTIONAL ACh RECEPTORS

The appearance of ACh hypersensitivity following denervation of adult skeletal muscle has been shown to result from turn-on of receptor biosynthesis (Brockes and Hall, 1975; Devreotes and Fambrough, 1976b). Just which aspect of the denervated state triggers renewed receptor synthesis is not known. However, it is clear that this turn-on can be reversed by direct stimulation of denervated muscle (Lømo and Rosenthal, 1972). Electrical stimulation of hypersensitive muscle in organ culture suppresses the biosynthesis and incorporation of receptor molecules into extrajunctional plasma membrane (Reiness and Hall, 1977; Card, unpublished observations, Fig. 9), while somewhat slowing (Hogan et al., 1976) or not changing (Card, 1977) the rate of extrajunctional ACh receptor degradation. Thus the disappearance of hypersensitivity during electrical stimulation results from decline in the number of extrajunctional ACh receptors due to degradation without balancing replacement. The mechanism underlying the depression of ACh receptor biosynthesis remains unknown.

FOR THE FUTURE

In the context of cholinergic mechanisms, the eventual goals in the study of receptor life history are to describe the mechanisms and the rates of production and destruction of receptors in developing and adult muscle and to determine all of the mechanisms regulating metabolic rates and spatial distribution of recep-

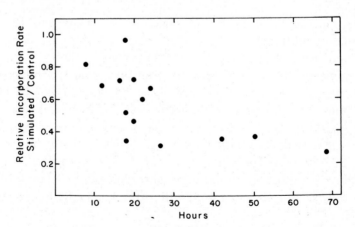

Fig. 9. Effect of electrical stimulation on biosynthesis and plasma membrane incorporation of ^{2}H-, ^{13}C-, ^{15}N-labelled ACh receptors in denervated rat extensor digitorum longus muscle in organ culture. These muscles contracted vigorously with full tetanic tension, in response to the electrical stimulation (100 Hz for 1 sec given every 80 sec). Relative biosynthetic rate was measured as the amount of ^{2}H-, ^{13}C-, ^{15}N-containing receptors (identified by velocity sedimentation sucrose gradients) per mg muscle weight in stimulated muscles divided by that measured in non-stimulated muscles cultured for the same time. Cultures were labelled with ^{2}H-, ^{13}C-, ^{15}N-amino acids for 7–9 h in experiments where stimulation was for 8–18 h. All longer experiments had labelling times of 11–12 h. Each point on the graph represents one experiment, involving two or three stimulated muscles and two or three control, non-stimulated muscles.

tor molecules. One of tne most elusive goals continues to be the assessment of the roles of factors other than muscle activity in regulation of receptor metabolism.

Several recent observations may prove pivotal to further progress. Apparently medium components derived from nervous tissue can alter the number and distribution of ACh receptors (Betz and Osborne, 1977; Cohen and Fischbach, 1977; Christian et al., 1978; Younkin et al., 1978). Receptor distribution is also sensitive to electric fields (Orinda and Poo, 1978). And receptor biosynthesis is accelerated by some drugs which may affect membrane structure and function (Shainberg et al., 1976). There is a promising variety of muscle and nerve-muscle preparations which appear favorable for further biophysical, biochemical and genetic analysis of receptor distribution, metabolism and regulation.

ACKNOWLEDGEMENT

Research in the authors' laboratory was supported in part by a grant from the Muscular Dystrophy Association. D.J.C. was a recipient of a Muscular Dystrophy Association prostdoctoral fellowship.

REFERENCES

Axelrod, D., Ravdin, P., Koppel, D.E., Schlessinger, J., Webb, W.W., Elson, E.L., and Podleski, T.R. (1976) Lateral motion of fluorescently labeled actylcholine receptors in membranes of developing muscle fibers. *Proc. nat. Acad. Sci. (Wash.)*, 73, 4594–4598.

Berg, D.K., and Hall, Z.W. (1974) Fate of α-bungarotoxin bound to acetylcholine receptors of normal and denervated muscle. *Science*, 194, 473–475.

Betz, W. and Osborne, M. (1977) Effect of innervation on acetylcholine sensitivity of developing muscle in vitro. *J. Physiol. (Lond.)*, 270, 75–88.

Brockes, J.P., and Hall, Z.W. (1975) Synthesis of acetylcholine receptor by denervated diaphragm muscle. *Proc. nat. Acad. Sci. (Wash.)*, 72, 1368–1372.

Card, D. (1977) Regulation of acetylcholine receptor metabolism by electrical stimulation. *Carnegie Inst. Wash. Yearbook*, 76, 17–22.

Chang, C.C., and Huang, M.C. (1975) Turnover of junctional and extrajunctional acetylcholine receptors of rat diaphragm. *Nature (Lond.)*, 253, 643–644.

Cohen, S.A., and Fischbach, G.D. (1977) Relative peaks of ACh sensitivity at identified nerve-muscle synapses in spinal cord-muscle cocultures. *Develop. Biol.*, 39, 24–30.

Devreotes, P.N., and Fambrough, D.M. (1975) Acetylcholine receptor turnover in membranes of developing muscle fibers. *J. Cell Biol.*, 65, 335–358.

Devreotes, P.N., and Fambrough, D.M. (1976a) Turnover of acetylcholine receptors in skeletal muscle. *Cold Spring Harbor Symp. quant. Biol.*, 40, 237–251.

Devreotes, P.N., and Fambrough, D.M. (1976b) Synthesis of the acetylcholine receptor by cultured chick myotubes and denervated mouse extensor digitorum longus muscles. *Proc. nat. Acad. Sci. (Wash.)*, 73, 161–164.

Devreotes, P.N., Gardner, J.M., and Fambrough, D.M. (1977). Kinetics of biosynthesis of acetylcholine receptor and subsequent incorporation into plasma membrane of cultured chick skeletal muscle. *Cell*, 10, 365–373.

Fambrough, D.M. (1977). Studies on the glycosylation of ACh receptors. *Carnegie Inst. Wash. Yearbook*, 76, 12–13.

Fambrough, D.M. (1979) Control of acetylcholine receptors in skeletal muscle. *Physiol. Rev.*, in press.

334

Fambrough, D.M., and Devreotes, P.N. (1978) Newly synthesized acetylcholine receptors are located in the Golgi apparatus. *J. Cell Biol.*, 76, 237–244.

Fambrough, D.M., Devreotes, P.N., and Card, D.J. (1977) The synthesis and degradation of acetylcholine receptors. In *Synapses*, G.A. Cottrell and P.N.R. Usherwood (Eds.), Blackie, Glasgow, pp. 202–236.

Fambrough, D.M., and Pagano, R.E. (1977) Positional stability of acetylcholine receptors at the neuromuscular junction. *Carnegie Inst. Wash. Yearbook*, 76, 28–29.

Fischbach, G.D., Berg, D.K., Cohen, S.A., and Frank, E. (1976) Enrichment of nerve-muscle synapses in spinal cord-muscle cultures and identification of relative peaks of ACh sensitivity at sites of transmitter release. *Cold Spring Harbor Symp. quant. Biol.*, 40, 347–357.

Gardner, J.M., and Fambrough, D.M. (1978) Properties of acetylcholine receptor turnover in cultured embryonic muscle cells. In *Maturation of Neurotransmission*, E. Giacobini, M. Giacobini, and A. Vernadakis (Eds.), International Society for Neurochemistry Satellite Symposium. Karger, Basel, pp. 31–40.

Hartzell, H.C., and Fambrough, D.M. (1973) Acetylcholine receptor production and incorporation into membranes of developing muscle fibers. *Develop. Biol.*, 30, 153–165.

Hogan, P.G., Marshall, J.M., and Hall, Z.W. (1976) Muscle activity decreases rate of degradation of α-bungarotoxin bound to extrajunctional receptors. *Nature (Lond.)*, 261, 328–330.

Lømo, T., and Rosenthal, J. (1972) Control of ACh sensitivity by muscle activity in the rat. *J. Physiol. (Lond.)*, 221, 493–513.

Merlie, J.P., Changeux, J.-P., and Gros, F. (1976) Acetylcholine receptor degradation measured by pulse chase labeling. *Nature (Lond.)*, 264, 74–76.

Orinda, N., and Poo, M. (1978) Electrophoretic movement and localization of acetylcholine receptors in the embryonic muscle cell membrane. *Nature (Lond.)*, in press.

Patrick, J., McMillan, J., Wolfson, H., and O'Brien, J.C. (1977) Acetylcholine receptor metabolism in a nonfusing muscle cell line. *J. Biol. Chem.*, 252, 2143–2153.

Reiness, C.G., and Hall, Z.W. (1977) Electrical stimulation of denervated muscles reduces incorporated of methionine into the ACh receptor. *Nature (Lond.)*, 268, 655–657.

Rotundo, R.L., and Fambrough, D.M. (1977) Kinetics of synthesis and release of acetylcholinesterase molecular forms in chick muscle culture. *Soc. Neurosci. Abstr.* III, p. 527.

Shainberg, A., Cohen, S.A., and Nelson, P.G. (1976) Induction of acetylcholine receptors in muscle cultures. *Pflügers Arch.*, 361, 255–261.

Younkin, S.G., Brett, R.S., Davey, B., and Younkin, L.H. (1978) Substances moved by axonal transport and released by nerve stimulation have an innervation-like effect on muscle. *Science*, 200, 1292–1295.

Accumulation of Acetylcholine Receptors at Nerve-Muscle Contacts in Culture

M.W. COHEN, M.J. ANDERSON *, E. ZORYCHTA ** and P.R. WELDON

Department of Physiology, McGill University, Montreal (Canada)

INTRODUCTION

The subsynaptic membrane at neuromuscular junctions in vertebrate skeletal muscle is specialized to respond to minute quantities of acetylcholine (ACh) and differs in this respect from the rest of the sarcolemma which is much less sensitive to the transmitter substance (Miledi, 1960; Feltz and Mallart, 1971; Dreyer and Peper, 1974; Kuffler and Yoshikami, 1975). This functional specialization reflects the fact that ACh receptors are present in the subsynaptic membrane in very high density, the order of magnitude being 10,000/ square μm, whereas elsewhere in the sarcolemma their density is about a thousand-fold lower (Barnard et al., 1971; Fambrough and Hartzell, 1972; Hartzell and Fambrough, 1972; Albuquerque et al., 1974; Fambrough, 1974; Fertuck and Salpeter, 1974, 1976; Burden, 1977a; Orkand et al., 1978).Even at relatively early stages of development when muscle fibres have a more substantial density of receptors along their entire length, receptor density is greatest at the developing neuromuscular junction (Diamond and Miledi, 1962; Bevan and Steinbach, 1977) and appears to be as high as at the adult neuromuscular junction (Burden, 1977a).

To study how ACh receptors become localized in the subsynaptic membrane during formation of the neuromuscular junction we have carried out experiments on cell cultures of neural tube (developing spinal cord) and myotomal muscle derived from 1-day-old embryos of the South African frog, *Xenopus laevis*. In this paper we summarize evidence, presented in detail elsewhere (Anderson et al., 1977; Anderson and Cohen, 1977), supporting the following conclusions: the nerve causes ACh receptors to accumulate along paths of nerve-muscle contact; this nerve-induced accumulation of ACh receptors is associated with functional innervation but can occur even when synaptic activity is blocked; it involves a process of redistribution whereby extrajunctional receptors change their location in the sarcolemma and aggregate at sites of nerve-muscle contact. We also present results showing that in contrast to the neural tube two other sources of nerve, sympathetic ganglia and dorsal root

---

* Present address: Neurobiology Department, The Salk Institute, LaJolla, CA (U.S.A.)
** Present address: Department of Pathology, McGill University, Montreal (Canada).

ganglia, do not induce the development of a high receptor density at sites of contact. These results suggest that the triggering of ACh receptor accumulation by neural tube neurites is due to a specific nerve factor.

MATERIALS AND METHODS

All experiments were carried out on cultures of myotomal muscle cells with or without nerve. One-day-old *Xenopus* embryos were skinned and treated with collagenase in order to isolate the myotomes and their associated neural tube from each other and from adjacent tissues. The isolated myotomes were then transferred to a calcium-magnesium-free solution of trypsin-EDTA to dissociate them into single cells or small clusters of cells. These in turn were plated in culture chambers whose floor consisted of a collagen-coated glass coverslip. For mixed cultures containing muscle as well as neural tube, the isolated neural tubes were likewise dissociated into small clusters of cells before being plated. Two other sources of nerve were also used in some cultures: sympathtic ganglia of *Xenopus* juveniles and dorsal root ganglia of *Xenopus* tadpoles. Like the neural tubes, these ganglia were treated with collagenase and with trypsin-EDTA before being added to the culture chambers. This treatment softened the ganglia but did not cause them to dissociate. Depending on their size the ganglia were cultured either whole or in fragments.

Cultures were maintained at room temperature in a medium consisting of L-15 diluted to 60% with water and Holmes' α-1 protein (0.2 μg/ml). For the first day of culture the medium also contained dialyzed horse serum (5%) in order to facilitate adhesion of the cells to the collagen substrate.

The distribution of ACh receptors on cultured myotomal muscle cells was examined by fluorescence and phase contrast microscopy after staining the receptors with fluorescent α-bungarotoxin (Anderson and Cohen, 1974; Anderson et al., 1977). For most experiments tetramethylrhodamine-labelled toxin was used in preference to fluorescein-labelled toxin because it gives brighter staining and fades less rapidly.

Cultures to be examined by electron microscopy were prepared and handled in the same way as described above except that the floor of the culture chamber consisted of a sheet of collagen-coated Aclar rather than glass (see Masurovrsky and Bunge, 1968). The cultures were fixed with 1.5% glutaraldehyde for 30—60 min, post-fixed with 1% osmium tetroxide, stained with 1% aqueous uranyl acetate, dehydrated in a graded series of ethanol, and embedded in Epon 812. Selected areas were sectioned with a Huxley ultramicrotome parallel to the floor of the culture chamber and examined with a Siemens Elsmiskop I electron microscope.

RESULTS

Description of muscle and neural tube cultures

The dissociated muscle cells attach to the collagen substrate and begin to elongate within a few hours after being added to the culture chamber. They begin to acquire striations during the first day and these increase in prominence over the next 3—4 days. During this period most of the yolk granules originally present in the cells are consumed and the cells grow in size, attaining lengths of up to 300 μm, but remain mononucleated. In all of these respects differentiation of the myotomal muscle cells in culture mimics that during the normal development of the embryos (Nieuwkoop and Faber, 1956; Hamilton, 1969; Muntz, 1975).

The growth of neurites and the formation of nerve-muscle contacts in cultures with neural tube was originally described by Harrison (1910) in his pioneering and classical study which firmly established the validity of the neurone theory. Briefly, by one day in culture neurites with active growth cones are seen extending from the neural tube cells. Many have already contacted muscle cells. The neurites may be as much as 2—3 μm in diameter and, as revealed by electron microscopy, are often composed of a few nerve fibres. They continue to grow during the next day in culture but by the third day most growth ceases and neurites begin to retract. As a result, after three days in culture fewer neurites and fewer contacts with muscle cells are seen.

As originally pointed out by Harrison (1910) some of the muscle cells which are contacted by neurites undergo spontaneous contractions whereas all of the non-contacted muscle cells remain quiescent. Muscle contractions can also be evoked by electrical stimulation of the appropriate neural tube cells. Both the spontaneous and neurally-evoked contractions are abolished by curare and by α-bungarotoxin indicating that they are due to cholinergic synaptic transmis-

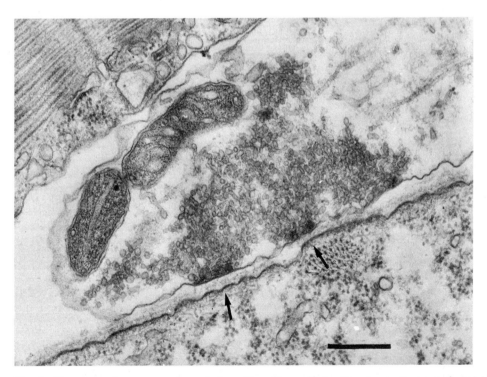

Fig. 1. Nerve-muscle synapse in a 1-day-old culture. Note the dense clusters of vesicles apposed to the axolemma, the basement membrane in the cleft, and the increased electron density of the underlying sarcolemma (arrows). Bar indicates 0.5 μm.

sion. Intracellular recordings have likewise confirmed that the myotomal muscle cells become functionally innervated in mixed cultures (Cohen, 1972).

Electron microscopy indicates that neuromuscular junctions develop in these cultures and that the sequence of development is much the same as during the normal development of the myotomes (see Kullberg et al., 1977). After 2–3 days in culture the nerve-muscle contacts display considerable variability in degree of ultrastructural differentiation. As in vivo, many appear rather immature but a few have the characteristics of relatively mature myotomal neuromuscular junctions, including aggregates of vesicles in close apposition to the junctional axolemma, a prominent basement membrane in the cleft, and a sarcolemma of increased electron density or thickness (Fig. 1).

Accumulation of ACh receptors at sites of nerve-muscle contact

Fluorescent staining of ACh receptors reveals that by 2 days in culture virtually all muscle cells not contacted by nerve have developed patches of high receptor density (see Fig. 5). These patches vary considerably in size, distribution and number. On the whole they appear to increase in size with the age of the culture, but even in 4- and 5-day-old cultures they are rarely more than 20 μm in their largest dimension and never more than 40 μm. The large patches often occur near the ends of the muscle cells or their processes, on the surface in contact with the culture dish. Another common location is in central regions of the cell, on the free surface. In addition to patches of high receptor density autoradiography with ^{125}I-labelled α-bungarotoxin reveals a widespread distribution of receptors whose mean density is some twenty-fold lower than the receptor density in the patches (Anderson et al., 1977). This widespread low density phase of ACh receptors is rather poorly resolved by fluorescent staining.

Quite different patterns of ACh receptor distribution are seen on many nerve-contacted muscle cells (Figs. 2, 3). In 2- and 3-day-old cultures fluorescent staining reveals that over 70% of the contacted cells have regions of high receptor density along the paths of contact (see Table I) and more than half of these cells have no patches of high receptor density elsewhere. The association of high receptor density with paths of nerve-muscle contact is always observed on muscle cells previously identified as being functionally innervated on the basis of spontaneous or neurally-evoked contractions (Anderson et al., 1977). There is considerable variability in the lengths of fluorescent stain along individual paths of nerve-muscle contact but of most interest is the fact that bands of stain sometimes extend for greater distances than the largest patches seen on non-contacted muscle cells. Such examples cannot be explained simply in terms of the neurites having contacted a pre-existing patch of high receptor density. Instead they indicate that the neural tube neurites can induce the development of a high density of ACh receptors at sites of contact.

This conclusion is further supported by experiments in which the patterns of fluorescent stain on individual cells are followed over a period of time. Figure 2 illustrates an example from such an experiment. In this case a 2-day-old culture was stained with fluorescent toxin and then rinsed only mildly so that a low concentration of fluorescent toxin remained in the culture. The first observa-

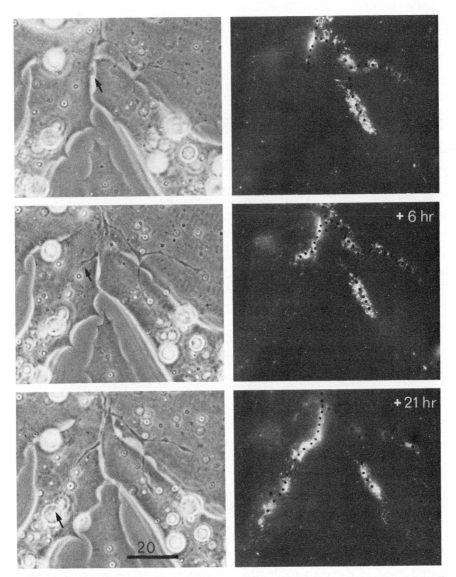

Fig. 2. Changes in the pattern of fluorescent stain on nerve-contacted muscle cells. The culture (2-days-old) was stained and rinsed mildly so that some fluorescent toxin was present throughout the period of observation. The same field was viewed with phase contrast optics (left) and with fluorescence optics (right). Note the neurite growth (arrows) between the first, second (+6 h) and third (+21 h) observations and the corresponding new appearance of fluorescent stain. Also note the loss of fluorescent stain along other paths of nerve-muscle contact. Black dots have been added to the fluorescence micrographs to indicate paths of nerve-muscle contact. Bar indicates 20 μ.

tion, made shortly afterwards, shows a field in which two of the muscle cells are clearly contacted by nerve and have fluorescent stain along the path of contact. A third muscle cell appears to be contacted along its edge and also has stain at this site. The next observation, made 6 h later, shows that the neurite has grown along this third muscle cell and that a high density of receptors has developed along the newly-formed contact. By the time of the third observation, 15 h later, the lengths of this nerve-muscle contact is even greater and, as

340

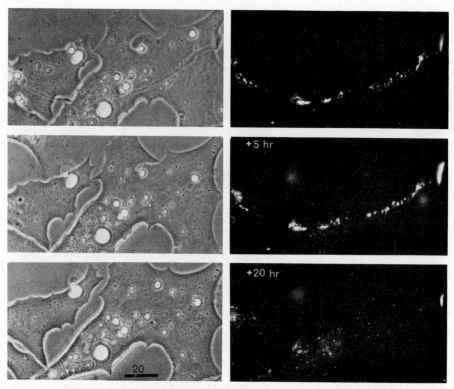

Fig. 3. Loss of fluorescent stain after nerve withdrawal. Procedures were the same as described for Fig. 2. Note that between the first and second (+5 h) observations the neurite withdrew and that by the third (+20 h) observation the fluorescent stain along the original path of nerve-muscle contact had for the most part disappeared. Culture was 2-days-old at the time of the first observation. Bar indicates 20 μm.

before, the region of newly-formed contact has acquired a high receptor density.

In following patterns of stain on nerve-contacted muscle cells one sometimes observes the converse result, the disappearance of fluorescent stain after spontaneous withdrawal of the neurite. Figure 3 shows such an example. The first observation shows a long length of fluorescent stain associated with the path of nerve-muscle contact. The next observation, 5 h later, shows that the neurite has withdrawn but the pattern of fluorescent stain is only marginally altered. However, by the third observation, 15 h later, there is very little stain left along the original path of contact although some new spots of stain have appeared elsewhere on the muscle cell. During the 20 h period in which the observations were made the culture contained a low concentration of fluorescent toxin, as in the experiment of Fig. 2, and no additional fluorescence was observed upon restaining the culture after the third observation.

The results, in addition to emphasizing the role of the nerve in inducing the development of a high receptor density at sites of contact, also raise the possibility that the maintenance of this local high density is initially dependent upon the continued presence of the nerve. However, it is also worth noting that the mere physical presence of the nerve is not a sufficient condition for maintaining the high receptor density. For example, in the experiment of Fig. 2

there was some disappearance of stain along one region of contact (the one on the right hand side) despite the continued presence of the nerve. Such observations suggest, not surprisingly, that the accumulation and maintenance of a high receptor density at sites of contact are dependent upon the state of the neurite and muscle cell.

Receptor accumulation occurs in the absence of synaptic or contractile activity

Myotomal muscle cells can be innervated even when they are cultured in the presence of a concentration (100 μg/ml) of curare which completely obliterates end-plate potentials (Cohen, 1972). Likewise, this high concentration of curare does not prevent the development of long bands of high receptor density along paths of nerve-muscle contact (Anderson et al., 1977). Thus the nerve-induced accumulation of ACh receptors is not triggered by synaptic potentials or muscle contraction or any of the associated changes in membrane permeability. Additional evidence in this regard is that ACh receptors also accumulate at sites of nerve-muscle contact in the presence of 5 μg/ml α-bungarotoxin (see below). This concentration of toxin blocks neurally evoked contraction within 2–3 min and saturates the ACh receptor toxin binding sites within 20 min (Anderson et al., 1977). Since activation of ACh receptors by transmitter must be entirely eliminated under these conditions it is clearly not essential for the nerve-induced accumulation of receptors. Instead these results indicate that the nerve-induced accumulation of receptors is triggered in some other way, by a factor which is either released from the neurites or associated with their surface membrane.

Dependence of receptor accumulation on nerve type

A question of obvious importance is whether all classes of nerve are able to induce accumulation of ACh receptors or whether this induction displays neural specificity. The latter possibility seemed likely in view of the fact that some of muscle cells contacted by NT (neural tube) neurites do not develop a high density of ACh receptors along the path of contact. A simple explanation of this result would be that the neural tube contains different types of nerve cells (see Spitzer, 1976) and that only some of these are competent to induce receptor accumulation. But other explanations, such as variability in the responsiveness of the cultured muscle cells, are not excluded by the observations presented so far. Experiments were therefore undertaken in which two other sources of nerve, sympathetic ganglia (SG) and dorsal root ganglia (DRG) were tested for their ability to induce accumulation of ACh receptors at sites of contact.

Explants of SG were cultured by themselves, usually for 4–6 days, in order to allow them to become firmly attached to the culture dish and to develop a reasonable outgrowth of neurites. Myotomal muscle cells were then plated in the same cultures. The SG neurites continued to grow and over the next 2–4 days contacted many muscle cells (Fig. 4). Explants of DRG adhered and grew more quickly than SG explants so that for these cultures the muscle cells were usually plated at the same time as the DRG explants or within the first 3 days.

342

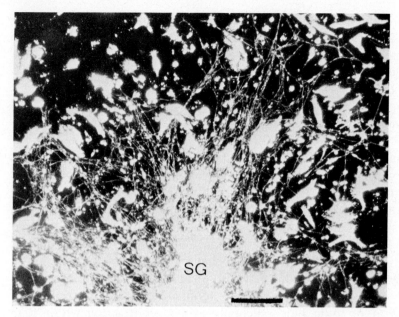

Fig. 4. Dark-field view showing scattered myotomal muscle cells and neuritic outgrowth from an explant of sympathetic ganglion (SG). The ganglion was in culture for 9 days and the muscle cells for 3 days. Bar indicates 300 μm.

In both types of culture ACh receptors were stained with fluorescent α-bungarotoxin and their distribution examined 2–4 days after plating the muscle cells.

In contrast to the muscle cells contacted by NT neurites, most of those contacted by SG or DRG neurites displayed no stain along the path of contact and had patches of stain elsewhere (Fig. 5). In the few examples where some stain was present along the path of contact it was never in the form of a characteristic band of stain as seen on many NT-contacted muscle cells but rather appeared like the patches that occur in non-contacted muscle cells. The results are summarized in Table I. Over 70% of NT-contacted muscle cells had some stain along the path of contact whereas the corresponding value for DRG-contacted muscle cells was 9%, and for SG-contacted cells it was 5%. In addition,

TABLE I

FLUORESCENT STAIN ALONG PATHS OF NERVE-MUSCLE CONTACT

Type of nerve	No. of cultures examined	No. of contacted muscle cells examined	Muscle cells with stain along path of contact (%) *
NT	9	438	73 ± 4
DRG	8	228	9 ± 3
SG	14	414	5 ± 1
SG + NT	6	131 (SG-contacted)	4 ± 3
		249 (NT-contacted)	77 ± 2

* Means and standard errors are based on values obtained in individual cultures.

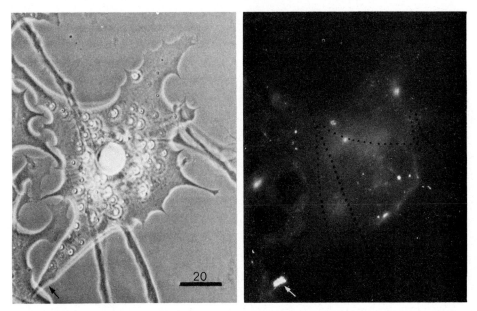

Fig. 5. Pattern of fluorescent stain on a muscle cell contacted by SG neurites. Note the general lack of stain along the paths of nerve-muscle contact. There is instead a bright patch of stain (arrow) near the end of one process of the muscle cell and several other smaller patches elsewhere on the cell. Age in culture; 7 days for SG; 3 days for muscle. Bar indicates 20 μm.

of those NT-contacted muscle cells that did have some stain along the path of contact only 37% had patches of stain elsewhere. In contrast almost all (over 95%) of the SG- and DRG-contacted muscle cells had patches of stain elsewhere even in those cases where there was some stain associated with the contact. These results suggest that SG and DRG neurites do not significantly influence the distribution of ACh receptors on the myotomal muscle cells.

Although it seemed unlikely, it could be argued that the DRG explants and SG explants modify the culture medium in some way which makes the myotomal muscle cells unresponsive to the neurites which contact them. This possibility was tested in cultures of SG explants by adding not only myotomal muscle cells but also neural tube cells. Contacts by NT neurites and SG neurites could easily be distinguished by following the neurites back to their source since SG explants were considerably larger and their neuritic outgrowth much more extensive than NT explants. As summarized in Table I the results obtained with these cultures were similar to those obtained when only one type of nerve explant was present. More than 70% of the NT-contacted muscle cells had stain along the path of contact whereas the corresponding value for SG-contacted muscle cells was only 4%. Clearly the ineffectiveness of SG neurites cannot be explained in terms of some modification of the culture medium. Instead these results indicate that the triggering of ACh receptor accumulation by neural tube neurites is due to a specific factor.

344

*Redistribution of receptors within the sarcolemma contributes
to receptor accumulation*

The lateral movement of surface membrane constituents has been well-established in recent years and has been found to contribute to phenomena such as capping whereby multivalent ligands such as antibodies and lectins cause their receptors to aggregate at one pole of the cell (for reviews see Bretscher and Raff, 1975; Edelman, 1976). To study whether the accumulation of ACh receptors at sites of nerve-muscle contact might involve their redistribution within the sarcolemma, experiments were carried out as follows. Myotomal muscle cells were cultured by themselves for 2–3 days and then stained with fluorescent α-bungarotxin. After several rinses the culture medium was replaced with one containing a high concentration (5 µg/ml) of unlabelled toxin and neural tube cells were added to the cultures. By this procedure only those receptors present in the sarcolemma prior to the addition of nerve would be stained, whereas any receptors incorporated into the membrane after the addition of the nerve would not be visible because unlabelled, rather than fluorescent, toxin would be bound to them. Observations were made 1–3 days after adding the neural tube cells. As in freshly stained cultures, a small but statistically significant number of examples were obtained in which the length of stain along paths of nerve-muscle contact was greater than the largest patches on non-contacted muscle cells (Anderson and Cohen, 1977). Accumulation of stained receptors at sites of nerve-muscle contact was also observed directly by following the same cell over a period of about one day (Anderson and Cohen, 1977). These experiments indicate that ACh receptors can change their position in the sarcolemma and aggregate at sites of nerve-muscle contact thereby contributing to the development of a high receptor density.

DISCUSSION

ACh receptors accumulate at sites of nerve-muscle contact in cultures of neural tube and myotomal muscle derived from *Xeonpus* embryos. This accumulation of receptors is induced by the NT neurites and is associated with functional innervation. Since ultrastructurally-differentiated nerve-muscle synapses also develop in these cultures it is most reasonable to assume that they are the sites of function and high receptor density, and that the accumulation of receptors is part of the overall process of neuromuscular synaptogenesis. The lack of effect of SG and DRG neurites on receptor distribution provides additional support that the NT-induced accumulation of ACh receptors is part of the normal process of synapse formation.

Evidence for nerve-induced accumulation of ACh receptors has recently been obtained in other studies as well. Frank and Fischbach (1977) have combined measurements of ACh sensitivity with focal external recordings of synaptic potentials in cultures of pectoral muscle cells and brachial spinal cord derived from chick embryos. By monitoring the same regions over a period of time they have found that new high levels of ACh sensitivity develop at sites of, and as a result of, innervation. Lømo and Slater (1976) have likewise reported nerve-induced development of high ACh sensitivity at ectopic synapses during reinnervation of rat soleus muscle. Also consistent with a process of neural induction is the finding that during normal development of the neuromuscular junction in the rat diaphragm (Bevan and Steinbach, 1977) and in *Xenopus* myotomes (Chow and Cohen, 1977) high densities of ACh receptors develop after the arrival of the growing nerve fibres.

Nerve-induced receptor accumulation occurs even when the receptors are entirely blocked and is therefore not dependent on synaptic or contractile activity or any of the associated changes in membrane permeability. The accumulation must be triggered in some other way, by a factor which is either released from the nerve or associated with its surface membrane. This factor appears to be specific inasmuch as DRG and SG neurites, unlike NT neurites, do not induce receptor accumulation. Furthermore, the rapidity with which receptor accumulation can occur after contact is made between NT neurites and muscle (see Fig. 2) suggests that the factor can act very shortly after the growing neurite comes into close proximity with the muscle cell. This raises the possibility that the triggering factor may participate in one of the primary interactions leading to synapse formation. In this context is should be recalled that skeletal muscle can be innervated not only by motor neurones but also by other cholinergic neurones (Landmesser, 1971, 1972; Bennett et al., 1973; Nurse and O'Lague, 1975; Hooisma et al., 1975; Betz, 1976; Obata, 1977; Schubert et al. 1977) and that at least one class of these "inappropriate" cholinergic neurones (chick ciliary ganglia) appears to be able to trigger receptor accumulation (Betz and Osborne, 1977). It will be of interest in future studies to try to assess further whether all cholinergic neurones have this ability and whether different neurones exhibit different capacities to induce receptor accumulation. The existence of such differences might help explain competitive interactions between different neurones for innervation of a muscle cell, such as occurs during normal development (Redfern, 1970; Bagust et al., 1973; Bennett and Pettigrew, 1974; Brown et al., 1976) and in response to experimental manipulation (Schmidt and Stefani, 1976; Bennett and Raftos, 1977; Yip and Dennis, 1977; Kuffler et al., 1977).

Our studies have indicated that the accumulation of ACh receptors at sites of nerve-muscle contact involves a process of redistribution whereby receptors change their position in the sarcolemma and aggregate at the sites of contact. Additional evidence that ACh receptors are mobile within the sarcolemma has also been provided by Axelrod et al. (1976) in experiments on cultured rat myotubes. These investigators stained ACh receptors with fluorescent α-bungarotoxin and examined the recovery of fluorescence in small areas after bleaching with high intensity excitation. Recovery of fluorescence after bleaching is attributed to the movement of unbleached, fluorochrome-labelled, receptors from adjacent regions into the bleached area. Very little recovery occurred after bleaching receptors within high density patches (hot spots) but significant recovery of fluorescence was observed when regions of lower receptor density were bleached. These results suggest that the receptors within high density patches are for the most part immobile whereas many of these which are present at lower density in the rest of the sarcolemma are mobile (Axelrod et al., 1976). Aggregation of receptors at sites of nerve-muscle contact is likewise most simply accounted for in terms of a loss of mobility.

At adult neuromuscular junctions ACh receptors have a half-life of at least several days (Berg and Hall, 1974, 1975; Chang and Huang, 1975). In contrast, extrajunctional receptors are much less stable and survive with a half-life of about one day (Berg and Hall, 1974, 1975; Chang and Huang, 1975; Devreotes and Fambrough, 1975; Merlie et al., 1976). This difference between extrajunctional receptors is apparent rather early in development; for example, it has been detected in neonatal rat muscle (Berg and Hall, 1975) and in muscle from 3-week-old chicks (Burden, 1977b). Of particular interest is the finding by Burden (1977a) that in muscle from younger chicks junctional receptors have the same short lifespan as extrajunctional receptors. Similar results have also been reported for junctional and extrajunctional receptors on chick myotubes in cultures (Frank and Fischbach, 1977). Despite the short lifespan of the junctional receptors in young chick muscle, junctional receptor density is maintained at the same high level as in the adult and it follows that receptors are added to the subsynaptic membrane at the same rate as they are degraded (Burden, 1977a). Such replenishment could occur by movement of extrajunctional receptors into the subsynaptic membrane or by preferential insertion of receptors directly into the subsynaptic membrane. Neither of these processes is mutually exclusive and both could operate in the formation as well as the maintenance of the high receptor density at sites of innervation.

SUMMARY

Nerve-induced accumulation of ACh receptors has been demonstrated to occur at sites of nerve-muscle contact in cell cultures of neural tube and myotomal muscle derived from *Xenopus* embryos. This acculumation of receptors is associated with functional innervation of the muscle cells and presumably occurs at the developing synaptic sites which have been observed in these cultures by electron microscopy. The accumulation can occur rapidly, within a few hours after nerve-muscle contact is made, and does not depend on synaptic or contractile activity. It involves a process of receptor redistribution whereby extrajunctional receptors aggregate at the site of contact. The neurites of two other sources of nerve, dorsal root ganglia and synpathetic ganglia, do not cause ACh receptors to accumulate at sites of muscle contact. It is suggested that the ACh receptor accumulation which is induced by neural tube neurites is due to the action of a specific neural factor.

ACKNOWLEDGEMENTS

We thank Mr. V. Vipparti for expert technical assistance. It is also a pleasure to express our gratitude to Professor Stephen Kuffler who introduced one of us some eleven years ago to the study of synapse formation in cultured explants of *Xenopus* neural tube and myotomal muscle. This work was supported by the Medical Research Council of Canada. Personal support (M.W.C.) from the Conseil de la Recherche en Santé du Québec is also gratefully acknowledged.

REFERENCES

Albuquerque, E.X., Barnard, A.E., Porter, C.W., and Warnick, J.E. (1974) The density of acetylcholine receptors and their sensitivity in the postsynaptic membrane of muscle endplates. *Proc. nat. Acad. Sci. (Wash.)*, 71, 2818–2822.

Anderson, M.J. and Cohen, M.W. (1974). Fluorescent staining of acetylcholine receptors in vertebrate skeletal muscle. *J. Physiol. (Lond.)*, 237, 385–400.

Anderson, M.J. and Cohen, M.W. (1977) Nerve-induced and spontaneous redistribution of acetylcholine receptors on cultured muscle cells. *J. Physiol. (Lond.)*, 268, 757–773.

Anderson, M.J., Cohen, M.W. and Zorychta, E. (1977) Effects of innervation on the distribution of acetylcholine receptors on cultured muscle cells. *J. Physiol. (Lond.)*, 268, 731–756.

Axelrod, D., Ravdin, P., Koppel, D.E., Schlessinger, J., Webb, W.W., Elson, E.L., Podleski, T.R. (1976). Lateral motion of fluorescently labeled acetylcholine receptors in membranes of developing muscle fibres. *Proc. nat. Acad. Sci. (Wash.)*, 73, 4594–4598.

Bagust, J., Lewis, D.M., and Westerman, R.A. (1973) Polyneuronal innervation of kitten skeletal muscle. *J. Physiol. (Lond.)*, 229, 241–255.

Barnard, E.A., Wieckowski, J. and Chiu, T.H. (1971) Cholinergic receptor molecules and cholinesterase molecules at mouse skeletal muscle junctions. *Nature (Lond.)*, 234, 207–209.

Bennett, M.R. and Pettigrew, A.G. (1974). The formation of synapses in striated muscle during development. *J. Physiol. (Lond.)*, 241, 515–545.

Bennett, M.R. and Raftos, J. (1977) The formation and regression of synapses during reinnervation of axolotl skeletal muscles. *J. Physiol. (Lond.)*, 265, 261–295.

Bennett, M.R., McLachlan, E.M., and Taylor, R.S. (1973) The formation of synapses in mammalian striated muscle reinnervated with autonomic preganglionic nerves. *J. Physiol. (Lond)*. 233, 501–517.

Berg, D.K. and Hall, Z.W. (1974) Fate of α-bungarotoxin bound to acetylcholine receptors of normal and denervated muscle. *Science*, 184, 473–475.

Berg, D.K. and Hall, Z.W. (1975) Loss of α-bungarotoxin from junctional and extrajunctional acetylcholine receptors in rat diaphragm in vivo and in organ culture. *J. Physiol. (Lond.)*, 252, 771–789.

Betz, W. (1976) The formation of synapses between chick embryo skeletal muscle and ciliary ganglia grown in vitro. *J. Physiol. (Lond.)*, 254, 63–73.

Betz, W. and Osborne, M. (1977) Effects of innervation on acetylcholine sensitivity of developing muscle in vitro. *J. Physiol. (Lond.)*, 270, 75–88.

Bevan, S. and Steinbach, J.H. (1977) The distribution of α-bungarotoxin binding sites on mammalian skeletal muscle developing in vivo. *J. Physiol. (Lond.)*, 267, 195–213.

Bretscher, M.S. and Raff, M.C. (1975) Mammalian plasma membranes. *Nature (Lond.)*, 258, 43–49.

Brown, M.C., Jansen, J.K.S., and Van Essen, D. (1976) Polyneuronal innervation of skeletal muscle in newborn rats and its elimination during maturation. *J. Physiol. (Lond.)*, 261, 387–422.

Burden, S. (1977a) Development of the neuromuscular junction in the chick embryo: the number, distribution, and stability of acetylcholine receptors. *Develop. Biol.*, 57, 317–329.

Burden, S. (1977b) Acetylcholine receptors at the neuromuscular junction: developmental change in receptor turnover. *Develop. Biol.*, 61, 79–85.

Chang, C.C., and Huang, M.C. (1975) Turnover of junctional and extrajunctional acetylcholine receptors of rat diaphragm. *Nature (Lond.)*, 253, 643–644.

Chow, I. and Cohen, M.W. (1977) Distribution of acetylcholine receptors in developing myotomes of *Xenopus laevis*. *Proc. Canad. Fed. biol. Soc.*, 20, 44.

Cohen, M.W. (1972) The development of neuromuscular connexions in the presence of (+)-tubocurarine. *Brain Res.*, 41, 457–463.

Devreotes, P.N. and Fambrough, D.M. (1975) Acetylcholine receptor turnover in membranes of developing muscle fibers. *J. Cell. Biol.*, 65, 335–358.

Diamond, J. and Miledi, R. (1962) A study of foetal and newborn rat muscle fibres. *J. Physiol. (Lond.)*, 162, 393–408.

Dreyer, F. and Peper, K. (1974) The acetylcholine sensitivity in the vicinity of the neuromuscular junction of the frog. *Pflügers Arch. ges. Physiol.*, 348, 273–286.

Edelman, G.M. (1976) Surface modulation in cell recognition and cell growth. *Science,* 192, 218–226.

Fambrough, D.M. (1974) Acetylcholine receptors: revised estimates of extrajunctional receptor density in denervated rat diaphragm. *J. gen. Physiol.* 64, 468–472.

Fambrough, D.M. and Hartzell, H.C. (1972) Acetylcholine receptors: number and distribution at neuromuscular junctions in rat diaphragm. *Science,* 176, 189–191.

Feltz, A. and Mallart, A. (1971) An analysis of acetylcholine responses of junctional and extrajunctional receptors of frog muscle fibres. *J. Physiol. (Lond.),* 218, 85–100.

Fertuck, H.C. and Salpeter, M.M. (1974) Localization of acetylcholine receptor by ^{125}I-α-bungarotoxin binding at mouse motor endplates. *Proc. nat. Acad. Sci. (Wash.),* 71, 1376–1378.

Fertuck, H.C. and Salpeter, M.M. (1976) Quantitation of junctional and extrajunctional acetylcholine receptors by electron microscope autoradiography after ^{125}I-α-bungarotoxin binding at mouse neuromuscular junctions. *J. Cell Biol.,* 69, 144–158.

Frank, E. and Fischbach, G.D. (1977) ACh receptors accumulate at newly formed nerve-muscle synapses in vitro. In *Cell and Tissue Interactions* (J.W. Lash and M.M. Burger, Eds.), Raven Press, New York, pp. 285–291.

Hamilton, L. (1969) The formation of somites in *Xenopus. J. Embryol. exp. Morphol.,* 22, 253–264.

Harrison, R.G. (1910) The outgrowth of the nerve fibre as a mode of protoplasmic movement. *J. exp. Zool.,* 9, 787–846.

Hartzell, H.C. and Fambrough, D.M. (1972) Acetylcholine receptors: distribution and extrajunctional density in rat diaphragm after denervation correlated with acetylcholine sensitivity. *J. gen. Physiol.,* 60, 248–262.

Hooisma, J., Slaff, D.W., Meeter, E., and Stevens, W.F. (1975) The innervation of chick striated muscle fibres by the chick ciliary ganglion in tissue culture. *Brain Res.,* 85, 79–85.

Kuffler, D., Thompson, W. and Jansen, J.K.S. (1977) The elimination of synapses in multiply-innervated skeletal muscle fibres of the rat: dependence on distance between end-plates. *Brain Res.,* 138, 353–358.

Kuffler, S.W. and Yoshikami, D. (1975) The distribution of acetylcholine sensitivity at the postsynaptic membrane of vertebrate skeletal twitch muscles: iontophoretic mapping in the micron range. *J. Physiol. (Lond.),* 244, 703–730.

Kullberg, R.W., Lentz, T.L. and Cohen, M.W. (1977) Development of the myotomal neuromuscular junction in *Xenopus laevis:* an electrophysiological and fine-structural study. *Develop. Biol.,* 60, 101–129.

Landmesser, L. (1971) Contractile and electrical responses of vagus-innervated frog sartorius muscles. *J. Physiol. (Lond.),* 213, 707–725.

Landmesser, L. (1972) Pharmacological properties, cholinesterase activity and anatomy of nerve muscle junctions in vagus-innervated frog sartorius. *J. Physiol (Lond.),* 220, 243–256.

Lømo, T. and Slater, C.R. (1976) Induction of ACh sensitivity at new neuromuscular junctions. *J. Physiol. (Lond.),* 258, 107P–108P.

Masurovsky, E.G. and Bunge, R.P. (1968) Fluoroplastic coverslips for long term nerve tissue culture. *Stain Technol.,* 43, 161–165.

Merlie, J.P., Changeux, J.P. and Gros, F. (1976) Acetylcholine receptor degradation measured by pulse chase labelling. *Nature (Lond.),* 264, 74–76.

Miledi, R. (1960) Junctional and extrajunctional ACh receptors in skeletal muscle fibres. *J. Physiol. (Lond.),* 151, 24–30.

Muntz, L. (1975) Myogenesis in the trunk and leg muscle during development of the tadpole of *Xenopus laevis. J. Embryol. exp. Morphol.,* 33, 757–774.

Nieuwkoop, P.D. and Faber, J. (1956) *Normal Table of Xenopus laevis (Daudin),* 2nd edn., Amsterdam, Nort Holland.

Nurse, C.A. and O'Lague, P.H. (1975) Formation of cholinergic synapses between dissociated sympathetic neurons and skeletal myotubes of the rat in cell culture. *Proc. nat. Acad. Sci. (Wash.),* 72, 1955–1959.

Obata, K. (1977) Development of neuromuscular transmission in culture with a variety of neurons and in the presence of cholinergic substances and tetrodoxin. *Brain Res.,* 119, 141–153.

Orkand, P.M., Orkand, R.K. and Cohen, M.W. (1978) Acetylcholine receptor distribution on *Xenopus* slow muscle fibres determined by α-bungarotoxin binding. *Neuroscience,* in press.

Redfern, P.A. (1970) Neuromuscular transmission in newborn rats. *J. Physiol. (Lond.)*, 209, 701–709.

Schmidt, H. and Stefani, E. (1976). Reinnervation of twitch and slow muscle fibres of the frog after crushing the motor nerves. *J. Physiol. (Lond.)*, 258, 99–123.

Schubert, D., Heinemann, S., Kidokoro, Y. (1977) Cholinergic metabolism and synapse formation by a rat nerve cell line. *Proc. nat. Acad. Sci. (Wash.)*, 74, 2579–2583.

Spitzer, N.C. (1976) Chemosensitivity of embryonic amphibian neurons in vivo and in vitro. *Proc. 6th Ann. Meet. Soc. Neurosci.*, 2, 204.

Yip, J.W. and Dennis, M.J. (1977) Suppression of transmission at foreign synapses in adult newt muscle involves reduction in quantal content. *Nature (Lond.)*, 260, 350–352.

Morphogenesis of the Cholinergic Synapse in Striated Muscle

JUHANI JUNTUNEN

Department of Neurology, University of Helsinki, SF - 00290 Helsinki 29 (Finland)

INTRODUCTION

The main purpose of this paper is to review the morphogenesis of the cholinergic synapse in mammalian striated muscle with special reference to some experimental approaches to the mechanism regulating this morphogenesis. Comprehensive reviews on trophic function of the nerve, a phenomenon also related to myoneural morphogenesis, have been published recently by Gutmann (1976) and Drachman (1976). The development of the cholinergic innervation in general has been thoroughly reviewed by Fambrough (1976). The majority of experimental studies on mammalian myoneural morphogenesis have been made on rats but our knowledge of the basic molecular events taking place during the myoneural morphogenesis has been considerably increased by the numerous studies on the development of myoneural junctions in chick and in amphibian larvae. Tissue culture techniques offer further important models for experimental studies on synaptogenesis. Since these subjects are touched on in other papers in this book, only limited references are given here. Although there may be a great diversity in the developmental events of different species, in the different muscles of one species and even in the different muscle fibres of one muscle, their morphological features are principally similar. The present paper deals mainly with the morphogenesis of the myoneural junction of rat, but for comparison, the corresponding human myoneural morphogenesis has also been touched on.

CHOLINESTERASE OF THE MYONEURAL JUNCTION

After the pioneering histochemical demonstration of cholinesterase activity in the myoneural junction by Koelle and Friedenwald (1949), the nature and distribution of cholinesterases have been the subject of a great number of investigations. It soon became apparent that acetylcholinesterase (AChE) is localized in the vicinity of the postsynaptic membrane of the myoneural junction. The exact localization of AChE has, however, caused much discussion. The question is whether the histochemical reaction products reflect the distribution of enzyme or merely the distribution of binding sites for the final

reaction products. The radioautographic techniques employed for the studies on the localization of the active site of AChE do, however, strongly support the postsynaptic localization of this enzyme (Salpeter, 1967).

A critical review on the characteristics of AChE histochemistry, with special emphasis on the technical sources of error involved in the different methods, has been made by Friedenberg and Seligman (1972). It is possible, by light microscopy, using selective inhibitors and substrates, to demonstrate by AChE histochemistry the postsynaptic infoldings of the myoneural junction (Eränkö and Teräväinen, 1967), a finding of particular importance for experimental studies on the development of myoneural junction. This histochemically demonstrable AChE in the myoneural junction probably represents the end-plate-specific form of this enzyme (Vigny et al. 1976), and it can usefully be employed in studies on the myoneural morphogenesis at the optical microscopic level.

APPEARANCE OF THE LOCAL AChE ACTIVITY IN THE MUSCLE

It is generally believed that a histochemically demonstrable AChE appears slightly after the neuromuscular contact has been established between the motor nerve and the muscle. On the basis of concomitant histological and electrophysiological studies, it seems that the appearance of this local AChE activity in the muscle coincides with the initiation of neuromuscular transmission (Bennett and Pettigrew, 1974). Although it is evident that developing or regenerating nerve terminals lie in the vicinity of the muscle before the formation of the functional cholinergic myoneural synapse, the neural influence on this development is not clear. The view that the embryonic muscle fibre is uniformly sensitive to iontophoretically applied acetylcholine before innervation occurs (Diamon and Miledi, 1962) has been questioned by recent studies using tissue culture and radioautographic techniques. These indicate that the clustering of acetylcholine receptors may occur independent of neural influence, but the significance of the clustering remains unknown (see e.g. Fischbach and Cohen, 1974; Fambrough et al., 1974; Bekoff and Betz, 1976). As judged from the appearance of AChE activity the formation of a primitive myoneural junction occurs in the tibialis anterior muscle on the 18th day of gestation on the rat (Teräväinen, 1968a) in the 10th gestational week in human embryos; junctions are found in the human intercostal muscles at 8.6 gestational weeks (Juntunen and Teräväinen, 1972). The time of formation of myoneural junctions varies not only in different species but also in different muscles of one animal owing to the time-course of the cervicocaudal and the cervicorostral closure and maturation of the neural tube. This developmental gradient is retained throughout the growth of the animal. The appearance of the AChE-positive myoneural junctions of the muscle during embryogenesis probably reflects the formation of a functional neuromuscular system, since at about the same time a primitive fusimotor innervation is established (Milburn, 1973), the concentration of end-plate-specific cholinesterase is doubled (Vigny et al., 1976) and end-plate potentials can be registered (Bennett et al., 1974).

MORPHOGENESIS OF THE POSTSYNAPTIC MEMBRANE
OF THE MYONEURAL JUNCTION

In contrast to the relatively simple morphological differentiation of the presynaptic axon terminal, i.e. accumulation of the synaptic vesicles and formation of the relatively thick presynaptic membrane, (see e.g. Lentz, 1969) more profound changes occur in the postsynaptic membrane during the morphogenesis of the myorneural junction (see e.g. Teräväinen, 1968b). A schematic illustration shows the successive stages of development during myoneural morphogenesis after elimination of inappropriate myoneural contacts (Fig. 1). Ultrastructural techniques usually reveal the detailed morphological changes of one axon terminal and its postsynaptic membrane, while a more general view of the morphogenesis of the myoneural junction can be achieved by histochemical techniques in which the localized AChE shows the outlines of the postsynaptic membrane and infoldings at a light microscopic level. At the time when a functional myoneural synapse is established, the axon terminals are apposed to a smooth muscle membrane, shown as a narrow band in side view (Figs. 1a, 2a and 3). Thereafter the area of the postsynaptic membrane expands, probably because of a slight separation of the axon terminals from each other (Figs. 1b and 2b). The most important stage in the morphogenesis of the postsynaptic membrane of the myoneural junctions is the formation of the characteristic postsynaptic infoldings, which occurs in the rat tibialis anterior muscle at the age of about 5 days, i.e. a week after the primitive myoneural junction is formed on day 18 of gestation. Similar postsynaptic infoldings were seen in the human tibialis anterior muscle at the age of 19.2 weeks (Juntunen and Teräväinen, 1972). First the axon terminals depress the muscle slightly, giving the myoneural junction a cup-like appearance in side view, and the primitive junctional folds are formed simultaneously with the increasing electron density of the postsynaptic membrane (Figs. 1c, 2c and 4). Thereafter, concomitant with the deepening of the postsynaptic infoldings, the area of myoneural junctions gradually expands due to ramification of axon terminals. Segmentation of the area of AChE activity occurs after the axon terminals are separated from each other (Figs. 1d–e, 2d–f). The histochemical appearance of a mature myoneural junction shows a complex area of AChE

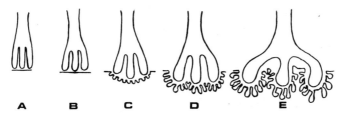

A   B   C   D   E

Fig. 1. Structural development of the myoneural junction: successive stages. The schematic representation is based on previously published data on the morphogensis of the myoneural junctions. The motor axon ramifies into many terminals, which are first apposed to a smooth muscle membrane (a). Subsequently, the postsynaptic area expands (b), a depression is formed and postsynaptic infoldings appear in it (c). The axon terminals are gradually separated (d) and each is surrounded by a folded portion of the postsynaptic membrane of the muscle fibre with concomitant deepening of the postsynaptic infoldings (e).

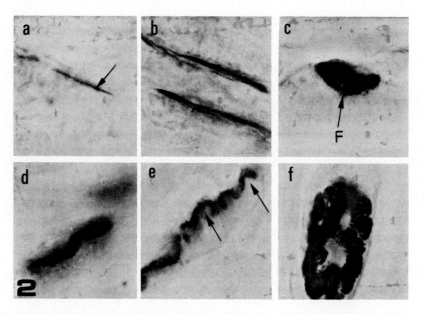

Fig. 2. Acetylcholinesterase (AChE) activity in the developing myoneural junction of the rat tibialis anterior muscle. Acetylthiocholine was used as the substrate and 10^{-5} M iso-OMPA as the inhibitor of non-specific cholinesterases. X900. (a) Primitive myoneural junction of newborn rat, completely devoid of postsynaptic infoldings. Arrow indicates postsynaptic membrane. (b) Widening of the area of myoneural AChE activity before formation of postsynaptic infoldings at the age of 4 days. Two primitive myoneural junctions in adjacent muscle fibres. (c) Myoneural junction in a one-week-old animal. Cup-like depression and postsynaptic infoldings (F) are clearly visible. (d) and (e) Subsequent morphological events: separation of terminal axons from each other (arrows in Fig. e) and deepening of postsynaptic infoldings. (f) Mature myoneural junctions can be seen in the tibialis anterior muscle of rat about three weeks after formation of first primitive acetylcholinesterase-positive myoneural junction.

activity, its form depending on the plane of section of the muscle (Fig. 2f). The increasing electron density of the walls of the postsynaptic infoldings in glutaraldehyde-osmium tetroxide fixed muscles during morphogenesis probably depends on the accumulation of protein particles in the postsynaptic membrane as revealed by freeze-etching techniques (e.g. Sandri et al., 1972).

Structurally mature myoneural junctions are seen in the rat tibialis anterior muscle at the age of 3—4 weeks, while the human tibialis anterior muscle shows relatively immature myoneural junctions at birth. The adult appearance is attained in human muscles during the first 4 postnatal years.

REGULATION OF THE MORPHOGENESIS OF THE CHOLINERGIC SYNAPSE IN THE STRIATED MUSCLE

There is convincing evidence that an intact innervation is necessary for the development of the myoneural junction (see e.g. Juntunen, 1974), and the morphogenesis of the myoneural junction is generally considered a suitable model for experimental studies on the trophic functions of the nerve (see Gutmann, 1976, for references). On the basis of such experimental studies, it

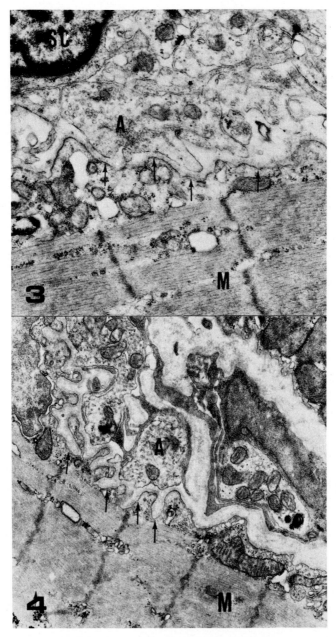

Fig. 3. Electron micrograph of myoneural junction of newborn rat. Axon terminal (A) is apposed to a smooth muscle (M) plasma membrane (arrows) devoid of any postsynaptic thickening or infoldings. (Routine glutaraldehyde-osmium tetroxide fixation). SC, Schwann cell. X14,760.

Fig. 4. Myoneural junction of the tibialis anterior muscle of a one-week-old rat. Postsynaptic infoldings are clearly visible (arrows). A, axon terminal; M, muscle. X9850.

seems that morphogenesis of the postsynaptic membrane is controlled by a developing or regenerating nerve alone; compression of the nerve at different ages of the animal delays the development of the myoneural junction by the time required for the nerve to regenerate. This is followed by normal myo-

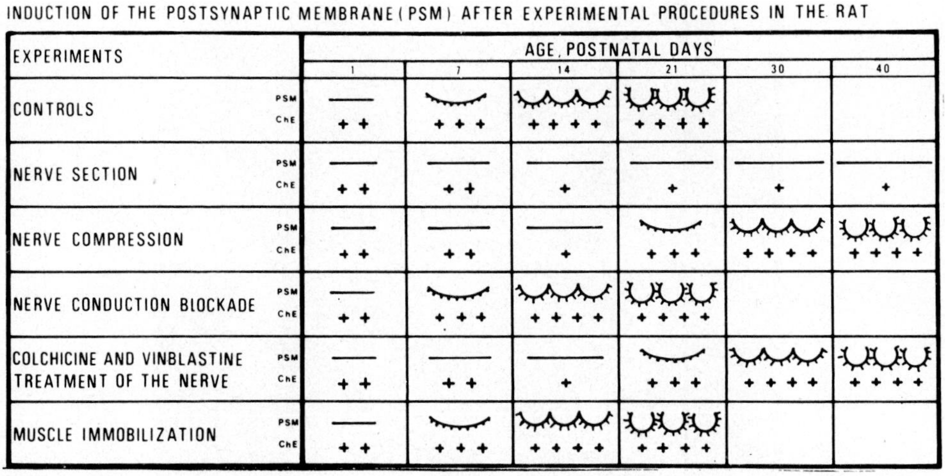

INDUCTION OF THE POSTSYNAPTIC MEMBRANE (PSM) AFTER EXPERIMENTAL PROCEDURES IN THE RAT

EXPERIMENTS		AGE POSTNATAL DAYS					
		1	7	14	21	30	40
CONTROLS	PSM	—	⌣	ᴧᴧᴧ	ᴴᴴᴴ		
	ChE	+ +	+ + +	+ + + +	+ + + +		
NERVE SECTION	PSM	—	—	—	—	—	—
	ChE	+ +	+ +	+	+	+	+
NERVE COMPRESSION	PSM	—	—	—	⌣	ᴧᴧᴧ	ᴴᴴᴴ
	ChE	+ +	+ +	+	+ + +	+ + + +	+ + + +
NERVE CONDUCTION BLOCKADE	PSM	—	⌣	ᴧᴧᴧ	ᴴᴴᴴ		
	ChE	+ +	+ + +	+ + + +	+ + + +		
COLCHICINE AND VINBLASTINE TREATMENT OF THE NERVE	PSM	—	—	—	⌣	ᴧᴧᴧ	ᴴᴴᴴ
	ChE	+ +	+ +	+	+ + +	+ + + +	+ + + +
MUSCLE IMMOBILIZATION	PSM	—	⌣	ᴧᴧᴧ	ᴴᴴᴴ		
	ChE	+ +	+ + +	+ + + +	+ + + +		

Fig. 5. Schematic illustration of the subsequent morphogenesis of the postsynaptic membrane (PSM) of the myoneural junction after experiments performed on rats at the age of one day. Intensity of histochemical cholinesterase (ChE) activity has been registered on the basis of visual estimation from − to ++++ (++++ = intensity of enzyme activity in adult myoneural junction; − = background activity). (The illustration is from the authors' M.D. thesis published in 1974).

neural morphogenesis (Teräväinen and Juntunen, 1968). Morphogenesis of the myoneural junction induced postnatally at any place on the muscle is similar to that occurring in normally developing muscle (Saito and Zacks, 1969; Juntunen and Teräväinen, 1970; Koenig, 1973), whereas permanent denervation results in the cessation of all myoneural morphogenesis. There is also a temporal correlation between the delay in myoneural morphogenesis and the reversible disappearance of axonal neurotubules caused by colchicine and vinblastine (Juntunen, 1973a). These drugs are known to inhibit axoplasmic transport (Ochs, 1972) without affecting the muscular activity itself (Albuquerque et al., 1972). Inhibition of the impulse-stimulated release of acetylcholine from nerve endings by prolonged nerve blockade with long-acting local anesthetic did not affect myoneural morphogenesis (Juntunen and Teräväinen, 1973). Immobilization of the target muscle by tenotomy (Juntunen, 1973b) had no effect either. These experiments, summarized schematically in Fig. 5 suggest that the nerve regulates the morphogenesis of the cholinergic synapse in striated muscle. The importance of axoplasmic flow for the maintenance of normal myoneural morphogenesis is evident, although the exact regulatory mechanism of this process still remains to be solved.

SUMMARY

The structural development of the cholinergic myoneural synapse is a relatively complicated phenomenon which can be studied by histochemical methods based on the demonstration of functionally important components. A histochemically demonstrable acetylcholinesterase (AChE) activity of the myoneural junction is localized in the postsynaptic membrane, revealing the structure of this membrane during the development. The motor axon first ramifies into many terminals which are apposed to a smooth muscle membrane. Subsequently, a depression is formed in which the postsynaptic infoldings appear. The axon terminals are gradually separated from each other and a folded postsynaptic membrane surrounds each terminal, thus forming the structurally complex mature myoneural junction. Morphogenesis of the cholinergic synapse in striated muscle appears to be, in the main, a similar phenomenon in different species; only the intervals between the different stages of development vary. On the basis of experimental studies, it looks as if developing or regenerating nerves control the morphogensis of the cholinergic synapse in the striated muscle by a mechanism not involved in the neurotransmission. This morphogenesis of the myoneural junction provides a suitable model for experimental studies on the trophic functions of the neurone.

REFERENCES

Albuquerque, E.X., Warnick, J.E., Tasse, J.R. and Sansone, F.M. (1972) Effects of vinblastine and colchicine on neural regulation of the fast and slow skeletal muscles of the rat. *Exp. Neurol.*, 37, 604–607.

Bekoff, A. and Betz, J. (1976) Acetylcholine hot spots: Development of myotubes cultured from aneural limb buds. *Science*, 193, 915–917.

Bennett, M.R., Florin, T. and Woog, R. (1974) The formation of synapses in regenerating mammalian striated muscle. *J. Physiol. (Lond.)*, 283, 79–92.

Bennett, M.R. and Pettigrew, A.G. (1974) The formation of synapses in striated muscle during development. *J. Physiol. (Lond.)*, 241, 515–545.

Diamond, J. and Miledi, R. (1962) A study of foetal and newborn rat muscle fibres. *J. Physiol. (Lond.)*, 162, 393–408.

Drachman, D.B. (1976) Trophic interactions between nerves and muscles: the role of cholinergic transmission (including usage) and other factors. In *Biology of Cholinergic Function*, A.M. Goldberg and I. Hanin (Eds), Raven Press, New York, pp 161–186.

Eränkö, O. and Teräväinen, H. (1967) Distribution of estrases in the myoneural junction of the striated muscle of the rat. *J. Histochem. Cytochem.*, 15, 399–403.

Fambrough, D., Hartzell, H.C., Rash, J.E. and Ritchie, A.K. (1974) Receptor properties of developing muscle. *Ann. N.Y. Acad. Sci.*, 228, 47–62.

Fambrough, D. (1976) Development of cholinergic innervation of skeletal cardiac and smooth muscle. In *Biology of Cholinergic Function*, A.M. Goldberg and I. Hanin (Eds.), Raven Press, New York, pp. 101–160.

Fischbach, G.D. and Cohen, S.A. (1974) Some observations on trophic interaction between neurons and muscle fibers in cells culture. *Ann. N.Y. Acad. Sci.*, 228, 35–62.

Friedenberg, R.M. and Seligman, A.M. (1972) Acetylcholinesterase at the myoneural junction cycochemical ultrastructure and some biochemical considerations. *J. Histochem. Cytochem.*, 20, 771–792.

Gutmann, E. (1976) In *Neurtrophic Relations*, E. Knobil, R.R. Sonnenschein and I.S. Edelman (Eds.), *Ann. Rev. Physiol.*, 38, 117–216.

Juntunen, J. (1973a) Effects of colchicine and vinblastine on neurotubules of the sciatic nerve and cholinesterases in the developing myoneural junction of the rat. *Z. Zellforsch.*, 142, 193–204.

Juntunen, J. (1973b) Morphogenesis of the myoneural junctions after immobilization of the muscle in the rat. *Z. Anat. Entwickl.-Gesch.*, 143, 1–12.

Juntunen, J. (1974) Morphological studies on the induction of the postsynaptic membrane of the myoneural junction. *Acta Inst. Anat. Univ. Helsinkiensis* Suppl. 7, 1–32 (Academic dissertation).

Juntunen, J. and Teräväinen, H. (1970) Morphogenesis of myoneural junctions induced postnatally in the tibialis anterior muscle of the rat. *Acta physiol. scand.*, 79, 462–468.

Juntunen, J. and Teräväinen, H. (1972) Structural development of myoneural junctions in the human embryo. *Histochemie*, 32, 107–112.

358

Juntunen, J. and Teräväinen, H. (1973) Effect of prolonged nerve blockage on the development of the myonerual junction. *Acta physiol. scand.*, 87, 344–347.

Koelle, G.B. and Friedenwald, J.S. (1949) A histochemical method for localizing cholinesterase activity. *Proc. Sco. exp. Biol. Med.*, 70, 617.

Koenig, J. (1973) Morphogenesis of motor end-plates 'in vivo' and 'in vitro'. *Brain Res.*, 62, 361–365.

Lentz, T.L. (1969) Development of the neuromuscular junction. I. Cytological and cytochemical studies of the neuromuscular junction of differentiating muscle in the regenerating limb of the newt Triturus. *J. Cell Biol.*, 42, 431–443.

Milburn, A. (1973) The early development of muscle spindles in the rat. *J. Cell Sci.*, 12, 175–195.

Ochs, S. (1972) Fact axoplasmic transport of materials in mammalian nerve and its integrative role. *Ann. N.Y. Acad. Sci.*, 193, 43–58.

Saito, A. and Zacks, S.I. (1969) Fine structure of neuromuscular junctions after nerve section and implantation of nerve in denervated muscle. *Exp. Molec. Pathol.*, 10, 256–273.

Salpeter, M.M. (1967) Electron microscope radioautography as a quantitative tool in enzyme cytochemistry. I. The distribution of acetylcholinesterase at motor endplates of a vertebrate twitch muscle. *J. Cell Biol.*, 32, 379–389.

Sandri, C., Akert, K., Livingston, R.B. and Moor, H. (1972) Particle aggregations at specialized sites of freeze-etched postsynaptic membranes. *Brain Res.*, 41, 1–16.

Teräväinen, H. (1968a) Carboxylic esterases in developing myoneural junctions of rat striated muscle. *Histochemie*, 12, 307–315.

Teräväinen, H. (1968b) Development of the myoneural junction in the rat. *Z. Zellforsch.*, 87, 249–265.

Teräväinen, H. and Juntunen, J. (1968) Effect of temporary denervation on the development of the acetylcholinesterase-positive structure of the rat myoneural junction. *Histochemie*, 15, 261–269.

Vigny, M., Koenig, J. and Rieger, F. (1976) The motor end-plate specific form of acetylcholinesterase: appearance during embryogenesis and re-innervation of rat muscle. *J. Neurochem.*, 27, 1347–1353.

Elimination of Inappropriate Nerve-Muscle Connections During Development of Rat Embryos

* M.J. DENNIS and ** A.J. HARRIS

* Department of Physiology, University of California Medical Center,
San Francisco, Calif. 94143 (U.S.A.)
and ** Department of Physiology, University of Otago Medical School,
Dunedin (New Zealand)

INTRODUCTION

Contemporary theories about the operation of the vertebrate brain rest on the premise that its functions reflect an orderly and genetically determined specificity of interconnection among its billions of nerve cells. Evidence for this sweepingly general hypothesis is of several kinds: as an extension of actual maps of interconnections between all the nerve cells in the brains of small invertebrates, for example Macagno et al. (1973) study of the central nervous system in different individuals cloned from the water flea, Daphnia; from the well defined and reproducible pattern of connections between different cell types making up functional units within parts of the mammalian brain, for example cerebellum (Eccles, Ito and Szentagothai, 1967), or visual cortex (Hubel and Wiesel, 1965); and at less precise level, different operations are carried out at particular locations within the brain, and these locations are the same in different individuals. The important question then arises as to how this immensely complex pattern of interconnection is formed during development of the brain. All the information required to organize this development must be coded in the DNA of a single diploid cell. Thus, it can be argued on the grounds of economy that this information should provide a program, rather than a list of characteristics for each individual brain cell. Experimental evidence supports this idea. Embryonic nerve cells placed in tissue culture are promiscuous in their relations with other cells, and form and receive synapses indiscriminately (Nelson, 1975). Developmental regulation of synapse formation depends to a large extent on ensuring that particular classes of neurones are formed at the right time and in the right place and their axons extended in the right direction; synapses are then formed with whatever cells and processes are available at that time. This being the case it seems not unlikely that some mistakes should be made. It was the purpose of the study described here to see if such mistakes could be identified, and then to see if there was any natural mechanism for recognising and eliminating them.

Previous investigations of this question have been of two kinds. One has been to attempt to account for the massive neuronal cell death which takes place during development, on the supposition that this may reflect survival of cells making proper connections, and death of cells making no or mistaken connec-

360

tions. The results of such studies have often been unclear, at least as far as this particular question is concerned. For example, cells which have made apparently proper connections may still die (Landmesser and Pilar, 1976), and preventing cells from making connections by removing the tissue with which they should connect may lead to a form of cell death with characteristics different from that of cells dying in normal control embryos (Chu-Wang and Oppenheim, 1978). Clarke and Cowan (1975) showed that cells in the isthmo-optic nucleus (ION) of the chick died if they were not part of a loop of cells interconnecting appropriate regions of retina, optic tectum, and ION; in order to survive, cells of the ION had to receive appropriate afferent connections as well as make appropriate efferent connections. The other kind of investigation has been deliberately to induce formation of inappropriate connections in post-embryonic animals and to follow their survival. It has been difficult to show deletion of inappropriate connections in adult vertebrates (Frank and Jansen, 1976) other than in urodele amphibia such as salamanders (Bennett and Raftos, 1977; Dennis and Yip, 1978). Here, reinnervation by their proper nerves caused muscles to lose an inappropriate innervation. These animals are neotenous, so that it is possible that their ability to disciminate between appropriate and inappropriate innervation reflects a general characteristic of vertebrate embryos which in mammals is lost some time before birth.

RESULTS

The experimental preparation

Our investigation was made on rat embryo intercostal muscles. We chose this preparation for its segmental innervation, which allowed easy definition of appropriate or inappropriate innervation. It proved to have the added advan-

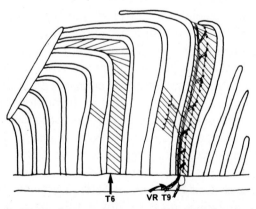

T6   VR T9

Fig. 1. Schematic illustration of normal pattern of innervation and of "mistakes" in rat embryo intercostal muscles. A hemi-ribcage is represented pinned out flat as for an experiment. The spinal column in split, and the cord removed, leaving ventral and dorsal roots, and dorsal root ganglia. Each nerve passes laterally just posterior to each rib, and nerve branches pass to the midpoints of muscle fibres and form end-plates there. In this case, when ventral roots T6 and T9 were stimulated, contractile responses were seen in segments appropriate for T5 and T8, respectively, as well as in those stimulated.

361

tage that the muscles were readily immobilised by pinning the ribs, so that microelectrode recordings were comparatively easy to make, and penetrations could be maintained for long periods. The preparation is illustrated schematically in Fig. 1, which shows the normal fields of innervation of ventral roots T6 and T9, together with two "mistakes" in innervation. "Mistakes" were defined simply as groups of muscle fibres that could be caused to contract by stimulating a ventral root supplying the intercostal muscle of a different segment.

The nature of mistakes in innervation

All mistakes occurred in a posterior-to-anterior direction. Most were one segment anterior to the ventral root supplying them; a few were two, three, or in a single case four segments anterior.

Rat embryo muscle fibres are multiply innervated (Redfern, 1970); we found that both appropriate and inappropriate synapses were placed near the midpoint of individual muscle fibres. This was done by making intracellular recordings from single muscle fibres in calcium-free Ringer solution. Calcium was applied focally by ejection from a micropipette, and appropriate or inappropriate ventral roots stimulated. If, after application of calcium, stimulation of an inappropriate root evoked an end-plate potential (EPP) in the muscle fibre, stimulation of the appropriate root always evoked a response (Fig. 2). Many muscle fibres were mapped along their whole lengths with this technique; responses to any form of nerve stimulation could be evoked at only one spot, near the midpoint of the fibre. Spatial resolution of this technique was about

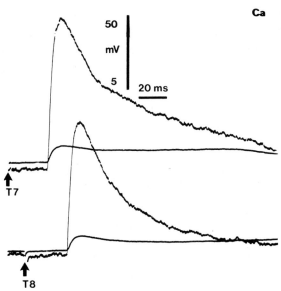

Fig. 2. Responses to appropriate and inappropriate nerve stimulation are evoked at the same site on single muscle fibres. The figure shows EPPs evoked in a single muscle fibre from an 18-day-old embryo, following stimulation of the appropriate nerve (T7) and an inappropriate one (T8). The preparation was in Ca^{2+}-free bathing solution, and a single endplate was activated by focal application of Ca^{2+} from a micropipette. If responses to inappropriate nerve stimulation could be evoked, responses to appropriate nerve stimulation could always be evoked, showing that nerve terminals co-existed at the same synaptic sites on single muscle fibres.

25 μm, the muscle fibres were 200–500 μm long at the time of the experiments.

Responses to inappropriate nerve stimulation were abolished if a fine scalpel blade was drawn along the rib between the appropriate and inappropriate muscle segments. Silver stained specimens prepared after physiological examination had fine bundles of neurites crossing the rib near regions where inappropriate responses had been recorded. The normal pattern of innervation of intercostal muscles is for the nerve to pass just posterior to the rib, and for bundles of neurites to branch off at regular intervals and form synapses near the midpoint of the muscle (Fig. 1). Mistakes apparently were due to such branches passing anteriorly into the next segment, instead of posteriorly into the appropriate muscle.

Electrical coupling

Embryo muscle fibres were powerfully electrically coupled one to another, so that in an 18-days-old embryo an EPP evoked in a single fibre could be recorded in neighbours at least 5 fibres distant. Coupling was reduced after this age, and by the time of birth became physiologically insignificant, although detectable at the level of a coupling ratio of a few per cent during the first post-natal week. The existence of this coupling invalidates attempts to perform statistical analyses of transmitter release in embryo muscles (Bennett and Pettigrew, 1974). Some embryo muscle fibres were found to be coupled to fibres in another segment, either anterior or posterior. When mapping mistakes in innervation by the technique of visually observing contraction this possibility was taken into account by cutting through the muscle fibres of successive appropriate segments near their posterior attachments, and thus depolarising them so that EPPs evoked by appropriate nerve stimulation could not be transmitted to anterior segments by electrical junctions. The problem did not arise when the technique of focal application of calcium was used to map mistakes.

Elimination of mistakes

The number of mistakes was estimated by counting in individual embryos the number of muscle segments in which there was a contractile response to stimulation of the ventral root supplying a different segment. No distinction was made as to the area of muscle involved. It was found that the probability of finding a mistake varied linearly with the length of each segment, i.e. the longer the intercostal muscle, the more likely it was to contain a mistake somewhere along its length. For this reason only the 8 larger segments, T3–T10 were used for this analysis. The results are presented in Fig. 3. Each point is an average taken from observations of several hundred muscles. For embryos of up to 18 days gestation, the likelihood of finding a mistake in any randomly chosen intercostal muscle was 30–40%. For embryos of 20 days gestation, neonates, and older animals, the probability was 10%. Thus the number of mistakes suffered a drastic reduction between day 18 and day 20 of gestation. No further reduction took place during the period in the second neonatal week when the juvenile state of multiple innervation of single muscle fibres was

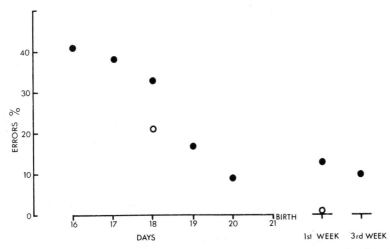

Fig. 3. Occurrence of mistakes in innervation. Each point represents the frequency with which mistakes in innervation were detected in single muscles examined in embryos and post-natal rats of different ages. Mistakes were detected by the contractile response to inappropriate nerve stimulation (filled circles) or by intracellular recording from muscle fibres sampled at fixed regular intervals (open circles). Each point is the average from 150–450 muscles (filled circles) or intracellular recordings from 140 fibres (18-day) and 760 fibres (post-natal) respectively.

reduced to the adult state of unitary innervation (Brown et al., 1976). Intracellular recordings from animals in their fourth week of age showed that inappropriately innervated muscle fibres were innervated by a single axon, and received no additional innervation.

CONCLUSIONS

The mechanisms by which mistakes are recognised, and by which they are deleted, are still under investigation. It is clear that they occur, and that some but not all are removed at a particular time in development. Their occurrence was completely random, both with regards to position and to number in any given animal. Thus they were not part of an orderly stage in development of nerve-muscle contacts. The time at which the number of mistakes was reduced coincides with the time Golgi tendon organs are formed in the muscles (Zelená and Soukup, 1976). Golgi tendon organs respond in a specific way to contraction of particular motor units, and project back, via an interneurone, to the motoneurones supplying those units (Binder et al., 1977). The simplest hypothesis currently being considered is that recognition and deletion of mistakes in this system is the same as in the isthmo-optic nucleus of the chick, as described by Clarke and Cowan (1976), that is, that nerve cells that are not a part of the loop,

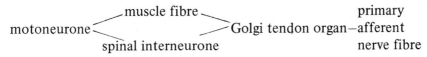

are functionally and perhaps physically deleted.

SUMMARY

"Mistakes" in innervation, whereby discrete groups of intercostal muscle fibres receive innervation from inappropriate ventral roots, are common in rat embryos of up to 18 days gestation. They are much less frequently seen in embryos of 20 days gestation, and their number in embryos of this age persists unchanged into adult life. There appears to be a critical period, during the time 18–19 days of gestation, in which a large proportion of mistakes is recognised and eliminated.

ACKNOWLEDGEMENTS

We thank the University of California for providing a Visiting Associate Professorship for A.J.H.; Dr. D.A. Riley for assistance with histology, and Nancy Johnson for technical help. This work was supported by N.I.H. grant no. 5R01 NS 10792, and the New Zealand Medical Research Council.

REFERENCES

Bennett, M.R. and Pettigrew, A.G. (1974) The formation of synapses in striated muscle during development. *J. Physiol. (Lond.)*, 241, 515–545.

Bennett, M.R. and Raftos, J. (1977) The formation and regression of synapses during the reinnervation of axolotl striated muscles. *J. Physiol. (Lond.)*, 265, 261–296.

Binder, M.P., Kroin, J.S., Moore, G.P. and Stuart, D.G. (1977) The response of Golgi tendon organs to single motor unit contractions. *J. Physiol. (Lond.)*, 271, 337–349.

Brown, M.C., Jansen, J.K.S. and Van Essen, D. (1976) Polyneuronal innervation of skeletal muscle in new-born rats and its elimination during maturation. *J. Physiol. (Lond.)*, 261, 387–422.

Chu-Wang, I. and Oppenheim, R.W. (1978) Cell death of motoneurones in the chick embryo spinal- cord. I. A light and electron microscopic study of naturally occurring and induced cell loss during development. *J. comp. Neurol.*, 177, 33–58.

Clarke, P.G.H. and Cowan, W.M. (1975) Ectopic neurones and aberrant connections during neural development. *Proc. nat. Acad. Sci. (Wash.)*, 72, 4455–4458.

Dennis, M.J. and Yip, J.W. (1978) Formation and elimination of foreign synapses on adult salamander muscle. *J. Physiol. (Lond.)*, 274, 299–310.

Eccles, J.C., Ito, M. and Szentagothai, J. (1967) *The Cerebellum as a Neuronal Machine*, Springer-Verlag, New York.

Frank, E. and Jansen, J.K.S. (1976) Interaction between foreign and original nerves innervating gill muscles in fish. *J. Neurophysiol.*, 39, 84–90.

Hubel, D.H. and Wiesel, T.N. (1965) Receptive fields and functional architecture in two nonstriate visual areas (18 and 19) of the cat. *J. Neurophysiol.*, 28, 229–289.

Landmesser, L. and Pilar, G. (1976) Fate of ganglionic synapses and ganglion cell axons during normal and induced cell death. *J. Cell Biol.*, 68, 357–374.

Macagno, E.R., Lopresti, V. and Levinthal, C. (1973) Structure and development of neuronal connections in isogenic organisms. Variations and similarities in the optic system of *Daphnia magna.. Proc. nat. Acad. Sci. (Wash.)*, 70, 57–61.

Nelson, P.G. (1975) Nerve and muscle cells in culture. *Physiol. Rev.*, 55, 1–61.

Redfern, P.A. (1970) Neuromuscular transmission in new-born rats. *J. Physiol. (Lond.)*, 209, 701–709.

Zelená, J. and Soukup, T. (1977) The development of Golgi tendon organs. *J. Neurocytol.*, 6, 171–194.

The Elimination of Polyneuronal Innervation of End-Plates in Developing Rat Muscles with Altered Function

JIŘINA ZELENÁ, F. VYSKOČIL and ISA JIRMANOVÁ

Institute of Physiology, Czechoslovak Academy of Sciences, 142 20 Prague (Czechoslovakia)

INTRODUCTION

It has been firmly established that focally innervated muscle fibres receive transitory polyneuronal innervation during development which is gradually reduced to a monoaxonal supply per end-plate (Redfern, 1970; Bagust et al., 1973; Bennett and Pettigrew, 1974; Brown et al. 1975, 1976; Jansen et al., 1975; Korneliussen and Jansen, 1976; Riley, 1976, 1977; Rosenthal and Taraskevich, 1977). However, the factors and mechanisms which bring about the reduction of the transient polyneuronal innervation are still not known. It has been postulated that the state of activity of the postsynaptic cell may control, in a retrograde manner, the selective stabilization of synapses during the process of maturation (Changeux et al., 1973). In developing rat skeletal muscles, tenotomy appeared to delay the regression of multiaxonal innervation, which had been interpreted as an indication that the state of activity of the nerve-muscle preparation regulates the evolution of its innervation (Benoit and Changeux, 1975).

In order to test the above hypothesis (Benoit and Changeux, 1975), we have examined the development of motor end-plates in the rat extensor digitorum longus (EDL) muscles after having changed the state of their activity (1) by removal of the synergic anterior tibialis muscles, which induced compensatory hypertrophy in the EDL; (2) by de-afferentation combined with a high transection of the spinal cord, which restricted active movements of the hind limbs. The reduction of multiaxonal innervation was studied by electrophysiological and morphological methods 9 and 14 days after birth, after either operation performed in neonatal and 1-day-old rats.

METHODS

Operations

(1) In the first group of animals, the tibialis anterior muscle was totally removed from the right side in new-born male rats under cold anaesthesia. The EDL was left intact in situ, and the skin was sutured. The contralateral EDL

muscles, and EDLs of 1—2 unoperated rats of each litter served as control.

(2) In the second group, 1-day-old male rats were operated under cold ana-esthesia. The spinal cord was exposed by laminectomy from the level of Th_8 caudally, and all lumbosacral dorsal roots were sectioned on both sides. In addition, the spinal cord was transected at the level of Th_{8-9}. In this manner, the motoneurones of the lumbosacral region were deprived of most of their afferent input. A hyperextension of both hind limbs developed and was main-tained from the first day after the operation onwards, and active movements of the hind limbs were thus restricted. Unoperated littermates served as control.

Electrophysiology

Nine and 14 days after the operation, EDL muscles were removed and immersed in oxygenated standard Liley solution, with a potassium content lowered from 5 to 2.5 mM, in order to ensure better nerve conductivity. (+)-tubocurarine was added into the bath in a concentration of 0.5×10^{-6} to 2×10^{-6} M. The standard microelectrode technique was used for recording end-plate potentials. The nerve was stimulated at its entry into the EDL with a double wire platinum electrode at a frequency of 0.5 Hz. As the intensity of stimulation was gradually increased, the amplitude of the end-plate potentials varied in a stepwise manner which made it possible to reveal the number of axons with different thresholds innervating each end-plate examined (cf. Vyskočil and Magazanik, 1977). Thirty to 60 end-plates were examined in each muscle.

Morphology

For light microscopy, EDL muscles of 9- and 14-day-old rats were fixed in a 2.5% paraformaldehyde solution overnight. Longitudinal sections 30—40 μm thick were cut with a cryostat microtome, stained by Karnovsky's method for end-plate acetylcholinesterase and impregnated with $AgNO_3$ according to a modified method of McIsaac and Kiernan (1974). The number of axons per end-plate was determined under the light microscope in 50—150 end-plates in each muscle.

For electron microscopy, muscles were fixed in a solution of 1% paraformal-dehyde and 1% glutaraldehyde in 0.4 M phosphate buffer for 2—3 h, post-fixed with 2% OsO_4 for 2 h, dehydrated and embedded in Durcupan. Regions con-taining end-plates were selected for ultrathin sectioning on transverse semithin sections stained with toluidine blue. Ultrathin sections were stained with 1% uranylacetate and 0.1% lead citrate, and examined with a JEM 100B electron microscope.

RESULTS

(1) Differences in the degree of end-plate maturation in experimental and control muscles

Motor end-plates were examined in hypertrophic, de-afferented and control muscles 9 and 14 days after birth. The number of monoaxonally and poly-

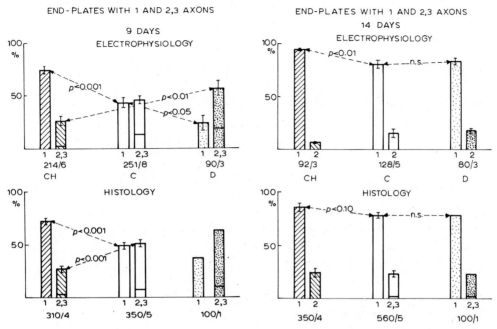

Fig. 1 (left). Percentage of motor end-plates with mono- and multiaxonal innervation in EDL muscles 9 days after birth, determined by electrophysiological (upper row) and histological (lower row) methods. Ordinate: percentage of end-plates. Abscissa: paired columns indicating mean % ± S.E.M. of monoaxonal (1) and multiaxonal (2, 3) end-plates in each muscle group; the 2nd column, if divided by a transverse line, comprises the percentage of end-plates with 2 (above the line) and 3 axons (below the line). CH (hatched columns), muscles undergoing compensatory hypertrophy; C (white columns), control muscles; D (dotted columns), de-afferented muscles. The number of end-plates and muscles investigated in each group is given below each pair of columns. Significant differences between muscle groups, calculated according to the t-test, are indicated at interrupted lines connecting the related columns; n.s., not significant.

Fig. 2 (right). Percentage of motor end-plates with mono- and multiaxonal innervation in EDL muscles 14 days after birth. For explanation see text to Fig. 1.

axonally supplied end-plates was determined by electrophysiological and morphological methods and the results were expressed as percentage of mono-axonal and polyaxonal end-plates in each muscle group. The summarized results are shown in Figs. 1 and 2. The results obtained in the two control groups for hypertrophic and de-afferented EDLs were not significantly different, and both control groups are pooled in the figures.

In EDL muscles undergoing compensatory hypertrophy, the muscle weight was increased by 70% on day 9 and by 25% on day 14 in comparison with the contralateral control muscles. The proportion of monoaxonal end-plates on day 9 was 70% higher in hypertrophic muscles than in control EDLs (Fig. 1). By day 14 the electrophysiologically revealed difference diminished to 15% (Fig. 2, Electrophysiology) and the small difference found by morphological method was no longer statistically significant (Fig. 2, Histology).

In de-afferented muscles in which active movements were greatly restricted, the proportion of monoaxonal end-plates was 50% lower than in the control muscles on day 9 (Fig. 1). The difference was lost by day 14, although

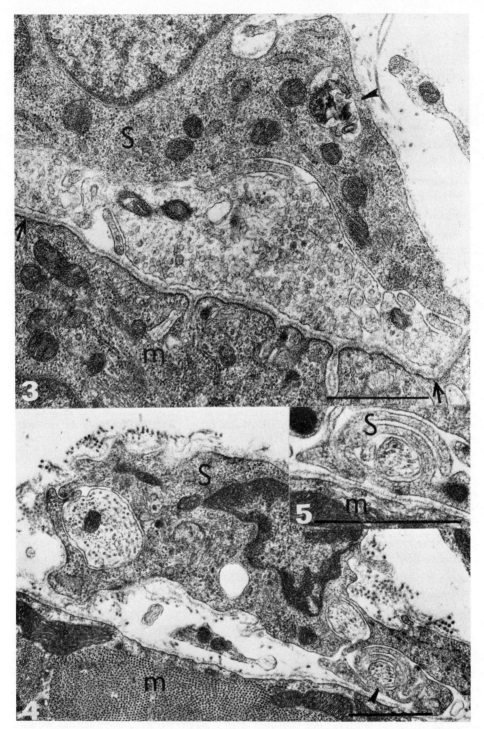

Fig. 3. Transverse section of an end-plate with 1 large and several small terminals in a shallow groove on the muscle fibre surface (m), from a de-afferented EDL muscle 14 days after birth. The postsynaptic membrane is thickened along the synaptic contact (between arrows) but not in the developing infoldings. Note that 2 tiny terminal profiles are not in contact with the basal lamina. A degenerated axonal profile (arrowhead) is seen engulfed within the cytoplasm of the Schwann cell (S) covering the end-plate.

Fig. 4. A Schwann cell (S) with 1 large and 2 small axonal profiles near an end-plate on the muscle fibre (m) seen below; from a hypertrophic EDL 14 days after removal of the anterior tibial muscle. The small axon (arrowhead) enwrapped by Schwann cell processes is assumed to be retracting from the end-plate.

Fig. 5. A larger magnification of the presumably retracting axon seen in Fig. 4, taken from another section. Note that the Schwann cell processes (S) are still in contact with basal lamina of the muscle fibre (m). Scale 1 μm.

de-afferented muscles had restricted movements during the whole investigated period.

If we take the percentage of monoaxonal end-plates as a measure of end-plate maturation, it can be concluded that the rate of maturation is accelerated in hypertrophic muscles and retarded in de-afferented muscles by day 9. In two week's time, however, the majority of end-plates becomes monoaxonal both in experimental and control muscles.

(2) Retraction of redundant axons – a probable mode of reduction of multiaxonal innervation

A sample of about 50 end-plates from control and experimental muscles was examined in the electron microscope with the aim of elucidating the question of whether or not the reduction of multiaxonal innervation occurred by degeneration of redundant axons and terminals. None of the investigated end-plates contained degenerating or degenerated terminals. However, occasional degenerated axonal profiles were observed in the end-plate region, encircled by Schwann cell processes or engulfed by the cytoplasm of Schwann cells covering the end-plate (Fig. 3). In muscles of 14-day-old animals, a small axonal profile was sometimes seen enwrapped by numerous Schwann cell processes, near a large-sized axon approaching the same end-plate (Figs. 4 and 5). Intramuscular nerve branches which mainly consisted of axons with incipient myelination by day 14, frequently contained one or two tiny axonal profiles on the outer circumference of the Schwann cells encircling the central axon. These ultrastructural findings suggest that terminals become detached and axons are retracted during the process of end-plate maturation.

DISCUSSION

A transient functional hyperinnervation of muscle fibres appears to be regular developmental feature preceding the establishment of monoaxonal contacts characteristic for mature neuromuscular junctions of focally innervated muscles. In the rat, polyneuronal innervation is being reduced during the first two weeks after birth, and the monoaxonal pattern is established between the second and third week postnatally (Redfern, 1970; Bennett and Pettigrew, 1974; Benoit and Changeux, 1975; Riley 1977; Rosenthal and Taraskevich, 1977).

In our experiments the process of reduction of multiaxonal innervation was found to be accelerated in rat EDL muscles undergoing compensatory hypertrophy. A similar effect has recently been observed in developing rat soleus muscles after chronic nerve stimulation (O'Brien et al., 1977). However, no difference was found between stimulated and unstimulated muscles by histological method (Östberg and Vrbová, 1977), which was explained by the assumption that non-functional axons still maintained their contact with the muscle fibre. It is possible that redundant axons are retracted from the end-plate with a delay, after their terminals have lost functional contact with the muscle fibre. In our study, the percentage of multiaxonal end-plates determined histologically was usually somewhat higher than that revealed electro-physiologically but the difference was small.

In de-afferented muscles of animals with restricted movement of hind limbs, elimination of multiaxonal innervation was found to be retarded, as was previously described after tenotomy (Benoit and Changeux, 1975).

Although chronic effects upon the neuromuscular activity of the model situation used in our experiments are not clearly defined (cf. Hník, 1956; Gutmann, 1972; Macková and Hník, 1973), it is apparent that activity influences the process of elimination of redundant axons. It should be noted, however, that the retardation effect observed after de-afferentation was only transient. The developmental programme of end-plate maturation — which is presumably genetically encoded — was eventually carried out without significant delay in de-afferented muscles with reduced active movements, i.e. by motoneurones chronically deprived of their afferent input.

As regards the mechanism of elimination of polyneuronal innervation during development, our findings contradict the conclusions of Rosenthal and Taraskevich (1977) that redundant terminals degenerate as in denervated adult muscles. Our results rather indicate that redundant terminals are eliminated by retraction, as has been suggested previously (Korneliussen and Jansen, 1976; Riley, 1977). Although degenerated axons are occasionally observed in the end-plate region, it is assumed that such degeneration occurs during the process of axonal retraction, when terminal parts of receding axons may be severed or pinched off and phagocytized by Schwann cells. The presence of tiny axonal profiles, found sometimes in intramuscular nerves outside the Schwann cells myelinating the growing axons, suggests that non-functional axons cease to grow, become atrophic and are retracted into persisting parent axons within the muscle. However, the problem concerning the factors affecting the detachment and retraction of supernumeraxy terminals requires further study.

SUMMARY

The elimination of polyneuronal innervation from developing end-plates was studied in rat EDL muscles with altered function (1) during compensatory hypertrophy and (2) after de-afferentation.

(1) The EDL muscles became overloaded and underwent compensatory hypertrophy after removal of the tibialis anterior muscles in new-born rats. The maturation of end-plates was speeded up: electrophysiological examination revealed that 74% end-plates already had a monoaxonal supply by day 9, and 94% were monoaxonally supplied by day 14, as compared with 43% and 82% monoaxonal end-plates in corresponding control muscles. Similar results were obtained by histological method, but the percentage of monoaxonal end-plates was as a rule somewhat lower: 72% monoaxonal end-plates were found in hypertrophic muscles on day 9 and 85% on day 14, as compared with 49% and 77% monoaxonal end-plates in control muscles.

(2) After bilateral section of lumbosacral dorsal roots and high transection of the spinal cord in 1-day-old rats, active movements were restricted and the maturation of end-plates in EDL muscles was retarded: only 24% and 37% end-

plates became monoaxonal by day 9 as determined by electrophysiological and histological method respectively. No difference was found between de-afferented and control muscles by day 14, when about 80% of all end-plates investigated already had a monoaxonal supply.

The maturation of motor end-plates thus appears to be speeded up by compensatory hypertrophy and slowed down by de-afferentation. The retardation observed after de-afferentation is only temporary: the majority of end-plates become monoaxonally supplied in two weeks time in both experimental and control muscles.

Since no degenerating or degenerated motor terminals were found in our sample of 50 end-plates investigated in the electron microscope, our findings support the assumption that redundant terminals become detached and axons retract during postnatal development. Occasional axons may, however, degenerate when they are trapped and severed during the process of retraction.

ACKNOWLEDGEMENTS

The authors wish to thank Mrs. Marie Sobotková, Mrs. Markéta Krupková, Mr. H. Kunz and Ing. V. Pokorný for their skillful technical assistance, and Dr. P. Hník for critical reading of the manuscript.

REFERENCES

Bagust, J., Lewis, D.M., Westerman, R.A. (1973) Polyneuronal innervation of kitten skeletal muscle. *J. Physiol. (Lond.)*, 229, 241–255.

Bennett, M.R., Pettigrew, A.G. (1974) The formation of synapses in striated muscle during development. *J. Physiol. (Lond.)*, 241, 515–545.

Benoit, P., Changeux, J.-P. (1975) Consequences of tenotomy on the evolution of multi-innervation in developing rat soleus muscle. *Brain Res.*, 99, 354–358.

Brown, M.C., Jansen, J.K.S. and Van Essen, D. (1975) A large-scale reduction in moto-neurone peripheral fields during postnatal development in the rat. *Acta physiol. scand.*, 95, 3–4A.

Brown, M.C., Jansen, J.K.S. and Van Essen, D. (1976) Polyneuronal innervation of skeletal muscle in new-born rats and its elimination during maturation. *J. Physiol. (Lond.)*, 261, 387–422.

Changeux, J.-P., Courrege, P.H. and Danchin, A. (1973) A theory of the epigenesis of neuronal networks by selective stabilization of synapses. *Proc. nat. Acad. Sci. (Wash.)*, 70, 2974–2978.

Gutmann, E. (1972) Comparative aspects of compensatory hypertrophy of skeletal and heart muscle. *Folia Fac. med. Univ. Comenianae Bratisl.*, 10, Suppl., 229–239.

Hník, P. (1956) Motor function disturbances and excitability changes following de-afferentation. *Physiol. bohemoslova*, 5, 305–315.

Jansen, J.K.S., Van Essen, D. and Brown, M.C. (1975) Formation and elimination of synapses in skeletal muscles of rat. *Cold Spring Harb. Symp. quant. Biol.*, 40, 425–434.

Korneliussen, H., Jansen, J.K.S. (1976) Morphological aspects of the elimination of poly-neuronal innervation of skeletal muscle fibres in newborn rats. *J. Neurocytol.*, 5, 591–604.

Macková, E., Hník, P. (1973) Compensatory muscle hypertrophy induced by tenotomy of synergists is not true working hypertrophy. *Physiol. bohemoslov.*, 22, 43–49.

McIsaac, E. and Kiernan, J.A. (1974) Complete staining of neuromuscular innervation with bromoindigo and silver. *Stain Technol.*, 49, 211–214.

O'Brien, R.A.D., Purves, R.D., Vrbová, G. (1977) Effect of activity on the elimination of multiple innervation in soleus muscles of rats. *J. Physiol. (Lond.)*, 271, 54–55P.

Östberg, A.J.C., Vrbová, G. (1977) Illustration of the disappearance of polyneuronal innervation of developing skeletal muscle. *J. Physiol. (Lond.)*, 271, 6–7P.

Redfern, P.A. (1970) Neuromuscular transmission in newborn rats. *J. Physiol. (Lond.)*, 209, 701–709.

Riley, D.A. (1976) Multiple axon branches innervating single end-plates of kitten soleus myofibers. *Brain Res.*, 110, 158–161.

Riley, D.A. (1977) Spontaneous elimination of nerve terminals from the end-plates of developing skeletal myofibers. *Brain Res.*, 134, 279–285.

Rosenthal, J.L., Taraskevich, P.S. (1977) Reduction of multiaxonal innervation at the neuromuscular junction of the rat during development. *J. Physiol. (Lond.)*, 270, 229–310.

Vyskočil, F., Magazanik, L.G. (1977) Dual end-plate potentials at the single neuromuscular junction of adult frog. *Pflügers Arch.*, 368, 271–273.

Competitive Interactions between Developing Cholinergic Neurones

LYNN T. LANDMESSER

Department of Biology, Yale University, New Haven, CT 06520 (U.S.A.)

INTRODUCTION

Normal development of most neurones, including the peripheral, cholinergic ones considered here, appears to require a number of interactions with other cells. Of primary inportance are both the pre- and postsynaptic elements with which the neuron synapses, and evidence supporting such interactions can be found in several recent reviews (Purves, 1976; Black, 1978; Varon and Bunge, 1978). Both in vivo and in vitro studies on autonomic ganglia have shown that even the basic decision as to what transmitter a neuron will synthesize, can be determined by environmental interactions (for review see Patterson, 1978).

The target tissue with which a neuron synapses has been shown to play an especially critical role. In several peripheral cholinergic systems early ablation of the peripheral target has been shown the result in the death of most neurons during a critical developmental period: the avian ciliary ganglion (Landmesser and Pilar, 1974a), the avian (Hamburger, 1958; Oppenheim et al., 1978) and amphibian (Prestige, 1967) spinal cord and the duck trochlear nucleus (Sohal et al., 1978). Of greater interest perhaps is the fact that in all of these systems, a large proportion of neurons normally dies during approximately the same period. This has given rise to the idea that neurons are dependent on same interaction with their peripheral target, and that they normally compete for some aspect of the peripheral target which is limited in supply; possibly synaptic sites or a trophic factor. Competitive interactions have also been implicated in the process of selective reinnervation in lower vertebrates (Mark, 1974) and in the ultimate sorting out of connections that result in specific innervation patterns both during re-innervation and initial development (Prestige and Willishaw, 1975; Yip and Dennis, 1976; Harris and Dennis, 1977).

This chapter will review previous studies carried out on the chick ciliary ganglion to document that critical interactions with the target tissue are necessary for survival and full maturation of these cholinergic ganglion cells. More recent results which suggest that ciliary neurons actively compete amongst themselves for some as yet undefined aspect of their synaptic target will also be presented (Burstein et al., 1977). Finally the potential role of competition in the establishment of specific innervation patterns will be considered for both the ciliary ganglion and the chick lumbosacral spinal cord.

DEPENDENCE OF DEVELOPING NEURONS ON THEIR PHERIPHERAL TARGETS

The chick ciliary ganglion has proven to be an especially useful model to study developmental interactions since it is relatively simple and is accessible to experimental manipulations at all stages of development. As depicted diagramatically in Fig. 1, the ciliary ganglion consists of two spatially segregated populations each consisting of about 1500 ganglion cells: the large diameter ciliary cells send their axons in the ciliary nerves to innervate the iris and ciliary muscles; the smaller choroid cells innervate the smooth vascular muscle of the choroid coat (Marwitt et al., 1971). Both populations of ganglion cells are synapsed upon by cholinergic preganglionic neurons and both make cholinergic synapses with their peripheral targets (Marwitt et al., 1971; Chiappinelli et al., 1976).

Following their early migration from the neural crest and the formation of a ganglion by stage 25 (stage according to Hamburger and Hamilton, 1951), the ganglion cells appear to undergo an autonomous period of differentiation when they are not dependent on the peripheral target. This was deduced by comparing the developmental sequence in ganglia which had had their peripheral targets removed (optical vesicle ablation at stages 19—20) with that in normal ganglia (Landmesser and Pilar, 1974a,b, 1976; Pilar and Landmesser, 1976).

First, proliferation was unaffected and ganglion cells differentiated in normal numbers. By comparing cell and axon counts, it was possible to determine that most, if not all, ganglion cells had sent out axons. Finally synapse formation onto the ganglion cells was unaltered, and typical functioning cholinergic synapses were formed on all ciliary cells at the normal time (stages 30—33).

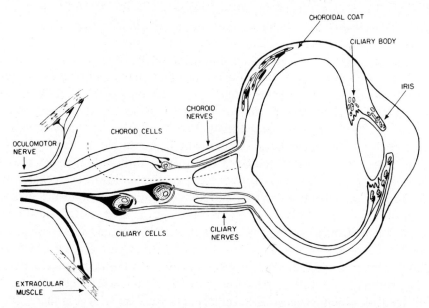

Fig. 1. A diagrammatic view of the chick ciliary ganglion. The two ganglion cell populations, ciliary and choroid project out separate nerves to innervate their peripheral targets within the eye. Each population is innervated by a separate class of preganglionic fibres which run in the oculomotor nerve with fibres to the extraocular muscles. (From Landmesser and Pilar, 1974a).

This implies that the ability of ganglion cells to synthesize and insert acetylcholine (ACh) receptors into their membrane in sufficient numbers to result in normal transmission, is independent of interactions with their peripheral target. Similar results showing a considerable degree of autonomous differentiation of neurons deprived of their peripheral targets have been obtained for the chick lumbosacral spinal cord motoneuron (Oppenheim et al., 1978) and for insect sensory neurons (Sanes et al., 1976).

Death of most of the peripherally deprived ganglion cells between stages 34–40, was the most striking manifestation of their dependence on the peripheral target. The cells died, even though they had differentiated and had functioning preganglionic synapses on them, (Landmesser and Pilar, 1974a), indicating the importance of the peripheral target in this system.

However, prior to actual ganglion cell degeneration, several more subtle effects of the peripheral target were noted. First, while virtually all of the

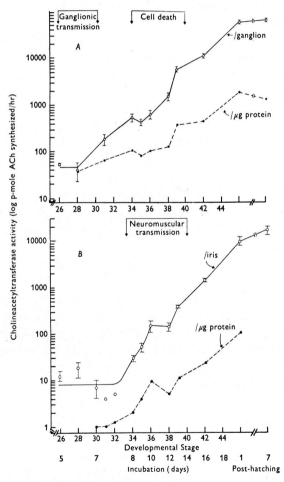

Fig. 2. ChAT activity in the ciliary ganglion (A) and iris (B) during different stages of development. In each graph the solid line plots ChAT activity per ganglion (A) or per iris (B), whereas the dashed line plots activity per microgram protein. Vertical bars represent mean ± S.E.M. of the mean. Periods during which synaptic transmission is established in ganglion and iris, as well as period of ganglion cell death are indicated. (From Chiappinelli et al., 1976).

periphally deprived ganglion cells appeared to send out axons as has been observed in other systems (Chu-Wang et al., 1978; Sohal et al., 1978), collateral branching was not as extensive as in the normal ganglion at stage 34, there being only 1.6 axons/cell compared to approximately 5 axons/cell in the control ganglion (Landmesser and Pilar, 1974a; Pilar et al., 1978). An interaction with the periphery also seems to trigger collateral sprouting in the trochlear nucleus (Sohal et al., 1978) and may be important in insuring complete innervation of the target.

Secondly, between stages 34—36, all ganglion cells in control ganglia undergo a rapid alteration in ribosomal organization: free ribosomes dominant at stage 34 become organized into polysomes and there is a marked increase in the amount of rough endoplasmic reticulum. These changes do not occur in peripherally deprived ganglion cells (Pilar and Landmesser, 1976). At least associated with these changes in time is an increase in the synthesis of two enzymes involved in cholinergic transmission, choline acetyltransferase (ChAT) and acetylcholinesterase (AChE) (Chiappinelli et al., 1976).

Fig. 2B shows that in the iris the level of the enzyme ChAT, which is entirely localized within the nerve terminals of the ciliary cells (Chiappinelli et al., 1976), increases approximately 100 fold between stages 34—36. Since ChAT is not synthesized in the nerve terminals, but is transported from the cell somas, this presumably indicates an increase in the synthesis of ChAT by ciliary cells in the ganglion (Fig. 2A). The level of ChAT also increases in the ganglion over the same period, but the presence of cholinergic preganglionic terminals complicates the interpretation. These observations have led to the suggestion that some interaction with the peripheral target, possibly synapse formation itself which occurs during this period (Landmesser and Pilar, 1974b), triggers the cell to increase their synthesis of transmission related enzymes (Chiappinelli et al., 1976). This hypothesis is thus far based only on a temporal coincidence of the two events, since no biochemical studies have been performed on peripherally deprived ganglia.

During the same period in which the peripherally deprived cells die, roughly half of the normal cells die as well (Landmesser and Pilar, 1974a). Substantial normal cell death during similar phases of neuronal development has been observed in a large number of cases (see Cowan, 1973 for review) although the role played by neuronal overproduction and subsequent degeneration is not yet clear. However, a number of observations on the ciliary ganglion suggest that the cells which die are not grossly defective in some genetic sense, since they are capable of proceeding through the basic steps of neuronal differentiation.

First, they all extend axons which reach the peripheral target. Thus all cells of the ciliary population can be labeled by exposing the iris and ciliary muscles to horse radish peroxidase (HRP) which is retrogradely transport back to the ganglion cell somas where it can be histochemically visualized (Pilar et al., 1978). Similar observations have been made in other systems where large numbers of neurons die (Clarke and Cowan, 1976; Chu-Wang et al., 1978; Sohal et al., 1978). Futhermore, they are receptive to synapse formation, since all ciliary ganglion cells have been shown to have received preganglionic synapses, which appear ultrastructurally normal and which continue to function until stage 34, just prior to cell death. Finally, all cells, even those which degenerate, undergo the alteration in ribosome complement, described earlier (Pilar and Landmesser, 1976).

One additional observation suggests that the ciliary cells which die may even have established some synapses with the peripheral target. In the curve of Fig. 2B, indicating ChAT activity in the iris, there is a definite dip which correlates with the cell death period. A reasonable interpretation of this fact, is that the nerve terminals of the cells which die during this period, contain relatively high levels of ChAT, compared with the levels at stage 32 which is prior to synapse formation. Increased amounts of ChAT in the nerve terminals is perhaps a biochemical indication of synapse formation. A similar dip in the curve depicting ChAT activity has been observed by Giacobini et al. (1973), in the chick hindlimb during a comparable period of motoneuron cell death (Hamburger, 1975).

These observations suggest that the cells which ultimately degenerate may not be irrevocably destined to die, there merely being insufficient target for all. It then should be possible to rescue some of the cells either by increasing the amount of peripheral target, or by decreasing the number of competing neurons. Support for this idea has already been obtained in the avian spinal cord using the former procedure (Hollyday and Hamburger, 1976) and experiments using the second procedure will now be described for the ciliary ganglion.

COMPETITION BETWEEN GANGLION CELLS FOR SURVIVAL

The ciliary population, considered thus far as a homogeneous group, actually consists of subgroups of cells which innervate different parts of the peripheral target. As diagramatically depicted in Fig. 3, it has been possible with electrophysiological studies (Pilar et al., 1978) to show that each of the three main ciliary nerve branches innervates a characteristic part of the target: the most lateral branch innervates the lateral third of the ciliary muscle and the iris constrictor (stippling); the middle branch, the middle third of the ciliary muscle (cross-hatching) and the medial branch, the medial third of the ciliary muscle (oblique lines); a small branch usually comes off the middle branch to innervate the iris dilator (solid line). Retrograde labeling with HRP of individual branches

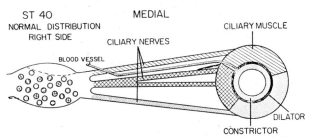

Fig. 3. Diagram showing the characteristic innervation territories of the three main branches of the ciliary nerve. Iris dilator and nerve innervating it are shown in black, Choroid cells are located in ganglion medial to dashed line and project out choroid nerves (not shown) medial to the last ciliary nerve. Ciliary cells projecting out any one branch are distributed throughout ganglion as indicated. In partial axotomy procedure, the two ciliary branches lateral to the blood vessel (stippled and cross-hatched) are cut, leaving branch 3 (oblique lines) intact. (From Landmesser and Pilar, 1976).

has shown, however, that the ganglion cells projecting out any one branch are distributed throughout the ganglion and have no obvious distinguishing characteristics.

Use was made of this characteristic pattern of innervation in order to determine whether ciliary cells actively competed with each other for survival. It was found that embryonic axotomy of all of the ciliary nerves at stage 32 resulted in rapid death of all ciliary cells, since this prevented interaction with the peripheral target during the critical period. We then used this technique to reduce the number of competing ganglion cells. In a series of right ciliary ganglia, axotomy at stages 32–34 of all branches except the one branch lying medial to the blood vessel (designated branch 3) was performed. The contralateral control ganglia served as controls. Following the period of cell death, pairs of ganglia were isolated and the number of ganglion cells projecting out the control and experimental branch 3 determined by retrogradely labeling the cells with HRP (Pilar et al., 1978). In this procedure, with only branch 3 left intact, the entire peripheral target was incubated in HRP.

Table I shows that following the period of cell death the number of branch 3 cells on the experimental side was nearly double the control number. In each of the five cases shown in Table I a large number of ganglion cells, ranging from 138–475, was rescued. Therefore cell death in this system is not inevitable and many cells can be rescued by decreasing the number of competing ganglion cells. However, it was not possible to save all the cells. By determining the number of branch 3 cells at stage 34, prior to cell death, it was possible to show that normally 69% of these cells die. On the axotomized side only 42% of the cells die so that 39% (or 259 cells on the average) were saved.

TABLE I
NUMBER OF BRANCH THREE GANGLION CELLS

	Control side St34	Control side St 39–40	Partial axotomy side St 39-40
		270	546
		241 **	716 **
		328	479
		380	518
	1067	230	484
	971	270 *	—
	807	395 *	—
Mean ± S.D.	948 ± 131	302 ± 66	549 ± 97

NUMBER OF AXONS IN BRANCH THREE

	Control side St 34	Control side St 39–40	Partial axotomy side St 39–40
		1292	1889
	4994	1232	1796
	4564	1254	2055
Mean ± S.D.	4779 ± 304	1259 ± 30	1913 ± 131

* These two values were obtained from normal unoperated embryos and fall within the same range of the contralateral control group.
** See next page.

A similar increase (approximately 700 axons/ganglion) over control values was reflected in the number of branch 3 axons determined from EM montages (Table I). However, the relative increase in axons (141%) was not as great as the increase in cell numbers (189%), so apparently the number of collaterals/cell was reduced in the experimental situation.

In summary, reducing the number of competing neurons substantially enhances the survival of the remaining cells. This suggests that normally cells compete with each other for some aspect of their peripheral target which is in limited supply. It is interesting to note that even when axotomy was performed as late as stage 34, just prior to the onset of cell death, a large number of cells could be rescued (double asterisk, Table I). Therefore, the presumed interaction with the peripheral target must take place rather late in development, and is at least temporally correlated with synapse formation. Hollyday and Hamburger (1976) were able to rescue chick lumbosacral motoneurons by adding a supernumerary limb, roughly doubling the amount of peripheral target. One can calculate from their data that on the average, 29% of the cells that normally would have died were saved. In the ciliary ganglion, partial axotomy which we estimated reduced the ganglion cell population by 63%, rescued 39% of the cells that would have died, a rather comparable result.

In interpreting these results as a rescue of cells, we are assuming that neuronal proliferation has been unaffected. Autoradiographic analysis showed that normally ciliary ganglion cell proliferation is over by stage 25 (Landmesser, unpublished observations) and the partial axotomies were not performed until stages 32–34. While it is therefore quite unlikely that cell proliferation was affected we can not yet completely rule out this possibility. Similar arguments supporting rescue of cells can be made from Hollyday and Hamburger's results (1976). Here motoneuron proliferation is complete by stage 24, but the number of cells on the supernumerary side does not differ from control values at stage 29. Only at the completion of the cell death period (stage 34) is there a difference between the two sides. Again, this is strong, although not conclusive evidence, that cell proliferation has not been affected, for it is possible that a second wave of proliferation could be induced on the experimental side.

COMPETITION NORMALLY RETARDS NEURONAL MATURATION

Competition not only results in the death of a large number of neurones, but actually delays the maturation of the surviving cells. It became apparent in comparing electron micrographic cross sections of branch 3 at stage 39 that not only was branch 3 larger on the partial axotomy side (Fig. 4, bottom) as would be expected from the larger number of axons, but that the axons in branch 3 were more mature. As can be better appreciated from the higher magnification insets on the right of Fig. 4, axons in the experimental branch (bottom) were larger and most had become individually encircled by Schwann cells. In contrast, axons in the contralateral control branch 3 (Fig. 4, top) were smaller and large numbers of naked axons were surrounded by single Schwann cells. In this case, mean axon diameter on the experimental side 0.74 ± 0.01 μm (mean $\pm$

380

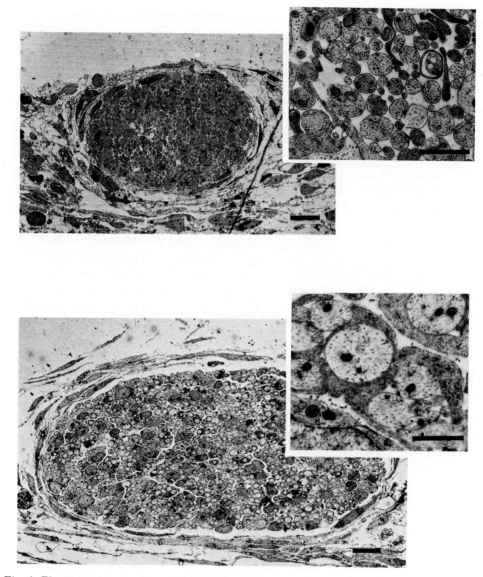

Fig. 4. Electron micrographic cross sections of branch 3 of the ciliary nerve from both sides of one stage 39 embryo. The low magnification views show that the control branch 3 (top) is considerably smaller than the contralateral branch on the partial axotomy side (bottom). Calibration bar 10 μm. The higher magnification insets, taken from the same sections show that the axons on the partial axotomy side (bottom) are larger and have been enwrapped by glial cells to a greater extent than control axons (top). Calibration bar 1 μm.

S.E.M.) was nearly double that of the control side 0.42 ± 0.01 μm. Conduction velocity was also measured in two additional pairs of nerves at stage 39 and on the experimental side was found to be increased by 150%, and 145% in the two cases.

Reduction of the ganglion cell population, therefore, not only rescues a number of cells, but allows the surviving cells to develop at a faster rate. Their axons increase in diameter and conduction velocity and become ensheated by glia, sooner than in the normal animal. This suggests that competition between neurons normally retards their rate of maturation.

IS COMPETITION INVOLVED IN THE ESTABLISHMENT OF SPECIFIC INNERVATION TERRITORIES?

Since each branch of the ciliary nerve normally innervates a restricted portion of the target, it was of interest to see if branch 3 would expand its peripheral field into the territory left denervated by axotomy of branches 1 and 2. Fig. 5 shows that in 12 different preparations, branch 3 did increase its peripheral territory, activating 284° ± 62 (mean ± S.D.) of the ring of ciliary muscle, compared to 157° ± 27 on the control side. In four cases it innervated the entire ciliary muscle and in $\frac{7}{12}$ cases the iris constrictor as well (Pilar et al., 1978). Since branch 3 invaded foreign territory in all cases, there is apparently nothing inherent in the properties of the ciliary muscles and iris that prevents innervation by inappropriate nerves. This suggests that some form of competition between nerve branches normally restricts each nerve, contributing to the specific innervation pattern observed.

Choroid neurons, in contrast, did not appear to innervate the ciliary muscles or iris. One explanation of the observations, is that there exists a hierarchy of mechanisms involed in the establishment of specific innervation patterns. At one level, there may be class to class matching based on specific recognition mechanisms, while within a class, competitive mechanisms may be employed.

It is interesting to compare these results with studies on the development of innervation patterns by chick lumbosacral motoneurons. In this system, each muscle is innervated by a coherent group of motoneurons which are located in a characteristic position and which project out specific spinal nerves (Landmesser and Morris, 1975; Lance Jones and Landmesser, 1978). Specific segments of the lumbosacral cord were deleted at very early embryonic stages (stages 14—16) prior to even the birth of lumbosacral motoneurons (Lance Jones and Landmesser, 1978). The projection patterns of the remaining segments of cord were electrophysiologically determined at stages 32—36 and were found to be entirely normal. Muscles that would have received their entire innervation from deleted segments, remained uninnervated and at later stages were highly atrophic.

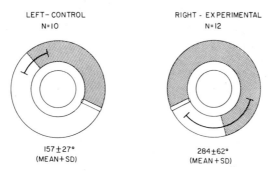

LEFT— CONTROL
N=10

RIGHT - EXPERIMENTAL
N=12

157±27°
(MEAN+SD)

284±62°
(MEAN+SD)

Fig. 5. Diagram of peripheral innervation territory of ciliary branch 3 in control and partial axotomy sides. Outer ring represents ciliary muscles, inner ring the iris constritor. Branch 3 normally activates somewhat more than $\frac{1}{3}$ of the ring of ciliary muscle, extending medially from the blood vessel (indicated by hollow cylinder). On partial axotomy side, the territory has been considerably extended.

Therefore, during initial innervation in this system, motoneurons are not prevented from innervating inappropriate muscles merely by competition from the appropriate nerve sources (Prestige and Willishaw, 1975). Some more specific recognition mechanisms are apparently used. Thus certain constraints which result in specific innervation patterns appear to operate at early embryonic times. These constraints may not operate at later developmental stages or during re-innervation in the adult. For it is well known that muscle nerve regeneration in higher vertebrates is generally not specific (Fambrough, 1975; Purves, 1976), and sprouting of foreign nerves into denervated muscle is common, especially in lower vertebrates (Yip and Dennis, 1976; for review see Fambrough, 1975 and Purves, 1976). These observations suggest that initial innervation may differ in important ways from re-innervation and denervation-induced sprouting, so that caution must be exercised in generalizing between the two situations.

SUMMARY

The importance of the peripheral target in neuronal development was documented in the chick ciliary ganglion. Early ablation of the peripheral targets for these cholinergic parasympathetic neurons altered their development in several ways: (1) While all neurons extended axons the number of axons colaterals was considerably reduced. (2) The ganglion cell somal complement of ribosomes retained an immature appearance and (3) all neurons died during a critical developmental period. Furthermore data was presented that suggests that some interaction with periphery may trigger the ganglion cells to increase their synthesis of enzymes involved in cholinergic transmission.

During normal development in this, as well as in many other neural systems, a large proportion of neurons die. Such neurons differentiate to a considerable degree and do not appear to be grossly defective. A large number of cells can be rescued from cell death by experimentally reducing the number of ganglion cells just prior to the critical period. This suggests that ganglion cells normally compete amonst themselves for some aspect of the peripheral target, resulting in death of many of them. Competition was also shown to normally retard the rate of neuronal maturation. Finally, competition between neurons was suggested to be involved in the establishment of specific innervation territories in the peripheral target of the ganglion. This was contrasted with the development of innervation territories by chick lumbosacral motoneurons, where competition was not required, and where some specific recognition mechanism between neuron and target was therefore suggested.

ACKNOWLEDGEMENTS

I would like to thank Mrs. Frances Hunihan for typing the manuscript. Some of the research reported in this chapter was supported by NIH grants NS10666 to L. Landmesser and NS10338 to G. Pilar.

REFERENCES

Black, I.B. (1978) Regulation of autonomic development. Ann. Rev. Neurosci., 1, 183–214.

Burstein, L.G., Pilar, G., and Landmesser, L. (1977) Competitive interactions between developing ciliary ganglion cells. *Soc. Neurosci. Abstr.*, 3, 442.

Chiappinelli, V., Giacobini, E., Pilar, G., and Uchimura, H. (1976) Induction of cholinergic enzymes in chick ciliary ganglion and iris muscle during synapse formation. *J. Physiol. (Lond.)*, 257, 749–766.

Chu-Wang, I.W., and Oppenheim, R.W. (1978) Cell death of motoneurons in the chick embryos spinal cord II. *J. comp. Neurol.*, 177, 59–86.

Clarke, P.G.H. and Cowan, W.M. (1976) The development of the isthmo-optic tract in the chick, with special reference to the occurrence and correction of developmental errors in the location and connections of the isthmo-optic neurons. *J. comp. Neurol.*, 167, 143–164.

Cowan, W.M. (1973) Neuronal death as a regulative mechanism in the control of cell number in the nervous system. *Development and Aging in the Nervous System*, Academic Press, New York, pp. 19–41.

Fambrough, D.M. (1976) Specificity of nerve–muscle interactions. In *Neuronal Recognition*, S.H. Barondes (Ed.), Plenum Press, New York, pp. 25–67.

Giacobini, G., Filogamo, G., Weber, M., Boquet, P., and Changeux, J.P. (1973) Effects of a snake α-neurotoxin on the development of innervated skeletal muscles in chick embryos. *Proc. nat. Acad. Sci. (Wash.)*, 170, 1708–1712.

Hamburger, V. (1958) Regression versus peripheral control of differentiation in motor hypoplasia. *Amer. J. Anat.*, 102, 365–410.

Hamburger, V. (1975) Cell death in the development of the lateral motor column of the chick embryo. *J. comp. Neurol.*, 160, 535–546.

Hamburger, V. and Hamilton, H.L. (1951) A series of normal stages in the development of the chick embryo. *J. Morph.*, 88, 49–92.

Harris, A.J. and Dennis, M.J. (1977) Deletion of "mistakes" in nerve–muscle connectivity during development of rat embryos. *Soc. Neurosci. Abstr.*, 3, 107.

Hollyday, M. and Hamburger, V. (1976) Reduction of the naturally occurring motor neuron loss by enlargement of the periphery. *J. comp. Neurol.*, 170, 311–320.

Lance Jones, C. and Landmesser, L. (1978) Effect of spinal cord deletions and reversals on motoneuron projection patterns in the embryonic chick hindlimb. *Soc. Neurosci. Abstr.*, 4, 118.

Landmesser, L. (1978) The organization of motoneuron pools supplying chick hindlimb muscles. *J. Physiol. (Lond.)*, in press.

Landmesser, L. and Morris, D.G. (1975) The development of functional innervation in the hindlimb of the chick embryo. *J. Physiol. (Lond.)*, 249, 301–326.

Landmesser, L. and Pilar, G. (1974a) Synapse formation during embryogenesis on ganglion cells lacking a periphery. *J. Physiol. (Lond.)*, 241, 715–736.

Landmesser, L. and Pilar, G. (1974b) Synaptic transmission and cell death during normal ganglionic development. *J. Physiol. (Lond.)*, 241, 737–749.

Landmesser, L. and Pilar, G. (1976) Fate of ganglionic synapses and ganglion cell axons during normal and induced cell death. *J. Cell Biol.*, 68, 357–374.

Mark, R.F. (1974) Selective innervation of muscle. *Brit. Med. Bull.*, 30, 122–126.

Marwitt, R., Pilar, G. and Weakly, J.N. (1971) Characterization of two cell populations in avian ciliary ganglion. *Brain Res.*, 25, 317–334.

Oppenheim, R.W., Chu-Wang, I.-W., and Maderdrut, J.L. (1978) Cell death of motoneurons in the chick embryo spinal cord. III. The differentiation of motoneurons prior to their induced degeneration following limb-bud removal. *J. comp. Neurol.*, 177, 87–112.

Patterson, P.H. (1978) Environmental determination of autonomic neurotransmitter functions. *Ann. Rev. Neurosci.*, 1, 1–17.

Pilar, G., and Landmesser, L. (1976) Ultrastructural differences during embryonic cell death in normal and peripherally deprived ciliary ganglia. *J. Cell Biol.*, 68, 339–356.

Pilar, G., Landmesser, L. and Burstein, L.G. (1978) Competition for survival among developing ciliary ganglion cells, submitted for publication.

Prestige, M.C. (1967) The control of cell number in the lumbar ventral horns during the development of *Xenopus laevis* tadpoles. *J. Embryol. exp. Morph.*, 18, 359–387.

Prestige, M.C. and Willishaw, D.J. (1975) On a role for competition in the formation of patterned neural connexions. *Proc. roy. Soc. Lond. B.* 190, 77–98.

384

Purves, D. (1976) Long-term regulation in the vertebrate peripheral nervous system. *Int. Rev. Physiol.*, 10, 125–177.

Sanes, J.R., Hildebrand, J.G., and Prescott, D.J. (1976) Differentiation of insect sensory neurons in the absence of their normal synaptic targets. *Dev. Biol.*, 52, 121–127.

Sohal, G.S., Weidman, T.A. and Stoney, S.D. (1978) Development of the trochlear nerve. Effects of early removal of periphery. *Exp. Neurol.*, 59, 331–341.

Varon, S., and Bunge, R.P. (1978) Trophic mechanisms in the peripheral nervous system. *Ann. Rev. Neurosci.*, 1, 327–361.

Yip, J.W., and Dennis, M.J. (1976) Suppression of transmission of foreign synapses in adult newt muscles involves reduction in quantal content. *Nature (Lond.)*, 260, 350–352.

Studies on Neurotrophic Regulation of Skeletal Muscle

S. THESLEFF

Department of Pharmacology, University of Lund, Lund (Sweden)

INTRODUCTION

Following denervation the extrajunctional ACh sensitivity of mammalian skeletal muscle fibres is enhanced by to the appearance of new receptors in the muscle membrane (Axelsson and Thesleff, 1959). At the same time tetrodotoxin (TTX) resistant action potentials appear (Redfern and Thesleff, 1971b). The rise in extrajunctional ACh sensitivity and the presence of TTX resistant spikes can be reversed by direct electric stimulation of the muscle (Drachman and Witzke, 1972; Lømo and Rosenthal, 1972). These findings have led to the suggestion that muscle activity per se may regulate the presence of extrajunctional ACh receptors and of TTX resistant spikes in the muscle membrane.

It has recently been reported that chronic poisoning of the rat extensor digitorum longus (EDL) muscle with botulinum toxin (BoTX) results in the appearance of fewer extrajunctional ACh receptors than are induced in this tissue by surgical denervation (Pestronk et al., 1976). Since BoTX poisoning produces complete muscle paralysis this suggests that trophic factors other than muscle activity participate in the regulation of ACh receptor numbers. In this study, a quantitative comparison was made of the influence of denervation and BoTX poisoning on the development of TTX resistant action potentials in rat muscle (see Mathers and Thesleff, 1978).

METHODS

The experiments were carried out on the extensor digitorum longus (EDL) muscle of adult male rats. *Clostridium botulinum* toxin type A was injected into the anterolateral region of the right hind leg superficial to the tibialis anterior and EDL muscles. Surgical denervation of the EDL muscle was performed under ether anaesthesia by sectioning the peroneal nerve at the knee.

At various times following BoTX poisoning or denervation the EDL muscle was removed under ether anaesthesia and mounted in an organ bath containing oxygenated saline at 29°C. TTX 10^{-6} M was routinely added to the bathing solution to block the TTX sensitive component of the muscle action potential. Action potentials were evoked and recorded by inserting two microelectrodes into the muscle fibre about 50 μm apart. Constant anodal current was used to hyperpolarize the membrane to -90 mV to ensure maximum rates of rise of the action potential (Redfern and Thesleff, 1971a).

RESULTS

Within 18 h, BoTX caused a complete paralysis of the leg which lasted for a period of up to 3 weeks. During the first two weeks after toxin injection the frequency of spontaneous miniature endplate potentials was reduced tenfold and evoked endplate potentials consisted of only a few quanta. In many fibres the nerve impulse failed to release transmitter.

Three days after surgical denervation, TTX resistant action potentials could be elicited from all fibres tested. In contrast muscle fibres examined 3 days after the administration of BoTX generally showed only delayed rectification in response to cathodic current and no action potentials were present. On the fourth postoperative day TTX resistant action potentials could be recorded in all denervated but only in about half of the BoTX poisoned muscle fibres tested. On the fourth and on all subsequent postoperative days the rate of rise was significantly lower in BoTX poisoned muscles as compared to denervated preparations (Fig. 1A). The amount of overshoot was also significantly smaller in BoTX poisoned muscles. The number of fibres which failed to produce an

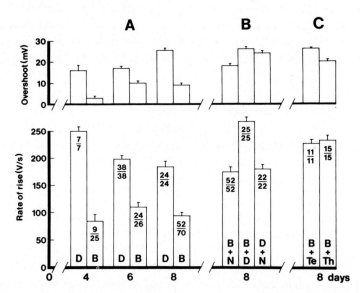

Fig. 1. Maximum rates of rise and overshoot of action potentials in the presence of TTX 10^{-6} M. The staple diagrams give means $\pm$ S.E.M. The figures within the staples show the number of fibers responding with an action potential over the number of fibres examined. (A) After 4, 6 and 8 days of denervation (D) or BoTX poisoning (B); (B) After 8 days of BoTX poisoning and (B + N) 2 days of α-neurotoxin or (B + D) 2 days of surgical denervation. (D + N) after 8 days of surgical denervation and 2 days of α-neurotoxin; (C) After 8 days of BoTX poisoning and 2 days of twice daily subcutaneous injections of an adrenergic β_2-receptor stimulating drug terbutaline (2-tert-butylamino-1-)3,5-dihydroxyphenyl)ethanol) 20 mg/kg body weight (Te) or a phosphodiesterase inhibitor, theophylline 75 mg/kg body weight (Th).

action potential was greater in the case of BoTX poisoned muscles at each of the postoperative days tested as shown in Fig. 1A.

The results suggested that in BoTX poisoning a trophic influence remains which suppresses the appearance of TTX resistant action potentials. Experiments were therefore made in which the muscles were denervated surgically on the sixth day after BoTX treatment. Both the rate of rise and the overshoot of the action potentials was increased by denervation as shown in Fig. 1B.

The possibility that the trophic influence of the nerve in BoTX poisoned muscles consisted of a residual acetylcholine release was tested by the administration of α-neurotoxin from *Naja naja siamensis* to BoTX poisoned rats. This neurotoxin blocks selectively and almost irreversibly the cholinergic receptor in skeletal muscle (Lester, 1972). Following the injection of 15 μg of the α-neurotoxins into the hind leg the rate of rise of the action potential and its amount of overshoot was the same as that seen following denervation as shown in Fig. 1B. The possibility that the α-neurotoxin acted directly on the mechanism responsible for spike generation was tested by administering the toxin to surgically denervated muscles. As shown by Fig. 1B this procedure failed to affect either the rate of rise or the overshoot of the action potential.

DISCUSSION

The results reported show that BoTX in a dose sufficient to cause complete paralysis was less effective than surgical denervation in inducing the appearance of TTX resistant action potentials. The lower potency of BoTX is unlikely to result from a systemic action of the toxin since injection of BoTX failed to alter the response of the muscle to denervation. We concluded therefore that a trophic influence supressing the appearance of TTX resistant action potentials and of extrajuctional ACh receptors (Pestronk et al., 1976) is released from the BoTX poisoned nerve.

It has recently been shown that only a fraction of the total ACh released from rat motor nerve endings can be accounted for by spontaneous and nerve evoked *quantal* release of transmitter. It has been proposed that the remaining release of ACh may occur as a leakage of transmitter from nerve (Katz and Miledi, 1977). BoTX reduces quantal transmitter release to a few per cent of normal but total ACh leakage only by one third (Brooks, 1954). It is therefore possible that a considerable release of ACh remains from BoTX poisoned nerves and that this leakage could have a trophic influence on muscle. The α-neurotoxin used in this study blocks specifically and with low reversibility ACh receptors in muscle. Our results show that the α-neurotoxin when applied to BoTX poisoned muscles enhances the appearance of TTX resistant action potentials in a way similar to surgical denervation. This finding suggests that an interaction between the ACh released from the BoTX poisoned nerve and its receptor in the muscle membrane has a neurotrophic influence which, in part, is responsible for the regulation of the appearance of changes associated with denervation in the muscle. In addition, the results emphasize the importance of using quantitative techniques when examining the role of various neurotrophic influences.

388

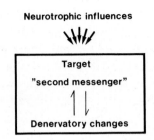

Neurotrophic influences

Target
"second messenger"

Denervatory changes

Fig. 2. For explanation, see text.

From the eminent research of Gutmann and his colleagues (Gutmann, 1976) we now know that a variety of neurotrophic actions serve to regulate muscle function. Factors such as nerve impulses, muscle use, unknown agents transported by axoplasmic flow, the release of acetylcholine, blood borne hormones such as testosterone, ATP etc. are all capable of exerting some degree of regulatory control of skeletal muscle function, however, their relative effectiveness varies from muscle to muscle and species to species. For instance, in the baboon, muscle use and disuse has a much less prominent role as a neurotrophic influence than in the rat (Gilliatt et al., 1977). Testosterone is effective only on certain target muscles (levator ani muscle of the female rat etc.). However, despite the great number of trophic influences examined to date no specialization has been detected and all factors seem to exert control over the same myogenic properties and to date, only quantitative differences have been demonstrated. Since a number of neurotrophic factors appear to be involved in regulating the same myogenic properties it might be of more importance to establish whether or not the muscle cell has a common target mechanism for all these influences instead of attempting to evaluate the relative "potency" and role of each neurotrophic factor. The possibility appears quite likely that the muscle cell has a "second messenger" which translates all the incoming neurotrophic signals into a regulating denervation/innervation signal (Fig. 2). Very little has so far been done to establish the presence of such a myogenic regulatory mechanism. Interestingly enough, however, it has been shown that the cyclic AMP concentration in skeletal muscle is increased very early after nerve section (Carlsen, 1975). It therefore appears possible that in muscle the adenyl cyclase-cyclic AMP system is neurotrophically controlled and that this system as pointed out by Greengard (1976), may control protein phosphorylation and thereby regulate the synthesis of specific proteins. The administration to BoTX poisoned rats of drugs, terbutaline or theophylline, which are potent stimulators of the adenyl cyclase-cyclic AMP system was shown to enhance the rate of rise of the TTX resistant action potential (Fig. 1C).

To conclude, it seems likely that we may learn more about neurotrophism by looking for a common target mechanism for neurotrophic influences in the muscle cell instead of trying to establish the nature and relative importance of each neurotrophic factor which affects the muscle.

SUMMARY

A quantitative comparison was made of the effects of the paralysis caused by the local injection of botulinum toxin type A with those of surgical denervation on the development of tetrodoxin (TTX) resistant action potentials in rat skeletal muscle. After surgical denervation TTX resistant action potentials were present in all fibres on the third day and the rate of rise and amount of overshoot reached peak values on the fifth day. Botulinum toxin poisoning caused complete paralysis, but failed to induce TTX resistant action potentials in all fibres. In fibres in which a TTX resistant spike could be evoked the average rate of rise was at all times only about half that in denervated fibres. Administration of the α-neurotoxin, which selectively blocks cholinergic receptors, to botulinum toxin poisoned animals resulted in the appearance of TTX resistant action potentials in all fibres and in a significant increase in their rate of rise and overshoot. The results show that despite causing complete paralysis botulinum toxin is less effective than surgical denervation in inducing denervatory changes in skeletal muscle. This suggests that the presence of the botulinum poisoned nerve prevents the appearance of the full blown signs of denervation. Since the influence of the botulinum poisoned nerve is blocked by α-neurotoxin the residual release of acetylcholine is thought to be responsible for this neurotrophic action. It is concluded that a great number of neurotrophic factors, among which is acetylcholine, suppress the appearance of denervatory changes in skeletal muscle. The relative importance of each neurotrophic influence varies quantitatively but not qualitatively between species and muscles studied. It is suggested that neurotrophic influences act on a common target mechanism ("second messenger") in the muscle cell which in turn regulates the appearance of the signs of denervation.

REFERENCES

Axelsson, J. and Thesleff, S. (1959) A study of supersensitivity in denervated mammalian skeletal muscle. *J. Physiol. (Lond.)*, 129, 178–193.

Brooks, V.B. (1954) The action of botulinum toxin on motor-nerve filaments. *J. Physiol. (Lond.)*, 123, 501–515.

Carlsen, R.C. (1975) The possible role of cyclic AMP in the neurotrophic control of skeletal muscle. *J. Physiol. (Lond.)*, 247, 343–361.

Drachman, D.B. and Witzke, F. (1972) Trophic regulation of acetylcholine sensitivity of muscle: effect of electrical stimulation. *Science*, 176, 514–516.

Gilliatt, R.W., Westgaard, R.H. and Williams, I.R. (1977) Acetylcholine-sensitivity of denervated and inactivated baboon muscle fibres. *J. Physiol. (Lond.)*, 271, 21P–22P.

Greengard, P. (1976) Possible role for cyclic nucleotides and phosphorylated membrane proteins in postsynaptic actions of neurotransmitters. *Nature (Lond.)*, 260, 101–108.

Gutmann, E. (1976) Neurotrophic relations. *Ann. Rev. Physiol.*, 38, 177–216.

Katz, B. and Miledi, R. (1977) Transmitter leakage from motor nerve endings. *Proc. roy Soc. Lond. B.*, 196, 59–72.

Lester, H.A. (1972) Blockade of acetylcholine receptors by cobra toxin: electrophysiological studies. *Molec. Pharmacol.*, 8, 623–631.

Lømo, T. and Rosenthal, J. (1972) Control of ACh sensitivity by muscle activity in the rat. *J. Physiol. (Lond.)*, 221, 443–513.

Mathers, D.A. and Thesleff, S. (1978) Studies on neurotrophic regulation of murine skeletal muscle. *J. Physiol. (Lond.)*, 282, 105–114.

Pestronk, A., Drachman, D.B. and Griffin, J.W. (1976) Effects of botulinum toxin on trophic regulation of acetylcholine receptors. *Nature (Lond.)*, 264, 787–789.

Redfern, P. and Thesleff, S. (1971a) Action potential generation in denervated rat skeletal muscle. I. Quantitative aspects. *Acta physiol. scand.*, 81, 557–564.

Redfern, P. and Thesleff, S. (1971b) Action potential generation in denervated rat skeletal muscle. II. The action of tetrodotoxin. *Acta physiol. scand.*, 82, 70–78.

Factors Affecting the Distribution of Acetylcholine Receptors in Innervated and Denervated Skeletal Muscle Fibres

ROSEMARY JONES

Department of Biochemistry, University of Birmingham, Birmingham B15 2TT
(United Kingdom)

INTRODUCTION

In normal adult skeletal muscles acetylcholine (ACh) receptors are apparent at the end-plate region only (Kuffler, 1943; Nastuk, 1953; Miledi, 1960b). Localisation of ACh receptors at the end-plate occurs gradually during post-natal development in mammals, for at birth the entire muscle fibre membrane is sensitive to ACh (Diamond and Miledi, 1962). During the first few weeks of the animal's life extrajunctional regions of the muscle fibre membrane become desensitized and sensitivity to ACh is confined to the postsynaptic membrane.

In developing muscles differentiation of junctional and extrajunctional regions of the muscle fibre membrane results not only from the elimination of receptors from the extrajunctional membrane but also involves an increase in sensitivity at the neuromuscular junction. End-plates of new born animals are relatively insensitive to the transmitter (Diamond and Miledi, 1962) or depolarising drugs (Maclagan and Vrbová, 1966) and sensitivity increases as the muscle matures (Jones and Vrbová, 1972). Both desensitization of the extrajunctional membrane and the development of high chemosensitivity at the neuromuscular junction require the presence of the motor nerve (Jones and Vrbová, 1972; Brown, 1975). The nerve is therefore important for the initial development of the different regions of the muscle fibre membrane, but once the neuromuscular junction has developed its properties apparently remain fixed, even if the motor nerve is subsequently removed (Frank et al., 1975).

The sensitivity of the extrajunctional membrane can, however, change dramatically, showing increased sensitivity following a variety of "insults" to the muscle. After denervation the entire muscle membrane again becomes sensitive to ACh (Axlesson and Thesleff, 1959; Miledi, 1960a) and muscle injury also results in the appearance of extrajunctional ACh receptors (Katz and Miledi, 1964; Redfern and Thesleff, 1971; Jones and Vrbová, 1974, Vyskočil and Gutmann, 1976).

The finding that hypersensitivity can be induced in innervated as well as denervated muscle fibres suggests that the presence of the nerve alone is not enough to control the distribution of chemosensitivity along the muscle fibre membrane. Several factors apparently interact to control muscle sensitivity,

392

and it is the nature of some of these interactions that will be discussed here.

In addition, the significance of changes in the muscle fibre membrane for the maintenance of normal nerve-muscle contact will be considered.

ROLE OF MUSCLE ACTIVITY IN THE CONTROL OF MUSCLE SENSITIVITY

That the motor nerve is important for ensuring the desensitization of the extrajunctional membrane is evidenced by the finding that denervated muscles remain hypersensitive until they become reinnervated (Miledi, 1960c). The presence of the motor nerve is also required to bring about the desensitization of the extrajunctional muscle fibre membrane in young muscles (Brown, 1975). There has been some debate, however, as to what aspect of the interaction between nerve and muscle is responsible for confining sensitivity to ACh to the end-plate region. It has been proposed that the nerve exerts a "trophic" influence, independent of impulse activity over the extrajunctional membrane (Miledi, 1960c; 1963) or that spontaneous release of small amounts of ACh by the nerve (Thesleff, 1960) keeps the extrajunctional membrane insensitive.

Recently, however, much evidence has been presented to show that it is the activity imposed on the muscle by its motor nerve that is important in confining sensitivity to the neuromuscular junction (Jones and Vrbová, 1970; 1971; 1974; Lømo and Rosenthal, 1972; Lømo and Westgaard, 1975).

Role of activity during development

The muscle fibres of newborn animals are sensitive to ACh over their entire length (Diamond and Miledi, 1962). Muscles of newborn rats removed from the body and placed in Krebs-Henseleit solution respond to ACh added to the solution with a contracture. When contractures to ACh are measured in muscles taken from rats at different ages it is found that as the muscle mature, progressively larger doses of ACh are required to produce a contracture. The decrease in sensitivity to ACh corresponds to the gradual desensitization of the extrajunctional muscle membrane (Fig. 1).

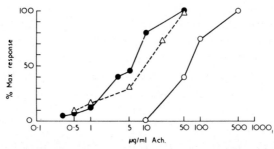

Fig. 1. The graph shows contractures in response to different doses of acetylcholine (ACh) of rat soleus muscles recorded in vitro plotted as a percentage of the maximum response for each muscle. Responses of a muscle from a 4-day-old rat (●) are compared with those of an 8-day-old rat (○) and those of an 8-day-old rat soleus muscle denervated at 4 days (△).

Desensitization of young muscle fibres is apparently due to the activity imposed on the muscle by the motor nerve, and electrical stimulation of young muscles, even for a few hours, reduces their response to ACh (Jones and Vrbová, 1970). Also, if a muscle is denervated soon after birth, desensitization of the extrajunctional membrane fails to occur (see Fig. 1). Denervated muscles of young rabbits were found to remain hypersensitive unless the activity normally imposed by the motor nerve was replaced by electrical stimulation via implanted electrodes (Brown, 1975).

Role of activity in adult muscles

Once the extrajunctional membrane has become desensitized during development it remains insensitive to the transmitter. If the motor nerve is cut, or the muscle injured, however, extrajunctional receptors reappear.

Thomson (1952) reported that following a period of electrical stimulation, denervated tibialis anterior muscles of rats showed smaller responses to ACh than unstimulated denervated muscles. Stimulation of denervated rat soleus muscles both in vitro and in vivo (Jones and Vrbová, 1970; 1971, 1974) was found to significantly reduce their response to ACh (Fig. 2).

That this is due to a reduction in extrajunctional membrane sensitivity to ACh was confirmed by Lømo and Rosenthal (1972). Further, it was found that the frequency of muscle stimulation was significant, fast patterns of activity desensitizing the muscle more effectively than slow ones (see Lømo and Westgaard, 1975). This finding affords an explanation for the observation that

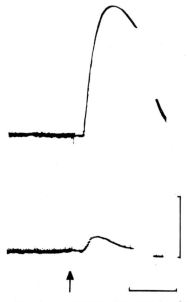

Fig. 2. Contractures of adult rat soleus muscles recorded in vitro in response to 500 μg/ml ACh given as indicated by the arrow; the upper trace shows the response of a muscle denervated 3 days previously and the lower trace the response of a 3 day denervated muscle that had been stimulated directly via implanted electrodes with trains of impulses of 40 Hz for 500 msec every 2 min, 7–8 h daily. The vertical calibration bar represents 5 g and the horizontal bar 10 sec.

394

in the slow soleus muscle of the rat, supplied by motoneurones with characteristically slow firing rates, low sensitivity to ACh can be detected outside the
end-plate region (Miledi and Zelená, 1966). This extrajunctional chemosensitivity disappears if the soleus muscle is reinnervated by a fast motor nerve
supplying high frequency impulses (Miledi et al., 1968). This offers further
evidence that not only activity per se, but the pattern of activity imposed on a
muscle by its motor nerve or by electrical stimulation is important in the
regulation of the distribution of ACh receptors.

NERVE DEGENERATION AND MUSCLE HYPERSENSITIVITY

Although extrajunctional sensitivity to ACh can be eliminated in both young
muscles and in denervated adult muscles by imposed electrical stimulation,
many experimental observations suggest that loss of activity following denervation cannot be the only factor leading to sensitivity of the extrajunctional
muscle fibre membrane. For instance denervation changes occur sooner if the
motor nerve is cut close to the muscle than when a long nerve stump is left
(Luco and Eyzaguirre, 1955; Harris and Thesleff, 1972), although in both cases
activity ceases at the same time. It is also known that nerve terminal degeneration occurs sooner when a short nerve stump is left (Miledi and Slater, 1968). It
has been suggested by Jones and Vrbová (1974) that two factors interact to
produce denervation hypersensitivity, loss of muscle activity coupled with an
inflammatory response to degenerating nerve tissue. Three lines of experimental evidence support this hypothesis:

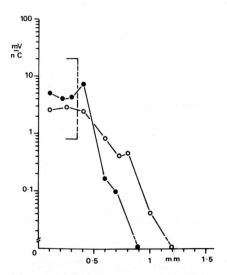

Fig. 3. Sensitivity of different regions of rat EDL muscle fibres to iontophoretically applied
ACh 3 days after placing a small piece of isolated nerve onto the surface of the muscle. The
broken line indicates the limit of the area covered by the piece of nerve. Sensitivity to ACh
(mV/nC) in the two muscle fibres mapped is highest in the area underlying the degenerating
nerve, falling to zero at distances approximately 1 mm away from the edge of the area under
the nerve explant. (Jones and Vyskočil).

Effects of degenerating nerve tissue on the sensitivity of the innervated muscle fibre membrane

If the response to degenerating nerve tissue has a role in the development of denervation hypersensitivity any situation in which degenerating nerve tissue is present should lead to increased membrane sensitivity. Vrbová (1970) tested this directly by placing a small piece of isolated nerve onto the extrajunctional surface of the tibialis anterior muscle in rabbits. She found that 5days after placing the piece of nerve onto the muscle the area of the extrajunctional membrane underlying the degenerating nerve was sensitive to applied ACh. Similar results were obtained with rat muscles onto which a piece of nerve had been placed (Jones and Vrbová, 1974); using iontophoretically applied ACh, Jones and Vyskočil (1975) found that the highest sensitivity was confined to the area underlying the piece of nerve (Fig. 3). Thus, even in an innervated muscle that remains active, degenerating nerve tissue can induce the appearance of extrajunctional ACh receptors. Histological sections through the muscle in the area where the piece of nerve has been placed show clear evidence of an inflammatory response (Blunt et al., 1975) and the phagocytic activity of cells invading the region can be demonstrated in electromicrographs (see Jones and Vrbová, 1974).

Partial denervation and muscle membrane properties

Another situation in which degenerating nerve tissue is present within a muscle is after partial denervation. Following section of one ventral root supplying muscles of the hind limb in rats many muscles undergo partial denervation (Hoffman, 1950; Edds, 1953). It was therefore interesting to see

TABLE I

RESPONSES OF DENERVATED AND PARTIALLY DENERVATED SOLEUS MUSCLES OF AN ADULT (180 g) RAT TO ELECTRICAL STIMULATION AND ACh 4 DAYS AFTER OPERATION

Stimulation	Muscle tension (g)	
	Partially denervated	Fully denervated
Electrical stimulation		
Motor nerve	7.7	0
Direct muscle stimulation	12.9	14.2
(Approx.) % denervation	40.3	100
Acetylcholine		
g/ml 1×10^{-6}	1.2	1.8
1×10^{-5}	2	2.1
1×10^{-4}	2.7	2.6

Measurement of contractions in response to electrical stimulation was performed in vivo under nembutal anaesthesia, and the maximum response obtained by soleus nerve stimulation using small shielded electrodes was compared with maximum response from direct muscle stimulation using large silver wire bipolar electrodes placed along either side of the muscle. Both muscles were then removed, placed in Krebs-Henseleit solution in the same organ bath and contractures in response to different doses of ACh added to the bathing medium were recorded.

TABLE II

TETRODOTOXIN RESISTANT ACTION POTENTIALS IN PARTIALLY DENERVATED MUSCLES

Days after operation	Total no. fibres	Denervated fibres	Innervated fibres	TTX Resistant fibres
4	37	12	25	13
5	13	2	11	7
6	20	7	13	19
7	36	6	30	27

Micropipettes were filled with 3M KCl and inserted into single muscle fibres in rat EDL muscles removed at 4–7 days after partial denervation

Fibres were assumed to be innervated if they displayed miniature end-plate potentials or if nerve stimulation resulted in an action potential. In a large number of the innervated muscle fibres tested, directly evoked action potentials could be elicited in the presence of TTX (10^{-6} M).

whether any changes occur in the innervated fibres of a partially denervated muscle as well as in the denervated ones. Muscles of one hind limb were partially denervated either by section of one ventral root (L4, L5) or by splitting the sciatic nerve in the thigh into two and crushing one part. Such procedures produced between 45 and 80% denervation of the ipsilateral soleus or extensor digitorum longus (EDL) muscle. However, when contractures of the muscles in response to ACh were compared with those of completely denervated muscles little difference could be detected (see Table I). This suggested that innervated muscle fibres had also undergone denervation-like changes. To explore this further, individual fibres in partially denervated muscles were examined using intracellular recording. Fibres were assumed to be innervated if they displayed miniature end-plate potentials or if nerve stimulation resulted in a muscle action potential. In addition to the appearance of extrajunctional acetylcholine receptors denervated muscle fibres also show tetrodotoxin (TTX) resistant action potentials (Redfern and Thesleff, 1971). Table II shows the results of experiments performed at different times after partial denervation in rat EDL muscles. It is clear that not only the denervated fibres developed TTX resistant action potentials, but many innervated fibres also showed this change (Jones and Ward, unpublished results). Similar changes in the properties of innervated muscle fibres in partially denervated muscles were reported by Cangiano and Lutzemberger (1977).

The inflammatory response in denervation

Direct experimental evidence that denervation induces an inflammatory response, and that this may be related to nerve terminal degeneration and the development of extrajunctional chemosensitivity, was obtained by Jones and Lane (1975). At various intervals after cutting the sciatic nerve in the thigh in rats, soleus and gastrocnemius muscles were removed, teased apart in culture medium and gently agitated on a catherine wheel. The resulting suspension was then filtered and centrifuged so that any mononuclear cells present in the medium were precipitated as a pellet from which dilutions for cell counts or cell smears were made. The number of mononuclear cells that could be isolated

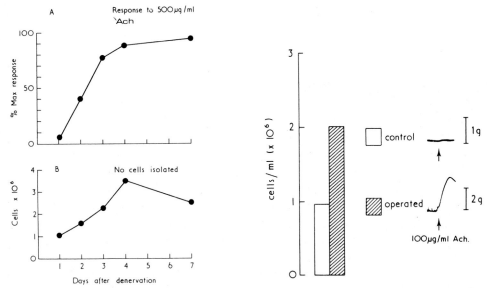

Fig. 4. (A) The graph shows responses of adult rat soleus muscles removed 1–7 days after denervation to 500 μg/ml ACh plotted as a percentage of the maximum response of a 7-day denervated muscle; (B) The number of mononuclear cells (cells × 10⁶/ml) that were isolated from rat soleus and gastrocnemius muscles is plotted against days after denervation.

Fig. 5. Contractures in response to 100 μg/ml ACh of a control adult rat soleus muscle and one examined 3 days after placing a small piece of nerve onto its surface are shown on the right hand side of the figure. On the left the number of mononuclear cells isolated from the operated muscle is compared with the control muscle.

from denervated muscles increased during the first few days after nerve section, the greatest relative increase being between days 2 and 4 and this corresponds well with the time course of the development of muscle hypersensitivity (Figs. 4a,b). Cell types found were monocytes, lymphocytes and polymorphonuclear leucocytes.

An increase in the number of mononuclear cells isolated from muscles on to which a piece of degenerating nerve had been placed was also found to correspond with increased sensitivity to ACh (Fig. 5).

INDUCTION OF SENSITIVITY CHANGES IN ADULT MUSCLES IN VITRO

It is clear that cellular infiltration and extrajunctional membrane sensitivity develop simultaneously following denervation or muscle injury, but it is less clear whether there is a direct causal link between the cellular response and increased membrane sensitivity. It has been suggested that in adult muscle fibres ACh receptors are present throughout the muscle fibre membrane, but are concealed or inactivated (Blunt et al., 1975). Substances released from infiltrating cells or from the injured or denervated muscle fibres themselves may act on the extrajunctional membrane to expose or activate latent receptors (Blunt et al., 1975).

The receptor molecule is thought to be a glycoprotein and in order to test whether enzymes that might act on receptors or on the muscle fibre membrane are present in denervated muscles, the responses of normal adult muscles incubated in homogenates prepared from denervated muscles were compared with those incubated in different phospholipases.

Advantage was taken of the fact that normal adult soleus muscles, which have some extrajunctional ACh receptors (Miledi and Zelená, 1966) respond to large doses of ACh applied in vitro with a contracture (Gutmann and Hanzlíková, 1966) and depolarisation.

Both soleus muscles were removed from adult rats, placed in Krebs-Henseleit solution and the average contracture or depolarisation in response to a test dose of ACh applied several times was obtained. One of the muscles was then incubated in extracts obtained from denervated muscles, cellular fractions from denervated or damaged muscles or with phospholipase C or D. Responses of incubated muscles to the same dose of ACh were then compared with control muscles. An increase in the contracture or depolarisation in response to ACh of incubated muscles was seen (Watson et al., 1976; Harbone et al., 1978).

Responses to ACh of incubated muscles increased by approximately 200% after 1–3 h incubation. This increase in sensitivity is much less than that seen after denervation or muscle injury where sensitivity to ACh increases many hundred-fold. However the development of full sensitivity following denervation or injury in vivo takes several days and it is possible that the relatively small increase in sensitivity in muscles incubated with homogenates or enzymes in vitro represents an early stage in the development of extrajunctional chemosensitivity.

MEMBRANE CHANGES AND MUSCLE INNERVATION

One interesting feature of muscles in which an inflammatory response has been induced is that changes in the motor innervation are also seen. It has long been known that nerve sprouting occurs in partially denervated muscles (Hoffman, 1950; Edds, 1953) and we have observed sprouting in partially denervated muscles at a time when changes in both innervated and denervated muscle fibres are apparent (see Table II and Fig. 6). Innervated fibres showing TTX resistant action potentials are not thought to be denervated fibres newly innervated by sprouts as the development of extrajunctional membrane changes precedes the first appearance of nerve sprouting. Nerve sprouting and degenerative changes have been observed in muscle onto which a piece of degenerating nerve has been placed although transmission at these synapses appears normal (Jones and Tuffery, 1973; see also Gordon et al., 1976).

Thus it may be that in the presence of an inflammatory reaction, changes in the nerve terminals are induced either through the direct action of substances released by invading cells on the nerve terminals, or indirectly by changes induced in the membrane of the muscle fibres which the axon terminals contact.

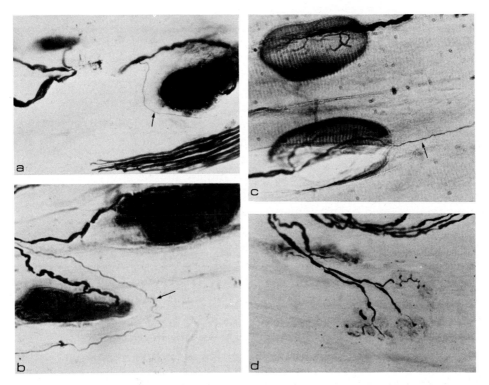

Fig. 6. (a, b and c) show examples of nerve sprouting (arrows) at end-plates in a 7-day partially denervated EDL muscle from an adult rat. (d) shows a group of degenerating (presumably denervated) end-plates in the same muscle. The 50–100 μm frozen sections were prepared using a modified combined silver and cholinesterase stain (Namba et al., 1967) (prepared in collaboration with Dr. Anna Östberg).

DISCUSSION

At birth the fibres of mammalian muscles are sensitive to ACh over their entire surface membrane. As the muscles mature ACh receptors become confined to the end-plate region (Diamond and Miledi, 1962). The motor nerve plays an essential role in the establishment and maintenance of the normal distribution of ACh receptors. Following section of the motor nerve adult muscle fibres again become sensitive to ACh over their entire membrane (Axelsson and Thesleff, 1959; Miledi 1960a) and desensitization of the extrajunctional muscle membrane fails to take place in denervated young muscles (Brown, 1975). There is now considerable evidence that it is the activity imposed on the muscle by its motor nerve that brings about desensitization of the extrajunctional muscle fibre membrane, and young or adult denervated muscles can be desensitized by imposed electrical stimulation (Jones and Vrbová, 1970; 1974; Lømo and Rosenthal, 1972; Brown, 1975; Lømo and Westgaard, 1975).

It has been suggested that the nerve governs the distribution of ACh receptors by suppressing receptor manufacture (Fambrough, 1970). It was proposed by Grampp et al. (1974) that following denervation release of this suppression leads to the appearance of extrajunctional ACh receptors. However there is a constant turnover of ACh receptors at the neuromuscular junction (Chang and Huang, 1975) and the normal muscle must manufacture receptors to incorporate them at the neuromuscular junction. Moreover the presence of the motor nerve cannot prevent the appearance of extrajunctional ACh receptors, for in innervated muscles subjected to injury, changes in the membrane similar to those seen after denervation are observed.

In the presence of degenerating nerve, or following denervation or muscle injury an inflammatory reaction occurs and extrajunctional ACh receptors appear. It was proposed by Blunt et al. (1975) that the appearance of extrajunctional ACh receptors following denervation or muscle injury was due to the exposure or activation of ACh receptors present at extrajunctional membrane by enzymes released by inflammatory cells.

However attempts to "expose" receptors in vitro have yielded only a very small increase in sensitivity of muscle to ACh (Watson et al., 1976) compared with that seen after injury or denervation.

Another possibility is that inflammatory reaction induces some modification in the muscle fibre membrane that allows the incorporation of receptors. Such receptors would then be subject to turnover (Chang and Huang, 1975) which in turn may stimulate increased receptor synthesis. Increased proteosynthetic activity following denervation has been reported (Grampp et al., 1972; see Brockes et al., 1975) and some of this activity has been shown to be associated with the appearance of new ACh receptors (Brockes et al., 1975).

The maintenance of the normal distribution of ACh receptors may then be explained by proposing that the end-plate becomes modified during development, or is constantly modified by substances released by the motor nerve or Schwann cells, so as to allow the incorporation of ACh receptors. At extrajunctional sites receptors are concealed or receptor incorporation is suppressed as a result of muscle activity. The appearance of extrajunctional receptors in the presence of an inflammatory reaction in innervated muscle fibres is short lasting (Jones and Vrbová, 1974) further indicating that muscle activity counteracts changes in the extrajunctional muscle fibre membrane that lead to increased chemosensitivity.

Disturbances of innervation observed in both partially denervated muscles or muscles onto which a piece of degenerating nerve has been placed (Jones and Tuffery, 1973) suggest that nerve sprouting may be induced by changes in innervated muscle fibres in response to an inflammatory reaction, or by the direct action of inflammatory cells on the nerve terminals. The many issues arising from the experiments briefly described require fuller analysis than can be given here, but these results serve to emphasise the influence of the environment within the muscle on nerve terminals and the establishement and maintenance of synaptic connections.

SUMMARY

During postnatal muscle development, ACh receptors, distributed at birth over the entire muscle fibre membrane, become confined to the end-plate. It is the activity imposed on the muscle by its motor nerve that brings about desensitisation of the extrajunctional membrane. In adults denervation or muscle injury induces changes in the extrajunctional muscle membrane leading to exposure or incorporation of ACh receptors. Muscle activity counteracts the effects of denervation or injury and extrajunctional receptors are eliminated in active muscles.

ACKNOWLEDGEMENTS

I would like to thank the Muscular Dystrophy group of Great Britain for financial support. Partial denervation experiments were carried out in collaboration with Dr. Anna Östberg and Dr. Martyn Ward.

REFERENCES

Axelsson, J. and Thesleff, S. (1959) A study of supersensitivity in denervated mammalian skeletal muscle. *J. Physiol. (Lond.)*, 147, 178–193.

Blunt, R.J., Jones, R. and Vrbová, G. (1975) Inhibition of cell division and the development of denervation hypersensitivity in skeletal muscle. *Pflügers Arch.*, 355, 189–204.

Brockes, J.P., Berg, D.K. and Hall, Z.W. (1975) The biochemical properties and regulation of acetylcholine receptors in normal and denervated muscle. *Cold Spring Harbor Sym. quant. Biol.*, Vol. XL, 253–262.

Brown, M.D. (1975) Activity as a factor in the postnatal development of chemosensitivity in rabbit skeletal muscles. In *Recent Advances in Myology*, W.G. Bradley, D. Gardner-Medwin and J.N. Walton (Eds.), *Proc. III Int. Congr. Muscle Dis.*, Excerpta Medica, Amsterdam, pp. 16–21.

Cangiano, A. and Lutzemberger, L. (1977) Partial denervation affects both denervated and innervated fibres in the mammalian skeletal muscle. *Science*, 196, 542–545.

Chang, C.C. and Huang, M.C. (1975) Turnover of junctional and extrajunctional acetylcholine receptors of the rat diaphragm. *Nature (Lond.)*, 253, 643.

Diamond, J. and Miledi, R. (1962) A study of foetal and newborn rat muscle fibres. *J. Physiol. (Lond.)*, 162, 393–408.

Edds, M.V. (1953) Collateral nerve regeneration. *Quar. Rev. Biol.*, 28, 260–276.

Fambrough, D.M. (1970) Acetylcholine sensitivity of muscle fibre membranes: mechanism of regulation by motorneurones. *Science*, 168, 372–373.

Frank, E., Gautvik, K. and Sommerchild, H. (1975) Cholinergic receptors at denervated mammalian motor end-plates. *Acta physiol. scand.*, 95, 66.

Gordon, T., Jones, R. and Vrbová, G. (1976) Changes in chemosensitivity of skeletal muscle as related to end-plate formation. *Progr. Neurobiol.*, 6, 103–136.

Grampp, W., Harris, J.B. and Thesleff, S. (1972) Inhibition of denervation changes in skeletal muscle by blockers of protein synthesis. *J. Physiol. (Lond.)*, 221, 743–754.

Gutmann, E. and Hanzlíková, V. (1966) Contracture responses of fast and slow mammalian muscles. *Physiol. Bohemoslov.*, 15, 404–414.

Harborne, A.J., Smith, M.E. and Jones, R. (1978) The effect of hydrolytic enzymes on the acetylcholine sensitivity of the skeletal muscle cell membrane. Pflügers Arch., 377, 147–153.

Harris, J.B. and Thesleff, S. (1972) Nerve stump length and membrane changes in denervated skeletal muscle. *Nature New Biol.*, 236, 60–61.

Hoffman, H. (1950) Local re-innervation in partially denervated muscle: A Histo-physiological study. *Aust. J. exp. Biol. med. Sci.*, 28, 383–397.

Jones, R. and Lane, J.T. (1975) Proliferation of mononuclear cells in skeletal muscle after denervation or muscle injury. *J. Physiol. (Lond.)*, 246, 60–61P.

Jones, R. and Tuffery, A.R. (1973) Relationship of end-plate morphology to junctional chemosensitivity. *J. Physiol. (Lond.)*, 232, 13–15P.

Jones, R. and Vrbová, G. (1970) Effect of muscle activity on denervation hypersensitivity. *J. Physiol. (Lond.)*, 210, 144–145P.

402

Jones, R. and Vrbová, G. (1971) Can denervation hypersensitivity be prevented? *J. Physiol. (Lond.)*, 217, 67—68P.

Jones, R. and Vrbová, G. (1972) Effects of denervation and reinnervation on the responses of kitten muscles to acetylcholine and suxamethonium. *J. Physiol. (Lond.)*, 222, 569—581.

Jones, R. and Vrbová, G. (1974) Two factors responsible for the development of denervation hypersensitivity. *J. Physiol. (Lond.)*, 236, 517—538.

Jones, R. and Vyskočil, F. (1975) An electrophysiological examination of the changes in skeletal muscle fibres in response to degenerating nerve tissue. *Brain Res.*, 88, 309—317.

Katz, B. and Miledi, R. (1964) The development of ACh sensitivity in nerve-free segments of skeletal muscle. *J. Physiol. (Lond.)*, 170, 389—396.

Kuffler, S.W. (1943) Specific excitability of the end-plate region in normal and denervated muscle. *J. Neurophysiol.*, 6, 99—110.

Lφmo, T. and Rosenthal, J. (1972) Control of ACh sensitivity by muscle activity in the rat. *J. Physiol. (Lond.)*, 221, 493—513.

Lφmo, T. and Westgaard, R.H. (1975) Further studies on the control of ACh sensitivity by muscle activity in the rat. *J. Physiol. (Lond.)*, 252, 603—626.

Luco, J.V. and Eyzaguirre, C. (1955) Fibrillation and hypersensitivity to ACh in denervated muscle. Effect of length of degenerating nerve fibres. *J. Neurophysiol.*, 18, 65—73.

MacLagan, J. and Vrbová, G. (1966) A study of increased sensitivity of denervated and reinnervated muscle to depolarising drugs. *J. Physiol. (Lond.)*, 182, 131—143.

Miledi, R. (1960a) Acetylcholine sensitivity of frog muscle fibres after complete or partial denervation. *J. Physiol. (Lond.)*, 151, 1—23.

Miledi, R. (1960b) Junctional and extrajunctional acetylcholine receptors in skeletal muscle fibres. *J. Physiol. (Lond.)*, 151, 24—30.

Miledi, R. (1960c) Properties of regenerating neuromuscular synapses in the frog. *J. Physiol. (Lond.)*, 154, 190—205.

Miledi, R. (1963) An influence of nerve not mediated by impulses. In *The Effect of Use and Disuse on Neuromuscular Functions*, E. Gutmann and P. Hník, (Eds.), Elsevier, Amsterdam, p. 576.

Miledi, R. and Slater, C.R. (1968) Electrophysiology and electronmicroscopy of rat neuromuscular junctions after nerve degeneration. *Proc. roy. Soc.* B 169, 289—306.

Miledi, R., Stefani, E. and Zelená, J. (1968) Neural control of acetylcholine sensitivity in rat muscle fibres. *Nature (Lond.)*, 220, 497—498.

Miledi, R. and Zelená, J. (1966) Sensitivity to acetylcholine in rat slow muscle. *Nature (Lond.)*, 210, 855—856.

Namba, T., Nakamura, T. and Grob, D. (1967) Staining for nerve fibre and cholinesterase activity in fresh frozen sections. *Amer. J. clin. Pathol.*, 47, 74—77.

Nastuk, W.L. (1953) Membrane potential changes at a single muscle end-plate produced by transitory application of acetylcholine with an electrically controlled microjet. *Fed. Proc.*, 12, 102.

Redfern, P. and Thesleff, S. (1971) Action potential generation in denervated rat skeletal muscle II. The action of tetrodoxin. *Acta physiol. scand.*, 82, 70—78.

Thesleff, S. (1960) Supersensitivity of skeletal muscle produced by botulinum toxin. *J. Physiol. (Lond.)*, 151, 598—607.

Thomson, J.D. (1952) The effect of electrotherapy on twitch time and ACh sensitivity in denervated skeletal muscle. *Amer. J. Physiol.*, 171, 773.

Vrbová, G. (1970) In Control of Chemosensitivity at the Neuromuscular Junction, R. Eigenmann (Ed.), *Proc. 4th Int. Congr. Pharmacol.*, Vol. III, Schwabe, Basel, pp. 158—169.

Vyskočil, F. and Gutmann, E. (1976) Control of ACh sensitivity in temporary unconnected ("decentralized") segments of diaphragm-muscle fibres of the rat. *Pflügers Arch.*, 367, 43—47.

Watson, J.E., Gordon, T., Jones, R. and Smith, M.E. (1976) The effect of muscle extracts on the contracture response of skeletal muscle to acetylcholine. *Pflügers Arch.*, 363, 161—166.

A Histological and Electrophysiological Study of the Effects of Colchicine on the Frog Sartorius Nerve-Muscle Preparation

E.M. VOLKOV [1], G.I. POLETAEV [1], E.G. ULUMBEKOV [2] and KH.S. KHAMITOV [3]

Departments of [1] General Biology and Biophysics, [2] Histology and [3] Physiology, Medical Institute, Kazan (U.S.S.R.)

INTRODUCTION

Denervation of muscle impairs neurotrophic control and causes a number of changes in the skeletal muscle and in the properties of the neuromuscular junction. However, the nature and mechanism of neurotrophic control are not clear. One of the approaches to the study of neurotrophic control mechanisms is to block axoplasmic flow by means of the mitostatic alkaloids, colchicine and vinblastine, obtained from *Colchicum autumnale* and *Vinca rosea* respectively. In mammals these alkaloids cause denervation-like changes in skeletal muscle without impairing neuromuscular transmission (Hofmann and Thesleff, 1972; Albuquerque et al., 1974; Fernandez and Ramirez, 1974) but their effects on amphibian skeletal muscle and neuromuscular junctions have not been established.

It has, however, been shown that changes which follow denervation in amphibian muscles present a number of peculiarities when compared with those in mammals. These include the absence of spontaneous muscle fibre fibrillation and resistance to tetrodotoxin (Nasledov and Thesleff, 1974). It therefore seemed of interest to investigate whether the application of colchicine to frog nerves interfered with axoplasmic transport, and whether it influenced muscle fibres and synaptic transmission. The object of the present research was to establish what effects the application of colchicine to the nerve had on electrophysiological and morphological characteristics of skeletal muscle fibres and the synaptic apparatus in frog and to compare these effects with those produced by nerve section.

METHODS

Experiments were carried out on the sciatic nerve — sartorius muscle preparation of *Rana ridibunda* and *Rana temporaria* in the autunm—winter period. In some animals a portion of a nerve (2—3 mm), innervating the muscle, was removed under ether anaesthesia. In other animals the same portion of the nerve was treated with colchicine solution (20 mM, Merck) as described previously (Volkov, 1977). Sympathetic denervation of the hind leg of the frog was performed by total unilateral extirpation of the paravertebral sympathetic chain. The animals were kept at room temperature. As controls, muscles or neuromuscular preparations of intact frogs or, in some experiments, contra-lateral muscles of experimental animals were used. Input resistance, length constant, the time constant and acetylcholine (ACh) sensitivity of muscle fibres were studied by a conventional technique (Henček and Zachar, 1965; Nasledov and Thesleff, 1974). Values for ACh sensitivity were calculated by correcting the membrane potential (MP) to a value of −90 mV (Castillo and Katz, 1954) and assuming the ACh-equilibrium potential to be −15 mV (Katz and Thesleff, 1957). Action potentials (AP), end-plate potentials (EPP) and miniature end-plate potentials (MEPP) were recorded. AP were evoked by passing depolarizing rectangular current pulses through a second microelectrode and hyperpolarizing the muscle fibre to a potential of −90 mV through the recording electrode. In a number of experiments the EPP quantum content (m) was determined (Martin, 1966). To block AP and muscle contractions the preparation was precurarized with (+)-tubocurarine chloride (1 μg/ml). Standard microelectrode techniques were used. The Ringer solution had the following composition: NaCl, 115 mM; KCl, 2.5 mM; CaCl$_2$, 1.8 mM; NaHCO$_3$, 3.0 mM, pH 7.2−7.4.

Nerve endings were impregnated with silver salts by the Gros-Bielschovsky method. Acetylcholinesterase (AChE) activity was demonstrated histochem-ically by the method of Gomori (1952) and succinic dehydrogenase (SDH) was shown in cross sections of muscle using nitroblue tetrazolium salt. Cross sec-tional areas of muscle fibres were determined by planimetric method.

RESULTS

Effect of colchicine application or nerve section on MP and other electrical properties of the muscle fibre membrane

A week after nerve section the MP, input resistance, time constant and length constant of muscle fibres did not differ from the control (Table I) but on the 13−15th day after denervation the MP, input resistance and time con-stant had decreased (Table I). The length constant remained unchanged, perhaps as a consequence of the concurrent change in fibre size. After col-chicine application to the nerve, the muscle fibre MP did not initially differ from the control but by the 13−15th day was clearly reduced (Table I). How-ever, the dynamics of changes of electric properties in the muscle membrane had some peculiarities. By 6−8 days after colchicine the input resistance and the time constant had become considerably less than in the controls. However at the 13−15th day this reduction in the resistance was less marked (Table I). The length constant remained unchanged both at 6−8 days and 13−15 days after colchicine application (Table I).

The effect of colchicine application or nerve section on muscle fibre sensitivity to ACh

6−8 days after nerve section an increase in the area of the muscle fibre sensitivity to ACh was noted and this enlarged progressively with time

TABLE I

PASSIVE ELECTRIC PROPERTIES AND ACETYLCHOLINE SENSITIVITY OF MUSCLE
FIBRE MEMBRANE IN M. SARTORIUS AFTER DENERVATION AND COLCHICINE
TREATMENT OF THE NERVE

Treatment (days)	Resting MP (mV)	Effective resistance (kΩ)	Time constant (msec)	Length constant (mm)	Sensitivity to Ach (mV/nC)	Width of the cholino-ceptive zone (μm)
Control	77 ± 1.8 (49)	226 ± 12 (42)	18.0 ± 1.1 (42)	1.7 ± 0.15 (17)	23.6 ± 9.3 (5)	350 ± 50 (5)
Denervation (6–8)	71 ± 2.9 (40)	229 ± 13 (25)	16.7 ± 1.9 (25)	1.6 ± 0.14 (18)	38.3 ± 7.8 (9)	900 ± 100 (9)
Denervation (13–15)	65 ± 2.7 (43)	180 ± 12 (24)	13.1 ± 0.9 (24)	1.8 ± 0.04 (16)	180 ± 30.2 (9)	1600 ± 100 (9)
Colchicine (6–8)	70 ± 3.8 (42)	131 ± 14 (33)	12.1 ± 0.2 (33)	1.6 ± 0.20 (16)	19.7 ± 21.2 (9)	1050 ± 150 (9)
Colchicine (13–15)	65 ± 2.4 (41)	180 ± 13 (28)	12.3 ± 0.8 (28)	1.8 ± 0.15 (15)	46.9 ± 16.1 (13)	1500 ± 110 (13)

The values are means ± S.E.M. Figures in brackets give the number of fibres examined.

Table I). At the same period the absolute senstivity to ACh of the former synaptic zone considerably increased (Table I). Colchicine application to the nerve also caused a widening of the ACh-sensitive zone and 13–15 days after colchicine nerve treatment the ACh-sensitive zone was the same size as that 13–15 days after surgical denervation (Table I). Thus, the rate of increase in the cholinoceptive zone after nerve section and after colchicine nerve treatment were the same. The mean value for the absolute sensitivity of the postsynaptic membrane to ACh had doubled 13–15 days after colchicine treatment of the nerve but statistically significant differences were not obtained because of the great variability of the results (Table I).

Peculiarities of muscle fibre AP generation following nerve section or colchicine treatment of the nerve

In mammals surgical denervation causes a decrease in the rate of rise of the muscle fibre AP (Albuquerque and Thesleff, 1968) whereas in amphibian muscles under similar conditions this parameter is unaltered by denervation (Nasledov and Thesleff, 1974). However, in this study first derivatives of the depolarising and repolarising phases of the muscle fibre AP were unchanged by denervation and colchicine treatment of the nerve (Table II). The percentage of fibres in which an AP could be evoked by indirect stimulation of the muscle after colchicine treatment of the nerve was unaltered. However, the amplitude and duration of the negative after-potential in muscle fibres, 7–14 days after colchicine nerve treatment were less than in controls (Table II). In the control muscles almost all the fibres generate a single AP (Table II) in response to a depolarising current impulse of 10 msec duration. In contrast, 13–15 days after nerve section or colchicine treatment, approximately one third of muscle fibres responded to a single current impulse with a train of AP (Table II).

TABLE II

ANOMALIES IN AP GENERATION IN FROG SARTORIUS MUSCLE FIBRES ON THE 13–15th DAY AFTER DENERVATION AND AFTER COLCHINE TREATMENT OF THE NERVE

	Control	Denervation	Colchicine application
Rate of rise of AP (V/sec)	147 ± 6.4	213 ± 5.7	204 ± 8.1
Rate of repolarisation of AP (V/sec)	97 ± 5.1	89 ± 2.3	97 ± 6.2
Rhythmic spike activity	3/63	19/53	15/38
After-potential amplitude (mV)	20 ± 0.6	–	15 ± 0.9 *

The values are means ± S.E.M. * Value significantly different from the control. "Rhythmic spike activity" gives the number of fibres in which a train of AP was evoked in response to a single current impulse, compared to the total number of fibres examined.

Morphometric and histochemical characteristics of muscle fibres after colchicine application or nerve section

The staining pattern of sartorius muscle fibres in cross section revealed by the SDH technique is extremely uneven. It is possible to single out three groups of muscle fibres: dark, intermediate and lightly stained (Volkov, 1977). By 13–15 days after muscle denervation the proportion of light fibres had increased and the number of dark and intermediate fibres had decreased (Table III), and in general the muscle had lost its typical mosaic pattern. At the same time the cross sectional areas of all three groups of muscle fibres decreased (Table III). It should be noted that the muscles, the nerves of which were subjected to colchicine treatment had, on average, 2–3 well stained fibres which were of a considerably greater diameter than the largest of the light fibres. Similar giant fibres sometimes also occur after nerve section.

Characteristics of induced and spontaneous ACh release in myoneural junctions after colchicine treatment of the nerve

No degenerative changes in motor nerve endings after colchicine nerve treatment were noted whereas after nerve section these degenerated. At 13–15 days after muscle denervation or colchicine nerve treatment the histochemically demonstrable AChE activity in the synaptic region was unchanged.

15 days after colchicine nerve treatment MEPP amplitude was decreased from $199 ± 6 \mu V$ in the controls to $131 ± 5 \mu V$ ($P < 0.001$) and MEPP frequency from $3.2 ± 1.1$ to $0.81 ± 0.1$ Hz ($P < 0.05$) on the 13–15th day; these numbers are means ± S.E.M. of results obtained on 48 fibres in 8 control muscles and on 50 fibres in 9 muscles after colchicine treatment. In previous studies we have determined the quantal content and the effect of repetitive stimulation on the EPP emplitude (Volkova et al., 1976). Following colchicine treatment the number of synapses in which the quantal content was 100 or less had increased at 13–15 days and at most synapses, irrespective of their original EPP quantal content, the initial potentiation of the EPP amplitude seen during stimulation at 100 Hz was quickly replaced by a deep depression.

407

TABLE III

CROSS SECTIONAL AREAS OF MUSCLE FIBRES AND PERCENTAGE OF FIBRES WITH DIFFERENT SUCCINIC DEHYDROGENASE (SDH) ACTIVITY IN SARTORIUS MUSCLE FROM CONTROL FROGS AND FROGS 15 DAYS AFTER AN EXPERIMENTAL PROCEDURE

Treatment	Groups of muscle fibres according to their SDH activity	Cross sectional areas of muscle fibres (in arbitrary planimetric units)		Percentage of fibres in each SDH activity class	
		Control	Experiment	Control	Experiment
Nerve section	Light	3.42 ± 0.07 (1927)	2.09 ± 0.06 (1607) *	38.81	50.39 *
	Intermediate	1.32 ± 0.04 (1367)	0.96 ± 0.02 (860) *	36.10	29.96
	Dark	0.67 ± 0.02 (948)	0.52 ± 0.01 (723) *	25.09	22.65
Nerve treated with 30 mM colchicine solution	Light	3.34 ± 0.07 (1663)	1.34 ± 0.05 (1927) *	46.06	57.41 *
	Intermediate	1.22 ± 0.04 (1281)	1.05 ± 0.02 (948) *	35.48	28.24
	Dark	0.84 ± 0.02 (666)	0.63 ± 0.02 (482) *	18.46	14.35
Sympathectomized	Light	3.50 ± 0.07 (1170)	3.61 ± 0.09 (1040)	48.18	45.35
	Intermediate	1.52 ± 0.03 (852)	1.60 ± 0.06 (874)	33.95	38.11
	Dark	0.78 ± 0.05 (434)	0.73 ± 0.04 (379)	17.87	16.54

The values are means ± S.E.M. * Values which are significantly different from the control ($P < 0.05$). The numbers of fibres examined are given in parentheses.

DISCUSSION

Colchicine treatment of the motor nerve enlarged the area of membrane sensitive to ACh, changed the passive electrical properties of the membrane, produced a repetitive response to a single direct stimulation, and caused partial atrophy of muscle fibres with a change in their histochemical profile. The same picture is seen in muscle fibres after denervation. In other words, colchicine nerve treatment produces changes in frog skeletal muscle that are similar to the denervation syndrome.

It must be emphasized that both spontaneous and induced release of ACh is preserved after colchicine treatment and the muscle retains its activity.

It is known that colchicine disturbs the fast axoplasmic transport (Kreutzberg, 1969; Jeffrey and Austin, 1973). Therefore denervation-like changes in frog muscle after colchicine treatment of the nerve can be explained by the disturbance of axoplasmic transport to the muscle of the substances which provide neurotrophic control.

As mentioned above the time-course of the development of some denervation-like changes in the muscle after colchicine application to the nerve is somewhat different from that of changes following nerve section. Thus the decreases in the input resistance and time constant occur more quickly after colchicine nerve treatment than after nerve section. However, the increase in absolute sensitivity to ACh is less, even though there is little difference in the final size of the cholinoreceptive zone.

In control experiments in which the nerve was treated with Ringer's solution, there was no decrease in the cross sectional areas nor any changes in the histochemical profile of muscle fibres (Volkov, 1977). Furthermore there were no changes in passive electrical properties of the membrane of muscle fibres directly submerged in 10 mM colchicine solution (Volkov et al., 1977). Thus, the effect of colchicine on the nerve cannot be explained by a possible direct action of colchicine on the membrane nor of the preparative surgery.

In control sartorius muscle the EPP amplitude, the quantal content of the EPP and the effect of repetitive stimulation vary greatly from end-plate to end-plate (Volkova et al., 1976). However, after colchicine application most of the synapses have a similar EPP quantal content of about 100 and EPP amplitude is rapidly depressed after only a short initial potentiation during repetitive stimulation. Colchicine application to the nerve, probably by disturbing axoplasmic transport, decreases the functional heterogeneity of frog sartorius neuromuscular synapse, and this may be related to the decrease in the heterogeneity of muscle fibres seen in histochemical tests (Volkov, 1977).

Denervation-like changes in muscle fibers after colchicine nerve treatment may be due to an action of colchicine not only on the motor nerves but also on sympathetic nerve fibres (Dahlström and Heiwall, 1975). The latter, according to some data, may have trophic effects on the muscle (Govyrin, 1967), however, after surgical sympathectomy we could not find any changes in either the muscle fibres or the function of the synapses.

From the results, we have come to the following conclusion. We believe that neurotrophic control of frog sartorius muscle and of its synaptic apparatus involves substances that are transmitted by axoplasmic flow along motor nerve fibres and, at the same time adequate motor activity plays some role.

SUMMARY

Electrophysiological and morphological characteristics of frog skeletal muscle fibres and anomalies of neuromuscular transmission after treatment of the motor nerve with colchicine and after surgical denervation have been studied. After colchicine treatment the cholinoceptive zone expands, the absolute sensitivity to ACh increases, the input resistance and membrane time constant decrease, rhythmic spike activity appears in response to a single stimulation, and partial atrophy and changes in the histochemical profile of muscle fibres occur. Similar changes develop in muscle fibres after surgical denervation. However, while nerve section causes the cessation of neuromuscular transmission, colchicine treatment does not affect it. However, there is a decrease in MEPP amplitude and frequency, an increase in the number of end-plates with a quantal content of less than 100 and an alteration in the response to repetitive stimulation. The regulation of muscle fibre membranes and neuromuscular transmission in the frog by substances transmitted by axoplasmic flow is discussed.

REFERENCES

Albuquerque, E.X. and Thesleff, S. (1968) A compartive study of membrane properties of innervated and chronically denervated fast and slow skeletal muscles of the rat. *Acta physiol. scand.,* 73, 471–480.

Albuquerque, E.X., Warnick, J.E., Sansone, F.M. and Onur, R. (1974) The effects of vinblastine and colchicine on neural regulation of muscle. *Ann. N.Y. Acad. Sci.,* 228, 224–243.

Castillo del, J. and Katz, B. (1954) The membrane change produced by the neuromuscular transmitter. *J. Physiol (Lond.),* 125, 546–565.

Dahlström, A. and Heiwall, P.O. (1975) Intra-axonal transport of transmitters in mammalian neurones. *J. Neural. Transm., Suppl.* 12, 97–114.

Fernandez, H.L. and Ramirez, B.V. (1974) Muscle fibrillation induced by blockage of axoplasmic transport in motor nerves. *Brain Res.,* 79, 385–395.

Govyrin, V.A. (1967) Trophic function of the sympathetic nerve of the heart and skeletal muscles. *Nauka (Leningrad),* 3–13 (in Russian).

Henček, M. and Zachar, J. (1965) The electrical constants of single muscle fibres of the crayfish (Astacus fluvitalis). *Physiol. Bohemoslov.,* 14, 297–311.

Hofmann, W.W. and Thesleff, S. (1972) Studies on the trophic influence of nerve in skeletal muscle. *Europ. J. Pharmacol.,* 20, 256–278.

Jeffrey, P.L. and Austin, L. (1973) Axoplasmic transport. *Progr. Neurobiol.,* 2, 205–255.

Katz, B. and Thesleff, S. (1957) On the factors which determine the amplitude of the "miniature end-plate potentials". *J. Physiol. (Lond.),* 137, 267–278.

Kreutzberg, G.W. (1969) Neuronal dynamics and axonal flow. IV. Blockage of intra-axonal enzyme transport by colchicine. *Proc. nat. Acad. Sci. (Wash.),* 62, 722–728.

Martin, A.R. (1966) Quantal nature of synaptic transmission. *Physiol. Rev.* 46, 51–66.

Nasledov, G.A. and Thesleff, S. (1974) Denervation changes in frog skeletal muscle. *Acta physiol. scand.,* 90, 370–280.

Volkov, E.M. (1977) Morphological and histological characteristics of frog skeletal muscle on application of colchicine to nerve. *Bull. Exp. Biol. Med.,* 83, 413–415 (in Russian).

Volkov, E.M., Nasledov, G.A., Poletaev, G.I. and Ulumbekov, E.G. (1977) The comparative characteristic of electrphysiological changes in the frog muscle fibre, after denervation and after blockade of the axoplasmic transport. *Sechenov Physiol. J. USSR*, 63, 1432—1437 (in Russian).

Volkova, I.N., Zefirov, A.L., Nikolsky, E.E. and Poletaev, G.I. (1976) Functional heterogeneity of the frog m. sartorius synapses. *Sechenov Physiol. J. USSR*, 62, 406—413 (in Russian).

The Effect of Altered Function of Dopaminergic Neurones on the Cholinergic System in the Striatum

H. LADINSKY and S. CONSOLO

Instituto di Ricerche Farmacologiche "Mario Negri", Via Eritrea, 62-20157 Milan (Italy)

INTRODUCTION

In mammalian brain, very high concentration of acetylcholine (ACh) and choline acetyltransferase (ChAT), markers for cholinergic terminals, is found in the caudate nucleus. The caudate receives projections from thalamic nuclei, from the substantia nigra, and from the entire extent of the cortex. None of these 3 major efferent pathways to the caudate are cholinergic. McGeer et al. (1971) first showed that electrolytic lesions placed in thalamic nuclei and in the ventral tegmental area as well as over large areas of the cortex did not alter the activity of ChAT or cholinesterase of the caudate. Butcher and Butcher (1974) confirmed and extended these results by showing that the ACh content of the neostriatum was unaltered by destruction of the known inputs to the neostriatum by various means. These findings suggested that the cholinergic system in the neostriatum is primarily organized within that structure.

The intracerebral injection of kainic acid, a rigid analogue of glutamate, into the rat striatum provided further evidence that cholinergic innervation in the neostriatum is primarily intrinsic, and may be localized in the Golgi type II interneurons that populate the region. Kainate apparently causes degeneration of neurones with cell bodies near the injection site, while sparing axons terminating in or passing through the region. After kainate injections, there is a profound decrease in the neurochemical markers for cholinergic neurons (Schwarcz and Coyle, 1977). The chemical lesion caused a 70% reduction in ChAT, ACh and the synaptosomal uptake of choline. It was suggested by Schwarcz and Coyle (1977) that the residual 30% level of ChAT may reflect survival of terminals from other cholinergic pathways afferent to the neostriatum. Thus, all of the ChAT destroyed may appertain to the neuronal cell bodies of the caudate which were nearly entirely destroyed by the kainic acid injection.

Consistent with these results, the injection of kainic acid left presynaptic dopaminergic nerve terminals intact, as shown by the finding that the striatal content of dopamine (DA) and the synaptosomal uptake of DA were unchanged; the activity of tyrosine hydroxylase, a marker for presynaptic catecholaminergic terminals, was actually increased (Schwarcz and Coyle, 1977). By contrast, striatal DA-sensitive adenylate cyclase, a putative receptor for DA, was decreased by more than 85% by kainic acid injection (Schwarcz and Coyle, 1977, Di Chiara et al., 1977), confirming that this enzyme is postsynaptic to dopaminergic terminals (Mishra et al., 1974). These data support the concept that the cyclase is located on the intrinsic cholinergic neurones.

Some evidence that dopaminergic nerve endings may form synapses in the striatum with cholinergic interneurones has been provided by the histochemical study of Hattori et al. (1976). These workers developed an immunohistochemical stain for ChAT and found a profusion of positively staining, medium sized, spiny cells, which are characteristic of interneurones, in the striatum of the guinea pig. Positive staining for ChAT was not found in the large striatal cells, which comprise only a small proportion of the striatal cell population and may be efferents. They also presented pictures of synaptic contacts between unidentifiable terminals and ChAT containing dendrites. After injection of 6-hydroxydopamine (6-OHDA), some degenerating dopaminergic nerve endings were seen to make contact with dendritic processes positively staining for ChAT. Hence, some synapses are probably formed of dopaminergic terminals and dendrites of cholinergic neurones.

FUNCTIONAL LINK BETWEEN DOPAMINERGIC AND CHOLINERGIC NEURONES IN THE STRIATUM

A functional link between the dopaminergic and cholinergic systems in the striatum has been demonstrated pharmacologically with the use of specific and

TABLE I

SUMMARY OF THE EFFECTS OF DOPAMINE-RELATED DRUGS ON THE ACh CONTENT IN RAT STRIATUM

Drug	Action on DA system	Effect on ACh content
Reserpine	Depletor	Decrease
Pimozide	Antagonist	Decrease
Chlorpromazine	Antagonist	Decrease
Clozapine	Antagonist	Decrease
Sulpiride	Antagonist	Decrease
Haloperidol, acute treatment	Antagonist	Decrease (see Table III)
Haloperidol, repeated treatment	Blocks and produces supersensitivity	Increase (see Table III)
6-OHDA, 9 days	Denervates and induces denervation supersensitivity	Increase (see Table II)
Apomorphine	Direct agonist	Increase (see Table II)
Piribedil	Direct agonist	Increase
Lisuride	Direct agonist	Increase
Bromocriptine	Direct and possibly indirect agonist	Increase
Picrotoxin	Indirect agonist	Increase (see Table IV)
d-Amphetamine	Indirect agonist	Increase
Amantadine	Indirect agonist	Increase
Nomifensine	Indirect agonist	Increase
l-Dopa	Indirect agonist	Increase
Alpha-methyl-p-tyrosine	Synthesis inhibitor	No effect
6-OHDA, 30 days	Denervates and induces supersensitivity	No effect (see Table II)

potent DA receptor agonists, antagonists, and DA depletors. The DA antagonists and depletors stimulate the release (Stadler et al., 1973) and the turnover (Trabucchi et al., 1974) of striatal ACh, and reduce steady-state ACh levels (McGeer et al., 1974; Sethy and Van Woert, 1974; Consolo et al., 1975; Guyenet et al., 1975; Rommelspacher and Kuhar, 1975); in contrast, direct- or indirect-acting DA agonists increase the steady-state level of ACh (Consolo et al., 1974; McGeer et al., 1974; Sethy and Van Woert, 1974; Guyenet et al., 1975; Ladinsky et al., 1975; Rommelspacher and Kuhar, 1975), and reduce its turnover (Trabucchi et al., 1975).

Table I shows such actions of a number of DA-related drugs on the rat striatal ACh level in relation to their action on the dopaminergic system, as studied in our laboratory. These results are consistent with the concept that inhibitory dopaminergic nerve endings are in apposition to cholinergic cells. It should be further pointed out that apomorphine, piribedil, haloperidol, and pictrotoxin appeared to have selective action on the ACh level in the striatum, and did not change the level in other major brain areas such as the mesencephalon, diencephalon, hippocampus, cerebral hemispheres, cerebellum and nc. accumbens (Ladinsky et al., 1974; Sethy and Van Woert, 1974; Guyenet et al., 1975; Consolo et al., 1977). Hence, the data suggest that the dopaminergic-cholinergic link is peculiar to the striatum and may be one of the features that characterize this brain region.

ACUTE LESIONS AND DA RECEPTOR SUPERSENSITIVITY SUPPORT THE CONCEPT OF INHIBITORY DOPAMINERGIC REGULATION OF INTRINSIC STRIATAL CHOLINERGIC NEURONES

The hypothesis of inhibitory regulation of cholinergic neurones by DA predicts that a decrease in striatal ACh content should occur when the striatum is devoid of dopaminergic input. Interruption of dopaminergic transmission by placement of radio frequency lesions into the median forebrain bundle reduced striatal ACh levels within 60 min (Agid et al., 1975; Rommelspacher and Kuhar, 1975). Thus, release of the cholinergic neurones from inhibitory dopaminergic regulation presumably produced disinhibition of the cholinergic neurones, an increased release of ACh and, in turn, a decrease in its level.

By contrast, no difference could be detected in the striatal ACh level on the side ipsilateral to the lesion following long-term complete degeneration of the dopaminergic nigrostriatal tract (Fibiger and Grewaal, 1974; Agid et al., 1975). This finding could indicate that a long-term adaptation of the striatal cholinergic neurones occurred (Fibiger and Greewaal, 1974; Agid et al., 1975; Consolo et al., 1978).

The long term adaptation of the striatal cholinergic neurones following complete degeneration of the dopaminergic pathway may be a consequence of the triggering of internal compensatory mechanisms generally described as denervation supersensitivity. It is known that a variety of surgical and pharmacological manipulations that interrupt synaptic transmission in central monoaminergic pathways can induce supersensitivity of postsynaptic receptors in response to the physiological agonist. Concerning the dopaminergic system,

414

unilateral degeneration of the ascending nigro-neostriatal pathway by 6-OHDA provokes an assymetric response to agents acting directly at the postsynaptic receptors, such as apomorphine, causing the animals to circle in a direction contralateral to the lesion (Ungerstedt and Arbuthnott, 1970). Bilateral injections of 6-OHDA into the caudate, or the nucleus accumbens septi, induce supersensitivity to apomorphine as manifested by enhanced stereotyped behavior or locomotor activity, respectively (Kelly et al., 1975).

Chronic receptor blockade with neuroleptics (Klawans and Rubovits, 1972; Tarsy and Baldassarini, 1973; Christensen et al., 1976), inhibition of transmitter synthesis with α-methyl-p-tyrosine (Costall and Naylor, 1973), or depletion of biogenic amines by acute and chronic reserpine treatment (Tarsy and Baldessarini, 1974), induce behavioural supersensitivity to the subsequent administration of dopamine, methylphenidate, or apomorphine.

It was thus postulated that the supersensitivity of the postsynaptic DA receptors should cause an increased sensitivity of the cholinergic system in the striatum to exogenous dopaminergic agonists. Thus, we tested whether denervation supersensitivity of dopaminergic receptors, elicited by unilateral lesion of the nigroneostriatal pathway with 6-OHDA or by repeated treatment with haloperidol, potentiates the action of apomorphine in increasing the level of ACh in the striatum.

DENERVATION SUPERSENSITIVITY

Nine days following unilateral lesion of the nigrostriatal projection, induced by infusion of 6-OHDA into the ventral tegmental area of rats, DA was depleted by more than 90% in the striatum of the ipsilateral side but its noradrenaline content was not altered. These animals responded to apomorphine with contralateral circling indicating that postsynaptic DA supersensitivity had developed. The results of the dose–response effect of apomorphine on the ipsilateral striatal ACh content 9 and 30 days after unilateral lesion are shown in Table II. At 9 days post-lesion, the basal level of ACh was significantly increased by about 25%. Because of the difference in basal levels, the dose–response effects of apomorphine were different in the vehicle-treated group as compared to the lesioned group. At doses of 0.2 and 0.35 mg/kg of the agonist, the level of ACh in the ipsilateal striatum was higher than in the vehicle-treated group. However, neither group responded significantly to a dose of 0.2 mg/kg of apomorphine, and the peak increase of around 60% was reached at a dose of about 0.5 mg/kg in both groups. A completely randomized 2×4 factorial design in fact indicated no interaction between the two groups and thus no significant supersensitivity response to apomorphine at this time.

Thirty days after lesion, the situation was quite different and a denervation supersensitivity response became evident. The basal level of ACh returned to the normal equilibrium state. The sensitivity to apomorphine was greater and the ACh level was increased by about 25% after a dose of 0.1 mg/kg (Table II). The maximum increase of around 60% was achieved at a dose of about 0.3 mg/kg in the lesioned group. The ACh level in the lesioned group was significantly higher than in the vehicle-treated group at all doses of apomorphine between

TABLE II

SUPERSENSITIVITY OF THE IPSILATERAL STRIATAL CHOLINERGIC NEURONS TO APOMORPHINE AT 9 AND 30 DAYS AFTER UNILATERAL LESION OF THE VENTRAL TEGMENTAL AREA WITH 6-OHDA

Dose of apomorphine (mg/kg, i.p., 30 min)	Ipsilateral striatal ACh (nmol/g wet wt ± S.E.)			
	9 days		30 days	
	Vehicle	6-OHDA	Vehicle	6-OHDA
Saline	30.3 ± 0.7 (23)	37.2 ± 1.2 * (23)	32.4 ± 1.0 (11)	32.9 ± 1.0 (14)
0.1	—	—	32.4 ± 1.1 (7)	40.1 ± 1.2 *,** (6)
0.2	33.1 ± 0.8 ** (6)	39.2 ± 1.5 * (10)	32.1 ± 1.1 (6)	43.5 ± 1.8 *,** (6)
0.35	40.6 ± 1.8 ** (9)	52.8 ± 1.4 *,** (8)	41.8 ± 1.6 ** (6)	51.8 ± 3.3 *,** (5)
0.5	47.2 ± 1.7 ** (10)	57.8 ± 2.0 *,** (9)	48.6 ± 1.7 ** (9)	50.7 ± 1.8 ** (4)
0.65 ***	50.2 ± 2.3 ** (6)	56.5 ± 1.8 (6)	—	—

The figures represent the mean and S.E.M. The number of animals is shown in brackets.
* $P < 0.01$ vs respective vehicle-treated group.
** $P < 0.01$ vs respective saline-treated group.

Statistics: completely randomized factorial design (2×4 for 9 days and 2×5 for 30 days): conventional analysis for unequal cell n's and Tukey's HSD test. 9 days, F interaction N.S.; 30 days, F interaction 5.0 (4; 54 df) $P < 0.01$.
*** Not included in statistical analysis but shows that the same plateau level is reached in both groups. ACh was measured by the radiochemical method of Saelens et al. (1970).

0.1 and 0.35 mg/kg. A completely randomized 2 × 5 factorial design revealed a highly significant interaction ($P < 0.01$). There was about a 3-fold increase in sensitivity to apomorphine in the lesioned group.

Neither the V_{max} of ChAT (4.7 μmole ACh synthesized $min^{-1} \cdot g$ protein^{-1}) nor the apparent Km's for choline (0.7 mM) or acetylcoenzyme A (0.06 mM) were affected in the ipsilateral striatum at 30 days after unilateral lesion. This result rules out modifications in activity of ChAT as being involved in the supersensitivity phenomenon described above.

DISUSE SUPERSENSITIVITY

Rats were treated with haloperidol (1 mg/kg) twice daily for 12 consecutive days. The experiment was performed at 48 h after the last dose to ensure complete elimination of the drug. At this time, apomorphine-induced stereotype was potentiated, indicating development of supersensitivity (Table III). At 48 h, the basal level of ACh in the striatum was significantly increased by about 20% (Table III). Supersensitivity to apomorphine was revealed by Anova 2 × 4 factorial analysis ($P < 0.05$). The sensitivity to apomorphine was increased about 2-fold.

Neither the V_{max} of striatal ChAT nor the Michaelis constants for choline or acetylcoenzyme A were altered by the repeated haloperidol treatment. As in the case of the unilateral chemical lesion, this treatment appears to produce supersensitivity of the cholinergic response secondarily to the induction of supersensitivity at the level of DA receptor.

Disuse supersensitivity induced by pharmacological receptor blockade occurs in the presence of intact presynaptic neurones. This situation might be responsible for the increase in the basal level of ACh following repeated

TABLE III

SUPERSENSITIVITY OF STRIATAL CHOLINERGIC NEURONS AND STEREOTYPE TO APOMORPHINE FOLLOWING REPEATED TREATMENT WITH HALOPERIDOL

Dose of apomorphine (mg/kg, i.p., 30 min)	Striatal ACh (nmol/g wet wt)		Stereotypy score	
	Vehicle	Repeated haloperidol administration	Vehicle	Repeated haloperidol administration
Saline	33.7 ± 1.1	40.8 ± 1.3 **		
0.2	35.6 ± 1.3	52.2 ± 1.7 *,**	0	0.25 ± 0.2
0.35	43.2 ± 1.6 *	55.6 ± 2.2 *,**	0	0.88 ± 0.1
0.5	46.0 ± 1.6 *	57.7 ± 1.4 *,**	0.25 ± 0.2	2.75 ± 0.3

Haloperidol 1 mg/kg twice daily for 12 days. The experiment was performed 48 h after the last dose.
* $P < 0.01$ vs respective saline-treated group.
** $P < 0.01$ vs respective vehicle-treated group.
Statistics: for biochemical data, Anova (2 × 4) factorial analysis and Tukey's test for unconfounded means. F interaction 3.4 (3; 56 df) $P < 0.05$; stereotypy, Smirnov's test.
The figures are the mean and S.E.M. of 8 animals per group.

TABLE IV

SUPERSENSITIVITY OF STRIATAL CHOLINERGIC NEURONS TO PICROTOXIN FOLLOWING REPEATED TREATMENT WITH HALOPERIDOL

Dose of picrotoxin (mg/kg i.p.)	Striatal ACh (nmol/g ± S.E.)	
	Vehicle	Repeated haloperidol
Saline	32.3 1.5 (8)	39.8 1.5 (8)
0.25	35.2 1.2 (8)	46.1 1.0 (12)
0.50	34.7 1.1 (8)	54.6 1.5 (12)
1.0	41.5 1.7 (8)	52.1 2.4 (8)
1.5 *	51.0 2.5 (8)	—
2.0 *	53.0 2.2 (8)	—

Statistics: completely randominzed factorial design (2 × 4); conventional analysis for unequal cell n's and Tukey's HSD test. F interaction 4.82 (3;56 df) $P < 0.01$.
* Not included in statistical analysis but shows that the same plateau level is reached in both groups.
The rats were killed 30 min after picrotoxin and 48 h after the last dose of haloperidol.

haloperidol treatment. Thus, after withdrawal of the neuroleptic drug, a decreased concentration of blocker exposes the sensitized DA receptors (Burt et al., 1977; Muller and Seeman, 1977) to endogenous DA which can lead to increased inhibitory tone and reduced ACh synthesis and release. This phenomenon has particular relevance to the movement disorder, tardive dyskinesia, which sometimes appears in patients when neuroleptic drug therapy is stopped (Van Woert, 1976).

This method of producing disuse supersensitivity be repeated haloperidol treatment is presently being used in our laboratory to determine whether drugs which increase the striatal ACh level do so through an eventual DA receptor stimulation. Increased sensitivity (a leftward shift) in the dose—response curve of the drug in rats made supersensitive is taken as evidence for DA receptor activity. The sensitivity of the increase in striatal ACh to picrotoxin was increased 3-fold (Table IV). This result supports a previous study (Ladinsky et

TABLE V

LACK OF SUPERSENTIVITY OF STRIATAL CHOLINERGIC NEURONS TO OXOTREMORINE FOLLOWING REPEATED TREATMENT WITH HALOPERIDOL

Dose of oxotremorine (mg/kg i.p.)	Striatal ACh (nmol/g ± S.E.)	
	Vehicle	Repeated haloperidol
Saline	32.3 ± 1.5 (12)	42.6 ± 1.2 (12)
0.05	32.7 ± 1.3 (10)	45.6 ± 1.8 (8)
0.1	51.6 ± 2.4 (8)	61.7 ± 2.5 (10)
0.2	60.1 ± 2.5 (8)	61.2 ± 2.3 (8)
0.4 *	62.0 ± 2.1 (3)	

Statistics: completely randomized factorial design (2 × 4): conventional analysis for unequal cell n's and Tukey's HSD test. F interaction NS.
* Not included in statistical analysis but shows that the same plateau level is reached in both groups.
The rats were killed 20 min after oxotremorine and 48 h after the last dose of haloperidol.

al., 1976) which showed that the increase in striatal ACh induced by picrotoxin was mediated by DA. We interpret these results as indicating that blockade of GABA receptors by picrotoxin released GABA-regulated dopaminergic neurones from inhibition, resulting in an increased release of DA which, in turn, inhibited intrinsic striatal cholinergic neurones.

On the other hand, there was no increased sensitivity in the dose—response curve of oxotremorine (Table V), the muscarinic agonist which effectively increases striatal ACh levels (Campbell and Jenden, 1970). Although the parenteral administration of oxotremorine was found to activate striatal dopaminergic neurones (Javoy et al., 1975), our data do not support involvement of such a mechanism in the increase in striatal ACh.

CONCLUSION

The concerted results of biochemical, pharmacological, and lesion studies have provided strong evidence for a link between two neuronal systems (dopaminergic and cholinergic) in the CNS. It appears from the data that dopaminergic neurones arising in the substantia nigra synapse with and exert inhibitory influence upon cholinergic neurones intrinsic to the neostriatum of the rat. Alteration of dopaminergic activity by drugs, lesions and possibly by mental disease can affect cholinergic activity in the neostriatum.

REFERENCES

Agid, Y., Guyenet, P., Glowinski, J., Beaujouan, J.C. and Javoy, F. (1975) Inhibitory influence of the nigrostriatal dopamine system on the striatal cholinergic neurones in the rat. *Brain Res.*, 86, 488–492.

Burt, D.R., Creese, I. and Snyder, S.H. (1977) Antischizophrenic drugs: Chronic treatment elevates dopamine receptor binding in brain. *Science*, 196, 326–328.

Butcher, S.G. and Butcher, L.L. (1974) Origin and modulation of acetylcholine activity in the neostriatum. *Brain Res.*, 71, 167–171.

Campbell, L.B. and Jenden, D.J. (1970) Gas chromatographic evaluation of the influence of oxotremorine upon the regional distribution of acetylcholine in rat brain. *J. Neurochem.*, 17, 1697–1699.

Christensen, A.V., Fjalland, B. and Møller Nielsen, I. (1976) On the supersensitivity of dopamine receptors, induced by neuroleptics. *Psychopharmacologia (Berl.)*, 48, 1–6.

Consolo, S., Ladinsky, H. and Bianchi, S. (1975) Decrease in rat striatal acetylcholine levels by some direct- and indirect-acting dopaminergic antagonists. *Europ. J. Pharmacol.*, 33, 345–351.

Consolo, S., Ladinsky, H., Bianchi, S. and Ghezzi, D. (1977) Apparent lack of a dopminergic-cholinergic link in the rat nucleus accumbens septi-tuberculum olfactorium. *Brain Res.*, 135, 255–263.

Consolo, S., Ladinsky, H. and Garattini, S. (1974) Effect of several dopaminergic drugs and trihexylphenidyl on cholinergic parameters in the rat striatum. *J. Pharm. Pharmcol.*, 26, 275–277.

Consolo, S., Ladinsky, H. Samanin, R., Bianchi, S. and Ghezzi, D. (1978) *Brain Res.*, 155, 45–54.

Costall, B. and Naylor, R.J. (1973) On the mode of action of apomorphine. *Europ. J. Pharmacol.*, 21, 350–361.

Di Chiara, G., Porceddu, M.L., Spano, P.F. and Gessa, G.L. (1977) Haloperidol increases and apomorphine decreases striatal dopamine metabolism after destruction of striatal dopamine-sensitive adenylate cyclase by kainic acid. *Brain Res.*, 130, 374–382.

Fibiger, H.C. and Grewaal, D.S. (1974) Neurochemical evidence for denervation super-sensitivity: The effect of unilateral substantia nigra lesions on apomorphine-induced increases in neostriatal acetylcholine levels. *Life Sci.,* 15, 57—63.

Guyenet, P.G., Agid, Y., Javoy, F., Beaujouan, J.C., Rossier, J. and Glowinski, J. (1975) Effects of dopaminergic receptor agonists and antagonists on the activity of the neo-striatal cholinergic system. *Brain Res.,* 84, 227—244.

Hattori, T., Singh, V.K., McGeer, E.G. and McGeer, P.L. (1976) Immunohistochemical localization of choline acetyltransferase containing neostriatal neurones and their rela-tionship with dopaminergic synapses. *Brain Res.,* 102, 164—173.

Javoy, F., Agid, Y. and Glowinski, J. (1975) Oxotremorine- and atropine-induced changes of dopamine metabolism in the rat striatum. *J. Pharm. Pharmacol.,* 27, 677—681.

Kelly, P.H., Seviour, P.W. and Iversen, S.D. (1975) Amphetamine and apomorphine responses in the rat following 6-OHDA lesions of the nucleus accumbens septi and corpus striatum. *Brain Res.,* 94, 507—522.

Klawans, H.L. Jr. and Rubovits, R. (1972) An experimental model of tardive dyskinesia. *J. neural. Trans.,* 33, 235—246.

Ladinsky, H., Consolo, S., Bianchi, S. and Jori, A. (1976) Increase in striatal acetylcholine by picrotoxin in the rat: Evidence for a gabergic-dopaminergic-cholinergic link. *Brain Res.,* 108, 351—361.

Ladinsky, H., Consolo, S., Bianchi, S., Samanin, R. and Ghezzi, D. (1975) Cholinergic-dopaminergic interaction in the striatum: The effect of 6-hydroxydopamine or pimozide treatment on the increased striatal acetylcholine levels induced by apo-morphine, piribedil and (+)-amphetamine. *Brain Res.,* 84, 221—226.

Ladinsky, H., Consolo, S. and Garattini, S. (1974) Increase in striatal acetylcholine levels in vivo by piribedil, a new dopamine receptor stimulant. *Life Sci.,* 14, 1251—1260.

McGeer, P.L., Grewaal, D.S. and McGeer, E.G. (1974) Influence of noncholinergic durgs on rats striatal acetylcholine levels. *Brain Res.,* 80, 211—217.

McGeer, P.L., McGeer, E.G., Fibiger, H.C. and Wickson, V. (1971) Neostriatal choline acetylase and cholinesterase following selective brain lesions. *Brain Res.,* 35, 308—314.

Mishra, R.K., Gardner, E.L., Katzman, R. and Makman, M.H. (1974) Enhancement of dopamine-stimulated adenylate cyclase activity in rat caudate after lesions in sub-stantia nigra: Evidence for denervation supersensitivity. *Proc. nat. Acad. Sci. (Wash.),* 71, 3883—3887.

Muller, P. and Seeman, P. (1977) Brain neurotransmitter receptors after long-term halo-peridol: Dopamine, acetylcholine, serotonin, α-noradrenergic and naloxone receptors. *Life Sci.,* 21, 1751—1758.

Rommelspacher, H. and Kuhar, M.J. (1975) Effects of dopaminergic drugs and acute medial forebrain bundle lesions on striatal acetylcholine levels. *Life Sci.,* 16, 65—70.

Saelens, J.K., Allen, M.M. and Simke, J.P. (1970) Determination of acetylcholine and choline an enzymatic assay. *Arch. int. Pharmacodyn.,* 186, 279—286.

Schwarcz, R. and Coyle, J.T. (1977) Striatal lesions with kainic acid: Neurochemical charac-teristics. *Brain Res.,* 127, 235—249.

Sethy, V.H. and Van Woert, M.H. (1974) Modification of striatal acetylcholine concentra-tion by dopamine receptor agonists and antagonists. *Res. Commun. chem. pathol. Pharmacol.,* 8, 13—28.

Stadler, H., Lloyd, K.G., Gadea-Ciria, M. and Bartholini, G. (1973) Enhanced striatal acetyl-choline release by chlorpromazine and its reversal by apomorphine. *Brain Res.,* 55, 476—480.

Tarsy, D. and Baldessarini, R.J. (1973) Pharmacologically induced behavioural supersensitiv-ity to apomorphine. *Nature New Biol.,* 245, 262—263.

Tarsy, D. and Baldessarini, R.J. (1974) Behavioural supersensitivity to apomorphine follow-ing chronic treatment with drugs which interfere with the synaptic function of cate-cholamines. *Neuropharmacology,* 13, 927—940.

Trabucchi, M., Cheney, D.L., Racagni, G. and Costa, E. (1974) Involvement of brain cholin-ergic mechansims in the action of chlorpromazine. *Nature (Lond.),* 249, 664—666.

Trabucchi, M., Cheney, D.L., Racagni, G. and Costa, E. (1975) "In vivo" inhibition of striatal acetylcholine turnover by L-DOPA, apomorphine and (+)-amphetamine. *Brain Res.,* 85, 130—134.

Ungerstedt, U. and Arbuthnott, G.W. (1970) Quantitative recording of rotational behavior in rats after 6-hydroxy-dopamine lesions of the nigrostriatal dopamine system. *Brain Res.,* 24, 485—493.

Van Woert, M.H. (1976) Parkinson's disease, tardive dyskinesia, and Huntington's chorea. In *Biology of Cholinergic Function,* A.M. Goldberg and I. Hanin (Eds.), Raven Press, New York, pp. 583—601.

PART VI

PATHOLOGY

The Immunopathological Basis of Acetylcholine Receptor Deficiency in Myasthenia Gravis

ANDREW G. ENGEL

Department of Neurology and Neuromuscular Research Laboratory, Mayo Clinic and Mayo Foundation, Rochester, Minn. 55901 (U.S.A.)

INTRODUCTION

The physiological hallmark of myasthenia gravis is the small amplitude of the miniature end-plate potential (MEPP) (Elmqvist et al., 1964; Lambert and Elmqvist, 1971). The number of transmitter quanta released by nerve impulse, or the quantum content of the end-plate potential, is essentially normal. It was originally assumed that the small MEPP in myasthenia gravis (MG) was due to a decreased amount of acetylcholine (ACh) in the quantum (Elmqvist et al., 1964). However, quantitative electron microscopy of the end-plate showed that the synaptic vesicles, which hold the transmitter quanta, were of normal size and number, and the structural integrity of the nerve terminals, although about 30% smaller than normal, was preserved. By contrast, there were degenerative changes in the postsynaptic region involving especially the terminal expansions of the junctional folds and the postsynaptic membrane was simplified (Engel and Santa, 1971). Subsequently, decreased α-bungarotoxin binding by myasthenic end-plates, suggesting a decreased number of available acetylcholine receptors (AChRs) was demonstrated by Fambrough et al. (1973). The induction in experimental animals immunized with AChR of a syndrome which clinically (weakness and easy fatigability on exertion), electrophysiologically (small MEPPs), pharmacologically (response to anticholinesterase drugs and increased curare sensitivity) and morphologically resembled human MG, has provided strong evidence that the acquired form of MG was caused by an auto-immune reaction to AChR (Patrick and Lindstrom, 1973; Lennon et al., 1975; Tarrab-Hazdai et al., 1975; Lindstrom et al., 1976b; Lambert et al., 1976; Heilbronn et al., 1976; Engel et al., 1976). Further, about 90% of patients with MG were found to have detectable circulating antibodies to human AChR (Lindstrom et al., 1976c), and both MG (Toyka et al., 1977) and the experimental disease (Lindstrom et al., 1976a) could be passively transferred to normal animals with immunoglobulin G (IgG) from affected donors. These findings implied (1) that the abnormal neuromuscular transmission in acquired MG was primarily due to AChR deficiency and (2) that this was caused by an

---
* This work was supported in part by NIH Grant 6277 and by a research Center Grant from the Muscular Dystrophy Association.

424

autoimmune attack on the postsynaptic region. However, direct morphological proof for these assumptions was not available.

ULTRASTRUCTURAL LOCALIZATION OF AChR IN MG

Our objectives were (1) to compare the ultrastructural localization of AChR at the normal and at the MG end-plate, (2) to observe any relationship between changes in end-plate fine structure and alterations in receptor distribution, (3) to express the results in quantitative terms and (4) to determine whether the alterations in receptor distribution could be correlated with changes in the MEPP amplitude. AChR was localized with peroxidase-labelled α-bungaro-toxin in external intercostal muscles of MG patients and non-weak control subjects (Engel et al., 1977b). Peroxidase was conjugated directly to α-bungaro-toxin by a modification of the method of Nakane and Kawaoi (1974). Thin strips of fresh muscle, intact from origin to insertion, were exposed to 4 × 10^{-7} M of labelled toxin in well oxygenated Tyrode solution for 2 h, rinsed in the same solution for 2 h and then fixed with glutaraldehyde. Small pieces of muscle containing end-plates were isolated from the fixed strips by needle dissection, reacted with Karnovsky's diaminobenzidine medium, post-fixed with osmium tetroxide and then embedded and sectioned for electron microscopy. The application of the labelled toxin to fresh muscle strips followed by glutar-

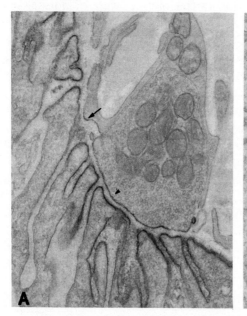
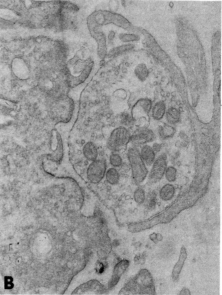

Fig. 1. External intercostal end-plate region from control subject (A) and from a case of moderately severe, generalized myasthenia gravis (B). In (A), acetylcholine receptor is associated with terminal expansions and deeper surfaces of the junctional folds; the pre-synaptic membrane (arrowhead) and Schwann cell membrane (arrow) facing crests of folds are lightly stained. In (B), the postsynaptic region is simplified and only a short segment of the simplified postsynaptic membrane reacts for the receptor. Presynaptic staining is barely discernible. A and B, ×18,500. (Reproduced from Engel et al., 1977b, by permission of Harcourt Brace Jovanovich, Inc.).

aldehyde fixation resulted in optimal preservation of fine structure and in excellent localization of AChR.

At the normal human end-plate AChR was localized on the terminal expansions of the junctional folds and sometimes also on the stalks of the folds (Fig. 1A). Less intense reaction was observed on the presynaptic membrane and occasionally on Schwann cell membranes where they faced reactive segments of the synaptic folds. In patients with moderately severe, generalized MG, most postsynaptic regions showed a marked decrease in AChR (Fig. 1B) and in some end-plate regions no AChR could be detected on any part of the postsynaptic membrane. In patients with predominantly bulbar MG, the amount and distribution of AChR appeared normal at some end-plate regions while in other regions it was decreased. In general, those postsynaptic regions that appeared to be the simplest showed the least reaction for AChR. Staining of the presynaptic membrane was reduced or absent when the junctional folds facing it were not reactive (Fig. 1B). For this reason, the presynaptic staining was considered to be secondary, at least in the large part, to diffusion of reaction product from the postsynaptic membrane.

To quantitate our observations, that segment of the postsynaptic membrane reacting for AChR was measured at each end-plate region and this value was normalized by dividing it by the length of the primary synaptic cleft. The ratio thus obtained was an index of the relative abundance of the postsynaptic AChR at a given end-plate region. For each patient, the mean AChR index was compared with the mean MEPP amplitude.

In 2 control subjects the AChR indices were remarkably similar. In all 4 patients the AChR indices were significantly decreased and these decreases were essentially proportionate to the decreases in the MEPP amplitudes. When the MEPP amplitude for each subject was plotted as the dependent variable against the corresponding mean AChR index as the independent variable, a highly significant linear correlation was observed. Figure 2 represents the findings in the first 4 myasthenic patients and in a control subject investigated in our laboratory. In 8 additional cases of MG and 3 additional control

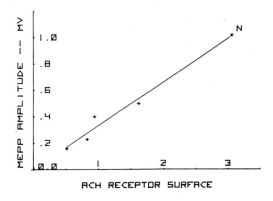

Fig. 2. Relationship between mean miniature end-plate potential amplitude and mean acetylcholine receptor index in control subject (N) and in four myasthenia gravis patients. The regression line is linear with a correlation coefficient of 0.988; $P < 0.001$. (Reproduced from Engel et al., 1977b, by permission of Harcourt Brace Jovanovich, Inc.).

426

subjects we found the same linear relationship between the AChR index and the MEPP amplitude.

This study clearly shows that the postsynaptic AChR available for α-bungarotoxin binding is reduced in MG. The fact that the decreased MEPP amplitude was proportionate to the AChR index demonstrates nicely that AChR deficiency is primarily responsible for the impaired neuromuscular transmission in MG. The greatest decrease in AChR occurred at those postsynaptic regions that were either the simplest or showed the most degenerative changes. This is consistent with the assumption that those portions of the postsynaptic membrane that bear AChR had been selectively destroyed.

LOCALIZATION OF IMMUNE COMPLEXES (IgG AND C3) AT THE MG ENDPLATE

Despite the massive evidence for an autoimmune pathogenesis of MG, immune complexes have never been directly demonstrated at the MG end-plate. We succeeded in localizing IgG at the MG end-plate utilizing peroxidase-labelled staphylococcal protein A (Engel et al., 1977a). Protein A binds to the Fc portion of human IgG subclasses 1, 2 and 4 and is relatively highly reactive toward human and rabbit IgG but is much less reactive to rat or goat IgG (Forsgren and Sjöquist, 1966; Kronvall and Williams, 1969; Biberfeld et al., 1975). The specificity and sensitivity of the reagent (and hence the intensity of the reaction at the end-plate versus the background staining) are further enhanced by protein A having an increased affinity for IgG molecules bound to antigenic sites (Kessler, 1976).

The third component of complement (C3) was localized with peroxidase-labelled anti-human C3 (Engel et al., 1977a). Immunoglobulin fraction of rabbit anti-human C3 serum, monospecific for human C3 by immunoelectrophoretic criteria, was obtained from Bio-Rad Laboratories.

The immunoreagents were labelled with peroxidase by a modification of the method of Nakane and Kawaoi (1974) as described by Engel et al. (1977a).

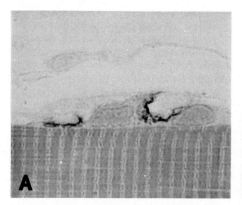

Fig. 3. Semi-thin resin sections showing light microscopic localization of IgG (A) and C3 (B) in a mild case of myasthenia gravis. The reaction is sharply localized to the upper region of the junctional sarcoplasm. Background staining is absent. A and B, ×1500. (Reproduced from Engel et al., 1977a, by permission.)

The labelled immunoreagents were applied in suitable concentrations to fresh muscle strips, which were well-rinsed before and after exposure to the reagents, followed by glutaraldehyde fixation. The fixed muscle strips were then treated the same way as those in the procedure for AChR localization. The method gave excellent localization of the immune complexes with minimal or absent background staining and optimal preservation of fine structure. In addition, the immunoreagents were also suitable for demonstrating immune complexes in fresh frozen sections and semi-thin resin sections by light microscopy (Fig. 3).

Both IgG and C3 were demonstrated at the end-plate in MG patients but not in control subjects. The reaction product was localized on the post-synaptic membrane as well as on degenerate material in the synaptic space (Figs. 4 and 5). In less severe cases of MG with relatively well preserved postsynaptic folds, abundant immune complexes were localized on the terminal expansions of the folds where the AChR is known to be located (Fig. 4A). In the more severely affected patients the postsynaptic region was generally more simplified, or showed more degenerative changes, and bound less abundant immune complexes (Fig. 4B). In each patient the relative abundance and pattern of distribution of IgG and C3 deposits were similar.

Omission of immunoreagents during the preparation of MG tissues for electron microscopy or of H_2O_2 from the diaminobenzidine medium, or the absorption of peroxidase-labelled staphylococcal protein A with human IgG, abolished the reaction at the MG end-plates. Absorption of peroxidase-labelled anti-human C3 with fresh human plasma abolished the reaction at the MG end-plates but absorption of this immunoreagent with human IgG was without effect.

The relative abundance of the immune complexes at the MG end-plates was estimated by measuring the length of the postsynaptic membrane reacting for IgG or C3 and dividing this value by the length of the primary synaptic cleft. The IgG and C3 indices thus obtained were compared with the m.e.p.p. amplitude in 5 patients with MG. The patients with higher IgG and C3 indices had higher m.e.p.p. amplitudes than those with lower indices. For any one patient, the IgG and C3 indices were not significantly different. Regression of the mean m.e.p.p. amplitude on the corresponding IgG and C3 index demonstrated a significant linear correlation of both indices with the MEPP amplitude ($P < 0.01$).

The less intense reaction for immune complexes in more severely affected patients was attributed to the smaller quantity of AChR remaining at their end-plates. The fact that the length of the postsynaptic membrane that bound immune complexes was proportionate to the MEPP amplitude was also consistent with this assumption, for we have also shown that the MEPP amplitude in MG patients is a linear function of the length of the postsynaptoc membrane reacting for AChR (Engel et al., 1977b). These observations clearly indicate that the deficiency of postsynaptic AChR is the primary reason for the reduction of the MEPP amplitude and that the binding of antibody to AChR probably only partially inhibits AChR function in MG.

The immuno-electron microscopic findings provide evidence for a destructive autoimmune reaction involving the postsynaptic membrane in MG and implicate the complement system in this reaction. In addition to having a destruc-

428

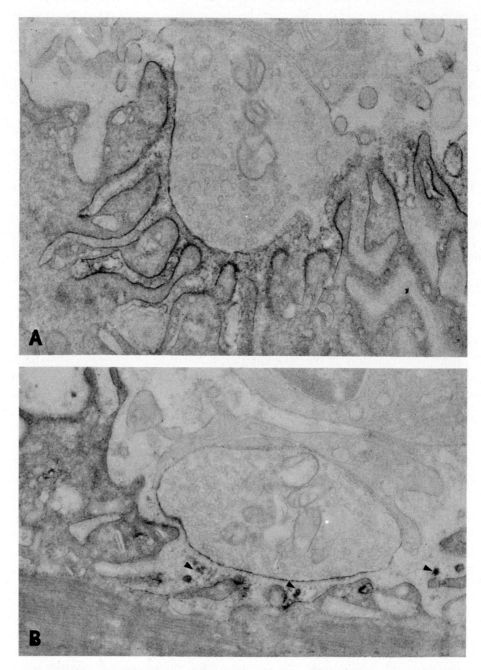

Fig. 4. Ultrastructural localization of IgG in mild (A) and more severe (B) case of myasthenia gravis. In (A), postsynaptic region is well preserved and terminal expansions of most folds bind IgG. In (B), postsynaptic region is highly simplified and junctional folds are small and sparse. IgG deposits occur on short segments of highly simplified postsynaptic membrane and on degenerate material in widened synaptic space (arrowheads). (A), ×29,500; (B), ×23,600. (Reproduced from Engel et al., 1977a, by permission.)

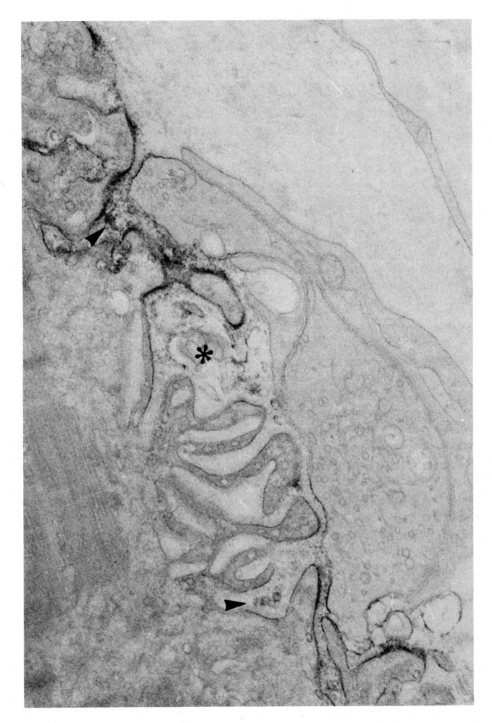

Fig. 5. IgG localization at a myasthenia gravis end-plate. The IgG deposits have a patchy distribution, occurring on some junctional folds but not on others and on debris in the synaptic space (arrowheads). In one region there is degeneration of the junctional folds (asterisk). Note reciprocal staining of presynaptic membrane where it faces reactive segments of postsynaptic membrane. ×33,900. (Reproduced from Engel et al., 1977a, by permission).

tive effect on the postsynaptic membrane, the anti-AChR antibodies could also impair neuromuscular transmission by interfering with AChR function: antibodies could allosterically hinder the attachment of ACh to AChR, or prevent ACh from opening ion channels, or decrease the conductance or open time of the ion channels. However, we have estimated that antibody binding in itself could not depress the MEPP amplitude by more than 40% (Engel et al., 1977a). Further, recent studies suggest that the conductance and open time of the acetylcholine-induced ion channels are normal in MG (Cull-Candy et al., 1978).

The antibodies could also induce modulation of AChR, consisting of its accelerated internalization and intracellular destruction. There is evidence for modulation of AChR by myasthenic immunoglobulin in cultured muscle fibers (Appel et al., 1977; Kao and Drachman, 1977; Heinemann et al., 1977). However, the AChR of cultured muscle fibers resembles the extrajunctional AChR of denervated muscle fibers which has a faster turnover rate than the junctional AChR (Heinemann et al., 1977; Chuang and Huang, 1975). The capacity for modulation of junctional and extrajunctional membranes might be different, and the in vitro system, where modulation plays an important role, lacks complement. The presence of complement as well as of antibody on the postsynaptic membrane in vivo might favor membrane lysis rather than modulation. Finally, if modulation were the only reason for the postsynaptic AChR deficiency in MG, one would not expect to see destructive changes at the endplate. However, such changes do exist (Engel et al., 1976; Engel et al., 1977a and b; Engel and Santa, 1971).

Regardless of the relative importance of the 3 different mechanisms that could impair neuromuscular transmission in MG — antibody-dependent complement-mediated lysis of the postsynaptic membrane, antibody-induced modulation of AChR, and antibody-dependent interference with AChR function — the process leading to failure of neuromuscular transmission is initiated by the immune complexes demonstrated at the MG end-plates.

ULTRASTRUCTURAL LOCALIZATION OF AChR AND IMMUNE COMPLEXES (IgG AND C3) IN EXPERIMENTAL AUTOIMMUNE MYASTHENIA GRAVIS

The findings in MG were further validated by parallel studies in experimental autoimmune MG (EAMG). Rats immunized with AChR were studied in the chronic phase of EAMG which closely resembles MG morphologically and electrophysiologically (Engel et al., 1976). AChR was localized in the EAMG animals with peroxidase-labelled α-bungarotoxin. Postsynaptic regions in these animals showed a marked decrease of AChR and were highly simplified (Fig. 6), resembling end-plates in patients with moderately severe, generalized MG (Engel et al., 1977b).

IgG was localized in chronic EAMG with rabbit anti-rat IgG followed with treatment with peroxidase-labelled protein A; and C3 with peroxidase-labelled rabbit anti-rat C3 (Sahashi et al., 1978). The distribution of immune complexes on the postsynaptic membrane and on degenerate material in the synaptic

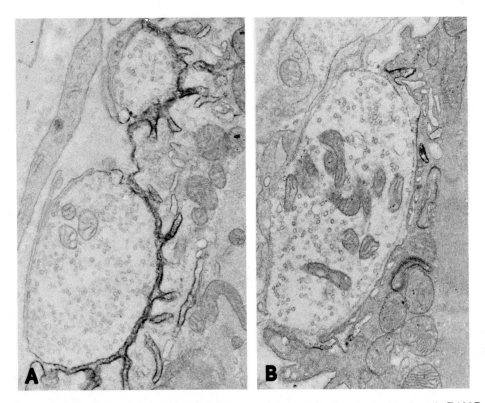

Fig. 6. Acetylcholine receptor localization in control rat (A) and in rat with chronic EAMG (B). In (B), postsynaptic region is highly simplified and only segments of it react for the receptor; presynaptic membrane is unstained except where it faces reactive segments of the postsynaptic membrane. (A), ×18,100; (B), ×16,100. (Reproduced from Engel et al., 1977b, by permission of Harcourt Brace Jovanovich, Inc.)

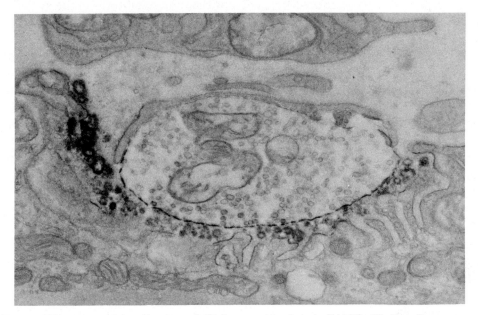

Fig. 7. Ultrastructural localization of C3 in rat with chronic EAMG. C3 deposits occur on globular, degenerated material in the synaptic space and on short segments of the post-synaptic membrane. There is also reciprocal staining of the presynaptic membrane. ×24,400.

432

space in EAMG closely resembled that in MG (Fig. 7). Further, in more severely affected animals the postsynaptic regions bound less abundant immune complexes and the abundance and distribution of IgG and C3 on the postsynaptic membrane were essentially identical. These data fully support the assumption that chronic EAMG is a valid model for MG.

SUMMARY

The synaptic dysfunction in myasthenia gravis (MG) is caused by a deficiency of the nicotonic postsynaptic acetylcholine receptor (AChR) protein. Ultrastructural localization of AChR with peroxidase-labelled α-bungarotoxin shows some postsynaptic regions devoid of AChR; in other regions the junctional folds display only faint traces of AChR, or only some segments of the postsynaptic membrane react for AChR; finally, in some postsynaptic regions there is no apparent decrease in AChR. In general, those postsynaptic regions which are the simplest or show the most degenerative changes show the greatest decrease of AChR. In MG cases of varying severity, morphometric estimates of the abundance of the postsynaptic membrane reacting for AChR correlate linearly with the miniature end-plate potential MEPP amplitude.

The AChR deficiency in acquired MG is caused by an autoimmune attack on the postsynaptic membrane. Immunoglobulin G (IgG) can be localized at the MG end-plate with peroxidase-labelled staphylococcal protein A, and the third component of complement (C3) can be demonstrated with peroxidase-labelled anti-human C3 under experimental conditions which give excellent ultrastructural localization, optimal preservation of fine structure and minimum background staining. Both IgG and C3 are found on segments of the postsynaptic membrane and on degenerated fragments of the junctional folds in the synaptic space. Immune complexes are more abundant in less severely affected patients than in the more severely affected patients. The abundance of the immune complexes is proportionate to the MEPP amplitude, and hence to the amount of AChR remaining at the end-plate. This indicates that the immune complexes act not merely by impairing AChR function but primarily by inducing a deficiency of AChR. The AChR deficiency can be caused by antibody-dependent complement-mediated lysis of the postsynaptic membrane or by antibody-induced modulation of AChR.

Parallel electron cytochemical and immuno-electron microscopic studies in chronic experimental autoimmune MG (EAMG) of the rat demonstrate a deficiency of the postsynaptic AChR and localize IgG and C3 at the end-plate. As in the human disease, the decrease in AChR is more marked and the immune complexes which bind to the postsynaptic membrane are less abundant in the more severely affected animals than in the less severely affected ones. These findings support the assumption that EAMG is a valid model of acquired MG.

REFERENCES

Appel, S.H., Anwyl, R., McAdams, M.W. and Elias, S. (1977) Accelerated degradation of acetylcholine receptor from cultured rat myotubes with myasthenia gravis sera and globulins. *Proc. nat. Acad. Sci. (Wash.)*, 74, 2130–2134.

Biberfeld, P., Ghetie, V. and Sjöquist, J. (1975) Demonstration and assaying of IgG antibodies in tissues and on cells by labelled staphylococcal protein A. *J. Immunol. Methods*, 6, 249–259.

Chuang, C.C. and Huang, M.C. (1975) Turnover of junctional and extrajunctional acetylcholine receptors of the rat diaphragm. *Nature (Lond.)*, 253, 643–644.

Cull-Candy, S.G., Miledi, R. and Trautmann, A. (1978) Acetylcholine induced channels and transmitter release at human end-plates. *Nature (Lond.)*, 271, 74–75.

Elmqvist, D., Hofmann, W.W., Kugelberg, J. and Quastel, D.M.J. (1964) An electrophysiological investigation of neuromuscular transmission in myasthenia gravis. *J. Physiol. (Lond.)*, 174, 417–434.

Engel, A.G. and Santa, T. (1971) Histometric analysis of the ultrastructure of the neuromuscular junction in myasthenia gravis and in the myasthenic syndrome. *Ann. N.Y. Acad. Aci.*, 183, 46–63.

Engel, A.G., Tsujihata, M., Lambert, E.H., Lindstrom, J.M. and Lennon, V.A. (1976) Experimental autoimmune myasthenia gravis: a sequential and quantitative study of the neuromuscular junction ultrastructure and electrophysiologic correlations. *J. Neuropathol. Exp. Neurol.*, 35, 569–587.

Engel, A.G., Lambert, E.H. and Howard, F.M., Jr. (1977a) Immune complexes (IgG and C3) at the motor end-plate in myasthenia gravis. *Mayo Clinic Proc.*, 52, 267–280.

Engel, A.G., Lindstrom, J.M., Lambert, E.H. and Lennon, V.A. (1977b) Ultrastructural localization of the acetylcholine receptor in myasthenia gravis and in its experimental autoimmune model. *Neurology (Minn.)*, 27, 307–315.

Fambrough, D.M., Drachman, D.B. and Satyamurti, S. (1973) Neuromuscular junction in myasthenia gravis: decreased acetylcholine receptors. *Science*, 182, 293–295.

Forsgren, A. and Sjöquist, J. (1966) "Protein A" from *S. aureus*. I. Pseudo-immune reaction with human γ-globulin. *J. Immunol.*, 97, 822–827.

Heilbronn, E., Mattsson, C., Thornell, L.-E., Sjöström, M., Stalberg, E., Hilton-Brown, P. and Elmqvist, D. (1976) Experimental myasthenia in rabbits: biochemical, immunological, electrophysiological, and morphological aspects. *Ann. N.Y. Acad. Sci.*, 274, 337–353.

Heinemann, S., Bevan, S., Kullberg, R., Lindstrom, J.M. and Rice, J. (1977) Modulation of acetylcholine receptor by antibody against the receptor. *Proc. nat. Acad. Sci. (Wash.)*, 74, 3090–3094.

Kao, I. and Drachman, D.B. (1977) Myasthenic immunoglobulin accelerates acetylcholine receptor degradation. *Science*, 196, 527–529.

Kessler, S.W. (1976) Cell membrane antigen isolation with the staphylococcal protein-A antibody absorbent. *J. Immunol.*, 117, 1482–1490.

Kronvall, G. and Williams, R.C. (1969) Differences in anti-protein A activity among IgG subgroups. *J. Immunol.*, 103, 828–833.

Lambert, E.H. and Elmqvist, D. (1971) Quantal components of end-plate potentials in the myasthenic syndrome. *Ann. N.Y. Acad. Sci.*, 183, 183–199.

Lambert, E.H., Lindstrom, J.M. and Lennon, V.A. (1976) End-plate potentials in experimental autoimmune myasthenia gravis in rats. *Ann. N.Y. Acad. Sci.*, 274, 300–318.

Lennon, V.A., Lindstrom, J.M. and Seybold, M.E. (1975) Experimental autoimmune myasthenia: a model of myasthenia gravis in rats and guinea pigs. *J. exp. Med.*, 141, 1365–1375.

Lindstrom, J.M., Engel, A.G., Seybold, M.E., Lennon, V.A. and Lambert, E.H. (1976a) Pathological mechanisms in experimental autoimmune myasthenia gravis. II. Passive transfer of experimental autoimmune myasthenia gravis in rats with anti-acetylcholine receptor antibodies. *J. exp. Med.*, 144, 739–753.

Lindstrom, J.M., Lennon, V.A., Seybold, M. and Whittingham, S. (1976b) Experimental autoimmune myasthenia gravis and myasthenia gravis: biochemical and immunochemical aspects. *Ann. N.Y. Acad. Sci.*, 274, 254–274.

Lindstrom, J.M., Seybold, M.E., Lennon, V.A., Whittingham, S. and Duane, D.D. (1976c) Antibody to acetylcholine receptor in myasthenia gravis: prevalence, clinical correlates, and diagnostic value. *Neurology (Minn.)*, 26, 1054–1059.

434

Nakane, P.K. and Kawaoi, A. (1974) Peroxidase-labeled antibody: a new method of conjugation. *J. Histochem. Cytochem.*, 22, 1084–1091.

Patrick, J. and Lindstrom, J. (1973) Autoimmune response to acetylcholine receptor. *Science,* 180, 871–872.

Sahashi, K., Engel, A.G., Lindstrom, J.M., Lambert, E.H. and Lennon, V.A. (1978) Ultrastructural localization of immune complexes (IgG and C3) at the end-plate in experimental autoimmune myasthenia gravis. *J. Neuropath. exp. Neurol.,* 37, 212–223.

Tarrab-Hazdai, R., Aharonov, A., Silman, A. and Fuch, S. (1975) Experimental autoimmune myasthenia induced in monkeys by purified acetylcholine receptor. *Nature (Lond.),* 256, 128–130.

Toyka, K.V., Drachman, D.B., Griffin, D.E., Pestronk, A., Winkelstein, J.A., Fishbeck, K.H., Jr. and Kao, I. (1977) Myasthenia gravis: study of humoral immune mechanisms by passive transfer to mice. *New Engl. J. Med.,* 296, 125–131.

The Role of Humoral Immunity in Myasthenia Gravis

ANNE D. ZURN and BERNARD W. FULPIUS

Department of Biochemistry, University of Geneva, CH-1211 Geneva 4 (Switzerland)

INTRODUCTION

In a synapse, neurotransmitters liberated from nerve endings diffuse within the narrow synaptic cleft and reach the postsynaptic receptors located at the top of the synaptic folds. When agonists or antagonists are injected into the blood stream, they have to cross the capillary wall and enter the cleft to act on synaptic transmission. This is certainly the case for pharmacological agents of molecular weights inferior to 1000 daltons such as dimethyltubocurarine and for small proteins of slightly higher molecular weights, about 10,000 daltons, such as the α-neurotoxins isolated from snake venoms. Very little is known, however, of the influence on the synaptic function, of larger molecules. A few toxins of molecular weights in the range of 150,000 daltons are known to interfere with synaptic function but do not act postsynaptically. Recent data from studies on myasthenia gravis (MG) and experimental autoimmune myasthenia gravis (EAMG) might give new insights into that question. In this relatively rare human disease and in its experimental model (animals immunized with acetylcholine receptor (AChR) purified from different sources) anti-AChR antibodies are present in the serum when the blockade of the neuromuscular transmission develops. Moreover, the number of AChR sites labelled with α-bungarotoxin (α-BgTX) is decreased (Fulpius, 1978). It could well be that these antibodies cause the block of neuromuscular transmission. In order to clarify this problem, we will review (i) the evidence favouring a role of anti-AChR antibodies in the synaptic dysfunction, (ii) data concerning the specific site of action of these antibodies, and (iii) data concerning their mechanism of action in the pathogenesis of MG.

Evidence favouring a role of anti-AChR antibodies in synaptic dysfunction

In order to cause synaptic dysfunction, molecules of the size of γ-immunoglobulins (IgG) such as anti-AChR antibodies have to leave the vascular compartment, diffuse into the extracellular space, enter the narrow synaptic cleft and reach the top of the postsynaptic folds. This was shown to be the case by Zurn and Fulpius (1976). These authors labelled the nerve terminal region of mouse diaphragms with covalent complexes made of IgG and ^{125}I-labelled α-BgTX. In addition, they showed that ^{125}I-labelled anti-α-cobratoxin antibodies injected into mice which had received sublethal doses of α-cobratoxin could reach the toxin molecules bound to the nerve terminals of the diaphragm. The ultimate proof of a pathogenic role of anti-AChR antibodies has to be given by the successful transfer of myasthenic symptoms with these antibodies. Transfer of anti-AChR antibodies occurs naturally in newborns from myasthenic mothers (Keesey et al., 1977). At birth, these infants, like their mothers, have high levels of anti-AChR antibodies in their serum and show transiently myasthenic symptons. These clinical manifestations disappear gradually together with the anti-AChR antibodies in their serum. In this context, one should mention the response of myasthenic patients to plasma exchange. This procedure, which removes humoral factors from the blood, produces a shortterm remission in myasthenic patients. In all patients studied, there is a marked inverse correlation between anti-AChR antibody titre and muscle strength (Newsom-Davis et al., 1978). Experimentally, myasthenic symptoms have been successfully transferred into mice (Toyka et al., 1977). A few days after injection of myasthenic IgG, these animals appeared weak, had a reduced amplitude of miniature end-plate potentials (MEPP), a reduced number of receptor sites and typical myasthenic electromyographic responses. The delayed development of reduced MEPP amplitude implies that the defect of neuromuscular transmission in MG is probably not due to a simple curariform block of the AChR site by an IgG antibody.

SITE OF ACTION OF ANTI-AChR ANTIBODIES

It has been reported that animals immunized with AChR denatured by heat treatment (Zurn, 1978), dissociation in sodium dodecyl sulphate (SDS) (Valderrama et al., 1976) or reduction followed by carboxymethylation (Bartfeld and Fuchs, 1977), do not develop autoimmune myasthenia. Accordingly, specific native antigenic sites on the AChR molecule are essential to induce the disease. The study of these is presently limited to one such site, the α-BgTX binding site. As shown by several authors, there exists a competition between anti-AChR antibodies and α-BgTX. The experiments reported were performed (a) with solubilized AChR or (b) with AChR in situ on membrane fragments, (c) on tissue sections, (d) on intact cells in culture, (e) on whole organs.

(a) Almon and Appel (1975) demonstrated 50% inhibition of α-BgTX fixation to soluble and purified AChR after incubation with myasthenic serum. Zurn and Fulpius (1976) showed inhibition of α-cobratoxin fixation on purified AChR from *Torpedo* with the serum of a rabbit immunized with *Torpedo* AChR. (b) Lefvert and Bergström (1977) demonstrated with myasthenic immunoglobulins an inhibition of α-BgTX binding to membrane fragments isolated from *Torpedo marmorata* electric organ. On an identical preparation, Zurn (1978) showed inhibition of α-cobratoxin fixation by the serum of a rabbit immunized with *Torpedo* AChR. This inhibition was less pronounced when compared to that measured on soluble AChR (Fig. 1). This finding favours the view that in addition to antibodies directed against the binding site,

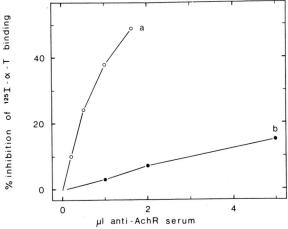

Fig. 1. Percentages of inhibition of ^{125}I-labelled α-cobratoxin binding produced with increasing quantities of anti-AChR serum on AChR-rich membrane fragments. (a), fragments solubilised with detergent; (b), fragments intact.

there are other antibodies which also inhibit toxin fixation. A more specific piece of evidence for the presence of antibodies directed against the toxin-binding site comes from the experiment illustrated in Fig. 2. In this case it is shown that α-BgTX inhibits ^{125}I-labelled anti-AChR antibody binding to *Torpedo marmorata* membrane fragments by 5% to 15%. According to Zurn (1978), the amount of ^{125}I-labelled anti-AChR antibodies competing for the toxin binding site represents only 0.15% of the total amount of $[^{125}I]IgG$ incubated. (c) Bender et al. (1975) showed inhibition of $[^{125}I]\alpha$-BgTX binding to human muscle sections with myasthenic sera. (d) Lindstrom (1976) showed that anti-eel AChR serum inhibits toxin binding to the eel electroplaque by 50%. Zurn and Fulpius (1977) showed inhibition of α-cobratoxin binding to isolated mouse diaphragms with rabbit anti-*Torpedo* AChR serum.

It is still debated whether or not α-BgTX and acetylcholine (ACh) binding sites are identical. According to Raftery et al. (1975) there are twice as many sites for the toxin as for ACh. It follows that the ACh binding site cannot be

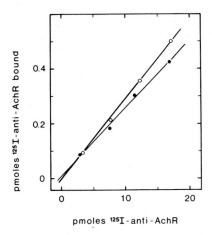

Fig. 2. ^{125}I-labelled anti-AChR antibodies bound to *Torpedo* AChR-rich membrane fragments in the presence of α-BgTX (4×10^{-6} M) (●———●) and in its absence (○———○).

studied adequately with ligands such as neurotoxins. Good indications, however, can be derived from electrophysiological experiments such as the following. Bevan et al. (1977) showed a decrease in the sensitivity of muscle cells in culture to iontophoretically-applied ACh in the presence of myasthenic serum. Several authors (Sugiyama et al., 1973; Patrick et al., 1973) showed a block of the electrophysiological response of the eel electroplaque to carbamylcholine with a rabbit anti-eel AChR serum. After incubation with a rabbit anti-*Torpedo* AChR serum, Green et al. (1975) observed on frog endplates a decrease in MEPP size resembling that seen in animals immunized with AChR.

According to these data, anti-AChR antibodies could act as curare-like agents to cause synaptic dysfunction. However the experiments reported do not give any indication concerning the mode of action of the anti-AChR antibodies.

MODE OF ACTION OF ANTI-AChR ANTIBODIES

Three mechanisms have been proposed (Engel et al. 1977): (1) A pure immunopharmacological blockade of AChR. This implies that antibodies compete for the ACh binding site. (2) An accelerated internalization and an intracellular degradation of AChR on binding of anti-AChR antibodies. (3) A partial destruction of the postsynaptic membrane consequent on the binding of antibodies to AChR and the subsequent activation of the complement reaction.

Experimental evidence for these mechanisms comprises the following: (1) A pure immunopharmacological blockade has been shown to occur in experiments in mouse diaphragm which were performed at 4°C and in the absence of complement (Zurn and Fulpius, 1977). In these conditions, the inhibition of toxin binding observed in the presence of anti-AChR antibodies cannot be due to internalization of the AChR-antibody complexes nor to a complement-mediated destruction of the post-synaptic membrane. (2) Evidence for the second mechanism has been brought forward by several authors (Anwyl et al., 1977, Appel et al., 1977; Bevan et al., 1977; Kao and Drachman, 1977) who have shown that myasthenic serum added to cultured muscle cells from different sources induces an accelerated rate of internalization and degradation of AChR. (3) Evidence for the third mechanism has been brought by Engel et al. (1977), who demonstrated the presence of IgG and C3 on the postsynaptic membranes of myasthenic neuromuscular junctions. More recently, Lennon et al. (1978) showed that it is not possible to induce EAMG by passive transfer of syngeneic antibodies or by active immunization, in rats depleted in complement by treatment with cobra venom factor.

In vivo, activation of the complement reaction upon binding of the antibodies to the antigen plays probably a major role. The contribution of the second mechanism to the synaptic dysfunction is questionable since the intrinsic rate of AChR turnover is much slower on the neuromuscular junction than in cultured muscle cells. A pure immunopharmacological blockade might play only a minor role in vivo since inhibition of toxin fixation on mouse diaphragms has been obtained only with sera containing high titers of anti-AChR antibodies, a condition allowing a minor subpopulation to become fully effective.

CONCLUSIONS

Recent data from studies on MG and EAMG indicate that anti-AChR anti-bodies play a major role in the impairment of neuromuscular transmission. Thus, it seems conceivable that antibodies directed against synaptic receptors can interfere with the normal function of synapses. This mechanisms might not be restricted to the neuromuscular junctions but could also take place in central synapses. Recent data (Fulpius et al., 1977; Lefvert et al., 1977) favour such a hypothesis in the case of myasthenia gravis.

SUMMARY

The authors review evidence favouring a pathogenic role of anti-acetyl-choline receptor (anti-AChR) antibodies in myasthenia gravis and discuss their site and mode of action. They conclude that in vivo complement mediated destruction of the postsynaptic membrane consecutive to binding of anti-AChR antibodies to AChR is the major factor causing synaptic dysfunction.

ACKNOWLEDGEMENTS

This work was supported by grants 3.551.75 and 3-157.77 of the Swiss National Science Foundation.

REFERENCES

Almon, R.R. and Appel, S.H. (1975) Interaction of myastenic serum globulin with the acetylcholine receptor. *Biochim. Biophys. Acta,* 393, 66–77.

Anwyl, R., Appel, S.H. and Narashi, T. (1977) Myasthenia gravis serum reduced acetyl-choline sensitivity in cultured rat myotubes. *Nature (Lond.),* 267, 262–263.

Appel, S.H., Anwyl, R., McAdams, M.W. and Elias, S. (1977) Accelerated degradation of acetylcholine receptor from cultured rat myotubes with myasthenia gravis sera and globulins. *Proc. natl. Acad. Sci. (Wash.),* 74, 2130–2134.

Bartfeld, D. and Fuchs, S. (1977) Immunological characterization of an irreversibly dena-tured acetylcholine receptor. *FEBS Lett.,* 77, 214–218.

Bender, A.N., Engel, W.K., Ringel, S.P., Daniels, M.P. and Vogel, Z. (1975) Myasthenia gravis: a serum factor blocking acetylcholine receptors of the human neuromuscular junction. *Lancet,* 1, 607–609.

Bevan, S., Kullberg, R.W. and Heinemann, S.F. (1977) Human myasthenic sera reduce acetylcholine sensitivity of human muscle cells in tissue culture. *Nature (Lond.),* 267, 263–265.

Engel, A.G., Lambert, E.H. and Howard, F.M. (1977) Immune complexes (IgG and C3) at the motor end-plate in myasthenia gravis. *Mayo Clin. Proc.,* 52, 267–280.

Fulpius, B.W., Fontana, A. and Cuénoud, S. (1977) Central nervous systems involvement in experimental autoimmune myasthenia gravis. *Lancet,* 2, 350–351.

Fulpius, B.W. (1978) The pathogenesis of myasthenia gravis. *Bull. Schweiz. Akad. med. Wiss.,* 34, 25–31.

Green, D.P.L., Miledi, R. and Vincent, A. (1975) Neuromuscular transmission after immunization against acetylcholine receptors. *Proc. roy. Soc. Lond. B.,* 189, 57–68.

Kao, I. and Drachman, D.B. (1977) Myasthenic immunoglobulin accelerates acetylcholine receptor degradation. *Science,* 196, 527–530.

440

Keesey, J., Lindstrom, J., Cokely, H. and Herrmann, Jr., C. (1977) Anti-acetylcholine receptor antibody in neonatal myasthenia gravis. *New Engl. J. Med.*, 296, 55.

Lefvert, A.K. and Bergström, K. (1977) Immunoglobulins in myasthenia gravis: effect of human lymph IgG 3 and F (ab')$_2$ fragments on a cholinergic receptor preparation from *Torpedo marmorata. Europ. J. Clin. Invest.*, 7, 115–119.

Lefvert, A.K. and Pirskanen, R. (1977) Acetylcholine receptor antibodies in cerebrospinal fluid of patients with myasthenia gravis. *Lancet*, 2, 351–352.

Lennon, V.A., Seybold, M.E., Lindstrom, J.M., Cochrane, C. and Ulevitsch, R. (1978) Role of complement in the pathogenesis of experimental autoimmune myasthenia gravis. *J. exp. Med.*, 147, 973–983.

Lindstrom, J. (1976) Immunological studies of acetylcholine receptors. *J. Supramolec. Struct.*, 4, 389–403.

Newsom-Davis, J., Pinching, A.J., Vincent, A. and Wilson, S.G. (1978) Function of circulating antibody to acetylcholine receptor in myasthenia gravis: investigation by plasma exchange. *Neurology*, 28, 266–272.

Patrick, J., Lindstrom, J., Culp, B. and McMillan, J. (1973) Studies on purified eel acetylcholine receptor and anti-acetylcholine receptor antibody. *Proc. nat. Acad. Sci. (Wash.)*, 70, 3334–3338.

Raftery, M.A., Bode, J., Vandlen, R., Michaelson, D., Deutsch, J., Moody, T., Ross, M.J. and Stroud, R.M. (1975) Structural and functional studies of an acetylcholine receptor. In *Protein–Ligand Interactions*, H. Sun and G. Blauer (Eds.), Walter de Gruyter, Berlin, pp. 328–355.

Sugiyama, H., Benda, P., Meunier, J.C. and Changeux, J.P. (1973) Immunological characterization of the cholinergic receptor protein from *Electrophorus electricus. FEBS Lett.*, 35, 124–128.

Toyka, K.V., Drachman, D.B., Griffin, D.E., Pestronk, A., Winkelstein, J.A., Fischbeck, Jr., K.H. and Kao, I. (1977) Myasthenia gravis. Study of humoral immune mechanisms by passive transfer to mice. *New Engl. J. Med.*, 296, 125–131.

Valderrama, R., Weill, C.L., McNamee, M. and Karlin, A. (1976) Isolation and properties of acetylcholine receptors from *Electrophorus* and *Torpedo. Ann. N.Y. Acad. Sci.*, 274, 108–115.

Zurn, A.D. and Fulpius, B.W. (1976) Accessibility to antibodies of acetylcholine receptors in the neuromuscular junction. *Clin. exp. Immunol.*, 24, 9–17.

Zurn, A.D. and Fulpius, B.W. (1977) Study of two different subpopulations of anti-acetylcholine receptor antibodies in a rabbit with experimental autoimmune myasthenia gravis. *Europ. J. Immunol.*, 8, 529–532.

Zurn, A.D. (1978) *Experimental autoimmune myasthenia gravis: possible role of anti-acetylcholine receptor antibodies in the appearance of the neuromuscular blockade.* M.D. Thesis. University of Geneva.

Humoral Immunity in Myasthenia Gravis: Relationship to Disease Severity and Steroid Treatment

R.J. BRADLEY, D. DWYER, B.J. MORLEY, G. ROBINSON, G.E. KEMP and S.J. OH

Neuroscience Program and Department of Neurology, The Medical Center, University of Alabama in Birmingham, Birmingham, AL 35294 (U.S.A.)

INTRODUCTION

Components of the nicotinic acetylcholine receptor (AChR) protein have been purified from several sources (Changeux et al., 1975) and antibodies directed against this protein have been discovered in the serum of myasthenia gravis (MG) patients (Almon et al., 1974; Aharonov et al., 1975; Lindstrom et al., 1976). Study of the AChR has been accelerated by the purification of a small basic protein, the so-called α-toxin from cobra venom which binds with high specificity and affinity to the AChR protein, blocking the effects of acetylcholine at the muscle endplate (Lee et al., 1972). Radioactive α-bungarotoxin (α-BgTX) has been used to quantify the AChR in situ at the endplate and after extraction of the protein from its membrane environment.

Fambrough et al. (1973) found a decreased number of receptors for α-BgTX in muscle from MG patients. Almon et al. (1974) showed that γ-immunoglobulin (IgG) from the serum of 5 out of 15 MG patients blocked α-BgTX binding to AChR extracts from denervated rat muscle. The use of immunocytochemistry to visualize α-BgTX binding has enabled Bender et al. (1975) to show that a serum factor from 44% of MG patients blocked α-BgTX binding to the endplate region of sections from normal human muscle. When sections of denervated human muscle were used the blocking effect was observed in 75% of MG patients. Mittag et al. (1976) found that serum from 2 out of 28 patients inhibited toxin binding to AChR extracts from denervated rat muscle. Similarly, Lindstrom et al. (1976) reported that serum from 6 out of 16 patients inhibited toxin binding to partially purified AChR from human muscle. Several investigators have used an anti-IgG immunoprecipitation assay to quantify the IgG in MG serum which binds to the α-[^{125}I]BgTX/AChR complex. This type of radioimmunoassay (RIA) has facilitated the diagnosis of a higher proportion of MG patients, namely 87% (Lindstrom et al., 1976); 85% (Mittag et al., 1976); 70% (Almon et al., 1976), and 92% (Monnier and Fulpius, 1977).

The application of these assays in the diagnosis of MG would be an improvement over currently accepted procedures including edrophonium testing and repetitive nerve stimulation. However, antibody assays must first be shown to have some meaningful correlation with the incidence of MG. If the correlation exists then it may shed light on the etiology of the disease, its prognosis and treatment. We have measured serum antibodies from a group of MG patients and controls using both α-BgTX inhibition and immunoprecipitation assay systems in all cases. Our studies have utilized a purified AChR from denervated rat muscle in order to maximize the reliability and sensitivity of the measurements. In addition, we have attempted to monitor the serum activity at different times in the same patient in order to find out if the antibody titre is directly related to the intensity of the disease.

METHODS

Sera from 27 patients with MG and 14 controls with various neuromuscular diseases were assayed for anti-AChR antibody. Diagnosis of MG was based on the clinical demonstration of muscle weakness and the unequivocal improvement of weakness to edrophonium or neostigmine. In 78% of these cases a decrementing evoked muscle potential to repetitive nerve stimulation was also documented.

α-[^{125}I]BgTX was radiolabelled by enzymatic iodination employing immobilized lactoperoxidase (Bradley et al., 1976). The specific activity of the monoiodinated product used in these studies was 40,000 Ci/mole.

AChR was purified from denervated rat muscle using a combination of affinity chromatography with α-toxin from the venom of *Naja naja Siamensis*, ion exchange chromatography and gel filtration (Kemp et al., 1978). The specific activity of the final product in our purification scheme has varied between 2.8–7.5 pM α-[^{125}I]BgTX/μg protein for many individual preparations. The AChR assay system was as previously described (Bradley et al., 1976) using ion exchange chromatography on CM-50 to separate free toxin from the α-[^{125}I]BgTX/AChR complex. All sera used in this study were frozen at −27°C and assayed at a later date.

Inhibition assays were carried out by incubating aliquots of AChR with increasing amounts of serum for 12 h at 4°C in a total volume of 0.2 ml. Excess α-[^{125}I]BgTX was then added and after incubation for 90 min at room temperature, the amount of complex was assayed. For the immunoprecipitation assay (RIA), AChR was incubated with excess α-[^{125}I]BgTX and the α-[^{125}I]BgTX/AChR complex was isolated on a small column of Sephadex CM-50 as in our standard AChR assay procedure. Aliquots of the radioactive complex were incubated with varying amounts of serum for 24 h at 4°C in a total volume of 0.2 ml. Sufficient goat anti-human IgG (Miles) was added to precipitate the total IgG in the assay and incubation was continued for 12 h at 4°C. The sample was then centrifuged and washed three times and finally counted. The data were expressed as pmoles of α-[^{125}I]BgTX/AChR complex bound per 1 of serum. The assay was carried out in the linear portion of the precipitation curve. For all assays, control tubes were included where α-[^{125}I]BgTX binding was inhibited by excess cold toxin or curare. This background was routinely subtracted.

RESULTS AND DISCUSSION

Our results agree well with the reports of others that only MG patients have serum antibodies directed against the AChR. Out of a total of 31 serum determinations from patients with myasthenia gravis 28 were significantly different from controls on either assay system. On the basis of the assay procedures presented here an accurate diagnosis could be made for 22 out of 25 or 88% of all MG patients identified by conventional means. In the case of patients with an Osserman classification (Osserman, 1958) of IIB or higher, 100% were positive, i.e. more than 2 standard deviations above the mean control value on at least one of the two assay systems (Table I). Only 65% sera from all MG patients showed nominal inhibition of α-[^{125}I]BgTX binding to the receptor and only 2 patients gave a maximum inhibition over 40% (Fig. 1).

One MG patient in this study (No. 26) had a positive antibody titre in the inhibition of α-[^{125}I]BgTX binding but did not show any detectable titre in the RIA. This suggests that in this patient the antibody competes with α-BgTX for AChR binding and does not bind to the receptor if α-BgTX is already bound. The binding isotherms for this patient (No. 26) and patient (No. 9) are shown in Fig. 1. These two sera gave the highest maximum percentage of toxin inhibition which have been observed with our toxin inhibition assay. The maximum inhibition was 68% for patient No. 26 and 44% for patient No. 9. In these two cases the active component in both assay systems was identified as IgG by anti-IgG immunoadsorption and anti-IgG immunoprecipitation. For the case of the most potent sera (No. 26) in our series an IgG concentration of approximately 5×10^{-8} M induced half maximal inhibition of α-BgTX binding in the assay system suggesting a highly specific interaction between the antibody and the AChR toxin binding site.

TABLE I

ANTI-RECEPTOR ANTIBODY TITRE IN MG PATIENTS AND CONTROLS

Group	(N)	pmol AChR by RIA	% Inhibition of BgTX	Diagnosis
Control	14	18 ± 12	1.6 ± 0.6	0
I	9	107 ± 40	4.3 ± 1.2	7
IIA	10	312 ± 141	8.3 ± 2.4	9
IIB	4	696 ± 396	25.8 ± 10.9	4
III & IV	2	1207 ± 381	32.5 ± 15.5	2
0	6	792 ± 344	8.9 ± 3.5	6

The anti-receptor assay for α-[^{125}I]BgTX inhibition and the RIA were carried out in a total of 27 MG patients classified according to Osserman (1958), and 14 controls with other neuromuscular diseases. Group 0 consisted of 6 patients who recently became symptom free but who all had a significant antibody titre. Data from 4 patients were included both while they had MG ($\geq$IIA) and when they were symptom free after treatment with prednisone plus an anticholinesterase. In addition, one patient was receiving only anticholinesterase and one was in remission without treatment. Antibody titres were expressed as pmol of AChR precipitated per litre of serum (±S.E.M.). Toxin inhibition values represent the maximum percentage inhibition of α-[^{125}I]BgTX binding in the AChR assay after incubation with MG serum.

444

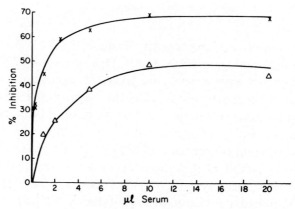

Fig. 1. Inhibition of α-[^{125}I]BgTX binding to AChR by increasing amounts of serum from two MG patients. The Osserman classification was IIB for patient No. 26 (X——X) and IV for patient No. 9 (△——△).

However, total inhibition of toxin binding was not observed in any of these patients. Almon and Appel (1976) first reported this failure to obtain complete inhibition of α-BgTX binding and it was suggested that the IgG blocked only one of two α-BgTX binding sites on the AChR macromolecule. The presence of a >66% inhibition from patient No. 26 might suggest the existence of more than two (possibly 3) α-BgTX binding sites per AChR molecule. In order to further investigate the failure of serum from patient No. 26 to show up in the RIA we ascertained by gel filtration the approximate molecular weight of the α-[^{125}I]-BgTX/AChR complex before and after incubation with this serum. As expected the molecular weight of the α-[^{125}I]BgTX/AChR complex did not increase in the case of patient No. 26 confirming that serum antibody did not bind (Fig. 2). However, in patient No. 9 an early peak is observed in the gel filtra-

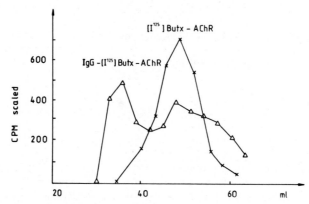

Fig. 2. Gel filtration of α-[^{125}I]BgTX—AChR complexes after incubation with a saturating amount of serum from patients No. 26 (X——X) and No. 9 (△——△). AChR was incubated with excess α-[^{125}I]BgTX for 1 h at room temperature, then passed through a pasteur pipette Sephadex CM-50 column to remove unbound α-[^{125}I]BgTX. The α-[^{125}I]BgTX—AChR complex was incubated for 12 h at 4°C with the myasthenic sera and then gel filtered on a 1.2 × 80 cm Sepharose 6B column. The common peak represents the α-[^{125}I]BgTX-AChR complex but an IgG—α-[^{125}I]BgTX—AChR is only found in the case of patient No. 9. The cpm. for patient No. 26 were scaled by a factor of 5 in order to compare the peak locations on this graph. α-BgTX is denoted as Butx in the graph.

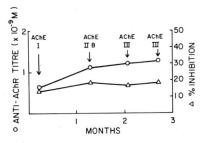 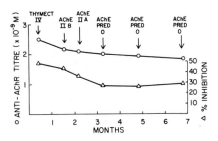

Fig. 3. Patient No. 21 (28-yr-old man) was treated with an anticholinesterase (AChE) throughout the course of antibody measurement. Anti-AChR antibody was assayed for maximum toxin binding inhibition and nanomoles of AChR precipitated per litre of serum during a three month period. The disease onset occured 1 month before our first measurement. A thymectomy 1 week before our last measurement showed hyperplasia. This patient demonstrates an increase in antibody as the disease becomes more severe as seen from the Osserman classification at each measurement.

Fig. 4. For patient No. 9 (20-yr-old female) anti-receptor antibody was assayed for maximum toxin binding inhibition and nanomoles AChR precipitated per litre of serum during a seven month period. The disease was first diagnosed at age 13. Thymectomy at age 15 showed hyperplasia. The current treatment and disease classification are given at each assay point. This figure shows that there is a decrease in antibody as the disease becomes less severe but the titre is still more than 40 times greater than control values when the patient becomes symptom free (stage 0) after prednisone (PRED) treatment.

tion profile indicating the increased molecular weight of the IgG α-[^{125}I]BgTX/ AChR-complex. Therefore, it would be advisable in the future to carry out both the toxin inhibition assay and the RIA in all cases as at least in this patient No. 26 the RIA was apparently insensitive due to α-BgTX blockade of IgG binding.

Within individual patients there was an excellent correspondence between the antibody titre and the increasing severity of the disease over time. An example of this is shown in Fig. 3. Our most remarkable result is that patients who become symptom free after steroid treatment (Group 0) still had very high antibody levels, well above many other patients who have the disease. In several patients for whom we measured antibody throughout the course of the disease, remission was not correlated with a significant reduction in antibody levels. In patient No. 9 there was a very slight reduction in titre at the return of normal strength but the RIA value was still higher than we have measured in any other patient with the disease (Fig. 4). The antibody level in this patient was still dramatically elevated even after 3 months in remission.

It is possible that the antibody present in the serum, although correlated with increasing severity of the disease, is in fact an epiphenomenon not necessarily responsible for the underlying pathogenesis. The remission induced by prednisone may be due to the suppression of some other feature of the immune system and the level of IgG may be coincidental. It has been shown that lymphocytes in MG patients are sensitized to AChR (Richman et al., 1976) and steroid-induced remission in MG is correlated with reduced lymphocyte sensitization (Abramsky et al., 1975). In this case it might be expected that serum anti-receptor IgG would remain high for some time after immune suppression. The role of autoantibody in MG could vary in different patients perhaps acting as a cholinergic modulator in some cases or alternatively supporting a form of antigenic modulation or cooperating in a cellular or complement induced attack at the endplate. Although the measurement of this antibody may be of great value in the diagnosis of MG patients our results would suggest that at least in some cases, an underlying cellular response might be dissociated from the production of specific antibody. This has already been determined in the case of experimental allergic encephalitis (Lisak et al., 1970).

It must also be considered that steroids might directly or indirectly interfere with an antibody mediated attack at the endplate. We carried out the RIA for anti-AChR IgG from MG patients in the presence of varying concentrations of prednisone to determine if antibody binding to the receptor was altered. We found that $\leq 10^{-4}$ M prednisone has no effect on the IgG—AChR interaction as measured in our two assay systems. Furthermore, prednisone did not inhibit the binding of α-BgTX to the AChR.

Recent evidence has shown that myasthenic IgG induces some features of MG when injected into mice (Toyka et al., 1977) and causes AChR modulation in cultured muscle cells with concomitant loss of receptor (Appel et al., 1977). However, there is good evidence that experimental autoimmune MG in rats induced by innoculation with AChR from electric fish, is mediated solely by the action of IgG in conjunction with complement activation (Lennon et al., 1978). The binding of the IgG to AChR at the endplate, is, therefore, likely to cause no direct electrophysiological effect. The role of the IgG would only be to target the endplate for complement activation and lysis. However, there is as yet no direct evidence that complement induced lysis occurs in MG although IgG and the C3 component of complement have been observed binding to the endplate of MG patients (Engel et al., 1977). It is already known that under certain conditions steroids will interfere with the complement induced lysis of tumor cells (Schlager et al., 1977). It is, therefore, conceivable that MG patients may become symptom free after treatment with steroids or other immunosuppressive drugs and still retain a very high titre of anti-receptor antibody. For example, it has been shown that there is no change in the level of anti-myelin antibody found in multiple sclerosis after long-term, high dose azathiaprine therapy (Wilkerson et al., 1977). The fact that steroids are used successfully in the treatment of MG may reflect a suppression of complement activity or some cellular component of the immune system. The role of suppressor and helper T-lymphocytes in MG should be studied in order to determine if the disease is associated with a defect in suppressor T cells, as has been suggested in certain animal models of autoimmunity (de Heer and Edgington, 1974). The remissions in MG sometimes observed during menstruation or pregnancy (Viets et al., 1942) may well be an indication of fluctuations in suppressor T cell activity.

In order to evaluate the specificity of the autoantibody we have repeated many of our assays using AChR from human muscle purified by the methods which we have used for rat AChR. We do not find a significant increase in sensitivity over AChR from denervated rat muscle. Our preliminary results in this respect do not warrant the use of AChR from human muscle which is difficult to obtain. Denervated rat muscle is a richer source of AChR and 5 rat legs can provide enough purified receptor to carry out both types of assay on more than 50 patients. The advantage of a highly purified and stable AChR preparation as used in our study may further increase the sensitivity of the assay, especially in the measurement of toxin binding inhibition.

It has recently been reported that anti-AChR antibody is found in the cerebrospinal fluid of MG patients (Lefvert and Pirskanen, 1977) and abnormal EEG has been measured in rabbits with experimental autoimmune MG induced by immunization with AChR from *Torpedo* (Fulpius et al., 1977). These findings suggest that the anti-AChR antibody might affect the central nervous system of MG patients. We have used Triton X-100 extracts and a purified α-BgTX binding protein from rat brain instead of muscle AChR in our assay systems. The antibody from MG patients does not interact with this rat brain protein and we do not find inhibition of toxin binding or precipitation of toxin-protein complexes. One explanation of this finding is that the brain and muscle receptor for α-BgTX may have different immunological properties.

SUMMARY

In order to determine the role of humoral factors in myasthenia gravis (MG) the anti-acetylcholine receptor antibody was measured in the sera of a series of patients and controls. A radioimmunoassay and a toxin-inhibition assay were used to assess the interaction of the autoantibody with receptor purified from denervated rat muscle. Antibody was detected in 88% of MG patients and was not present in controls. In some patients the antibody was assayed at different time during the course of the disease and after the drug-induced remission of symptoms. Antibody titre was found to increase with the progressive development of muscle weakness but remained very high after muscle strength returned to normal during steroid treatment. It is suggested that autoantibody may be necessary but is not sufficient to account for the pathogenesis of MG.

ACKNOWLEDGEMENTS

This work was supported in part by grants 5T 32 MH-14286 and NS-11356 from the NIH.

REFERENCES

Abramsky, O., Aharonov, A., Teitelbaum, D., Fuchs, S. (1975) Myasthenia gravis and acetylcholine receptors: effect of steroids in clinical course and cellular immune response to acetylcholine receptors. *Arch. Neurol.*, 32, 684–687.

Aharonov, A., Tarrab-Hazdai, R., Abramsky, O., Fuchs, S. (1975) Humoral antibodies to acetylcholine receptors in patients with myasthenia gravis. *Lancet,* ii, 340–342.

Almon, R.R., Andrew, C.G., Appel, S.H. (1974) Serum globulin in myasthenia gravis inhibition of α-bungarotoxin binding to acetylcholine receptors. *Science*, 186, 55–57.

Almon, R.R., Appel, S.H. (1976) Serum acetylcholine-receptor antibodies in myasthenia gravis. *Ann. N.Y. Acad. Sci.*, 274, 235–243.

Appel, S.H., Anwyl, R., McAdams, M.W., Elias, S. (1977) Accelerated degradation of acetylcholine receptor from cultured rat myotubes with myesthenia gravis sera and globulins. *Proc. nat. Acad. Sci. (Wash.)*, 74, 2130–2134.

Bender, A.N., Ringel, S.P., Engel, W.K., Daniels, M.P., Vogel, F. (1975) Myasthenia gravis: a serum factor blocking acetylcholine receptors of the human neuromuscular junction. *Lancet*, i, 607–609.

Bradley, R.J., Howell, J.H., Romine, W.O., Carl, G.F., Kemp, G.E. (1976) Characterization of a nicotinic acetylcholine receptor from rabbit skeletal muscle and reconstitution in planar phospholipid bilayers. *Biophys. biochem. Res. Commun.*, 68, 577–584.

Changeux, J.P., Benedetti, L., Bourgeois, J.P., Brisson, A., Cartaud, J., Devaux, P., Grunhagen, H.H., Moreau, M., Popot, J.L., Sobel, A., Weber, M. (1976) Some structural properties of the cholinergic receptor protein in its membrane environment relevant to its functions as a pharmacological receptor. *Cold Spring Harbor Symp. quant. Biol.*, 40, 211–230.

De Heer, D.H., Edgington, T.S. (1974) Clonal heterogeneity of the anti-erythrocyte autoantibody responses of NZB mice. *J. Immunol.*, 113, 1184–1189.

Engel, A.G., Lambert, E.H., Howard, F.M. (1977) Immune complexes (IgG and C3) at the motor end-plate im myasthenia gravis: ultrastructural and light microscopid localization and electrophysiological correlations. *Mayo Clin. Proc.*, 52, 267–280.

Fambrough, D.M., Drachman, D.B., Satyamurti, S. (1973) Neuromuscular junction in myasthenia gravis: decreased acetylcholine recpetors. *Science*, 182, 293–295.

Fulpius. B.W., Fontana, A., Cuenoud, S. (1977) Central-nervous-system involvement in experimental autoimmune myasthenia gravis. *Lancet*, i, 350–351.

Kemp, G.E., Morley, B.J., Robinson, G., Bradley, R.J. (1978) Purification and characterization of the nicotinic acetylcholine receptor from normal and denervated muscle, submitted for publication.

Lee, C.Y. (1972) Chemistry and pharmacology of polypeptide toxins in snake venoms. *Ann. Rev. Pharmacol.*, 12, 265–286.

Lefvert, A.K., Pitskanen, R. (1977) Acetylcholine-receptor antibodies in cerebrospinal fluid of patients with myasthenia gravies. *Lancet*, i, 351–352.

Lennon, V.A., Seybold, M.E., Lindstrom, J.M., Cochrane, C., Ulevitch, R. (1978) Role of complement in the pathogenesis of experimental autoimmune myasthenia gravis. *J. exp. Med.*, 147, 973–983.

Lindstrom, J.M., Lennon, V.A., Seybold, M.E., Whittingham, S. (1976) Experimental autoimmune myasthenia gravis: biochemical and immunological aspects. *Ann. N.Y. Acad. Sci.*, 274, 254–274.

Lisak, R.P., Falk, G.A., Heinz, R.G., Kies, M.W., Alvord, E.C., Jr., (1970) Dissociation of antibody production from disease suppression in the inhibition of allergic encephalomyelitis by myelin basic protein. *J. Immunol.*, 104, 1435–1445.

Mittag, T.,-Kornfeld, P., Tormay, A., Woo, C. (1976) Detection of anti-acetylcholine receptor factors in serum and thymus from patients with myasthenia gravis. *New Engl. J. Med.*, 294, 691–694.

Monnier, V.M., Fulpius, B.W. (1977) A radioimmunoassay for the quantitative evaluation of anti-human acetylcholine receptor antibodies in myasthenia gravis. *Clin. exp. Immunol.*, 29, 16–22.

Osserman, K.E. (1958) *Myasthenia gravis*. Grune and Stratton, New York.

Richman, D.P., Patrick, J., Arnason, B.G.W. (1976) Cellular immunity in myasthenia gravis: response to purified acetylcholine receptor and autologous thymocytes. *New Engl. J. Med.*, 294, 694–698.

Schlager, S.I., Ohanian, S.H., Boros, T. (1977) Kinetics of hormone-induced tumor cell resistance to killing by antibody and complement. *Cancer Res.*, 37, 765–770.

Toyka, K.V., Drachman, D.B., Griffin, D.E., Pestrouk, A., Winkelstein, J.A., Fishbeck, K.H., Kao, I. (1977) Myasthenia gravis: study of humoral immune mechanism by passive transfer to mice. *New Engl. J. Med.*, 196; 125–131.

Viets, H.R., Schwab, R.S., Brazier, M.A.B. (1942) The effect of pregnancy on the course of myasthenia gravis. *Arch. Neurol. Psychiat.*, 47, 1080–1084.

Wilkerson, L.D., Lisak, R.P., Zweiman, B., Silberberg, D.H. (1977) Antimyelin antibody in multiple sclerosis: no change during immunosuppression. *J. Neurol. Neurosurg. Psychiat.*, 40, 872–875.

Acetylcholine in Intercostal Muscle from Myasthenia Gravis Patients and in Rat Diaphragm after Blockade of Acetylcholine Receptors

P.C. MOLENAAR, R.L. POLAK, R. MILEDI, S. ALEMA *, A. VINCENT
and J. NEWSOM-DAVIS

(P.C.M.) Department of Pharmacology, Sylvius Laboratories, Leiden University Medical Centre, 2333 AL Leiden, (R.L.P.) Medical Biological Laboratory/T.N.O., Rijswijk-Z.H. (The Netherlands); (R.M. and S.A.) Department of Biophysics, University College London, London WC1E 6BT and (A.V. and J.N.D.) Department of Neurological Science, Royal Free Hospital, London NW3 2QG (United Kingdom)

INTRODUCTION

Myasthenia gravis (MG) is a disease in which neuromuscular transmission to skeletal muscle is impaired. In their study on myasthenic muscle Elmqvist et al. (1964) found that MEPPS — caused by the release of single quanta of acetylcholine (ACh) from motor nerve terminals — were abnormally small. More recently, it was discovered that ACh receptors (AChRs) are reduced in number at myasthenic end-plates (Fambrough et al., 1973; Green et al., 1975a). Further, an immunoglobulin has been demonstrated in the sera of myasthenic patients which binds to solubilized AChRs in vitro (Almon et al., 1974; Mittag et al., 1976; Lindstrom et al., 1976). These findings indicate that MG is an autoimmune disease in which the sensitivity of the muscle fibre membrane to ACh is reduced by a specific antibody. In fact, immunization of animals against AChR protein purified from various sources leads to MG-like symptoms (see for instance: Patrick and Lindstrom, 1973; Sugiyama et al., 1973; Heilbronn and Mattson, 1974; Green et al., 1975b) presumably by cross reaction of antibodies raised against AChR with endogenous receptors in striated muscle.

In our studies we have investigated whether the reduction in the number of AChRs in MG is accompanied by changes in the synthesis and release of ACh. We have also investigated rat diaphragms, whose AChRs were blocked in vitro by α-bungarotoxin (Chang and Lee, 1963; Lee et al., 1967; Miledi and Potter, 1971), or by immunization with AChR purified from *Torpedo* electroplax (Patrick and Lindstrom, 1973; Green et al., 1975b; Alema et al., 1978).

Parts of this work have been reported elsewhere (Ito et al., 1976; Miledi et al., 1978; Ito et al., 1978b; Cull-Candy et al., 1978b).

---

* Present address: Laboratory of Cellular Biology C.N.R., Roma (Italy).

METHODS

Human muscle. The biopsies were performed under general anaesthesia except in a few cases (congenital MG) in which local anaesthesia was used. Parasternal intercostal muscle was obtained from 15 patients with MG during thymectomy. In all "control" patients, in 5 patients with congenital MG and in

TABLE I

ACETYLCHOLINE CONTENT OF HUMAN INTERCOSTAL MUSCLE

Patient	ACh content fresh muscle $(pmol \cdot g^{-1})$	MEPP size (mV)	Age (y)	Sex	Clinical state (b)
Acquired myasthenia gravis					
1	719 (1)	0.59 [c]	28	M	IV
2	500 ± 73 (2)	0.73 [d]	27	M	II B
3	360 ± 45 (2)	0.26	18	M	II B
4	100 ± 15 (2)	0.19	27	F	II B
5	280 ± 37 (6)	0.23 [c]	18	M	III
6	290 ± 33 (6)	0.44 [c]	54	M	IV
7	340 ± 55 (2)	0.26	54	F	IV
8	710 ± 127 (3)	0.33 [c]	36	F	II B
9	280 ± 16 (4)	0.17	48	M	II B
10	460 ± 53 (2)	0.24 [c]	41	M	II A
11	210 ± 33 (3) [a]	0.18 [c]	27	M	IV
12	360 ± 19 (2) [a]	0.15 [c]	27	M	IV
13	840 (1)	0.25 [c]	32	F	II A
Mean ± S.E.M. [e]	420 ± 61				
Congenital myasthenia gravis					
1	160 ± 16 (4)	0.07 [c]	13	M	II B
2	130 ± 4 (4)	–	25	M	II B
3	130 ± 16 (4)	0.15 [c]	17	M	II A
4	240 ± 22 (3)	0.33 [c]	13	M	II A
5	300 ± 44 (2)	0.46 [c]	17	M	II B
Mean ± S.E.M. [f]	190 ± 34				
Controls					
1	250 ± 25 (2)	1.03 [c]	56	M carcinoma	
2	120 ± 10 (2)	0.70	56	M carc. of bronchus	
3	170 ± 8 (2)	0.64	64	M carc. of bronchus	
4	190 ± 40 (2)	0.64	41	M achalasia	
5	240 ± 60 (6)	0.94 [c]	63	M carcinoma	
6	150 ± 16 (6)	1.06 [c]	65	M carcinoma	
7	180 ± 33 (2)	–	57	M carc. of bronchus	
8	120 ± 12 (2)	–	57	M carc. of bronchus	
9	290 ± 20 (4)	–	61	M neurofibroma	
Mean ± S.E.M.	190 ± 20				

ACh was determined in 1–6 groups of tendon-to-tendon muscle bundles; each group normally consisted of 3–4 bundles (40–100 mg).
[a] Lateral biopsy; [b] clinical state in MG was classified ocular (I); mild generalized (II A); moderate generalized (II B); acute (III); severe acute (IV); severe chronic (V); for control patients the diagnosis is stated (cf. Ito et al., 1976, Table I).
[c] MEPP amplitude in neostigmine bromide (5 μM); [d] after DFP (10 μM).
[e] Significantly different from control ($P_2 < 0.005$); [f] significantly different from acquired MG ($P_2 < 0.005$); Welch's multiple test.

2 cases of MG, external intercostal muscle which had been taken through a lateral incision, was used instead of parasternal intercostal muscle. Muscle from "control" patients – without clinical signs of muscle disease – was obtained during thoracotomy. For clinical details, see Table I.

Rat diaphragm. AChR was purified from the electric organ of *Torpedo marmorata* by affinity chromatography either on a curare-column (Green et al., 1975b) or on a cobratoxin-Sepharose derivative (Alema et al., 1978). Wistar or PVG/C (hooded) rats (150–200 g) were injected subcutaneously on multiple sites with AChR (75–150 μg) in complete Freund adjuvant; animals were injected two to three times at three weeks intervals and killed about three weeks after the last injection.

Hemidiaphragms were dissected together with the phrenic nerve. MEPPs were recorded from portions of muscle from the same animal as used for the estimations of ACh.

ACh estimation. ACh was extracted from human and rat muscle and subsequently estimated by mass fragmentography; deuterated ACh was used as an internal standard (for details see: Miledi et al., 1977).

RESULTS

Human muscle

ACh content. We investigated two types of the disease: acquired and congenital MG; in congenital MG antibodies against AChR have not been found (Cull-Candy et al., 1978b). As shown in Table I muscle from acquired MG patients contained about twice as much ACh (420 pmol/g) as muscle from control patients (190 pmol/g) or muscle from patients suffering from congenital MG (190 pmol/g). The difference was statistically significant not-withstanding a considerable variation between different patients or even between bundles of one biopsy. In one patient with Eaton–Lambert syndrome *, the muscle contained 190 pmol/g ACh (not shown in Table I), which is within the range of control values. Table I further shows that the MEPP amplitudes were reduced in muscles from patients suffering from acquired as well as from congenital MG.

Release of ACh. To measure the release of ACh it is necessary to prevent its enzymic hydrolysis; for this purpose DFP was used. However, when cholinesterase activity is blocked, muscles may show spontaneous "fasciculations", due to the generation of nerve impulses in some terminals (Masland and Wigton, 1940). Since the liberation of ACh by such activity might contribute appreciably to the apparent spontaneous release of ACh from the muscle into the medium, we measured the ACh released in the presence and in the absence of TTX, which is known to block nerve impulse activity without interfering with spontaneous ACh release (cf. Katz and Miledi, 1967).

---

* Also named myasthenic syndrome, since it is characterized by muscular weakness; it is often associated with carcinoma.

452

TABLE II

SPONTANEOUS RELEASE OF ACh FROM HUMAN INTERCOSTAL MUSCLE

	Acetylcholine release (pmol $\cdot$ g^{-1} $\cdot$ min^{-1})	
	Control	Acquired MG
No TTX	1.7 ± 0.5 (4)	2.8 ± 1.1 (4)
With TTX (1.6 μM)	1.4 ± 0.6 (3)	0.9 ± 0.1 (4)

Muscle bundles (100–200 mg) were incubated in Ringer (for details see legends of Fig. 1 and Table I). The rates of spontaneous ACh release were averaged in each case from four incubation periods. The spontaneous release, though varying considerably between different patients, remained at about the same level in each individual case (values are somewhat different from those in Table II, Ito et al., 1976, due to additional data).
Mean ± S.E.M. with number of patients in brackets.

As shown in Table II, TTX seemed to reduce the spontaneous release of ACh from human muscle, but this reduction was not significant. Moreover, no significant difference was observed between the amounts of ACh released spontaneously from myasthenic and control muscle. However, the situation was different when the release of ACh was stimulated by raising the KCl concentration in the incubation medium. In the absence of TTX, KCl initially produced a significantly higher release of ACh from myasthenic than from control muscle. During prolonged incubation the ACh release from myasthenic muscle decreased with time, whereas that from control muscle increased (Fig. 1A). Surprisingly, in the presence of TTX the KCl induced release of ACh from myasthenic muscle was equal to that from control muscle, and distinctly less than in the absence of TTX (Fig. 1B).

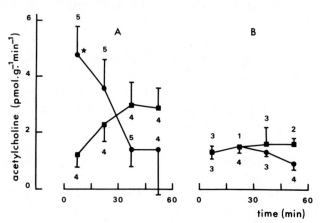

Fig. 1. KCl-induced release of ACh in intercostal muscle from acquired MG (circles) and control (squares) patients. Muscle bundles (100–200 mg) were incubated at about 20°C in 2 ml Ringer (pH 7.3) consisting of (in mM): NaCl 116; KCl 4.5; NaHCO$_3$ 23; NaH$_2$PO$_4$ 1; MgSO$_4$ 1; CaCl$_2$ 2; glucose 11; DFP 0.02. The medium was kept under an atmosphere of 95% O$_2$ and 5% CO$_2$. After 90 min (at time = 0 in the figure) the KCl was raised to 50 mM and, subsequently, the ACh was collected in 15-min periods. The release of ACh into the medium in the presence of 50 mM KCl has been corrected for the spontaneous release of ACh in normal Ringer which was determined in two preceding periods (not shown). Incubation was in the absence (A) or in the presence (B) of 1.6 μM TTX. Means ± S.E.M. with number of patients in brackets. ★ Value significantly different from control ($P_2 < 0.05$, Wilcoxon's Rank Sum Test).

TABLE III

EFFECT OF ACUTE AND CHRONIC RECEPTOR BLOCKADE ON ACh RELEASE AND CONTENT IN RAT DIAPHRAGM

	Evoked release [a] (3/sec, %) [b]	Evoked release [a] (50 mM-KCl, %) [b]	Spontaneous release (%) [b]	Content (%) [b]	MEPP amplitude (%) [b]
α-Bungarotoxin	170 ± 38 (2)	200 ± 18 (11) [c]	100 ± 13 (14)		Undetectable
Immunization with AChR	170 ± 26 (9) [c]	170 ± 26 (12) [c]	110 ± 15 (14)	100 ± 6 (12)	37 ± 1.7 (17) [c]

Rat hemi-diaphragms were incubated at about 20°C in 5 ml Ringer containing 20 μM DFP. After 90 min the preparations were rinsed and the incubation was continued for 60 min either in normal Ringer under supramaximal stimulation of the nerve at 3/sec or in Ringer containing 1.6 μM TTX and 50 mM KCl. When added, α-bungarotoxin (5 μg/ml) was present in the medium during the first 60–90 min of incubation. ACh was collected in 15- or 30-min periods. In the experiments with rats immunized against AChR-protein, littermates, most of which had been sham-immunized with Freund's adjuvant, served as controls. In the experiments with α-bungarotoxin the contralateral hemidiaphragms were used for controls. MEPPs were recorded in fibres form diaphragms not treated with DFP.

Means ± S.E.M. with number of animals in brackets.

[a] Evoked release of ACh was corrected for spontaneous release in two preceding 30-min periods.

[b] Percentages of the control values; the controls were (mean values): 1.0 (3/sec), 3.2 (50 mM KCl) and 0.4 pmol · min^{-1} (spontaneous release), 300 pmol (tissue content) and 0.7 mV (MEPP amplitude).

[c] Significantly different from controls (P_2 < 0.05, Student's t-test).

Synthesis of ACh. Since KCl stimulates the release of ACh, one might expect it to cause a reduction of the ACh content of the tissue. However, it was found that after incubation with 50 mM KCl the ACh content was virtually unchanged or even increased. Apparently synthesis must have occurred in the tissue, replacing the amounts of ACh lost to the medium. In the absence of TTX the synthesis (change in tissue ACh + ACh released into the medium) was about 600 pmol/g, both in myasthenic and control muscle.

Rat muscle

In isolated rat diaphragms after treatment with α-bungarotoxin end-plate potentials and MEPPs were completely abolished. On the other hand, in diaphragms from rats after immunization with AChR the MEPPs were not abolished, but their amplitude was reduced (see Table III). Furthermore, neuromuscular transmission was practically unaffected in immunized Wistar rats, though weakness was sometimes observed in PVG/C rats at a later stage of immunization (weak rats have not been used for the experiments in Table III). As shown in Table III both α-bungarotoxin and immunization with AChR caused about a 2-fold increase in the amount of ACh released under the influence of electrical (3/sec) or chemical (50 mM KCl) stimulation. Receptor blockade by α-bungarotoxin or by anti-AChR antibodies did not alter the spontaneous release of ACh, nor did it affect the ACh content of the muscles.

DISCUSSION

It was not possible to obtain control intercostal muscle from the same (parasternal) site as in MG patients; we therefore used external intercostal muscle taken from a lateral site in patients who were often ill and who were generally older than the MG patients. If we accept these limitations, the present findings appear to contradict the suggestion that a defect in transmitter synthesis or release is the cause of transmission failure in MG (see for instance: Desmedt, 1958; Elmqvist et al., 1964). In fact, we found that the ACh content of myasthenic muscle was double that of control muscle (see also Ito et al., 1976). This difference does not appear to be due to a difference in the number of end-plates in myasthenic and control muscle, which was about 15000/g of wet tissue in both types of muscle. Probably it was also not caused by the treatment with anticholinesterase drugs given to MG patients since increased ACh levels were also found in two patients who did not receive such drugs (patients Nos. 6 and 9 in Table I). Moreover, it should be noted that Chang et al. (1973) found that the ACh content of rat diaphragm was not altered after 7-days' treatment with neostigmine.

The severity of the clinical condition in MG patients was not clearly related to the ACh content of their muscles. A similar lack of correlation has been found between the clinical condition and MEPP amplitudes (Ito et al., 1978b; this paper), the number of AChRs or the plasma levels of anti-AChR antibodies (Lindstrom et al., 1976; Ito et al., 1978a).

It has been demonstrated (Miledi et al., 1977) that frog muscle contains a non-neural, probably muscular, store of ACh, apart from a larger store situated in the nerve terminals. It is possible that human muscle also contains a non-neural store of ACh, and that it is this store alone which is increased in MG. However, preliminary experiments in which the fibres were divided into end-plate containing and end-plate free segments, suggest that the increased total content of ACh of myasthenic muscle is due to an increase in the ACh content of the nerve terminals.

The finding that the high rate of KCl-induced release of ACh from myasthenic muscle is not sustained for 60 min although the ACh content of the muscle has not decreased at the end of this period, suggests that the ACh in the preparation is present in at least two compartments: a small compartment in which the ACh seems to be available for release and a larger one containing ACh which is unavailable for release; exhaustion of the small pool might not become apparent when the total ACh content is measured. An alternative possibility is that the release mechanism becomes "fatigued" during stimulation with KCl.

The finding that the KCl-induced release of ACh from myasthenic muscle was poorly sustained, is probably not related to the "decremental response" in MG since this is due to a phenomenon occurring at both normal and myasthenic end-plates. It is clinically evident in MG only because of the reduced sensitivity of the muscle membrane to ACh. Whether the exhaustion of ACh release is related to the aggravation of muscle weakness in MG during exercise, remains to be seen.

The finding that TTX depressed the KCl-induced release of ACh from myasthenic muscle suggests that Na^+-influx resulting from depolarization by K^+ may in some way influence ACh release, for instance by mobilizing internal calcium ions or by accelerating the synthesis of ACh, which is known to be dependent on Na^+ ions (Quastel, 1962; Birks, 1963; Bhatnagar and MacIntosh, 1967).

Our finding that the ACh content of muscle from patients with congenital MG (whose symptoms are indistinguishable from acquired MG) is within the normal range, and about half that of myasthenic muscle, raises the question whether this type of the disease is fundamentally different from acquired MG. This view is supported by the finding that congenital MG patients do not have elevated anti-AChR antibody titres (Cull-Candy et al., 1978b) and that their condition is not improved by total plasma exchange (Pinching et al., 1976), a procedure which brings about a considerable improvement in most patients with acquired MG (Newsom-Davis et al., 1978; Dau et al., 1977).

In diaphragms from immunized rats the MEPP amplitude was decreased to almost the same extent as in human myasthenic muscle. In spite of this practically all the fibres responded with an action potential to nerve stimulation. This contrasts with human myasthenic muscle and immunized rabbit muscle in which nerve impulses fail to trigger action potentials in a proportion of the fibres (cf. Green et al., 1975b). This difference may be due to a higher safety margin for neuromuscular transmission in rats than in rabbits or human muscle (see also Lambert et al., 1976), but it could also be that there is a more adequate adaptation (e.g. increase of ACh release, cf. Table III) to the loss of AChRs in the rat than in the other species.

456

Experimental MG in animals induced by immunization against AChR-protein has been proposed as an animal model for MG (see for review: Vincent, 1978). Notwithstanding several differences between the results obtained with human and rat muscle, we found that the evoked release of ACh was increased in muscles from both myasthenics and immunized rats. In either preparation the increase may have been due to an adaptation process compensating for the chronic loss of AChRs. In this connection it is of interest to note that Cull-Candy et al. (1978a) found that the number of quanta released by an action potential at low calcium concentrations was about six times normal in myasthenic intercostal muscle.

The increase in the evoked release of ACh from the rat diaphragm caused by α-bungarotoxin could also result from a compensatory mechanism which, of course, need not be identical to that in the immunized preparation: in the experiments with α-bungarotoxin the AChRs were blocked acutely and completely, in the immunized rats chronically and partially.

SUMMARY

ACh was extracted from intercostal muscles from patients with acquired and congenital myasthenia gravis (MG) and from control patients with no clinical signs of muscle disease. ACh was estimated by mass fragmentography. It was found that the concentration of ACh was about two times higher in muscle from acquired MG than from congenital MG or control patients. Muscle from acquired MG and control patients was also incubated in Ringer with 50 mM KCl in order to stimulate the release of ACh. During the first 15 min of incubation with KCl more ACh was released from myasthenic than from control muscle, but this difference was not sustained on prolonged incubation. Further, it was found that tetrodotoxin depressed the amount of ACh released into the medium in the presence of 50 mM KCl. Diaphragms from normal rats were treated with α-bungarotoxin, or taken from animals which had been immunized against the ACh receptor protein from *Torpedo marmorata*. It was found that from these preparations KCl released about twice as much ACh as from control diaphragms.

It is suggested that in myasthenic muscle and in diaphragms from immunized rats there is an adaptation process in the transmitter release compensating for the chronic loss of ACh receptors. If there is such a mechanism, it might be more efficient in rat than in man.

ACKNOWLEDGEMENTS

We are grateful to the patients for their collaboration and to Mr. M.F. Sturridge and Mr. J.R. Belcher for their help in obtaining the biopsies, to Drs. Y. Ito, S.G. Cull-Candy and O.D. Uchitel for measuring MEPPs and to Mrs. P. Braggaar-Schaap for technical assistance. Financial support by the Medical Research Council (to R.M.), the foundations I.B.R.O. and PROMESO (to P.C.M.) and FUNGO-ZWO (to R.L.P. and P.C.M.) is gratefully acknowledged.

S.A. was a recipient of a long-term fellowship from the European Molecular Biology Organization.

REFERENCES

Alema, S., Miledi, R. and Vincent, A. (1978) Nicotinic acetylcholine receptors and the role of specific antibodies in experimental myasthenia neuromuscular block. In *Organ Specific Immunity*, P.A. Miescher (Ed.), Menarini series on immunopathology, Vol. I, Schwabe, Basel, in press.

Almon, R.R., Andrew, C.G. and Appel, S.H. (1974) Serum globulin in myasthenia gravis: inhibition of α-bungarotoxin binding to acetylcholine-receptors. *Science*, 186, 55–57.

Bhatnagar, S.P. and MacIntosh, F.C. (1967) Effect of quaternary bases and inorganic cations on acetylcholine synthesis in nervous tissue. *Canad. J. Physiol. Pharmacol.*, 45, 249–268.

Birks, R.I. (1963) The role of sodium ions in the metabolism of acetylcholine. *Canad. J. Biochem. Physiol.*, 41, 2573–2597.

Chang, C.C. and Lee, C.Y. (1963) Isolation of neurotoxins from the venom of Bungarus multicinctus and their modes of neuromuscular blocking action. *Arch. Int. Pharmacodyn.*, 144, 241–257.

Chang, C.C., Chen, T.F. and Chuang, S.-T. (1973) Influence of chronic neostigmine treatment on the number of acetylcholine receptors and the release of acetylcholine from the rat diaphragm. *J. Physiol. (Lond.)*, 230, 613–618.

Cull-Candy, S.G., Miledi, R. and Trautmann, A. (1978) Acetylcholine-induced channels and transmitter release at human end-plates. *Nature (Lond.)*, 271, 74–75.

Cull-Candy, S.G., Miledi, R., Molenaar, P.C., Newsom-Davis, J., Polak, R.L., Trautmann, A. and Vincent, A. (1978), in preparation.

Dau, P.C., Lindstrom, J.M., Cassel, J.K., Denys, E.H., Shev, E.E. and Spitter, L.E. (1977) Plasmapheresis and immunosuppressive drug therapy in myasthenia gravis. *New Engl. J. Med.*, 297, 1134–1140.

Desmedt, J.E. (1958) Myasthenia-like features of neuromuscular transmission after administration of an inhibitor of acetylcholine synthesis. *Nature (Lond.)*, 182, 1973–1974.

Elmqvist, D., Hofmann, W.W., Kugelberg, J. and Quastel, D.M.J. (1964) An electrophysiological investigation of neuromuscular transmission in myasthenia gravis. *J. Physiol. (Lond.)*, 174, 417–434.

Fambrough, D.M., Drachman, D.B. and Satyamurti, S. (1973) Neuromuscular junction in myasthenia gravis: Decreased acetylcholine receptors. *Science*, 182, 293–295.

Green, D.P.L., Miledi, R., Perez de la Mora, M. and Vincent, A. (1975a) Acetylcholine receptors. *Phil. Trans. roy. Soc. Lond. B*, 270, 551–559.

Green, D.P.L., Miledi, R. and Vincent, A. (1975b) Neuromuscular transmission after immunization against acetylcholine receptors. *Proc. roy. Soc. Lond. B*, 189, 57–68.

Heilbronn, E. and Mattson, Ch. (1974) The nicotinic cholinergic receptor protein: improved purification method, preliminary amino acid composition and observed auto-immuno response. *J. Neurochem.*, 22, 315–317.

Ito, Y., Miledi, R., Molenaar, P.C., Vincent, A., Polak, R.L., van Gelder, M. and Newsom-Davis, J. (1976) Acetylcholine in human muscle. *Proc. roy. Soc. Lond. B*, 192, 475–480.

Ito, Y., Miledi, R., Molenaar, P.C., Newsom-Davis, J., Polak, R.L. and Vincent, A. (1978a) Neuromuscular transmission in myasthenia gravis and the significance of anti-acetylcholine antibodies. In *Biochemistry of myasthenia gravis and muscular dystrophy*, R. Marchbanks and J. Lunt (Eds.), *Academic Press*, pp. 89–109.

Ito, Y., Miledi, R., Vincent, A. and Newsom-Davis, J. (1978b) Acetylcholine receptors and electrophysiological function in myasthenia gravis. *Brain*, 101, 345–368.

Katz, B. and Miledi, R. (1967) Tetrodoxin and neuromuscular transmission. *Proc. roy. Soc. Lond. B*, 167, 8–22.

Lambert, E.H., Lindstrom, J.M. and Lennon, V.A. (1976) End-plate potentials in experimental autoimmune myasthenia gravis in rats. *Ann. N.Y. Acad. Sci.*, 274, 300–318.

Lee, C.Y., Tseng, L.F. and Chiu, T.H. (1967) Influence of denervation on localization of neurotoxins from elapid venoms in rat diaphragms. *Nature (Lond.)*, 215, 1177–1178.

Lindstrom, J.M., Lennon, V.A., Seybold, M.E. and Whittingham, S. (1976) Experimental

458

autoimmune myasthenia gravis and myasthenia gravis: biochemical and immuno-chemical aspects. *Ann. N.Y. Acad. Sci.,* 274, 254–274.

Masland, R.L. and Wigton, R.S. (1940) Nerve activity accompanying fasciculation produced by neostigmine. *J. Neurophysiol.,* 3, 269–275.

Miledi, R. and Potter, L.T. (1971) Acetylcholine receptors in muscle fibres. *Nature (Lond.),* 233, 599–603.

Miledi, R., Molenaar, P.C. and Polak, R.L. (1977) An analysis of acetylcholine in frog muscle by mass fragmentography. *Proc. roy. Soc. Lond. B,* 197, 285–297.

Miledi, R., Molenaar, P.C. and Polak, R.L. (1978) α-Bungarotoxin enhances transmitter 'released' at the neuromuscular junction. *Nature (Lond.),* 272, 641–643.

Mittag, T., Kornfeld, P., Tormay, A. and Woo, B. (1976) Detection of anti-acetylcholine receptor factors in serum and thymus from patients with myasthenia gravis. *New. Engl. J. Med.,* 294, 691–694.

Newsom-Davis, J., Pinching, A.J., Vincent, A. and Wilson, S.G. (1978) Functioning of circulating antibody to acetylcholine receptor in myasthenia gravis investigated by plasma exchange. *Neurol. (Minneap.),* 28, 266–272.

Patrick, J. and Lindstrom, J. (1973) Autoimmune response to acetylcholine receptor. *Science,* 180, 871–872.

Pinching, A.J., Peters, D.K. and Newsom-Davis, J. (1976) Remission of myasthenia gravis following plasma exchange. *Lancet,* ii, 1373–1376.

Quastel, D.M.J. (1962) Ph.D. Thesis, McGill University, Montreal, Quebec.

Sugiyama, H., Benda, P., Meunier, J.-C. and Changeux, J.-P. (1973) Immunological charac-terisation of the cholinergic receptor protein from *Electrophorus electricus. FEBS Lett.,* 35, 124–128.

Vincent, A. (1978) Experimental autoimmune myasthenia gravis. In *Neuroimmunology,* F.C. Rose (Ed.), Blackwell, Oxford, in press.

Short Review of Progress in Experimental Myasthenia Gravis, Bearing on the Pathogenesis of Myasthenia Gravis

EDITH HEILBRONN

Unit of Biochemistry, National Defence Research Institute, Department 4, S-172-04 Sundbyberg (Sweden)

INTRODUCTION

This article summarizes some recent work on experimental autoimmune myasthenia gravis (EAM), and includes some new results on human myasthenia gravis (MG) from the Stockholm–Uppsala–Umeå myasthenia gravis group.

ROLE OF HUMORAL ACETYLCHOLINE RECEPTOR ANTIBODY

In EAM destruction of the postsynaptic part of the motor end-plate, which occurs during the inflammatory stage of the induced disease and leads to a simplification of this membrane area, is caused by antibodies to the used immunogen, the nicotinic acetylcholine receptor (nAChR). It is possible, as suggested by Martinez et al. (1977), that macrophage-associated cytophilic antibodies with nAChR specificity are formed. Another antibody triggered mechanism of importance in the pathogenesis of EAM (and of MG) may be an internalization of functional receptors, as recently observed in muscle cell cultures (Bevan et al., 1977; Anwyl et al., 1977). It is now established that the number of α-neurotoxin binding sites at the motor end-plate and the amplitude of miniature end-plate potentials are reduced in MG and in EAM (for recent reviews see Heilbronn and Stålberg, 1978; Drachman, 1978a,b). Membrane destruction, as well as further antibody attack may change autoreceptors and cause them to break immunotolerance which would lead to the formation of autoreceptor antibodies.

The destruction of the motor end-plate membrane and the decrease in the number of functional receptors triggered by receptor antibodies causes the characteristic muscle weakness in EAM. In MG, it is not entirely clear what causes membrane destruction, although antibody-triggered receptor degeneration probably occurs. In later stages of the diseases, remaining receptors at now simplified postsynaptic membranes may be more or less blocked by autoreceptor antibodies. The occurrence of such antibodies and of complement C3 has been demonstrated by Engel et al. (1977a). Transfer experiments with IgG from myasthenics administered to EAM-rabbits have shown that such IgG can cause acetylcholine (ACh) receptor blockade within hours, as inferred from increased muscle decrement i.e. the progressive decline of muscle action potentials evoked by repetetive stimulation of the motor nerve (Mattsson et al., 1977). However such a process may not occur readily at a healthy motor end-plate, where long-term presence of receptor antibodies may be necessary to first cause some of the earlier mentioned antibody effects. For example, babies of myasthenic mothers can have humoral receptor antibodies without any clinical symptoms (Lefvert et al., 1978).

SPECIFICITY OF ANTIBODIES. RECEPTOR SUBUNITS

The nicotinic ACh receptor is a macromolecule, probably with several poly-peptide chains, each of which contains carbohydrate. The chains (subunits) can be separated from each other after reduction and SDS treatment. They have been shown to be less antigenic then the native receptor macromolecule, but their antibodies crossreact with each other (Uddgård et al., 1977) and probably also with the receptor macromolecule (Lindstrom et al., 1978). EAM-like symptoms following immunization with receptor subunits have not been observed in rabbits, but Lindstrom et al. (1978) report some effects in rats, where the amount of extractable muscle ACh receptor was decreased. The nAChR subunits have different immunogenic capacity. It is possible that the nAChR, itself, can induce a number of receptor antibodies which may be directed towards different immunological sites. In fact, of circulating anti-bodies present in MG patients, about 8% seem to prevent α-neurotoxin binding to nAChR (Fischer and Heilbronn, unpublished; Zurn and Fulpius, 1977), whereas the rest bind to nAChR without blocking toxin binding. The amount of each different kind of receptor antibody could vary from MG-patient to MG-patient (and from EAM-animal to EAM-animal), and this variation could be partly responsible for the poor correlation sometimes found between titers of humoral nAChR antibodies and the severity of the disease in different patients. However, a rough correlation between these parameters seems to exist: for example, Lefvert et al. (1978) observed that MG patients with thymomas and patients in stage IV of the disease (according to the Oossermann-Oosterhuis classification) always have high titers of circulating receptor antibodies (see also Bradley et al., this volume, p. 441). However it might be of interest to differentiate between antibodies with varying immunological specificity in measurements of receptor antibody titers in MG; this should be possible when methods used for the preparation of receptor subunits are improved.

DETERMINATION OF ANTIBODY TITER IN PATIENTS

At present, total receptor antibody titer is usually determined. For this purpose it is best to use human skeletal muscle receptor, or at least mammalian skeletal muscle receptor, as it has been shown (Uddgård and Heilbronn, 1976) that the crossreactivity between electric tissue receptor and that from skeletal muscle is very low, and that the electric tissue receptor is unsuitable for anti-body determination in EAM and MG (Lindstrom et al., 1976). Antibody titers

TABLE I

TABLE I

COMPARISON OF NEUROPHYSIOLOGICAL AND IMMUNOCHEMICAL METHODS IN THE DIAGNOSIS OF PATIENTS SUSPECTED OF HAVING MYASTHENIA GRAVIS

Number of patients tested	Number of patients with abnormal values		
	nAChR antibodies	Single fiber electromyography	Decrement in distal or proximal muscle
27	16	23	14
4 *	3	2 (border line)	

* Localized MG suspected.

are generally determined using radioimmunoassay technique in which α-toxin-labelled solubilized human skeletal muscle receptor-antibody complexes are precipitated with anti-IgG antibody.

A comparison between electrophysiological parameters used in MG diagnosis and antibody titers has been made in a number of Swedish patients (Table I). It was found that determination of antibody titers is a useful and sensitive additional diagnostic tool (see also Lefvert et al., 1978).

PASSIVE TRANSFER OF EAM

The role of the antibody in the pathogenesis of EAM and MG is further illustrated by the demonstration that passive transfer of EAM can be achieved by repeated injection of IgG from MG-patients into CB-mice (Toyka et al., 1977); the most obvious effect was a reduced number of functional receptor sites and a reduced amplitude of miniature end-plate potentials. Clinical manifestations of muscle weakness are not always observed in this kind of experiment: for example, in healthy rabbits, transfer of MG IgG did not induce clinical or electrophysiological EAM symptoms (Hammarström et al., 1978).

ROLE OF THYMUS. OCCURRENCE OF MYOPATHY

The role of the thymus in EAM and MG is not clear. Receptor (nAChR) antibodies seem to crossreact with some factor in the normal and myasthenic human thymus, and that tissue has a very low, but saturable α-neurotoxin binding capacity that is reduced by tubocurarine (Törnquist and Heilbronn, unpublished). These observations suggest the presence of some thymic nAChR sites, and Engel et al. (1977b) reported evidence that ACh receptors exist in thymus epithelial cells. Recent experiments on normal and thymectomized CB-mice injected with electric tissue nAChR have shown that a thymus-dependent antibody response is promptly produced; however clinical and electrophysiological signs of EAM seldom occur, and even when they do, their induction requires repeated challenges (Hammarström et al., to be published). More regularly, however, sarcotubular aggregates are observed, and these often have inner tubules and amorphous material or electron dense cores; these changes only occur in type II muscle fibers (Thornell et al., 1977). These morphological changes may be specific for EAM-animals and in particular mice, although we have observed membrane proliferation in rabbits (Thornell et al., 1976). These observations might be relevant to the specificity of the receptor antibody and also to a possible connection between the end-plate and the sarcoplasmic system of the muscle fibre.

DEVELOPMENT OF NEW DRUGS FOR MG

EAM-animals may be used to test the effect of new drugs with therapeutic potential in MG. Recently, 4-aminopyridine, a compound known to increase the release of ACh from the nerve ending (Molgo et al., 1975), was shown to relieve muscle decrement in EAM-rabbits (Heilbronn et al., 1977); the effect lasted for several hours. 4-Aminopyridine also restores neuromuscular transmission in rats paralysed by botulinum toxin (Lundh et al., 1977a), and has been tested in patients with Eaton—Lambert syndrome (Lundh et al., 1977b).

CONCLUSIONS

Recent work on MG and EAM confirms the autoimmune nature of MG. Humoral receptor antibodies play an essential role in the pathogenesis of the diseases, and may act by different mechanisms to cause: (a) postsynaptic membrane destruction, (b) acetylcholine receptor degeneration and (c) direct block of receptors remaining in a sick end-plate. Receptor antibodies are probably directed against a number of receptor sites and their specificity for receptors may sometimes not be absolute; antibody-induced changes selective for type II muscle fibers have been observed in mice. More than one type of receptor antibody may cause EAM and may be present in MG.

REFERENCES

Anwyl, R., Appel, S.M. and Narahashi, I. (1977) Myasthenia gravis serum reduces acetylcholine sensitivity in cultured rat myotubes. *Nature (Lond.)*, 167, 262—263.
Bevan, S., Kullberg, R.W. and Heinemann, S.F. (1977) Human myasthenic sera reduce acetylcholine sensitivity of human muscle cells in tissue culture. *Nature (Lond.)*, 267, 263—265.
Drachman, D.B. (1978a) Myasthenia gravis I. *New Engl. J. Med.*, 298, 136—142.
Drachman, D.B. (1978b) Myastenia gravis II. *New Engl. J. Med.*, 298, 186—193.
Engel, A.G., Lambert, E.H. and Howard, I.M. (1977a) Immune complexes (IgG and C3) at the motor end-plate in myasthenia gravis. *Mayo Clin. Proc.*, 82, 267—280.
Engel, W.K., Trotter, J.L., McFarlin, D.E. and McIntosh, Ch.L. (1977) Thymic epithelial cell contains acetylcholine receptor. *Lancet*, 1, 1310—1311.
Hammarström, L., Heilbronn, E., Hilton-Brown, P., Lefvert, A.K., Matell, G., Mattsson, C., Smith, E., Stålberg, E. and Uddgård, A. (1978) Pathological mechanisms in myasthenia gravis: Transfer of myasthenic symptoms to rabbits with human IgG. *Exp. Neurol.*, in press.
Heilbronn, E., Mattsson, C. and Stålberg, E. (1977) Effect of 4-aminopyridine in experimental autoimmune myasthenic rabbits. *Excerpta Med. Int. Congr. Ser.*, No 427, p. 105.

Heilbronn, E. and Stålberg, E. (1978) The pathogenesis of myasthenia gravis. *J. Neurochem.*, 31, 5–11.

Lefvert, A.K., Bergström, K., Mattell, G., Osterman, P.O. and Priskanen, R. (1978) Determination of acetylcholine receptor antibody in myasthenia gravis: Clinical usefulness and pathogenetic implications. *J. Neurol. Neurosurg. Psychiat.*, 41, 394–403.

Lindstrom, J.M., Lennon, V.A., Seybold, M.E. and Whittingham, S. (1976) Experimental autoimmune myasthenia gravis and myasthenia gravis: Biochemical and immunochemical aspects. *Ann. N.Y. Acad. Sci.*, 274, 254–274.

Lindstrom, J., Einarsen, B. and Merlia, J. (1978) Immunization of rats with polypeptide chains from Torpedo acetylcholine receptor causes an autoimmune response to receptors in rat muscle. *Proc. nat. Acad. Sci., (Wash.)*, 75, 769–773.

Lundh, H., Leander, S. and Thesleff, S. (1977a) Antagonism of the paralysis produced by botulinum toxin in the rat. *J. Neuron. Sci.*, 32, 29–43.

Lundh, H., Nilsson, O. and Rosén, I. (1977b) 4-aminopyridine – a new drug tested in the treatment of Eaton – Lambert Syndrome. *J. Neurol. Neurosurg. Psychiat.*, 40, 1109–1112.

Martinez, R.D., Tarrab-Hazdai, R., Aharonov, A. and Fuchs, S. (1977) Cytophilic antibodies in experimental autoimmune myasthenia gravis. *J. Immunol.*, 118, 17–20.

Mattsson, C., Hammarström, L., Smith, E., Lefvert, A.K., Matell, G. Stahlberg, E. and Heilbronn, E. (1977) Transfer of myasthenic symptoms to rabbits with IgG from human patients. *Excerpta Med. Int. Congr. Ser.*, No. 427, p. 274.

Molgo, M.J., Lemeignan, M. and Lechat, P. (1975) Modifications de la libération du transmetteur à la jonction neuromusculaire de grenouille sons l'action de l'amino-4 pyridine. *C. R. Acad. Sci. (Paris) Ser.* D, 281, 1637–1639.

Thornell, L.E., Sjöström, M., Mattsson, C. and Heilbronn, E. (1976) Morphological observations on motor end-plates in experimental myasthenia in rabbits. *J. neurol. Sci.*, 29, 389–410.

Thornell, L.E., Hammarström, L., Smith, E., Mattsson, C. and Heilbronn, E. (1977) Sarcotubular aggregates in acetylcholine receptor immunized mice. *Excerpta Med. Int. Congr. Ser.*, No. 427, p. 105.

Toyka, K.V., Drachman, D.B., Griffin, D.E., Pestronk, A., Winkelstein, Y.A., Fischbeck, K.H. and Kao, I. (1977) Myasthenia gravis: study of humoral immune mechanisms by passive transfer to mice. *New. Engl. J. Med.*, 269, 125–131.

Uddgård, A. and Heilbronn, E. (1976) Comparison of chemical and immunological properties of acetylcholine receptors from Torpedo electric organ and mammalian muscle. *Abstr. 10th Int. Congr. Biochem.*, No. 12-7-009, p. 560.

Uddgård, A., Elfman, L. and Heilbronn, E. (1977) Isolation and biochemical and immunological comparison of acetylcholine receptor subunits from different animals. *Excerpta Med. Int. Congr. Ser.*, No. 427, p. 107.

Zurn, A.D. and Fulpius, B.S. (1977) Study of two different subpopulations of anti-acetylcholine receptor antibodies in a rabbit with experimental autoimmune myasthenia gravis. *Europ. J. Immunol.*, 7, 529–532.

The Effects of Inorganic Lead on Cholinergic Transmission

ALAN M. GOLDBERG

Department of Environmental Health Sciences, The Johns Hopkins University,
School of Hygiene and Public Health, Baltimore, MD 21205 (U.S.A.)

INTRODUCTION

Our present understanding of the biologic effects of lead is largely based on the study of individuals suffering from overt lead poisoning and on the experience gained in the management of persons exposed to lead through the occupational setting. The exposure to lead in these groups is considerably greater than that in the normal population encountering lead via food sources and environmental pollution.

About one of four young children who survive an attack of acute encephalopathy, as a result of lead poisoning, suffer severe and permanent neurological consequences. Byers and Lord (1943) have delineated the nature and severity of central nervous system injury that follows lead poisoning in early childhood. The severe forms of acute lead encephalopathy, including convulsive episodes, are fortunately becoming less common; the subtle neurological deficits and mental impairment which might include sensory loss and a decrease in the I.Q. are more difficult to assess. The children appear to suffer from motor incoordination, lack of sensory perception, impaired learning, short attention spans and are generally easily distracted (National Academy of Science, 1972). These behavioral changes are difficult to distinguish from the changes seen after organic brain damage or environmental deprivation.

ANIMAL MODELS OF LEAD EXPOSURE

The administration of inorganic lead during critical periods of development can lead to behavioral dysfunction in experimental animals. Behavioral changes have been described in mice (Silbergel and Goldberg, 1973), rats (Saureoff and Michaelson, 1973; Shih et al., 1976; Overman, 1977) and monkeys (Allen et al., 1974). The major behavioral effects observed have been changes in spontaneous motor activity, agressiveness, arousal, patterning of behavior, and vocalization.

One criticism raised against many experimental studies has been that the results on intact animals are largely based on experiments where high doses of lead, on an absolute scale have been administered. Actually, since the sensitivity of different animal species to lead is very different, the lead exposure necessary in experimental work also varies. It is apparent from the literature that one cannot use blood levels of lead in different species to compare the severity of exposure. Rodents with blood levels of lead exceeding 100 μg% do not present symptoms of overt intoxication, while man with the same blood level of lead is clearly intoxicated. Therefore, to suggest that exposure in any study is high or low can only be argued on the basis of species sensitivity.

Various laboratories use different regiments of lead administration. Generally, rat or mouse pups are exposed to lead through the mother's milk. While most investigators initiate lead exposure on the day pf parturition, others expose their animals on day two or three of life. In certain studies the mothers have been exposed to lead for their lifetime and the f2 generation, also exposed to lead, are the subjects of the investigation.

As indicated above, data on behavioral changes produced by the exposure to lead are not consistent between laboratories. Reports from two laboratories describe either hyperactivity or overactivity in the mouse (Silbergeld and Goldberg, 1973; Maker et al., 1975). With the rat, hyperactivity has been reported from at least five different laboratories (Sauerhoff and Michaelson, 1973; Golter and Michaelson, 1975; Shih et al., 1976; Grant et al., 1975; Overmann, 1977). Some studies from the same laboratories, however, as well as from other laboratories have not revealed this effect (Sobotka et al., 1975; Modak et al., 1975; Grant et al., 1976; Krehbiel et al., 1978). To evaluate the apparent contradiction, several features of these studies should be taken into account. First, the route and level of exposure have in most cases been different. Second, the methods of measuring the behavioral changes have not been standardized. Third, the nutritional status of the animals has not been controlled as regards the intake of calories and nutritional factors such as other trace metals. Fourth, the litter size and thus the nutrition and the psychological factors have not been uniform. With this lack of uniformity it is difficult to compare directly the studies that have been reported. However, it may be noted that the attenuated or paradoxical pharmacological effects of amphetamines in lead exposed animals, originally observed in mice (Silbergeld and Goldberg, 1974) have been reported in the rat independently of the appearance of hyperactivity (Shih et al., 1976; Sobotka et al., 1975). The effects on the cholinergic system have also been consistent, independent of the species studied.

In our recent studies, we have developed a model of lead exposure that does not retard growth or produce undernourishment. The protocol has worked well with rats and mice; other species have not been tested. On the day of parturition the mothers receive 5 mg/ml lead acetate in their drinking water. The mothers of control animals receive tap water. Litters are reduced to 3 pups each within 24 h of birth. Neonatal animals are thus exposed to lead through the milk of their mothers from postnatal days 1 through 21. They are weaned at 28 days and group-housed by sex. Animals exposed to lead using this protocol show no growth retardation.

EFFECTS OF LEAD ON THE CHOLINERGIC SYSTEM

Peripheral system

Kostial and Vouk (1957) demonstrated that lead added in vitro blocked ganglionic transmission in the superior cervical ganglion of the cat and produced a decrease in the release of acetylcholine (ACh); this effect of lead appeared to be related to the interaction of lead and calcium. Manalis and Cooper (1973), using the frog sartorius muscle preparation, demonstrated that lead increases miniature end-plate potential frequency and decreases end-plate potential amplitude. Goldberg and van den Bercken (unpublished) confirmed these observations but observed that in the hemidiaphragm of the rat, the end-plate potential amplitude was decreased but minature end-plate potential frequency was unchanged. Similar results in the rat hemidiaphragm have been reported by Bornstein and Pickett (1977). Silbergeld et al., (1974) demonstrated that in vitro lead can decrease the force of contraction of the diaphragm when the phrenic nerve is stimulated. Further, they demonstrated that the force of contraction was not impaired when the muscle was stimulated directly. The ability of lead to decrease the end-plate potential amplitude and the force of muscle contraction cannot be explained by a postsynaptic inhibition since lead dose not alter the postsynaptic sensitivity to ACh (Silbergeld et al., 1974). All of these results are consistent with the suggestion that lead added in vitro decreases the release of ACh.

Steady-state levels

Steady-state levels of choline and ACh have been measured in mice and rats exposed to lead by Carroll et al. (1977) and Shih and Hanin (1978). These groups failed to observe any changes in either compound. On the other hand, Modak et al. (1975) did observe a small but significant decrease in the levels of ACh in the diencephalon of the rat.

Enzymes

No changes in the activity of choline acetyltransferase have been observed after exposure to lead but cholinesterase activity was depressed by approximately 20% in all studies (Sobotka et al., 1975; Modak et al., 1975; Carrol et al., 1977). In some of these studies the decrease reached statistical significance. Whether this is a biologically important phenomenon is difficult to determine.

Release and turnover

Carroll et al. (1977) studied the potassium induced release of ACh from brain tissue. The question that was being asked was whether inhibition of ACh release by lead in vitro was a general property of lead. ACh release was induced by exposing the brain minces to elevated levels of potassium (35 mM) for periods up to 1 h. As previously shown, the release of ACh was linear for at least 1 h and was calcium dependent. In control mice the release of ACh was in the range of 300 nmol/g wet wt · h. In lead exposed animals the release was decreased by about 50%. These results demonstrate that chronic lead exposure inhibits the potassium induced release of ACh from brain tissue. Presumably, the impairment of ACh release is a direct effect of lead and not a secondary consequence of the lead exposure.

Shih and Hanin (1978) have presented evidence that lead exposure in the rat results in a decrease in the turnover rate of ACh. They exposed rats to lead acetate from birth. In their study there was a decrease in growth rate of the lead exposed animals. At approximately 45 days of age, control and lead treated rats were administered radiolabelled choline phosphate, a precursor of choline and ACh. Specific activities of ACh were not altered, but the specific activity of choline was significantly elevated in all brain areas studied as compared to age-matched controls. The in vivo ACh turnover rate in cortex, hippocampus, midbrain, and striatum was decreased by 33–45%. These findings provide evidence for an inhibitory effect of lead exposure on central cholinergic function in vivo.

Pharmacology

Many drugs produce unexpected pharmacological responses in lead exposed animals. Methylphenidate and amphetamine produced a paradoxical or attenuated response in lead treated animals. Although these drugs are generally thought of as aminergic stimulants they have many additional effects not related to their catecholaminergic action. Amphetamines have been shown to increase ACh release from brain. In attempting to see whether methylphenidate also had a cholinergic action we did the following experiment (Carroll et al., 1977). Lead exposed animals were treated with 40 mg/kg methylphenidate and killed 2 h later. The brains from control, lead exposed, and lead exposed animals given methylphenidate were removed and the potassium induced release of ACh measured. As in previous experiments, lead exposed animals had a decreased release of ACh; this decrease was completely reversed in the methylphenidate-treated animals. These data are consistent with those of Shih et al., (1976 and 1978) who previously reported evidence suggesting a cholinergic effect of methylphenidate. Collectively, these data suggest a cholinergic link in the mechanisms of action of methylphenidate.

SUMMARY

Low levels of inorganic lead chronically administered during critical periods of development result in behavioral alteration and neurochemical changes. Among the changes seen consistently are alterations in central and peripheral cholinergic function. Administration of several drugs that influence cholinergic metabolism influences the behavioral effects of lead. In the peripheral nervous system, lead in vitro has been shown to impair cholinergic function in several different experimental models. Lead inhibits the release of ACh from the superior cervical ganglion of the cat during stimulation of the preganglionic fibers, decreases the size of the end-plate potential in the frog and rat neuromuscular preparations, and reduces the force of muscle contraction during nerve stimulation. The pharmacological evidence obtained in studies on whole animals and direct measurements of ACh release and turnover indicate that the release of ACh in the central and peripheral nervous systems is impaired by both acute in vitro and by chronic in vivo exposure to lead.

ACKNOWLEDGEMENTS

The studies reported from the author's laboratory were supported by NIEHS grant 00454 and 00034.

REFERENCES

Allen, J.R., McWay, P.J. and Suomi, S.J. (1974) Pathological and behavioral changes in rhesus monkeys exposed to lead. *Environ. Health Perspect.*, 7, 239–246.

Bornstein, J.C. and Pickett, J.B. (1977) Some effects of lead ions on transmitter release at rat neuromuscular junctions. *Soc. Neurosci.*, 3, 370.

Byers, R.K. and Lord, E.E. (1943) Late effects of lead poisoning on mental development. *Amer. J. Dis. Childh.*, 66, 471–494.

Carroll, P.T., Silbergeld, E.K., and Goldberg, A.M. (1977) Alterations of central cholinergic function in lead-induced hyperactivity. *Biochem. Pharmacol.*, 26, 397–402.

Golter, M. and Michaelson, I.A. (1975) Growth, Behaviour and brain catecholamines in lead-exposed neonatal rats: a reappraisal. *Science*, 178, 359–361.

Grant, L.D., Howard, J.L., Alexander, S. and Krigman, M.R. (1975) Low level lead exposure: behavioral effects. *Environ. Health Perspect.*, 10, 267.

Grant, L.D., Breese, J.L., Howard, J.L., Krigman, M.R. and Mushak, P. (1976) Neurobiology of lead-intoxication in the developing rat. *Fed. Proc.*, 35, 503.

Kostial, K. and Vouk, V.B. (1957) Lead ions and synaptic transmission in the superior cervical ganglion of the cat. *Brit. J. Pharmacol.*, 12, 219–222.

Krehbiel, D., David, G., LeRoy, L. and Bowman, R. (1978) Absence of hyperactivity in lead-exposed developing rats. *Environ. Health Perspect.*, 18, 147–157.

Maker, H.S., Lehrer, G.M. and Silides. D.J. (1975) The effect of lead on mouse brain development. *Environ. Res.*, 10, 79–91.

Manalis, R.S. and Cooper, G.P. (1973) Presynaptic and postsynaptic effects of lead at the frog neuromuscular junction. *Nature (Lond.)*, 243, 354–355.

Modak, A.T., Weintraub, S.T. and Stavinoha, W.B. (1975) Effect of chronic ingestion of lead on the central cholinergic system in rat brain regions. *Toxicol. appl. Pharmacol.*, 34, 340–347.

National Academy of Sciences (1972) *Lead, Airborn Lead in Perspective*.

Overmann, S.R. (1977) Behavioral effects of asymptomatic lead exposure during neonatal development in rats. *Toxicol. appl. Pharmacol.*, 41, 459–472.

Sauerhoff, M.W. and Michaelson, I.A. (1973) Effect of inorganic lead on monoamines in brain of developing rat. *Pharmacologist*, 15, 165.

Shih, T.M. and Hanin, I. (1978) Effects of chronic lead exposure on levels of acetylcholine and choline and on acetylcholine turnover rate in rat brain areas in vivo. *Psychopharmacol.*, in press.

Shih, T.M., Khachaturina, Z.S., Barry III, H. and Hanin, I. (1976) Cholinergic mediation of the inhibitory effect of methylphenidate on neuronal activity in the reticular formation. *Neuropharmacol.*, 15, 55–60.

Shih, T.M., Khachaturina, Z. and Hanin, I. (1978) Involvement of both cholinergic and catecholaminergic pathways in the central action of methylphenidate: a study utilizing lead-exposed rats. *Psychopharmacol.*, in press.

Silbergeld, E.K. and Goldberg, A.M. (1973) A lead-induced behavioral disorder. *Life Sci.*, 13, 1275–1283.

470

Silbergeld, E.K. and Goldberg, A.M. (1974) Lead-induced behavioral dysfunction. An animal model of hyperactivity. *Exp. Neurol.*, 42, 146–157.

Silbergeld, E.K., Fales, J.T. and Goldberg, A.M. (1974) The effect of inorganic lead on the neuromuscular junction. *Neuropharmacol.*, 13, 795–801.

Sobotka, T.J., Brodie, R.E. and Cook, M.P. (1975) Psychophysiologic effects of early lead exposure. *Toxicol.*, 5, 175–191.

PART VII

ABSTRACTS

Development of Cholinergic Enzymes in the Brain of
Normal and Hypertensive Rats

J. BAGJAR, V., HRDINA and V. GOLDA

Purkyně Medical Research Institute and Department of Neurosurgery, Charles University,
Hradec Králové (Czechoslovakia)

There are many mechanisms involved in the development of hypertension. We studied the changes of acetylcholinesterase (AChE, EC 3.1.1.7) and choline acetyltransferase (ChAT, EC 2.3.1.6) multiple molecular forms (MMF) in normal and spontaneously hypertensive rats (SHR) during development. Soluble supernatant and an extract from the microsomal fraction of the brain were separated by polyacrylamide gel electrophoresis and AChE and ChAT MMF were detected (Bajgar, J. and Patočka, J., Collect. Czechoslow. chem. Commun., 42, 2723–2727, 1977; Franklin, G.I., J. Neurochem., 26, 639–641, 1976). Total activities of both enzymes were higher in SHR in comparison with controls. ChAT was separated into 3 MMF, numbered according to their electrophoretic mobility. The highest activity of MMF 3 with the lowest mobility was detected in the microsomal fraction, while in the supernatant the activities of MMF of ChAT were distributed equally. The increase of ChAT in SHR was caused by an increase in ChAT MMF 3 in both microsomal and soluble fractions. The other forms (1, 2) were not changed in SHR in comparison with controls. AChE activity was separated into 4 MMF. The highest activity was observed in the microsomal fraction for MMF 3 and 4, while in the supernatant these activities were lower. The increase of AChE activity in the microsomal fraction SHR was caused by an increase in AChE MMF 3 and 4, while MMF 1 and 2 were not changed in comparison with controls. The development of activities of AChE MMF in the soluble fraction was similar in SHR and control rats. The results indicate that cholinergic mechanisms can be involved in the development of hypertension and that the MMF of AChE and ChAT are not of the same physiological importance.

Release of Acetylcholine from a Vascular Perfused Hemidiaphragm – Phrenic
Nerve Preparation

G. BIERKAMPER

Department of Environmental Health Sciences, Division of Toxicology, The Johns Hopkins
University, Baltimore, MD 21205 (U.S.A.)

A method for studying the release of endogenous acetylcholine (ACh) has been developed using the isolated perfused rat hemidiaphragm-phrenic nerve (Bierkamper, G. and Goldberg, A., J. Electrophysiol. Tech., 1978, in press). Briefly, the left hemidiaphragm with 2 cm of intact phrenic nerve is dissected from 300–400 g rats and placed in 6 mM HEPES buffer (pH 7.4, $24 \pm 2°C$) containing (in mM): 114 NaCl, 3.5 KCl, 2.5 $CaCl_2$, 1.0 $MgCl_2$ and 11 glucose. After cannulation of the diaphragmatic vein, the preparation is mounted in a closed oxygenated chamber. The nerve and muscle are then perfused vascularly using a syringe-type infusion pump (30–50 μl/min). Infusion of a permeating dye indicates that the entire nerve and muscle are perfused in about 2 min. Diisopropylfluorophosphate (DFP 1.5×10^{-5} M) is added to the buffer to inhibit acetylcholinesterase (AChE). Greater than 98% inhibition has been verified by direct assay of AChE after homogenization of the muscle. Following either a period of rest (spontaneous release) or stimulation (7 Hz) of the nerve, the perfusate is collected and ACh is measured by a sensitive enzymatic radioassay. Spontaneous release was 0.84 ± 0.07, 0.95 ± 0.08 and 1.2 ± 0.07 pmol/min in the presence of 0, 30, and 60 μM choline, respectively. Mean release during 60–70 min of stimulation (total –

spont = stim) was 5.13 ± 0.62, 6.75 ± 0.07, and 9.78 ± 1.67 pmole/min with 0, 30 and 60 μM choline, respectively. The results clearly indicate that exogenous choline levels can significantly influence the amount of ACh released from the neuromuscular junction. Efflux of endogenous choline has also been measured simultaneously in some preparations. After 1 h of equilibration, efflux was approximately 100 pmole/min. This amount decreased slowly over time (15%/h) at rest, but dropped quickly during the first 20 min of stimulation, indicating the utilization of choline for ACh synthesis. The ability to perfuse this preparation vascularly not only extends viability to beyond 6 h, but also allows optimum control for the application and study of pharmacological and toxicological agents.

Cholinergic Mechanisms in the Visual System of Rat

V. BIGL, H. WENK, U. MEYER and H.-J. LÜTH

Paul Flechsig Brain Research Institute, Karl Marx University, Leipzig, and Institute of Anatomy, Humboldt University Medical School, Berlin (G.D.R.)

The rich cholinergic innervation of subcortical visual structures is not associated with the primary optic projection (Bigl, V. and Schober, W., Exp. Brain Res., 27, 211—219, 1977) but ACh is regarded as the transmitter released from reticular connections to the lateral geniculate nucleus (LGN). A direct monosynaptic AChE-containing pathway from the mesencephalic reticular formation (MRF) to the LGN has been described in the rat (Shute, C.C.D. and Lewis, P.R., Brain 90, 497—520, 1967), but could not be confirmed by recent horseradish peroxidase studies (Lüth, H.-J., Seidel, J. and Schober, W., Acta Histochem., 60, 91—102, 1977). To provide quantitative data on the structural basis of cholinergic modulation of visual input, this communication describes biochemical and quantitative histochemical results on changes in ChAT and AChE activities in some visual structures after partial or total deafferentation. The results indicate: (i) The cholinergic fibres found within the optic nerve are partly of central and partly of retinal origin. (ii) About 30% of ChAT activity of the LGN seems to be related to afferents from the MRF. Lesions placed in this structure at the level of Nc. ruber induce bilateral loss, and lesions at the level of the sensory root of the Vth nerve, uni(ipsi)lateral loss of AChE and ChAT activity in the LGN (dorsal and ventral part) and in the superior colliculus. (iii) The large proportion of ChAT remaining in the LGN seems to be localized in intrinsic neurones. The facilitatory influence of MRF stimulation on synaptic transmission in the LGN has been shown to be due to disinhibition rather than to postsynaptic excitation (Singer, W., Physiol. Rev., 57, 386—420, 1977) but, on the other hand, LGN cells are excited by microiontophoretically applied ACh. The synaptic mechanism by which ACh acts in LGN transmission thus remains to be unravelled.

Low Ionic Strength Aggregation and Collagenase Sensitivity of Some Acetylcholinesterase Molecular Forms from Avian and Mammalian Tissues

S. BON and J. MASSOULIÉ

Ecole Normale Supérieure, 75005 Paris (France)

Acetylcholinesterase (AChE) from avian and mammalian tissues can be separated by sedimentation analysis into several molecular forms, in a manner similar to the separation of AChE forms from fish electric organs (Hall, Z., J. Neurobiol., 4, 343—361, 1973; Vigny, M. et al., FEBS Lett. 69, 277—280, 1976). The avian and mammalian heavy molecular forms (16—20 S) become insoluble in low salt conditions, while most of the enzyme remains soluble. Several lines of evidence demonstrate that this phenomenon reflects ionic interactions with polyanionic compounds, very similar to those which are involved in the low-salt aggregation of the tailed molecular forms of AChE from the electric organ (Bon, S., Cartaud, J. and Massoulié, J., Eur. J. Biochem., 85, 1–14, 1978): (a) both phenomena are ionic strength dependent; (b) they are equally abolished by acetylation of the enzyme; and (c) polyanions interact with the isolated enzymes in a similar way.

We have found that the tailed forms of AChE from *Electrophorus* are modified by collagenase in a characteristic manner and that the collagen-like tail is responsible for the aggregation of the enzyme. Exactly parallel effects can be observed with the mammalian heavy forms: their sedimentation coefficient is increased and they become low-salt soluble, while the other molecular forms remain unmodified. It seems therefore possible that the mammalian and avian heavy forms possess a collagen-like element similar to the tail of *Electrophorus* AChE. In agreement with this idea, the apparent Stokes radii obtained by gelfiltration chromatography reveal a markedly larger asymmetry for the heavy forms than for the lighter ones. The tails could participate in the multimolecular ionic interactions which are probably involved in the attachment of AChE to the cell coats.

Exo-endocytosis at Neuromuscular Junctions during Intense Quantal Release of Neurotransmitter

B. CECCARELLI, F. GROHOVAZ and W.P. HURLBUT

Department of Pharmacology, CNR Center of Cytopharmacology, University of Milano, 20129 Milano (Italy)

Freeze-fracture technique was used to study frog neuromuscular junctions that had been treated with black widow spider venom, K^+ rich solutions or that had been stimulated indirectly. In the presynaptic membranes of nerve terminals that had been treated with venom for 8–15 min numerous dimples (E face) or protuberances (P face) were found immediately adjacent to the double row of particles that line the active zones. In contrast no dimples or protuberances were found in the nerve terminals that had been treated for 1 h and which were depleted of synaptic vesicles. Numerous dimples or protuberances were also found in the presynaptic membranes of nerve terminals that had been stimulated indirectly or treated with 20 mM K^+. However, under these conditions the dimples or protuberances were not confined to the double rows of particles, but were scattered all over the presynaptic membrane. Dimples or protuberances were still found after 1 h soaking in 20 mM K^+, showing that extensive recycling of synaptic vesicles had occurred. Recycling in K^+ rich solutions was confirmed by thin section EM and by experiments with extracellular tracers.

When muscles are soaked in Ca^{2+}-free solutions with 1 mM EGTA, the active zones become disorganized and isolated remnants of the double rows are found dissociated from the active zones. When venom was applied under these conditions dimples or protuberances still occurred mainly near the remnants of the double rows, even when these had migrated from their usual locations opposite the postjunctional folds. Thus, when transmitter release is stimulated and recylcing is vigorous, dimples or protuberances occur all over the presynaptic membrane whereas when transmitter release is stimulated, but recycling is impaired, as in venom treated terminals, dimples or protuberances occur only near the particles. These results show that quantal release of neurotransmitter occurs by exocytosis and that the sites of exocytosis are predetermined by the large intramembranous particles that normally line the edges of the active zone. On the other hand endocytosis of synaptic vesicles can occur either at the sites of exocytosis or nearby, at regions between the active zones.

Supported by MDA Grant awarded to B. Ceccarelli.

Relationship between the Steady-State Choline Level in Plasma and the Acetylcholine Level in Rat Brain Areas

S. CONSOLO, H. LADINSKY and R. GOMENI

Istituto di Ricerche Farmacologiche 'Mario Negri', 20157 Milan (Italy)

Female CD_1 rats weighing 200–220 g were given 100 mg/kg i.p. of choline iodide. The animals were killed at 20 and 40 min either by near-freezing or by microwave irradiation to the head (4.5 sec, 1.3 kW). Acetylcholine (ACh) and choline (Ch) were measured radio-enzymatically (Saelens, J.K., et al., Arch. int. Pharmacodyn. Ther., 1986, 279–286, 1970). At 20 min, the Ch level was increased in the striatum (20%) and hippocampus (35%) and returned to normal at 40 min. The ACh level was jot changed in either area.

ACh and Ch content in brain areas were determined when the plasma Ch level was raised 15- and 30-fold above normal and maintained in the steady state. A method for rapidly achieving any desired steady-state plasma Ch concentration (SSC) by two consecutive infusion rates was used (Wagner, J.G., Clin. Pharmacol. Ther., 16, 691–700, 1974). Ch was infused into the jugular vein of awake rats via an indwelling catheter inserted the day before the experiment. Initially, a constant-rate infusion at the faster rate, Q_1 is given over 20 sec, then the rate is abruptly changed to a lower rate, Q_2 by which steady-state plasma levels are reached and maintained.

A Q_1 of 4.4 μmol $\cdot$ kg^{-1} $\cdot$ sec^{-1} and a Q_2 of 0.37 μmol $\cdot$ kg^{-1} $\cdot$ sec^{-1} gave a calculated SSC of Ch of 390 nmol Ch $\cdot$ ml^{-1} and a measured SSC of 456.5 $\pm$ 13.3 nmol $\cdot$ ml^{-1} of plasma taken from the carotid artery at 100, 200 and 400 sec after the start of the infusion. At this SSC, (30 times above normal), the Ch content was unchanged in the striatum of rats killed by near freezing, while increases of 20–30% were found in the brain stem, hippocampus and the rest of the hemispheres.

ACh levels were unchanged in the brain stem and telencephalon but increased in the hippocampus (21%) and striatum (34%).

The prior administration of atropine sulfate (5 mg/kg, i.p., 30 min) prevented the increase in striatal ACh induced by Ch infusion.

When lowering the SSC of Ch to 217.0 $\pm$ 6.3 nmol $\cdot$ ml^{-1} (15 times above normal) (Q_1 = 2.2; Q_2 = 0.18 μmol $\cdot$ kg^{-1} $\cdot$ sec^{-1}), no changes in ACh content were observed in the hemispheres, striatum and brain stem. The Ch levels were increased only in the striatum and hippocampus.

The unusually high plasma SSC of Ch needed to raise central ACh stores cannot nearly be achieved in rats by Ch-supplemented diet or by the i.p. administration of Ch. In this extreme condition ACh appears to be increased not because its synthesis is increased, but rather by a negative feedback due to a direct cholinergic agonist property of Ch. The latter hypothesis is in accord with the lack of effect of Ch on the turnover of ACh (Eckernäs, S.-A. et al., Acta physiol. scand., 101, 404–410, 1977) in striatum and cortex.

Acetylcholinesterase Forms in Dystrophic Chickens. Normal Axonal Transport in Peripheral Nerves, Abnormal Concentration in Fast Twitch Muscles and Plasma

L. DI GIAMBERARDINO *, J.Y. COURAUD, J. LYLES *, I. SILMAN and E.A. BARNARD

*Department of Biochemistry, Imperial College, London SW 7 (England) and
* Département de Biologie, CEA de Saclay, 91190 Gif-sur-Yvette (France)*

Axonal transport and tissue concentration of AChE forms were studied here in the dystrophic chicken (line 413, Davis, California, maintained at Imperial College, London). Four acetylcholinesterase (AChE) molecular forms have been described in normal chickens: L_1, L_2, M and H, with respective sedimentation coefficients of approximately 5, 7, 11 and 20 S. L_1 and L_2 are conveyed slowly down the axon while M and H move rapidly and in both directions (Di Giamberardino L. and Couraud, J.Y., Nature (Lond.), 271, 170–172,

1978). In the present study, axonal transport of AChE forms in the sciatic and brachial nerves of dystrophic chickens was investigated by cutting or ligating the nerve and studying the accumulation rate of the enzyme forms on both sides of the lesion. The results of this analysis showed no differences in the accumulation rate of AChE and its forms in either sciatic or in brachial nerve, suggesting that the mechanism of axonal transport is normal in this line of genetically dystrophic chickens. Assays of AChE concentration in several tissues (optic lobe, ciliary ganglia, heart, plasma, muscles) of normal and dystrophic chickens showed that only fast twitch muscles and plasma had an abnormally high concentration of enzyme activity, as reported in the Table I.

TABLE I

AChE CONCENTRATION (nmol AcThCh split/min/mg wet weight)

	Normal	Dystrophic
Biceps *	0.81	1.91
Pectoral *	0.74	3.75
PLD *	0.59	3.38
Plasma	0.07	0.27

* AChE was extracted in 1 M NaCl, 1% Triton-X-100, 1 mM Tris, pH 7. AChE was assayed with acetylthiocholine in the presence of 10^{-7} M.

Comparison of the molecular forms of AChE in the pectoral muscle and in plasma of dystrophic and control chickens revealed pronounced differences in the patterns observed on sucrose gradients. The pectoral muscle extracts were made from fresh tissue in the presence of pepstatin and EGTA, which were added in order to retard proteolytic modification of the intact molecular forms. Gradient fractions were analysed after preincubation with 10^{-4} M iso-OMPA, to inhibit pseudocholinesterase activity. In normal chickens the H form, believed to be associated with the end-plate, is the predominant species, and only small amounts of lighter molecular forms can be detected. In the pectoral muscle of dystrophic chickens the overall elevation of AChE is mainly due to a large increase in the light forms, L_1 and L_2, although the H form is also elevated. In both cases the M form is present to only a minor extent. In plasma the higher concentration of AChE is mostly due to the elevation of the M form and to a minor degree of the L_2 form.

Supported by the Muscular Dystrophy Association (U.S.A.) and the Muscular Dystrophy Group of Great Britain and by EMBO.

Biochemical Effects of Presynaptic Snake Neurotoxins on Cholinergic Nerve Terminals

M.J. DOWDALL

Abteilung Neurochemie, Max-Planch-Institut für biophysikalische Chemie, Postfach 968, 3400 Göttingen (F.R.G.)

Taipoxin (TPX), notexin (NTX) and β-bungarotoxin (β-BgTX) are snake neurotoxins which block transmitter release at peripheral cholinergic synapses. Although the mechanism of this blockade is unknown recent studies have shown that these three toxins are potent inhibitors of high affinity choline transport (HAChT) by isolated nerve terminal sacs (T-sacs) from Torpedo electric organ (Dowdall, M.J., et al., Nature (Lond.), 269, 700–702, 1977). These toxins consist of at least one polypeptide chain of molecular weight 13,500 (13.5 K) and in each case this has sequence homology with purified phospholipase A_2. NTX is a single chain toxin, TPX a ternary complex of three 13.5 K chains whilst in β-BTX the 13.5 K chain is covalently linked by S–S bridges to an 8 K chain. *Enhydrina* myotoxin and *Notechis* II-5 are single chain, 13.5 K toxins which also block HAChT. For single chain toxins there is a good correlation between toxicity and blockade of HAChT, but inhibitory potency on HAChT is not directly related to phospholipase activity. Modification of a catalytically important histidine residue in the 13.5 K chain using *p*-bromophenylacyl bromide produces toxins with little or no phospholipase A_2 activity. TPX modification results in a considerable lowering of neurotoxicity but only a small change of inhibition of HAChT suggesting that physical interaction of this toxin with the nerve terminal is sufficient for blockade of HAChT. Other evidence shows that the toxins exert their effect by interacting with a common "target" site on the nerve terminal membrane. HAChT by synaptosomes from mammalian brain and *Sepia* optic lobes is quite insensitive to these toxins except in the case of TPX. Thus toxin target sites are either absent of modified (occult sites?) in central cholinergic nerve endings. The ubiquitous HAChT system is therefore probably not the primary target site for these toxins. The observation that TPX blocks adenosine uptake by *Torpedo* nerve terminals also supports this view.

Involvement of Thiamine in Synaptic Transmission

L. EDER, Y. DUNANT, M. BAUMANN and L. SERVETIADIS

Département de Pharmacologie, Ecole de Médicine, Genève (Switzerland)

The relationship of thiamine (vitamin B_1) and its phosphoric esters to cholinergic neurotransmission was examined using the electric organ of *Torpedo marmorata* (Eder, L., et al., Nature (Lond.), 264, 186–188, 1976).

The total amount of thiamine and its esters is 117 ± 51 nmol/g as compared to acetylcholine (ACh) which is 590 ± 77 nmol/g of electrogenic tissue. Thiamine, its monophosphate, diphosphate and triphosphate esters represent 37%, 25%, 6% and 32% of the total respectively; thiamine triphosphate is most abundant in the isolated cholinergic nerve endings. When incubated with increasing concentrations of external [^{14}C]thiamine (from 10^{-5} to 10^{-3} M) the vitamin is phophorylated with an apparent K_m of 2×10^{-4} M and approximate V_{max} of 0.76 nmol $\cdot$ g^{-1} $\cdot$ min^{-1}.

Electrophysiological and radiochemical methods were used to measure the effects of external thiamine or its analogue, oxythiamine, on synaptic transmission. Thiamine (5×10^{-3} M) first caused an increase in the amplitude of the discharge, followed by a subsequent decrease. Oxythiamine (10^{-4} M) induces a long lasting increase both in amplitude and duration of the discharge. These effects were found to be due to modifications in the amount of ACh released, as measured directly.

Histofluorescence analysis of thiamine shows that a strong fluorescence reaction is produced in the axons and nerve terminals and only a weak response takes place in the electroplaque cells (Eder, L., et al., J. Neurocytol., in press). From these observations we conclude that thiamine is involved in ACh metabolism and release at the presynaptic nerve endings.

Interactions of cholinergic drugs, GABA and Cyclic GMP in Picrotoxin-Induced Convulsions

V.P. GEORGIEV and N.D. LAMBADJIEVA

Institute of Physiology, Bulgarian Academy of Sciences and Sofia University, Sofia (Bulgaria)

The effects of cholinergic drugs and their interactions with GABA and cyclic GMP were studied by measuring the convulsive-seizure threshold (PCST) by timing the length of an intravenous infusion of the GABAergic antagonist picrotoxin required to produce a seizure.

The experiments were performed on albino mice and test drugs were administered intracerebroventricularly.

It was found that oxotremorine (10 μg/mouse, 10–30 min) and physostigmine (2, 5 and 10 μg/mouse, 10, 20, 40 and 70 min) significantly increased the PCST. GABA (75 and 100 μg/mouse, 15 min) also increased PCST but weakened the effect of oxotremorine and potentiated that of physostigmine.

PCST was increased at the 15th min after cGMP administration in a dose of 150 μg/mouse, afterwards it gradually decreased. PCST was lowered after cGMP in a dose of 20 μg/mouse. The effect of oxotremorine on PCST was potentiated and that of physostigmine weakened by cGMP (150 μg/mouse).

The following conclusions are drawn: (1) Cholinergic mechanisms may interfere with the convulsant action of the GABA antagonist picrotoxin; (2) GABA interacts with cholinergic mechanisms and thus may change the threshold for picrotoxin-induced seizures; (3) cGMP interferes with the action of picrotoxin; (4) However, cGMP also interacts with cholinergic mechanisms and may thus change the threshold for picrotoxin-induced seizures.

Cytochemical Studies of Enzymes Located at the Neuromuscular Junction in the Course of Denervation

JOLANTA GODLEWSKA-JĘDRZEJCZYK

Department of Histology and Embryology, Institute of Biostructure, Medical Academy, Warsaw (Poland)

Although the fast, significant decrease of the activity of muscle cholinesterase (ChE) seen between 1 and 3 days after motor nerve transection in mammals is well documented (Guth, L., Brown, W.C. and Watson, P.K., Exp. Neurol., 18, 443–452, 1967), the neural regulation of the end-plate concentration of the active sites of these enzymes is still obscure. The dynamics of change in the concentration of end-plate active sites of ChE in the soleus and plantaris muscles of rats was investigated in the present study after nerve transection, in relation to short and long nerve stumps.

In attacking this quantitative problem the autoradiographic method of using an irreversible labelled inhibitor of these enzymes, [3H]diisopropylfluorophosphate ([3H]DFP) was applied ((Barnard, E.A., Rymaszewska, T. and Wieckowski, J., (1971) in Cholinergic Ligands, (Triggle, D.J., Moran, J.F. and Barnard, E.A., Eds.) pp. 175–200, Academic Press, New York)). For the discrimination of acetylcholinesterase (AChE) active centres, the specific reactivator of this enzyme, pyridine-2-aldoxime methiodide, was used, in conjunction with [3H]DFP, whereas total ChE (i.e. AChE + pseudo-ChE) concentration was estimated by the protection of the active centres of these enzymes from [3H]DFP by their reversible inhibitor, eserine, in the grain-counting measurements over the end-plates. A significant decrease in the concentration of active sites of AChE at the end-plates in both muscles was observed 8 h after the sciatic nerve had been cut at the level of the sciatic foramen. Differences in the concentration of this enzyme caused by cutting the nerve at proximal and distal levels were observed (the latter giving a slower decline in AChE) during the first two days of the denervation process, in either type of end-plate. A decrease in the concentration of pseudo-ChE was seen between 3 to 5 days after the operation, and this was independent of the length of the nerve stump.

The changes in the concentration of AChE as early as 8 h after cutting the nerve close to the muscle, and the delay in this decrease when a long stump remained, indicate some form of trophic influence of the nerve on the maintenance of this enzyme at the motor end-plate.

Blood Cholinergic Parameters as a Potential Index of Central Cholinergic Function

I. HANIN, U. KOPP, C. NEVAR, D.G. SPIKER, J.F. NEIL and D.J. KUPFER

Western Psychiatric Institute and Clinic, Department of Psychiatry,
University of Pittsburgh School of Medicine,
Pittsburgh, PA 15261 (U.S.A.)

Experiments have been conducted in our laboratories to establish whether it is feasible to utilize red blood cell (RBC) and plasma choline (Ch) as an index of brain cholinergic function in vivo (Shih, T.-M. et al., Neurosci. Abstr. 3, 322, 1977). An animal model has been developed which has provided encouraging results justifying the initiation of parallel studies of RBC and plasma Ch profile in human subjects.

Our studies to date in human subjects indicate that: (1) within an individual, RBC and plasma Ch levels are maintained at an unusually constant level for months, provided external conditions and dietary factors are unchanged; (2) tremendous interindividual differences, on the other hand, do exist in RBC Ch levels, and they have been shown to span over a 20-fold range (Hanin, I., et al., Cholinergic Mechanisms and Psychopharmacology, D.J. Jenden, Ed., Plenum, New York, pp. 181–195, 1978); (3) these differences among individuals may furthermore be related in some manner to differences in the behavioral state of these individuals. Unipolar depressed, drug-free patients were shown to have significantly higher RBC Ch levels (51.7 ± 18.4 nmol/ml; n = 15) than normal controls (10.2 ± 1.0 nmol/ml; n = 10). Plasma Ch levels, at the same time, were similar in both groups; (4) finally, treatment of depressed patients with the antidepressant, amitriptyline (Elavil®), had no effect on RBC and plasma Ch levels, implying that endogenous factors play a major role in controlling the levels of RBC and plasma Ch in various individuals.

We believe that the higher observed levels of RBC Ch in some of our depressed patients are related in some very specific manner to the concept of a definitive involvement of cholinergic mechanisms in affective disease states. The nature and extent of this relationship are currently under further investigation in our laboratories.

Supported by NIMH grant No. MH 26320.

Ultrastructure of the Cholinergic Synapse Revealed by Quick-Freezing

JOHN E. HEUSER and SHELLEY R. SALPETER

Department of Physiology, University of California School of Medicine,
San Francisco, CA (U.S.A.)

This abstract outlines the use of quick-freezing to elucidate the ultrastructure of the cholinergic synapse and to capture the structural changes that underlie synaptic transmission. The machine developed for this purpose (Heuser, J.E., et al., Cold Spring Harbor Symp. quant. Biol., 40, 17–24, 1976), is designed to stimulate nerve-muscle preparations, record the resulting end-plate potentials, and at precisely timed intervals thereafter to freeze the tissues almost instantaneously by plunging them onto a pure copper block cooled to $4°$K with liquid helium. High speed resistance and capacitance measurements at the time of impact indicate that the surface of the muscle freezes within one msec.

Quick-freezing of the frog neuromuscular junction has been used to reconstruct the sequence of events that occurs during presynaptic nerve secretion. First, it has established that transmitter release is brought about by exocytosis of synaptic vesicles, by showing that such exocytosis occurs at exactly the same time as transmitter release, and that one synaptic vesicle opens for each quantum that is discharged (Fig. 1) (Heuser, J.E., et al., J. Cell Biol., in press). Second, quick-freezing has shown that discharged vesicle membrane begins to be retrieved from the plasma membrane within a few seconds, by a specialized form of endocytosis known as "coated vesicle" formation. These sorts of timing experiments add further evidence in favor of the original idea of Heuser and Reese (J. Cell Biol., 57, 315–344, 1973)

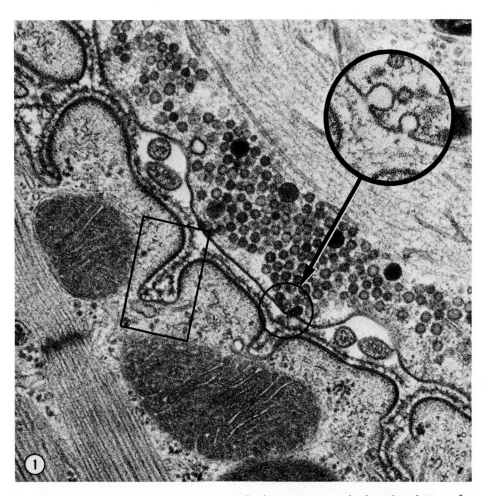

Fig. 1. Standard electron microscopy view of a frog neuromuscular junction that was frozen so fast that ice crystals are invisible, and then was OsO₄ fixed while still frozen, to obtain optimum preservation of membrane specializations and synaptic vesicles that characterize the cholinergic synapse. The magnified image shows what we take to be synaptice vesicle exocytosis caught by quick-freezing a nerve just at the moment it secreted acetylcholine in response to one nerve impulse. X40,000.

that synaptic vesicle membrane is recycled and used over again to make a new generation of synaptic vesicles (Heuser, J.E. and Reese, T.S., in The Nervous System, Handbook of Physiology, Vol. 1, Ed. E. Kandel, Amer. Physiol. Soc., Chapter 8, 1977).

The method of quick-freezing has also been used to stabilize the molecular architecture of the postsynaptic membrane at the cholinergic synapse (Heuser J.E. and Salpeter, S.R., J.Cell Biol., in press). The acetylcholine receptors have been visualized by etching water away from the surface of frozen electrocytes, the postsynaptic cells in the *Torpedo* electric organ (Fig. 2). The view thus obtained has illustrated that the acetylcholine receptors are arranged in a highly ordered lattice (Fig. 3). The forces which stabilize this lattice are unknown, but deep-etching reveals other aspects of the matrix around the postsynaptic membrane which may be involved. For one, it reveals the presence of a dense filamentous network in the cytoplasm beneath the postsynaptic membrane, which attaches to it periodically (Figs. 2, 3) and could possibly serve a role in immobilizing the receptors. Also, deep-etching reveals the structure of the basal lamina in the synaptic cleft (Figs. 1–3) and indicates its important function in cholinergic transmission, which is that of binding acetylcholinesterase and interconnecting the pre- and postsynaptic membranes.

480

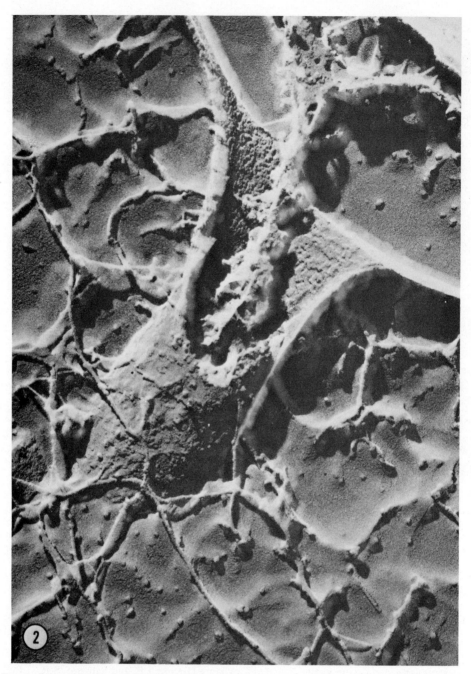

Fig. 2. View of a trumpet-shaped invagination of the postsynaptic membrane in the fish electrocyte cholinergic synapse, analogous to the muscle of folds at the neuromuscular junction (Boxed in Fig. 1). Before replication with platinum and carbon, this freeze fracture was freeze-dried for 3 min at $-100°C$ to remove water and bring structures inside and outside the membrane into view: the filamentous net work in the cytoplasm beneath, and the white "barbed-wire" looking basal lamina in the synaptic cleft above. Note: the basal lamina characteristically extends down into the center of the invagination, as it does at the neuro-muscular folds in Fig. 1. $\times 120,000$.

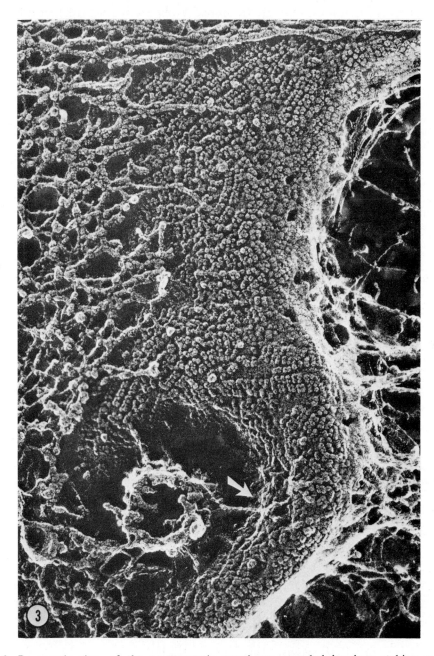

Fig. 3. Panoramic view of the postsynaptic membrane revealed by deep etching, rotary shadowing, and photographic reversal (Heuser, J.E. and Salpeter, S.R., J. Cell Biol., in press). With this technique, the platinum deposits which coat every surface irregularity look white, producing a 3-D perspective like that obtained from scanning electron microscopy. To the left of this figure, the lace-like basal lamina lies above the membrane and obscures it from view; and below, it assumes a ring-like appearance as it extends down into a dark postsynaptic invagination. Here and there, strands of the basal lamina extend out and attach to the membrane with claw-like feet (arrow). In the center of the figure, the basal lamina has been broken away to reveal the true external surface of the postsynaptic membrane with its characteristic clusters and linear arrays of "donuts" which are thought to be the acetylcholine receptor molecules. To the right, the membrane has been cleaved clear through, and deep-etching has revealed the underlying "trabecular meshwork" of cytoplasmic filaments which gird this membrane from beneath. ×175,000.

482

Yawning: A Central Cholinergic Response under Monoaminergic Modulation

B. HOLMGREN and R. URBÁ-HOLMGREN

*Department of Neurophysiology, Centro Nacional de Investigaciones Cietíficas,
La Habana (Cuba)*

Yawning is a cholinergic response in the sense that it occurs when the balance created by the release of a number of neurotransmitters is overcome by an excess of cholinergic activity.

Intraperitoneal injections of physostigmine or pilocarpine in neonatal rats induce stereotyped yawning; the effect declines towards the middle of the second week of life. In infant rats yawning is blocked by scopolamine, thus suggesting that muscarinic receptors are involved.

Phenoxybenzamine, a blocker of α-adrenergic receptor activity, increases significantly physostigmine-induced yawning thus suggesting that noradrenergic influences are inhibitory. A strong facilitatory influence of serotoninergic mechanisms on yawning is indicated by a 3- to 4-fold potentiation of physostigmine-induced yawning obtained with Lu 10-171, a potent and selective serotonin uptake inhibitor.

In summary, cholinergically-induced yawning in infant rats is presented as a simple experimental model of a central cholinergic response which is subject to negative and positive modulating influences by noradrenergic and serotoninergic mechanisms.

Control of Transmitter Release during Repetitive Activation of Motor Nerve Terminals

M.A. KAMENSKAYA

*Department of Human and Animal Physiology, Moscow State University,
117234 Moscow (U.S.S.R.)*

During repetitive activation of motor nerve terminals transmitter release at neuromuscular junctions is facilitated in amphibia but depressed in mammals. Experiments have been

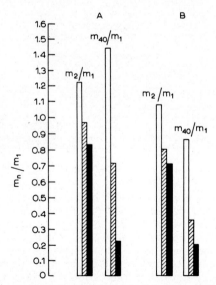

Fig. 1. The ratios of the quantal content of the 2nd (m_2) end-plate potential (EPP) to the 1st (m_1) and of the 40th (m_{40}) to the 1st in a train of 40 stimuli at 50 Hz. Open columns, frog sartorius msucle; hatched columns, mouse diaphragm; filled columns, human intercostal muscle. A and B, synapses with a mean quantal content of, respectively, 60 and 150 for a single EPP.

performed to test whether the development of facilitation or depression depends on the number of transmitter quanta released by single nerve impulses. Nerve terminals with a high or a low quantal content have been selected in frog, murine and human skeletal muscles and the number of quanta released from them by repetitive impulses has been investigated (Fig. 1). The results indicate that facilitation in amphibia and depression in mammals cannot be explained by a different quantal content of the first end-plate potential in the trains in the two classes of animals. It is suggested that the depression in mammals is due to (1) restricted store of available transmitter in the nerve terminals and (2) the lack of a reserve of latent release sites, with all release sites being involved in the response to the first impulse in the train.

Multiple Molecular Forms of Acetylcholinesterase

P. KÁSA and L. LEHOTAI

Central Research Laboratory, Medical University, Szeged (Hungary)

The molecular structure of acetylcholinesterase (AChE, EC 3.1.1.7) and its significance in neuronal function has been studied extensively. In the present work the multiple forms of AChE in the central and peripheral nervous system of the rat at different stages of postnatal development have been studied with polyacrylamide gel electrophoresis (PAGE), subcellular fractionations and optical and electronmicroscopic histochemistry.

It has been shown with PAGE that in the spinal cord and cerebellar cortex three bands are present which are subject to change during postnatal development. At birth, band No. 1 (fast migrating) predominates; later, band No. 2, and in the adult, band No. 3 (slow migrating) assume the greatest importance.

In order to correlate the electrophoretic forms of AChE in different tissues (spinal cord, cerebellar cortex, spinal ganglia and ischiadic nerve) at different developmental stages, subcellular fractionations and histochemical investigations were carried out. Subcellular fractionation ($100,000 \times g$; 60 min) of the tissues investigated shows the crude morphological localization of AChE activity. It was found that bands 1 and 2 correspond to a supernatant fraction, while band 3 was located in membranes. It was shown histochemically that the different molecular forms of AChE were associated with different subcellular structures of the neurones; AChE of bands 1 and 2 could be localized in the cytoplasm, rough endoplasmic reticulum and Golgi lamellae, while the enzyme in band 3 was incorporated into surface membranes. The electron histochemical study showed that the AChE is pentameric in both the cytoplasm and on the outer surface of the neurolemma.

Tissue Differences in the Effect of Atropine on
the Evoked Release of Acetylcholine

H. KILBINGER

Department of Pharmacology, University of Mainz, D-6500 Mainz (F.R.G.)

The effect of atropine on the release of acetylcholine (ACh) from different peripheral parasympathetically innervated tissues was investigated.

Myenteric plexus. The longitudinal muscle-myenteric plexus preparation of the guinea-pig was incubated in eserine-containing Tyrode solution. The ACh release evoked by high K^+ (45 or 108 mM) or by the nicotinic drug dimethylphenylpiperazinium (DMPP) (10 μM) was increased by atropine (0.1–10 μM) in a concentration-dependent fashion. Muscarinic agonists (oxotremorine; propargylester of arecaidine) prevented the facilitatory effect of atropine on ACh release. These results suggest that the ACh release from the myenteric plexus is regulated via presynaptic inhibitory muscarinic receptors.

Heart. Isolated guinea-pig hearts were perfused with Tyrode solution in the presence of eserine. The ACh overflow during a 2-min perfusion with 108 mM K^+ was not changed by atropine. Likewise, atropine did not modify the ACh output caused by 2-min perfusions of isolated chicken hearts with high K^+ (45 or 108 mM) or with DMPP (300 μM) either in the absence or presence of eserine.

Urinary bladder. Strips of the guinea-pig urinary bladder were incubated in eserine-containing Tyrode solution. The release of ACh evoked by a 10-min incubation of the strips with 108 mM K^+ was not changed by atropine. It is concluded that only the ACh release from the guinea-pig myenteric plexus can be modulated via presynaptic muscarinic-receptors. Apparently, the terminals of the postganglionic parasympathetic nerves of guinea-pig and chicken hearts and of guinea-pig urinary bladder are not equipped with inhibitory muscarinic-receptors.

Influence of Neurones and Contractile Activity on Acetylcholinesterase and Acetylcholine Receptors in Muscle Cell Cultures

J. KOENIG [1] and M. VIGNY [2]

[1] *Unité Biologie et Pathologie neuromusculaires (U 163), 75005 Paris and Laboratoire de Neurocytologie, Université P. et M. Curie, 75005 Paris and* [2] *Laboratoire de Neurobiologie, Ecole Normale Supérieure, 75230 Paris (France)*

The influence of neurones and of muscle contractility on the synthesis of the 16 S (synaptic) (Hall, Z., J. Neurobiol., 4, 343, 1973; Vigny, M., Koenig, J. and Rieger, F., J. Neurochem., 27, 1347, 1976) molecular form of AChE, the focalisation of this enzyme and of ACh receptors was investigated in cell cultures of muscle, mixed or not with spinal cord neurones. Tetrodoxin was used to block muscle contractile activity. We find:

(1) Muscle cells devoid of 16 S activity before the plating (13–14 day-old embryos) are unable to synthesize this form of AChE if neurones are not added to the cultures. With muscles already contacted by axons in vivo, the synthesis of 16 S AChE continues in vitro in the absence of neurones. Thus, the synthesis of this peculiar, "synaptic" form of AChE is induced under neural influence (Koenig, J. and Vigny, M., Nature, (Lond.), 271, 75, 1978).

(2) Focalisation of AChE does not occur in mixed cultures in the presence of tetrodoxin or in myotubes cultivated without neurones, even if they contract spontaneously. It is concluded that focalisation of AChE is dependent both on neurones and on muscle contractile activity, which is also suggested by experiments in vivo (Lømo, T. and Slater, M., in Synaptogenesis, Gif Lectures in Neurobiology, 1976, p. 930).

(3) Focalisation of ACh receptors is only neurone-dependent; inhibition of the myotube contractions does not prevent it.

In conclusion, in our conditions of culture, muscle contractile activity exerts different effects on the 3 synaptic markers studied.

Miniature End-Plate Potentials are Composed of Subunits

M.E. KRIEBEL, F. LLADOS and D.R. MATTESON

Upstate Medical Center, Syracuse, NY 13210 (U.S.A.)

Miniature end-plate potential (MEPP) amplitude histograms from frog and mouse muscle cells show a number of peaks that are integral multiples of the smallest peak (SMEPP). Kriebel, M.E. and Gross, C.E. (J. Gen. Physiol., 64, 85–103, 1974) found that in unstressed frog cells, SMEPP composed only 2–5% of the total but formed a distinct peak 1/7 that of the major peak. They also showed that repetitive nerve stimulation, heat and/or calcium challenges and nerve degeneration increased the percentage of SMEPPs and decreased the mean MEPP amplitude. MEPP distributions from mouse show SMEPPs to be 1/12 that of the mean amplitude. Colchicine reduced the mean 3–5-fold at which time MEPP distributions showed multiple peaks (Kriebel, M.E., Llados, F. and Matteson, D.R., J. Physiol. (Lond.), 262, 553–581, 1976). Botulinum toxin abolished larger MEPPs after the generation of as few as 200 MEPPs at which time SMEPPs and small multiple MEPPs were generated, but at very low frequencies. In comparison, black widow spider venom (Kriebel, M.E. and Stolper, D.R., Am. J. Physiol., 229, 1321–1329, 1975) or $LnCl_2$ challenges also left mainly SMEPPs but after depletion of vesicles. The number of MEPPs in the smaller peaks follows Poisson statistics which suggests a "drag effect". The variance in time-to-peak of MEPPs composing the 2nd and 3rd peaks was found to be larger than that of SMEPPs and larger MEPPs which also suggests a "drag effect". The number of MEPPs in the larger peaks appears to be binomially distributed. These observations suggest a subunit synchronizing mechanism. Since the intervals between the smaller peaks (drag mechanism) are the same as those between the larger peaks (synchronizing mechanism), both release mechanisms utilize the same subunit.

Choline Uptake, Acetylcholine Synthesis and Release in Synaptosomes from Hypoxic Rat Brain

H.J. KSIĘŻAK

Medical Research Centre of the Polish Academy of Sciences, 00-784 Warsaw (Poland)

In brain under hypoxia the general fall in pO_2 is followed by the leakage of K^+ which is first detectable in the cortex (Silver, I., Adv. exp. Med. Biol., 78, 299–317, 1977). The aim of this study was to determine effect of hypoxic depolarization in the synaptosomal fraction from rat brain after submission to 30 min of hypoxia (7% O_2 in N_2).

It was found that the V_{max} of Na^+-dependent high affinity choline (Ch) uptake was enhanced by 30% in hypoxic synaptosomes as compared to the control (79 and 60 pmol/mg of protein/4 min, respectively) without altering the K_m values. Also the percentage of [^{3}H]Ch that was converted to [^{3}H]ACh was slightly increased in hypoxic preparation. Taking into account that depolarization agents have similar effect in vitro (Kuhar, M.J., Molec. Pharmacol., 12, 1082–1090, 1976) our observations may reflect depolarization in vivo due to hypoxia.

High K^+ (35 mM KCl)-stimulated [^{3}H]ACh release was maximal during the first 10 min of incubation at 37°C and was Ca^{2+}-dependent in synaptosomal preparations from hypoxic and control brain. Spontaneous and K^+-stimulated [^{3}H]ACh release from synaptosomes from hypoxic brain, taken as a fraction of total labelled ACh, was not different as compared to the control. It appears that hypoxia does not lead to irreversible damage of the presynaptic mechanisms involved in these processes.

Synaptosomes from hypoxic brain showed a 2-fold increase over the control in [^{3}H]ACh formation after 30 min of incubation with [6–^{3}H]glucose in low K^+ medium. Incubation of these synaptosomes in high K^+ medium reduced markedly (by 67%) the incorporation of the radioactivity into ACh. However, differences between synaptosomes from hypoxic and normal brain in [U–^{14}C]glucose and [2–^{14}C]pyruvate decarboxylation were insignificant in both low and high K^+ medium. Thus, it seems that hypoxia has no irreversible influence on the glucose oxidative metabolism when studied in vitro. Our results indicate that acetyl-CoA supply was not a limiting factor in ACh synthesis from [6–^{3}H]glucose as well as from [^{3}H]Ch. It appears that enhanced Na^+-dependent Ch uptake is the main rate limiting factor in this process.

An increased rate of Na^+-dependent Ch accumulation as well as of ACh synthesis in synaptosomes from hypoxic brain suggests that the level of endogenous ACh pool has been lowered as a result of depolarization.

Drug-Induced Decremental Oscillations following Spike Potential in Cat Sympathetic Ganglion

J. MACHOVÁ

Institute of Experimental Pharmacology, Slovak Academy of Sciences, 88105 Bratislava (Czechoslovakia)

The aim of this study was to analyse drug-induced decremental oscillatory potentials (DOP) following spike potential in cat superior cervical ganglion in situ. Compound action potentials induced by supramaximal preganglionic stimulation (0.1–0.5 Hz) were recorded from the surface of the ganglion. Drug-induced asynchronous discharge (AD) was recorded from the internal carotid nerve. Drugs were administered i.a. into the ganglionic vascular bed. Muscarine-like drugs, 4-(*m*-chlorophenylcarbamoyloxy)-2-butynyl trimethylammonium chloride (McN-A-343), *N*-benzyl-3-pyrrolidyl acetate methobromide (AHR-602) and pilocarpine, and histamine and bradykinin induced DOP following spike potential occurring at the time of late negative wave (LN wave) onset. These drugs evoked slowly developing AD of low amplitude. Time of onset and duration of AD were similar to those of DOP. Atropine, which decreases LN wave amplitude, depressed DOP induced by McN-A-343; however it did not inhibit histamine induced DOP. The ganglion depolarizing drug isoprenaline, which did not induce AD, evoked an increase in LN wave amplitude without DOP. Nicotine induced a decrease in LN wave amplitude and no DOP were observed. These results indicate that drugs inducing DOP following spike potential have the common characteristic of causing AD by stimulation of non-nicotinic receptor sites in cat superior cervical ganglion. It is suggested that a process coincident with LN wave is responsible for the appearance of DOP by modulating discharges originating at non-nicotinic sites.

Modulation of Acetylcholine Turnover Rate by Neuropeptides

D. MALTHE-SØRENSEN and P. WOOD

Norwegian Defence Research Establishment, Division for Toxicology, N-2007 Kjeller (Norway) and Laboratory of Preclinical Pharmacology, St. Elizabeth Hospital, Washington DC 20032 (U.S.A.)

It is generally believed that by measuring dynamic parameters such as the turnover rate (TR) of a putative neurotransmitter in a given brain area, inferences can be made on the functional activity of a biochemically defined population of neurons. In cholinergic neurons, classical experiments performed by MacIntosh, Canad. J. Biochem. Physiol., 41, 2555–2571, 1963) on the cat superior cervical sympathetic ganglion, established that electrical stimulation of the preganglionic nerve results in an increase of acetylcholine (ACh) synthesis and release, without any modification of the ACh content in the ganglion. In the present study evidence is presented that changes in the functional activity of septal cholinergic

TABLE I

THE EFFECT OF PEPTIDES ON TR_{ACh}

Treatment	Cortex	Striatum	Hippocampus
Saline	0.93 ± 0.051	9.3 ± 1.52	3.5 ± 0.91
TRH	2.00 ± 2.25 *	6.4 ± 0.81	4.4 ± 0.83
Somatostatin	0.98 ± 0.052	7.4 ± 0.83	6.8 ± 0.33 *
Substance P	1.10 ± 0.35	7.2 ± 0.91	1.7 ± 0.25 *
α-MSH	0.92 ± 0.22	8.5 ± 0.74	6.8 ± 0.52 *

* $P < 0.01$ compared to saline treated animals.
TR_{ACh} is expressed in nmoles/mg protein/h.

neurones are reflected by corresponding changes in the TR_{ACh} in the hippocampus. Electrical stimulation of the medial septum increased the TR_{ACh} in the hippocampus by approx. 100% and acute and chronic lesions of fimbria decreased the TR_{ACh} by more that 70%. Neither acute nor chronic lesion of the fimbria induced changes in the TR_{ACh} of cortex. In view of the increasing interest in neuropeptides as possible neurotransmitters or modulators we investigated the effect of several neuropeptides on the cholinergic mechanisms in the brain. The peptides were injected intraventricularly or locally through a chronically implanted cannula. Data on the changes observed in TR_{ACh} after intraventricular injections of different peptides are given Table I.

Local injection of substance P in the septum specifically decreased the TR_{ACh} in hippocampus, while local septal injections of somatostatin and α-MSH had no effect in hippocampus. The nonuniform actions of the various peptides on TR_{ACh} in various brain structures are indicative of specific functions for these peptides and suggest that they are also able to modulate ACh metabolism and presumably cholinergic mechanisms in various brain structures. The presence of these peptides in synaptosomes and their ubiquitous distribution in the CNS all favour a neurotransmitter and neuromodulator function.

Source of Choline for Acetylcholine Synthesis in the Brain

STEPHEN P. MANN

A.R.C. Institute of Animal Physiology, Babraham, Cambridge (United Kingdom)

The supply of choline (Ch) to the brain has been discussed by Ansell, G.B. and Spanner, S. (in Cholinergic Mechanisms and Psychopharmacology, D. Jenden, Ed., Plenum Press, New York, pp. 431–445). These authors reviewed evidence for the loss of Ch from the brain and its replacement by Ch from the phospholipids. Estimations of free Ch in the brain vary but may be as low as 8.7 nmol $\cdot$ g^{-1} (Mann, S.P. and Hebb, C., J. Neurochem., 28, 241–244, 1977). The enzymes capable of releasing Ch from phospholipids and their metabolites may be of importance with regard to the supply of choline for the synthesis of acetylcholine (ACh). Two of these enzymes have now been examined. The presence of glycerophosphophorylcholine diesterase and its distribution in the brain has already been reported (Mann, S.P., Experientia, 31, 1256–1258, 1975).

The base-exchange system (Porcellati, G., et al., J. Neurochem. 18, 1395–1417, 1971) and alkaline phosphatase have now been investigated in the brain of the rats. The base-exchange system was found to be stimulated by Ca^{2+} and to have an activity of 2.1 nmol $\cdot$ g^{-1} $\cdot$ h^{-1} for both incorporation and release of Ch from phospholipids at pH 9.0. The system was evenly distributed in all fractions obtained by subcellular fractionation of the cerebrum. Ch incorporation and release were both inhibited by hemicholinium-3 (HC-3). Phospholipase D from plants was examined and had an optimum pH of 5.6; HC-3 did not inhibit this enzyme.

Alkaline phosphatase activity was assayed radiometrically and was found in all brain regions; the highest activity (38 μmol $\cdot$ g^{-1} $\cdot$ h^{-1}) was in the cerebrum at an optimum pH of 9.0. The enzyme had a K_m for choline phosphate of 1.06 mM. Subcellular fractionation showed high concentrations in fractions P_2 and P_3 (29 and 31%); 47% of the activity in P_2 was found in the synaptosome-rich (B) fraction. Activity was also found in fractions D (vesicles, 12%), F, G, H (synaptosome ghosts, 19, 14 and 27%) and I (mitochondria, 8%), obtained after osmotic rupture of the P_2 fraction.

These results indicate that there exist, within the brain, enzymes capable of making good the loss of Ch from the brain by releasing Ch from phospholipids and their metabolites. Ultimately, therefore, the Ch used in ACh synthesis may originate from the phospholipids of the brain.

Regeneration of Rat Neuromuscular Synapses

S. MANOLOV

Regeneration Research Laboratory, Bulgarian Academy of Sciences, 1431 Sofia (Bulgaria)

The regeneration of neuromuscular synapses was studied in the rat plantaris muscle over a period of 1 to 280 days after a crush lesion of the sciatic nerve. The initial signs of regeneration of the neuromuscular synapses were observed within 3 weeks of the operation. Thin axonal sprouts accompanied by large Schwann processes rich in organelles invaded some synaptic grooves. A considerable number of them were completely enveloped by Schwann cell processes and were situated at some distance from the postsynaptic membrane. Other terminals which penetrated deeper into the synaptic groove lost the Schwann processes from their surface and so established contact with the postsynaptic membrane. These terminals did not cover all the foldings of the subneural apparatus, and possessed a few synaptic vesicles, a single mitochondrion and a clear matrix. Single coated vesicles were rarely observed between the synaptic vesicles.

The Schwann cell processes accompanying the nerve terminals were filled with filaments and a larger than normal number of free ribosomes and vesicles were situated round the hypertrophied Golgi zones. At the same time the subsynaptic sarcoplasm was rich in vesicles located mainly round the distal parts of the subneural foldings. The free ribosomes, mitochondria and filaments beneath the subneural apparatus were augmented.

The axonal terminals began to grow in the second postoperative month and gradually filled the synaptic groove almost completely. The synaptic vesices increased in number and thickenings, with vesicular aggregations, appeared along the presynaptic membrane. The Schwann cell processes became thinner but were still rich in organelles.

After the third postoperative month all neuromuscular synapses were reinnervated and appeared completely normal.

Effect of Calcium Ionophores on Acetylcholine Output from Rat Brain Slices

P. MANTOVANI and G. PEPEU

Department of Pharmacology, University of Florence, Florence (Italy)

Acetylcholine (ACh) output from cortical slices incubated in Krebs solution was quantified by bioassay. The calcium ionophores BrX-537A and A23187 enhanced ACh output; BrX-537A exerted its maximal effect, a 6-fold increase, at a concentration of 1.8 μM; A23187 caused a 3-fold increase at 58 μM.

Raising Mg^{2+} to 9.3 mM doubled the peak effect of both ionophores. In Mg-free media the effect of both ionophores was reduced but the spontaneous ACh output was larger. In Ca-free media the effect of A23187 was abolished, and that of BrX-537A strongly reduced even in the presence of 9.3 mM Mg^{2+}. BrX-537A exerted some effect also when EDTA was added to the Ca-free medium. Raising Ca^{2+} to 5 mM increased the spontaneous ACh output but decreased the ionophore effect. Hyoscine (0.26 μM), but not the ionophores, further enhanced ACh output stimulated by KCl (25 mM). Hyoscine did not enhance ACh output stimulated by the ionophores The addition of hemicholinium-3 (0.01 mM) did not abolish the effect of the ionophores but suppressed the enhancing action of Mg^{2+}.

It is concluded that A23187 stimulates ACh output by transporting extracellular Ca^{2+} into cholinergic nerve endings. The effect of BrX-537A does not depend only on Ca^{2+}. The magnitude of the effect of both ionophores is modulated by ions and drugs affecting the synthesis and spontaneous output of ACh.

The Mechanism of the Trophic Action of Acetylcholine

A.M. MUSTAFIN

Kazakh State University, Alma-Ata (U.S.S.R.)

The transmitter acetylcholine is thought to exert a trophic function, but the mechanism has not been discovered yet. It could be carried out by modulating the sodium and potassium activated ATP-ase (Na^+, K^+-ATPase).

In our experiments acetylcholine 10^{-7} M increased Na^+, K^+-ATPase activity of frog sciatic nerves and brain as well as that of a membrane preparation containing Na^+, K^+-ATPase. The cholinolytic (+)-tubocurarine decreased Na^+, K^+-ATPase activity of the frog sciatic nerves and abolished the increase in activity caused by electrical stimulation.

Biochemical Characterization of Synaptic Vesicles Isolated from Guinea-Pig Cerebral Cortex

Á. NAGY, Z. RAKONCZAY, T. FARKAS and L. KESZTHELYI

Institute of Biochemistry and Biophysics, Biological Research Center of Hungarian Academy Sciences, Szeged (Hungary)

Synaptic vesicles with high ACh or catecholamine content were prepared from guinea-pig cerebral cortex (Nagy, Á., et al., J. Neurochem., 29, 449, 1977). The homogeneity and the purity of the two vesicle populations have been checked by their transmitter contents, investigation of marker enzymes and electronmicroscopic analysis.

More than 50% of the total vesicular proteins were strongly membrane bound. SDS electrophoretic protein patterns consisted of about 10 major and several minor bands. Actin, tubulin, synaptin, chromogranin A, dopamine-β-hydroxylase, ATPase and a few unidentified proteins could be determined on the basis of their molecular weights. 5–6 PAS stained components were also identified. It was shown that both vesicle fractions posses divalent cation dependent ATPases.

The lipid composition of vesicle membranes was different from that of the microsomal fraction. There was a higher level of phosphatidylcholine and sphingomyelin. The amount of phospholipid/mg of protein increased about 10 times by the purification on a glass bead column. There was a great similarity in the fatty acid composition between synaptic vesicular and other internal membranes, but certain polyenoic acids were undetectable in vesicular membranes.

Proton induced X-ray emission analysis was performed to determine P, S, K, Ca, Fe, Ni, Cu and Zn ion content of the vesicle fractions.

Neuronal Control in Tonic Muscle Fibres

G.A. NASLEDOV and T.L. RADZYUKEVICH

Sechenov Institute of Evolutionary Physiology and Biochemistry, Leningrad (U.S.S.R.)

Denervation in frog tonic muscle fibres is known to induce action potential (AP) generation (Miledi, R. et al., J. Physiol. (Lond.), 217, 737–751, 1971). The spread of acetylcholine (ACh) sensitivity after denervation in fibres of this type was also observed (Nasledov, G.A. and Thesleff, S., Acta physiol. scand., 90, 370–380, 1974).

In order to investigate the role of axoplasmic transport in these events, colchicine (10 mM) was applied briefly (30 min) to the nerve branches supplying pyriformis or cruralis muscles of frogs. Extrajunctional cholinergic sensitivity appeared in tonic fibres both after

490

colchicine treatment and after denervation. AP generation, however, did not occur after colchicine treatment. Moreover, extrajunctional ACh sensitivity, and not AP generation, was also observed in fibres of the contralateral muscle after colchicine treatment.

The results suggest that colchicine blocks a component of axoplasmic transport which suppresses extrajunctional ACh sensitivity (and presumably is moved by fast axonal flow), but not the component which may be responsible for the suppression of the AP mechanism. The velocity of transport of the trophic factor that controls AP suppression in innervated tonic fibres was calculated from experiments in which the nerve was cut at various distances from the muscle. The rate obtained was 2.2 mm/day which agrees with data obtained by Schalow, G. and Schmidt, H. (Pflüg. Arch., 355, Suppl. R57, 1975), and corresponds with the slow component of axonal transport (Bisby, M.A., Gen. Pharmacol., 7, 387–393, 1976).

We conclude that the AP mechanism in tonic fibres is controlled by axonal flow of substances which differ from those controlling the ACh receptors. Our results support the data indicating a systemic action of colchicine (Cangiano, A. and Fried, J.A., J. Physiol. (Lond.), 265, 63–84, 1977).

Relation between Acetylcholine Turnover and High Affinity Choline Uptake in Discrete Brain Regions

A. NORDBERG [1] and A. SUNDWALL [2]

Department of Pharmacology [1], University, S-751 23 Uppsala and Department of Clinical Pharmacology [2], Huddinge Hospital, S-141 86 Huddinge (Sweden)

The relation between the turnover rate of acetylcholine (ACh) in various brain areas in vivo and the high affinity choline uptake (HACU) in vitro has been studied in experiments on mice. Linear correlation was found between the turnover rat of ACh and HACU when different part of the brain (medulla, midbrain, striatum, hippocampus and cerebral cortex) were compared. After the administration of oxtremorine (0.5 mg of the base/kg body weight) was the turnover rate of ACh in all brain regions diminished to a greater extent than was the HACU. Pretreatment with atropine (5 mg/kg) prevented the effect of oxotremorine on the turnover rate of ACh and on HACU in all brain regions except the midbrain. Pretreatment with methylatropine (5 mg/kg) prevented the effect of oxotremorine in the striatum, but not in other brain regions. No obvious relationship has been found between the turnover rate of ACh and the HACU in various brain regions on the one hand and between the number of the muscarinic binding sites (as detected by the binding of quinuclidinyl benzilate to the synaptosomes) on the other hand.

Regional Biosynthesis of Acetylcholine in the Brain of Rat Following Chronic Treatment with Barbital

A. NORDBERG [1] and G. WAHLSTRÖM [2]

Department of Pharmacology [1], University, S-751 23 Uppsala and Department of Pharmacology [2], University, S-901 87 Umeå (Sweden)

Rats received a solution of sodium barbital as their only drinking fluid either for 33 weeks, or for a shorter period followed by a period of abstinence. The following five groups of rats were investigated: C, control; B, barbital until sacrifice; A3, abstinent for 3 days; A12, abstinent for 12 days; A30, abstinent for 30 days. Abstinance convulsions were recorded.

The biosynthesis of [3H]acetylcholine ([3H]ACh) from a tracer dose of [3H]choline was higher in the cerebellum + medulla + midbrain of groups B (+22%) and A3 (+54%), and also in the hippocampus + cortex of group A3 (+23%) as compared with group C. In the striatum no significant effect of barbital treatment or abstinence on [³H]ACh synthesis was found in any group. In a recent study (Europ. J. Pharmacol., 43, 237–242, 1977) we found a marked decrease in endogenous ACh in the striatum on the 3rd and the 12th day of barbital abstinence. In the present experiments the level of ACh in the striatum was still diminished on the 30th day of abstinence. The specific radioactivity of [³H]ACh was increased in all brain regions of abstinent rats in comparison with controls. This finding suggests that the turnover rate of ACh in the brain is increased during abstinence after chronic barbital treatment. The specific binding of the muscarinic antagonist quinuclidinyl benzilate to crude synaptosomal fractions from the striatum and from the midbrain + medulla + cerebellum was increased in rats abstinent for 3 days.

Cytochemical Localization of Choline Acetyltransferase

W. OVTSCHAROFF

Department of Anatomy, Histology and Embryology, Medical Academy, Sofia (Bulgaria)

The cytochemical localization of choline acetyltransferase in rat spinal cord, substantia nigra, caudate nucleus and cerebral cortex have been studied. Four adult Wistar-rats were used for this study. The cytochemical reaction was carried out according to Kása et al. (Nature (Lond.), 226, 812–816, 1970). No reaction product was visible when acetyl-coenzyme A or choline chloride and acetyl-coenzyme A were omitted from the incubation medium, but only a slight decrease of the reaction product was seen after removal of choline chloride alone from the medium. In experiments with the complete incubation medium the reaction product was located in the cytoplasm of terminals, as well as on the outer surface of the clear synaptic vesicles but it was never located in the boutons with small dense core vesicles. Unlike Kása P., et al. (Nature (Lond.), 226, 812–816, 1970) and Kása, P., (Progr. Brain Res., 34, 337–344, 1977) I found the product of enzyme activity in axon terminals containing round and flattened clear vesicles. Some intraterminal mitochondria were also enzyme positive. Choline acetyltransferase in the axonal endings was more frequently observed in the axonal endings in the anterior horn of the spinal cord. The product of enzyme activity was also seen in the neuronal perikarya and dendrites, associated with the outer surface of rough endoplasmic reticulum, vacuolar structures and mitochondria. The extraterminal enzyme localization was more clearly seen in the caudate nucleus than in the other areas of rat central nervous system that were examined.

The Effect of Long-Time Curarization on the Frog Neuromuscular Junction

M. PÉCOT-DECHAVASSINE and A. WERNIG

Laboratoire de Cytologie, Université P. et M. Curie, Paris (France) and Max-Planck-Institut für Psychiatrie und Neurophysiologie, München (F.R.G.)

Wernig and Stöver (Pflügers Arch., 368, R32, 1977) recently reported that frog muscle fibers show extrasynaptic sensitivity to ACh after four weeks treatment with paralysing doses of (+)-tubocurarine. However they did not observe correlative physiological changes in the process of transmitter release.

Electronmicroscopic investigation shows that morphological changes do appear in some neuromuscular junctions of treated frogs. Although many synapses remain normal, others show both regressive and regenerative features.

Significant signs of regression are well differentiated synaptic gutters either without any nerve terminal or partially occupied by shrinked nerve terminals.

Signs of regeneration consist of nerve sprouts either growing into empty preexisting synaptic gutters or inducing new synaptic sites.

Regressive and regenerative features are sometimes observed within a synapse while other regions of the same synapse remain normal.

Similar changes also occur (though much less frequently) in untreated winterfrogs and in very lean frogs.

These results suggest that the endplates in paralysed muscles and in muscles of low activity are in a state of continual remodeling; the nerve terminals successively withdraw from and reoccupy certain areas of the synaptic gutters or form new junctional sites from new sprouts.

Cholinergic (Nicotinic) Receptor and Lipids: Lipid Composition of Receptor-Rich Membranes Purified from *Torpedo marmorata* Electric Organ and Interaction of the Purified Receptor Protein with Monolayers of Pure Lipids

J.L. POPOT, R.A. DEMEL, A. SOBEL, L.L.M. VAN DEENEN and J.P. CHANGEUX

Laboratoire de Neurobiologie Moléculaire, Institut Pasteur, Paris (France) and Biochemical Laboratory, State University, Utrecht (The Netherlands)

Membrane fragments rich in cholinergic (nicotinic) receptor protein were purified from the electric organ of *Torpedo marmorata* by an improved method which yields preparations with a specific activity of ca. 4000 nmol α-toxin binding sites/g protein. Their lipid content is characterised by a high cholesterol/phospholipid ratio (0.9 mol/mol, i.e. 40% w/w), large amounts of phosphatidylethanolamine (40% of the phospholipids), phosphatidylcholine (37%) and phosphatidylserine (17%), and the near absence of sphingomyelin and cardiolipin ($\leqslant 1\%$). The fatty acid chains have a mean length of 18.7 carbon atoms, on the average, 2.0 double bonds per chain. The long, poly-unsaturated, docosahexaenoic acid ($C_{22:6}$) represents 25% of the chains; it is most abundant in the phosphatidylethanolamine and phosphatidylserine fractions.

Solubilised receptor was purified from these fragments and the concentration of sodium cholate lowered by dialysis to 0.01% (w/v). When this preparation was injected under a lipid monolayer, an increase of surface pressure developed, which was not observed with the detergent alone nor in the absence of the lipid film. When covalently-radiolabelled receptor preparations were injected at a constant surface pressure, the radioactivity recovered with the film was proportional to the increase in the area. It was concluded that the pressure or area increases are due to the penetration of the cholinergic receptor protein into the lipid film. Incorporation into films formed from varous pure lipids showed that the protein interacts more readily with cholesterol than with ergosterol, phosphatidylcholine, or other phospholipids. Its affinity is also higher for long-chain phosphatidylcholines than for short-chain ones. The degree of unsaturation and fluidity of the phosphatidylcholine films are of secondary importance. A detailed report of these experiments has been published (Popot, J.-L., et al., Europ. J. Biochem., 85, 27–42 (1978)).

The Effect of Hemicholinium-3 on the Turnover of Acetylcholine in the Auerbach Plexus of Guinea-Pig Ileum and in the Vas Deferens of Rat

G.T. SOMOGYI

Department of Pharmacology, Semmelweis University of Medicine, H-1085 Budapest (Hungary)

Isolated vas deferens of rat and longitudinal muscle strip of guinea-pig ileum was mounted in an organ bath of 2 ml and superfused (0.4 ml/min) by eserinised Krebs solution. Acetylcholine (ACh) was assayed biologically on guinea-pig ileum. In these experiments the synthesis of ACh was blocked by hemicholinium-3 (HC-3), and the amount of ACh released and its tissue level was measured. The release of ACh from guinea-pig ileum reached a peak value ($952.3 \text{ pmol} \cdot \text{g}^{-1} \cdot \text{min}^{-1}$), followed by a gradual decrease to an equilibrium state (568.4

pmol $\cdot$ g^{-1} $\cdot$ min^{-1}) during continuous high frequency stimulation. In the vas deferens, the ACh release reached a stable level (295.8 pmol $\cdot$ g^{-1} $\cdot$ min^{-1}). HC-3 did not inhibit the instantaneous release of ACh from either vas deferens or guinea-pig ileum, but completely prevented the formation of the equilibrium state. 50% of the ACh content was still present in both tissues treated with HC-3 when the stimulation induced release was abolished. The resting release from guinea-pig ileum was also reduced after HC-3 treatment. It is concluded that 50% of the ACh content of the tissues is not available for release. The turnover rate was calculated for both preparations on the basis of the released and the tissue ACh. The turnover rate was negative in the guinea-pig ileum (-37.1 nmol $\cdot$ g^{-1}), i.e. this amount of ACh was lost (decomposed) both in stimulated and resting preparations. This negative turnover rate was not detected in rat vas deferens, where the tissue level of ACh is relatively low (tissue content: 8.88 nmol $\cdot$ g^{-1}). It is concluded that backward reaction of choline acetyltransferase (ChAT) is responsible for the lost ACh in guinea-pig ileum, because the incomplete inhibition of ChE was excluded. However, in rat vas deferens, where the cytoplasmic level of ACh might be low, the reverse reaction of ChAT may not occur.

Pharmacologically Induced Stimulus-Bound Repetition (SBR) in Frog Sympathetic Ganglion

S. ŠTOLC

Institute of Experimental Pharmacology, Slovak Academy of Sciences, 881 05 Bratislava (Czechoslovakia)

The SBR is repetitive activity following a single orthodromic stimulus observed in frog sympathetic ganglion exposed to cholinesterase inhibitors (Riker, W.K. and Guerrero, S., J. Pharmacol. exp. Ther., 163, 54–63, 1968), cesium (Riker, W.K., et al., Life Sci., 13, 1069–1075, 1973), benzothiadiazines (Riker, W.K. et al., Pharmacologist, 16, 234, 1974), polyene antibiotic cyanein (Štolc, S. and Mačička, O., Českoslov. Fysiol., 25, 274, 1976), or ethanol (Montoya, G.A., et al., J. Pharmacol. exp. Ther., 200, 320–327, 1977). Our experiments carried out on the 8th sympathetic ganglion of frogs, *R. esculenta* or *R. temporaria,* seem to support the following conclusions: (1) It is improbable that, in the SBR induced by neostigmine, cesium, or cyanein, a cholinergic mechanism plays an essential role, because (a) the SBR has been induced by cesium or by cyanein which supposedly do not inhibit any cholinesterase, (b) the SBR-inducing potency of cesium was remarkably depressed by potassium as well as by anti- or orthodromic tetanus, and (c) the SBR has never been enhanced by an increase in stimulation rate. (2) The site of action of SBR-inducing drugs is most probably at the presynaptic terminals, because (a) the duration of neostigmine-induced SBR was not changed by an antidromic tetanus, while following orthodomic one it was remarkably prolonged and (b) the rate of development of the cesium-induced SBR was not accelerated by continuous stimulation of the preganglionic nerve as were other parameters of the postsynaptic action potential. (3) The SBR is likely connected with repolarization processes, because (a) there is a competitive relation between cesium and potassium, (b) cyanein, which depresses a membrane permeability ratio for potassium and sodium in skeletal muscle (Štekauerová, V. and Štolc, S., Českoslov. Fysiol., 25, 273–274, 1976) reveals the SBR-inducing potency, and (c) both cesium and cyanein prolong the duration of the repolarization phase of action potential in frog nerves consisting of C-axons.

Kinetics of Synthesis, Storage and Release of Acetylcholine in the Myenteric Plexus of the Guinea-Pig

J.C. SZERB

Department of Physiology and Biophysics, Dalhousie University, Halifax, B3H 4H7 (Canada)

Acetylcholine (ACh) release but not turnover in myenteric plexus (MP) has been extensively studied. The rate of [^{3}H]ACh formation from [^{3}H]choline was followed in

494

superfused MP preparations. Without stimulation [³H]ACh formation exceeded at least 3-fold the rate of [³H]ACh release. TTX and MnCl₂, which depress release, did not change the rate of [³H]ACh formation, while stimulation at 0.1 or 16 Hz, which increased ACh output 3- or 20-fold, increased [³H]ACh formation only in the first 9 min of superfusion. This suggests that a large fraction (over 50%) of ACh turns over spontaneously and stimulation enhances the turnover of only a small pool. During superfusion with HC-3 without stimulation the storage of [³H]ACh depended on the conditions of labelling: after labelling without stimulation it declined while stores formed during stimulation were maintained, suggesting that spontaneous formation of ACh occurs in a pool different from that activated by stimulation; [³H]ACh formed in the absence of stimulation or during 0.1 Hz stimulation could be released by both 0.1 and 16 Hz stimulation, but [³H]ACh formed during 16 Hz stimulation could be released only by 16 Hz stimulation, indicating that some cholinergic neurons respond only to high frequency stimulation. Supramaximal 0.1 Hz stimulation caused a biphasic outflow of [³H]ACh: from a fast emptying small and a slowly emptying large pool. Submaximal stimulation released only from the fast pool. 1 μM morphine reduced supramaximal twitches by about 50% and reduced the rate of efflux from the fast pool but not from the slow pool. Twitches and [³H]ACh produced by submaximal stimulation were completely suppressed by morphine. The morphine sensitive low threshold pool is about 1% of the total ACh. The MP is a complex preparation which contains: (1) a large spontaneously turning over pool; (2) a smaller heterogenous pool in terminals; and (3) a very small low threshold pool of ACh highly sensitive to morphine.

ATP-Citrate Lyase in the Cholinergic System of Rat Brain

A. SZUTOWICZ, W. ŁYSIAK and S. ANGIELSKI

*Department of Clinical Biochemistry, Institute of Pathology, Medical Academy,
Gdańsk (Poland)*

High activity of ATP citrate lyase (citrate cleavage enzyme, CCE) varying from 18 to 30 μmol/h · g of tissue, was shown in brains of several different species. It was much higher than the activity of choline acetyltransferase (ChAT) in the brain. Subcellular fractionation showed that over 70% of the activities of both CCE and ChAT in secondary fractions is found within synaptosomes, where they are preferentially located in the synaptoplasm. Activities of these enzymes in cerebellum were many times lower than in the brain regions having a higher density of cholinergic neurones. The CCE activity in cerebrum did not change significantly during animal maturation, while in cerebellum it decreased threefold during this period. Data is presented that indicate a particular link of CCE with cholinergic neurones in brain. It can be calculated that about 80% of CCE and only 10% of the activity of other enzymes of acetyl-CoA metabolism in the whole synaptosomal fraction is found inside cholinergic nerve endings.

The net citrate synthesis by synaptosomes utilizing either pyruvate or glucose was reduced by the CCE activator Mg–ATP and then increased by its specific inhibitor (−)-hydroxycitrate. Bromopyruvate, in a concentration that inhibits pyruvate utilization by about 75% (0.5 mM) resulted in 98% decrease in citrate production. Both inhibitors also caused a decrease in ACh synthesis by about 45 and 70%, respectively. The metabolic flux of citrate through the mitochondrial membrane and the CCE reaction was calculated to be only few times higher than the rate of ACh synthesis in cholinergic synaptosomes. These data suggest that the inhibition of ACh synthesis might be brought about either by an impairment of citrate cleavage in the synaptoplasm, or by a decrease in citrate efflux from the intrasynaptosomal mitochondria.

Postsynaptic Muscarinic Receptor and Sodium-Pump Activity

T.L. TÖRÖK and E.S. VIZI

Department of Pharmacology, Semmelweis University of Medicine, Budapest (Hungary)

Earlier it was shown by one of us (Vizi, E.S., J. Physiol. (Lond), 267, 261–280, 1977) that the amount of acetylcholine (ACh) liberated from the nerve terminals is in part controlled by the activity of (Na–K)-activated ATPase activity present in them. Present findings indicate that the response of the post-synaptic membrane to cholinergic agents is also affected by the Na^+-pump since during enhanced electrogenic sodium transport carbachol fails to produce the normal depolarization and contraction of smooth muscle.

The membrane potential of the guinea-pig taenia coli smooth muscle was measured by the sucrose gap technique (Bülbring, E. and Tomita, T., Proc. roy. Soc. B., 172, 89–102, 1969) and the intracellular ionic content by emission spectroscopy (Vizi, E.S., J. Physiol. (Lond.), 267, 161–280, 1977).

Under steady state conditions the coupling ration of the Na^+-pump was calculated to be 3Na : 2K : 1ATP. Carbachol (5.5×10^{-5} M) depolarized the smooth muscle cells and increased the Na_i^+ content from 35.1 ± 1.5 to 129.9 ± 12.7 mmol $\cdot$ kg^{-1} wet wt. In Ca^{2+}-free (EGTA, 0.1 mM) solution the depolarization was abolished and the Na_i^+ content was 31.6 ± 5.4 mmol $\cdot$ kg^{-1} wet wt. On washing, the carbachol induced depolarization was followed by a hyperpolarization and the muscle relaxed. Both these effects were inhibited in K^+-free, Na^+-free or Ca^{2+}-free (EGTA, 0.1 mM) media and by the application of ouabain (2×10^{-5} M) or cold-shock. Both effects were enhanced when Cl_0^- was replaced by the impermeant anion benzene-sulphonate. In K^+-free solution the membrane potential decreased from 56.9 to 35.1 mV and the Na^+ content increased from 42.2 ± 1.7 to 91.0 ± 6.8 mmol $\cdot$ kg^{-1} wet wt. Restoration of the K^+ caused a hyperpolarization. During this period E_m was 75.8 mV and clearly exceeded the calculated value of E_K 67.8 mV and Na_i^+ diminished to 48.8 ± 7.1 mmol $\cdot$ kg^{-1} wet wt. Ouabain (2×10^{-5} M) completely prevented the hyperpolarization observed during $(K^+)_0$ restoration. In the presence of ouabain the $(Na^+)_i$ content was 65.5 ± 4.0 mmol $\cdot$ kg^{-1} wet wt. Readmission of $(K^+)_0$ to tissues which had been kept previously in K^+-free solution abolished the depolarizing and contracting actions of carbachol.

Effects of Scopolamine and Pilocarpine on the Activity of Acetylcholinesterase in Rat Brain Synaptosomal Fractions

L. VENKOV and N. IANCHEVA

Regeneration Research Laboratory, Bulgarian Academy of Sciences, Sofia (Bulgaria)

In this work we studied the effect of scopolamine and pilocarpine on the activity of acetylcholinesterase (AChE) in isolated synaptosomal fractions of rat brain. Scopolamine was found to act as an inhibitor of AChE: at a concentration of 0.25×10^{-2} M, it exhibited competitive inhibition, whereas at 0.5×10^{-2} and 10^{-2} M the inhibition was of the mixed type. The following inhibition constants were found for scopolamine: at 0.25×10^{-2} M, $K_i = 1.5 \times 10^{-3}$ M; at 0.5×10^{-2} and 10^{-2} M, $K_i = 3.2 \times 10^{-3}$ M. The scopolamine concentrations used did not have any effect on the substrate inhibition by acetylcholine (ACh). At all concentrations, the Hill coefficient was 0.32. Pilocarpine was shown to inhibit AChE in a non-competitive way at 10^{-5} to 10^{-4} M; at 10^{-4} to 10^{-3} M the inhibition was of the mixed type. K_i was 0.36×10^{-3} M for all pilocarpine concentrations examined and the Hill coefficient was 0.27. At higher concentrations ($10^{-3}–10^{-2}$ M) pilocarpine provided a marked protection of AChE against substrate inhibition by ACh. This effect was not accompanied by an activation of AChE.

Non-Specific Cholinesterase and Acetylcholinesterase Molecular Forms in Rat Tissues

M. VIGNY, V. GISIGER and J. MASSOULIÉ

Laboratoire de Neurobiologie, Ecole Normale Supérieure, 75230 Paris (France)

The existence in vertebrate tissues of multiple molecular forms of acetylcholinesterase (AChE) differing in their sedimentation coefficients, has been well documented. We have observed that "non-specific" cholinesterase (ChE), which is present in rat superior cervical ganglion and muscle together with AChE, also exists in a series of molecular forms. The properties of these molecular forms closely parallel those of AChE. For example their sedimentation coefficients are 3.6 S, 6.3 S, 10 S and 16.6 S for AChE, and 4.2 S, 7 S, 10.8 S and 17 S for ChE. Every AChE form appears to possess its ChE counterpart, and except for the heaviest form, the relative abundance of the corresponding components is practically identical. The solubility of the corresponding molecular forms in different extraction media also appears very similar.

We asssume that the homologous AChE and ChE molecular forms represent molecules of similar quaternary structure. However, in spite of the structural similarity implied by the existence of parallel molecular forms in the two enzyme systems, we observed no immunological cross-reactivity between them.

A Search for the Factor in Peripheral Nerves Which Induces Acetylcholine Sensitivity of Skeletal Muscle

F. VYSKOČIL and I. SYROVÝ

Institute of Physiology, Czechoslovak Academy of Sciences, 142 20 Prague (Czechoslovakia)

A piece of peripheral nerve placed on the extrajunctional area of skeletal muscle induces the sensitivity to acetylcholine (ACh) within several days (Vrbová, G., J. Physiol. (Lond.), 191, 20P, 1967; Jones, R. and Vyskočil, F., Brain Res., 88, 309–317, 1975). One possibility is that the nerve contains a factor which can trigger the appearance of ACh receptors.

The fractionation of the homogenised rat sciatic nerve was therefore made and individual fractions were built into silastic concave plates which were placed on the exposed surface of the innervated m. extensor digitorum longus and left in situ for 3–4 days. After dissection, the ACh sensitivity of the muscle fibres was tested in the extrajunctional area with microelectrodes. It was found that the fraction remaining in the dialysis sac after 24 h of dialysis against water induced no ACh sensitivity, whereas the concentrated eluate evoked sensitivity ranging in different superficial fibres from 6 to 50 mV/nC.

The irradiation of the eluate with UV light for 10 min abolished the sensitivity-inducing potency completely.

Very low sensitivity (about 0.3 mV/nC) was induced by plates containing 0.1% w/w of trypsin and no sensitivity appeared after loading the silastic plates (20 mg) with NaCl (3.7 mg). The low molecular fraction obtained after dialysis was further fractionated on Sephadex G25 and two Eolin-positive peaks (mol. wt. approx. 10,000 and 8000) were obtained. Both fractions induced ACh sensitivity when built into plates. It is very probable that the peripheral nerve contains low molecular, probably peptide-like substance(s) with the ability of inducing extrasynaptic ACh sensitivity.

Cholinergic Projections to the Cortex of the Rat Brain

H. WENK, V. BIGL and U. MEYER

Institute of Anatomy, Humboldt University, Berlin and Paul Flechsig Brain Research Institute, Karl-Marx-University, Leipzig (G.D.R.)

Some cortical brain areas of rats were investigated after stereotaxic lesions of several structures of the basal forebrain. Choline acetyltransferase (ChAT) activity was estimated biochemically, and acetylcholinesterase (AChE) by both quantitative histochemical and biochemical methods.

After subtotal lesions of the medial septal nuclei, of the lateral preoptic area and of the substantia innominata the activities of AChE and ChAT in several cortical areas descreased to moderate or low residual values indicating direct cholinergic projections to paleo-, archi- and neocortical parts of the brain. The cholinergic fibres to paleocortical fields originate in the magnocellular nucleus of the lateral preoptic region. The archicortical hippocampal formation is supplied with cholinergic fibres arising in the medial septal nuclei. The medial limbic cortex receives its cholinergic innervation from the nucleus of the diagonal band of Broca while the neocortical fields are supplied with cholinergic fibres from scattered cell groups situated in the substantia innominata of the basal forebrain. These cells are possibly homologous to the nucleus basalis of Meynert described principally in primates. The cells of origin occupy surprisingly small areas but because of the enormous preterminal ramifications they give rise to a widespread monosynaptic and ipsilateral projection. This is remarkably similar to the arrangement of monoamine-containing neurones of the brain stem which provide the monoaminergic innervation of the forebrain. It seems, therefore, that the cortical neurones are subject to two antagonistic influences, the cholinergic system and the monoaminergic system, which modulate and control the specific pattern of excitation in the cortex in a manner resembling the control function exerted by the sympathetic and para-sympathetic system in the periphery.

Differences in Transmitter Release at Different Synapses of the Frog Sartorius Muscle

L.N. ZEFIROV and A.L. ZEFIROV

Kazan State V.I. Ulyanov –Lenin University, Kazan (U.S.S.R.)

Microelectrode recording from Mg^{2+}-blocked end-plates was used to analyse changes in quantum content (m), probability of release (p) and available transmitter store (n) in frog sartorius muscle fibres during repetitive nerve stimulation at 10 Hz. Measurements were made on the large fibres from the inner surface of the muscle, and on the small fibres on the outer surface (Zefirov, A.L., et al., Bull. exp. Biol. Med., 1978, in press).

In large fibres repetitive nerve stimulation caused first an increase, and then a sharp fall in m. In small fibres, the initial increase was smaller and the subsequent decrease was very gradual. In small fibres the maximum rate of transmitter release was about half that of large fibres.

Evaluation of binomial release parameters (Bennett, M.R. and Florin, T., J. Physiol. (Lond.), 238, 93–107, 1974) showed that n was smaller in small fibres (range 5–60) than in large fibres (range 30–160). During repetitive stimulation p increased steadily in large fibres, whereas it showed little change in small fibres.

It is suggested that these differences are related to the different length of synaptic contact in large and small fibres (Kuno, M., et al., J. Physiol. (Lond.), 213, 545–556, 1971).

Author Index

Subject Index

502